ADVANCES IN EXPERIMENTAL MEDICINE AND BIOLOGY

Recent Volumes in this Series

A Continuation Order Plan is available for this series. A continuation order will bring delivery of each new volume immediately upon publication. Volumes are billed only upon actual shipment. For further information please contact the publisher.

FUMONISINS IN FOOD

FUMONISINS IN FOOD

Edited by

Lauren S. Jackson

Food and Drug Administration
National Center for Food Safety and Technology
Summit-Argo, Illinois

Jonathan W. DeVries

General Mills, Inc.
Minneapolis, Minnesota

and

Lloyd B. Bullerman

University of Nebraska–Lincoln
Lincoln, Nebraska

PLENUM PRESS • NEW YORK AND LONDON

Library of Congress Cataloging-in-Publication Data

Fumonisins in food / edited by Lauren S. Jackson, Jonathan W. DeVries,
and Lloyd B. Bullerman.
    p.   cm. -- (Advances in experimental medicine and biology ; v.
392)
    "Proceedings of an American Chemical Society Symposium on
Fumonisins in Food, held April 2-7, 1995, in Anaheim, California"-
-T.p. verso.
    Includes bibliographical references.
    ISBN 0-306-45216-2
    1. Fumonisins--Toxicology--Congresses. 2. Corn as food-
-Contamination--Congresses. 3. Feeds--Contamination--Congresses.
I. Jackson, Lauren S. II. DeVries, Jonathan W. III. Bullerman,
Lloyd B. IV. Symposium on Fumonisins in Food. V. Series.
RA1242.F86F86 1996
615.9'54--dc20   95-26382
                                                         CIP

Proceedings of an American Chemical Society Symposium on Fumonisins in Food,
held April 2–7, 1995, in Anaheim, California

ISBN 0-306-45216-2

© 1996 Plenum Press, New York
A Division of Plenum Publishing Corporation
233 Spring Street, New York, N. Y. 10013

10 9 8 7 6 5 4 3 2 1

PREFACE

The contents of this book are the proceedings of the ACS symposium, "Fumonisins in Food," which was held April 4-6, 1995, at the American Chemical Society National Meeting in Anaheim, CA. This symposium, which was international in scope, brought together researchers from diverse backgrounds in academia, government, and industry. Thirty-three speakers discussed topics ranging from the analysis of fumonisins to toxicology and regulatory aspects.

The fumonisins became the spotlight of mycotoxin research in 1988, when researchers at the South African Medical Research Council isolated and structurally characterized the fumonisins. Since 1988, there has been an explosion in the numbers of papers dealing with fumonisin-related topics. The interest in the fumonisins has arisen for several reasons. First, fumonisins are found in measurable concentrations in corn grown throughout the world. Second, these compounds have been implicated as the causative agents in a variety of naturally occurring animal diseases. Finally, there is speculation that fumonisins may in part be responsible for the high incidence of esophageal cancer in regions of the world in which corn is the staple grain.

The purpose of this book is to provide the most comprehensive and current information on the topic of fumonisins in food. Part One of the book reviews the history of the fumonisins and their world-wide occurrence in food. Analytical methods for identifying and quantifying fumonisins in foods and feeds are reviewed in Part Two of the book. Part Three characterizes the fungal species that produce fumonisin and describes genetic and biochemical aspects of fumonisin production. Part Four is devoted to the toxicological research that is being conducted throughout the world. Part Five reviews the effects of processing on fumonisins and possible methods for reducing the fumonisin content of food. The final part of the book gives a perspective on risk assessment and how it is being used in making regulatory decisions related to the fumonisins.

ACKNOWLEDGMENTS

The symposium organizers would like to thank the ACS Division of Agricultural and Food Chemistry for their approval and financial support of the fumonisin symposium. We express special appreciation to the session chairs, Dr. Susanne Keller, Dr. Robert Eppley and Dr. William Allaben for their excellent job in organizing their prospective sessions. Most especially, the speakers and contributing authors are gratefully acknowledged for their contributions to the symposium and this book. We also express our gratitude to the following sponsors. Without their financial support, a symposium of this magnitude would certainly not have been possible.

- AMERICAN CORN MILLERS FEDERATION
- CARGILL ANALYTICAL SERVICES
- CORN PRODUCTS (Division of CPC International)
- CORN REFINERS ASSOCIATION
- DEKALB GENETICS
- FRITO-LAY
- GENERAL MILLS
- KELLOGG'S
- MEDALLION LABORATORIES
- ROMER LABS
- WATERS CHROMATOGRAPHY
- WOODSON TENENT LABORATORIES

Lauren S. Jackson

Jonathan W. DeVries

Lloyd B. Bullerman

CONTENTS

Part Five: Effects of Processing on Fumonisins

Part Six: Regulatory Aspects of Fumonisins

FUMONISINS: HISTORY, WORLD-WIDE OCCURRENCE AND IMPACT

Walter F. O. Marasas

Programme on Mycotoxins and Experimental Carcinogenesis (PROMEC)
Medical Research Council, P O Box 19070
Tygerberg 7505 South Africa

ABSTRACT

The history, toxicological effects, world-wide natural occurrence and impact of the fumonisins, food-borne carcinogenic mycotoxins produced by *Fusarium moniliforme*, are reviewed from the original description of the fungus in 1881 to the present. Following the isolation and characterization of fumonisin B_1 and B_2 and the publication of the first 3 papers on fumonisins by South African researchers in 1988, the interest in these compounds increased dramatically during 1989 and 1990 because of numerous outbreaks of mycotoxicoses in animals associated with the 1989 corn crop in the USA. Major advances made during this period were published in approximately 49 papers from 1988 to 1991. During the period 1992 to 1994, there was an explosion in the literature on fumonisins and at least 212 papers were published. The information contained in the more than 260 papers on fumonisins published to date is reviewed with respect to toxicological effects, association with animal and human diseases, and world-wide natural occurrence in corn and corn-based feeds and foods. Impact of the fumonisins is addressed with respect to their implications for human and animal health, risk assessment and establishment of tolerance levels.

INTRODUCTION

The fungus *Fusarium moniliforme* Sheldon is one of the most prevalent seedborne fungi associated with corn (*Zea mays L*) intended for human and animal consumption throughout the world (Marasas *et al.*, 1984a). The fumonisins, food-borne carcinogenic mycotoxins, were first isolated from cultures of *F. moniliforme* strain MRC 826 at the South African Medical Research Council by Gelderblom *et al.* (1988a). The structures of the fumonisins were elucidated in collaboration with scientists at the Council for Scientific and Industrial Research, Pretoria (Bezuidenhout *et al.* 1988).

Fumonisins in Food, Edited by L. Jackson *et al.*
Plenum Press, New York, 1996

The history and toxicological effects, world-wide natural occurrence in corn and corn-based foods and feeds and impact of fumonisins on human and animal health will be discussed in this paper.

FUMONISINS: HISTORY

1881 - 1988

Saccardo (1881) described *Oospora verticillioides* on corn implicated in pellagra in Italy. Butler (1902) reproduced leukoencephalomalacia (LEM) in horses with naturally contaminated moldy corn in the USA. Sheldon (1904) described *Fusarium moniliforme* on corn implicated in moldy corn toxicosis in farm animals in the USA. Wilson & Maronpot (1971) reproduced LEM in a donkey with a pure culture of *F. moniliforme* from corn in Egypt. Kellerman *et al.* (1972) induced a hepatotoxic syndrome in horses with a pure culture of *F. moniliforme* from corn in South Africa. Marasas *et al.* (1976) reproduced LEM in horses with a pure culture of *F. moniliforme* from corn in South Africa. Marasas *et al.* (1981) demonstrated that *F. moniliforme* is significantly more prevalent in corn from a high incidence area of human esophageal cancer (EC) than a low incidence area. Kriek *et al.* (1981) induced pulmonary edema in pigs with a pure culture of *F. moniliforme* from corn in South Africa. Marasas *et al.* (1984b) induced liver cancer in rats with a pure culture of *F. moniliforme* from corn in South Africa. Gelderblom *et al.* (1988b) demonstrated cancer promoting activity of pure cultures of *F. moniliforme* in rat liver in a short-term cancer initiation/ promotion bioassay.

1988 - 1991

1988 - Year of Isolation and Characterization. Gelderblom *et al.* (1988a) isolated fumonisins B_1 (FB$_1$) and B$_2$ (FB$_2$) from cultures of *F. moniliforme* MRC 826.

Bezuidenhout *et al.* (1988) elucidated the structures of FB$_1$ and FB$_2$.

Marasas et al. (1988) induced a syndrome resembling LEM in a horse by intravenous injection of FB$_1$.

1989 - Year of Confirmation, Excitation, Expectation and Preparation. Laurent *et al.* (1989a) isolated and characterized two new mycotoxins from *F. moniliforme*: macrofusin and micromonilin. The structure of macrofusin proved to be identical to fumonisin B$_1$.

Laurent *et al.* (1989b) confirmed the induction of a syndrome resembling LEM in a horse by intravenous injection of FB$_1$.

Voss *et al.* (1989) detected FB$_1$ and FB$_2$ in two corn samples associated with field outbreaks of LEM in horses, but did not quantify levels.

1990 - Year of the Great American Outbreaks of LEM in Horses and PES in Pigs, Analytical Methods and Natural Occurrence. Great American Outbreaks. Corn screenings of the 1989 USA crop caused widespread outbreaks of Leukoencephalomalacia (LEM) in horses and Pulmonary Edema Syndrome (PES) in pigs in the USA during the fall of 1989 and winter of 1990 (Harrison *et al.*, 1990; Ross *et al.*, 1990; Wilson *et al.*, 1990).

Year of the Horse. Kellerman *et al.* (1990) reproduced the pathognomonic brain lesions of LEM in two horses by oral dosing of FB$_1$. Thus the causative role of FB$_1$ in equine LEM was finally proven conclusively.

Year of the Pig. Harrison *et al.* (1990) reproduced PES characterized by pulmonary edema and hydrothorax in a pig by intravenous injection of FB_1.

Analytical Methods. Analytical methods for the detection and quantification of FB_1 and FB_2 in corn were developed and these methods were used to determine naturally occurring levels for the first time:

- Analysis by HPLC of the maleyl derivative of FB_1 with ultraviolet detection and of the fluorescamine derivative with fluorescence detection (Sydenham *et al.*, 1990a).
- Analysis by HPLC with pre-column derivatization with *o*-phthaldialdehyde (OPA) and fluorescence detection with detection limits: 50 ng/g FB_1 and 100 ng/g FB_2 (Shephard *et al.*, (1990).
- Analysis by hydrolysis and GC/MS with sensitivity: low ppm (Plattner *et al.*, 1990).

Natural Occurrence. Naturally occurring levels of FB_1 and FB_2 in corn associated with field outbreaks of LEM in horses were reported by Shephard *et al.* (1990); Plattner *et al.* (1990); and Wilson *et al.* (1990). One sample of South African commercial mixed feed contained 8.85 μg/g FB_1 and 3.00 μg/g FB_2. (Shephard *et al.*, 1990). Two samples of USA corn contained 12-130 μg/kg FB_1 and <1 - 17 μg/g FB_2 (Plattner *et al.*, 1990). Three samples of USA corn screenings contained 37-122 μg/g FB_1 and 2-23 μg/g FB_2 (Wilson *et al.*, 1990).

Naturally occurring levels of FB_1 and FB_2 in home-grown corn intended for human consumption from a high incidence area of EC in Transkei, South Africa were reported by Sydenham *et al.* (1990 a,b). Visibly healthy kernels and *Fusarium*-infected kernels selected by hand from one sample of home-grown corn contained 44 and 83 μg/g FB_1, respectively (Sydenham *et al.*, 1990a). Visibly healthy and moldy home-grown corn samples from a high incidence area of EC in Transkei contained significantly higher levels of both FB_1 (3.45 - 46.9 μg/g) and FB_2 (0.9 - 16.3 μg/g) than corresponding samples from a low incidence area (Sydenham *et al.*, 1990b).

Naturally occurring levels of FB_1 and FB_2 in commercial corn were reported by Shephard *et al.* (1990). Three samples of South African commercial corn contained FB_1 at levels from 0.4 - 4.4 μg/g and FB_2 at 0.1 - 1.3 μg/g.

Three publications during 1990 claimed to be the first report of naturally occurring levels of fumonisins in corn:

- Sydenham *et al.* (1990a): "This is the first conclusive report of the natural occurrence of FB_1 in corn". J. Agric. Food Chem. 38(1): 285 - 290. Published January 1990, accepted August 14, 1989.
- Shephard *et al.* (1990): "This publication reports, for the first time, a quantitative and sensitive HPLC method for the simultaneous determination of FB_1 and FB_2 in naturally contaminated corn and mixed feed". This method was used to determine FB_1 and FB_2 levels in a feed sample associated with an outbreak of LEM in South Africa . J. Liquid Chromat. 13(10): 2077-2087. Published June 1990, accepted March 15, 1990.
- Plattner *et al.* (1990): "Our report provides the first data on the fumonisin level in corn samples eaten by horses that died from LEM". Mycologia 82(6): 698-702. Published November- December 1990, accepted July 5, 1990.

1991 - Year of New Fumonisins, New Producers and Sphingolipids. New Fumonisins. Cawood *et al.* (1991) isolated and characterized two new fumonisins, FB_3 and FB_4, from cultures of *F. moniliforme* MRC 826. The N-acetyl derivatives FA_1 and FA_2 which were referred to by Bezuidenhout *et al.* (1988), were also shown to be minor metabolites of the fungus.

New Fumonisin Producers. Thiel *et al.* (1991a) confirmed the report by Ross *et al.* (1990) that *F. proliferatum* produces FB_1 and FB_2 and reported the production of FB_1 and FB_2 by *F. nygamai.*

Sphingolipids. Wang *et al.* (1991) reported that fumonisins are structurally similar to sphingosine and are the first known naturally occurring inhibitors of sphingolipid biosynthesis.

Year of the Rat. Gelderblom *et al.* (1991) induced liver cancer in rats by feeding a diet containing 50 mg/kg FB_1 for 18-26 months.

FUMONISINS: TOXICOLOGICAL EFFECTS

Toxicological Effects of FB_1

- Causes Leukoencephalomalacia (LEM) in Horses
- Causes Pulmonary Edema Syndrome (PES) in Pigs
- Is Hepatotoxic in Rats
- Is Hepatocarcinogenic in Rats
- Is a Cancer Promotor and Initiator in Rat Liver
- Is NOT Mutagenic or Genotoxic
- Is Cytotoxic to Mammalian Cell Cultures
- Is Phytotoxic
- Inhibits Sphingolipid Biosynthesis
- Is Associated with Esophageal Cancer (EC) Risk in Humans

FB_1 Causes Leukoencephalomalacia (LEM) in Horses. A syndrome resembling LEM was induced by intravenous injection of 276 to 680 mg FB_1 per horse (Marasas *et al.*, 1988; Laurent *et al.*, 1989). The pathognomonic brain lesions of LEM were induced by oral dosing of 8417 to 8925 mg FB_1 per horse by Kellerman *et al.* (1990).

Wilson *et al.* (1992) induced LEM by feeding a horse naturally contaminated corn screenings containing a total of 4519 mg FB_1. These authors also induced mild transient clinical signs and mild histopathological brain lesions in horses fed a diet containing 8 ppm FB_1. Ross *et al.* (1993) induced LEM by feeding ponies naturally contaminated corn screenings containing 40.56 to 51.46 g FB_1.

Naturally occurring levels of FB_1 (<1-160 µg/g), FB_2 (<1-49 µg/g) and FB_3 (<1-2.6 µg/g) in corn, corn screenings and corn-based mixed feeds associated with field outbreaks of LEM were reported by Plattner *et al.* (1990); Shephard *et al.* (1990); Wilson *et al.* (1990); Ross *et al.* (1991a,b); Thiel *et al.* (1991b); Ross et al. (1992); Sydenham *et al.* (1992a,b); Wilson *et al.* (1992); Caramelli *et al.* (1993); and Ross *et al.* (1993).

The mean level of total fumonisins in 14 corn samples associated with LEM was 10.8 µg/g according to Thiel *et al.* (1991b). Ross *et al.* (1991b) concluded that "feed with >10 µg/g FB_1 is not safe to feed to horses" while Wilson *et al.* (1992) concluded that there should be concern for feeding diets containing 8 ppm to horses.

FB₁ Causes Pulmonary Edema Syndrome (PES) in Pigs. Pulmonary edema and hydrothorax characteristic of PES were induced in two pigs fed corn screenings containing 155 mg/kg FB_1 and in one pig by intravenous injection of 11.3 mg FB_1 (Harrison *et al.*, 1990; Colvin & Harrison, 1992).

Haschek *et al.* (1992) induced PES in pigs by feeding corn screenings containing 166 mg/kg FB_1 and 48 mg/kg FB_2 and by intravenous injection of 72 mg FB_1. Osweiler *et al.* (1992) induced PES in pigs by feeding corn screenings containing 92 mg/kg FB_1 and 28 mg/kg FB_2.

Naturally occurring levels of FB_1 from <1 - 330 µg/g and FB_2 from <1 - 48 µg/g in corn screenings associated with outbreaks of PES in the USA during 1989/90 were reported by Harrison *et al.* (1990); Ross *et al.* (1991a); Colvin & Harrison (1992); Haschek *et al.* (1992); Osweiler *et al.* (1992); Ross *et al.* (1992); Riley *et al.* (1993a); and Motelin *et al.* (1994).

FB₁ Is Hepatotoxic in Rats. The principal pathological change in rats treated with FB_1 in short-term toxicity tests (21-33 days) was progressive toxic hepatitis characterized by hepatocellular necrosis, bile duct proliferation and fibrosis, identical to that induced by the hepatotoxic culture material of *F. moniliforme* MRC 826 (Gelderblom *et al.*, 1988a).

The liver was the main target organ in rats fed a diet containing 50 mg FB_1/kg in a long-term (26 months) experiment by Gelderblom *et al.* (1991). Pathological changes in the liver were characterized by a chronic toxic hepatitis that progressed to cirrhosis, cholangiofibrosis, the development of regenerative nodules and terminated in hepatocellular carcinoma and cholangiocarcinoma. These changes are identical to those induced by the hepatotoxic and hepatocarcinogenic culture material of *F. moniliforme* MRC 826 (Gelderblom *et al.*, 1991).

Voss *et al.* (1993) reported that FB_1 is hepatotoxic in rats fed a diet containing 150 mg/kg for 4 weeks and is also nephrotoxic at 15-50 mg/kg.

FB₁ Is Hepatocarcinogenic in Rats. FB_1 (not less than 90% pure) is hepatocarcinogenic and causes primary hepatocellular carcinoma and cholangiocarcinoma in rats at a dietary concentration of 50 mg/kg fed for 18-26 months (Gelderblom *et al.*, 1991). Ten out of 15 FB_1-treated rats developed primary hepatocellular carcinoma and metastases to the lungs and kidneys were present in four of these rats.

FB₁ Is a Cancer Promotor and Initiator in Rat Liver. FB_1 Is a cancer promotor in a short-term cancer initiation- promotion assay with diethylnitrosamine (DEN)-initiated rats and the induction of GGT-positive foci as endpoint (Gelderblom *et al.*, 1988a). This bioassay was used in the original isolation of FB_1 by Gelderblom *et al.* (1988a).

FB_1 is also a cancer initiator as evidenced by the induction of resistant hepatocytes in rat liver (Gelderblom *et al.*, 1992a,b; 1994). Although FB_1 is a complete carcinogen in rat liver, it is a poor cancer initiator requiring prolonged exposure (21 days) to a relatively high dietary level (250 mg/kg). Thus a time and dosage dependent threshold level exists for cancer initiation by FB_1 (Gelderblom *et al.*, 1994).

At a dietary level of 1000 mg/kg for 21 days, only the FB fumonisins (FB_1, FB_2 and FB_3) initiated cancer in rat liver whereas the N-acetylated analogues FA_1 and FA_2 and the hydrolysis products AP_1, AP_2 and TCA did not (Gelderblom *et al.*, 1993). Thus the free amino group and the intact molecule are required for cancer initiation.

FB₁ Is not Mutagenic or Genotoxic. FB_1 is not mutagenic in the Salmonella mutagenicity test (Gelderblom & Snyman, 1991; Park *et al.*, 1992). Thus *F. moniliforme* MRC

826 produces both FB_1 which is carcinogenic but non-mutagenic, and Fusarin C which is mutagenic but non-carcinogenic.

FB_1 does not induce unscheduled DNA synthesis in isolated rat hepatocytes (Norred *et al.*, 1991a), and is not genotoxic in the *in vivo* and *in vitro* DNA repair assays in rat primary hepatocytes (Gelderblom *et al.*, 1992a; Norred *et al.*, 1992a,b).

FB_1 Is Cytotoxic to Mammalian Cell Cultures. FB_1 is cytotoxic to baby hamster kidney cells at 1 µg/µl (La Grenade & Bean 1990), but is not cytotoxic to primary rat hepatocytes at concentrations as high as $10^{-2}M$ (Norred *et al.*, 1991a).

FB_1 is cytotoxic to mammalian cells and the most sensitive cell lines were rat hepatoma H4TG and dog kidney epithelial MDCK with IC_{50} values of 4 and 2.5 µg/ml, respectively (Shier *et al.*, 1991; Mirocha *et al.*, 1992a).

FB_1 is cytotoxic to primary rat hepatocytes at very high concentrations (350 µM) (Gelderblom *et al.*, 1992b), but not at 250 µM, whereas it is cytotoxic to pig kidney epithelial cells ($LCC-PK_1$) at concentrations > 35 µM (Norred *et al.*, 1992a,b; Yoo *et al.*, 1992).

In rat hepatoma H4TG and dog kidney epithelial MDCK cells, the FB fumonisins (FB_1, FB_2, FB_3) are cytotoxic at concentrations of 50 µg/ml or less and FB_2 is more cytotoxic than FB_1 and FB_3 (Abbas *et al.*, 1993). The N-acetylated analogues FA_1 and FA_2 are not cytotoxic. However, the hydrolysis products AP_1 and AP_2 have similar or greater cytotoxic activity than the FB parent compounds (Abbas *et al.*, 1993).

In primary rat hepatocytes FB_1, FB_2 and FB_3 are cytotoxic at high concentrations (1000-2000 µM) and FB_2 is more cytotoxic than FB_1 and FB_3 (Gelderblom *et al.*, 1993). The FA analogues FA_1 and FA_2 are much less active and TCA is not cytotoxic. However, the hydrolysis products AP_1 and AP_2 are more cytotoxic than the FB parent compounds (Gelderblom *et al.*, 1993). Cawood *et al.* (1994) also found that in primary rat hepatocytes FB_1 and FB_2 are cytotoxic at a high concentration of 300 µM and FB_2 is more cytotoxic than FB_1.

In turkey lymphocytes FB_1 and FB_2 are cytotoxic at concentrations of 1.4 and 0.4 µg/ml, respectively, i.e. FB_2 is 3 to 4-fold more cytotoxic than FB_1 (Dombrink-Kurtzman *et al.*, 1994).

FB_1 is cytotoxic to chicken macrophages at concentrations as low as 0.5 µg/ml (Qureshi & Hagler, 1992). This suggests that FB_1 may cause immune suppression in chickens.

FB_1 Is Phytotoxic. FB_1 is phytotoxic and causes rapid death of intact jimsonweed (*Datura stramonium*) plants at a concentration of 2.5 mg/50 ml and soft rot of excised leaves within 24 hours at 2.5 µg/100 µl (Abbas *et al.*, 1991). "This is the first report of the application of this fungus and fumonisin B_1 as a method of weed control" (Abbas *et al.*, 1991). Abbas & Boyette (1992) and Abbas *et al.* (1992) confirmed the phytotoxicity of FB_1 to jimsonweed and a wide variety of other weeds and crop plants at concentrations from 10 - 200 µg/ml. Abbas *et al.* (1991) concluded that "One possible objection to the use of Fumonisin B_1" (as a herbicide) "is a recent report by Marasas *et al.*[6] (sic) that long-term feeding may be carcinogenic. Also, the fumonisin used was not pure and may have been contaminated with other carcinogens".

FB_1 is structurally similar to TA-toxin produced by *Alternaria alternata* f.sp. *lycopersici* and produces identical genotype-specific necrotic symptoms on detached leaves of resistant (Asc) and susceptible (asc) tomato lines (Gilchrist *et al.*, 1992; Mirocha *et al.*, 1992a). TA-toxin has a 20-fold higher specific activity to the asc genotype (20 nM) than FB_1 (400 nM).

Vesonder *et al.* (1992 a,b) reported that FB_1 is phytotoxic to duckweed (*Lemna minor*) and reduces growth as well as the ability to synthesize chlorophyll at 0.7 µg/ml. At this concentration TA-toxin also inhibits chlorophyll synthesis but does not reduce growth to the same extent as FB_1. Tanaka et al. (1993) found that both FB_1 and TA-toxin are phytotoxic to duckweed at concentrations as low as 0.04 μM. At 1 μM significant reductions in growth are also caused by FB_2 and FB_3, whereas AP_1 has moderate activity and FA_1, FA_2 and AP_2 are not phytotoxic.

Lamprecht et al. (1994) confirmed that FB_1 and TA are phytotoxic to detached tomato leaves and cause identical genotype-specific leaf necrosis at concentrations as low as 0.1 μM. The FB analogues (FB_1, FB_2, FB_3) and TA cause significantly more necrosis than the FA analogues (FA_1 and FA_2) and the hydrolysis products AP_1 and AP_2, whereas TCA is not phytotoxic. FB_1 and TA also cause dose-dependent reductions in growth of corn seedlings and are more phytotoxic than FB_2 and FB_3 at a concentration of 10 μM (Lamprecht et al., 1994).

FB_1 Inhibits Sphingolipid Biosynthesis. FB_1 is structurally similar to sphingosine and inhibits the activity of ceramide synthase which leads to a change in the sphinganine:sphingosine ratio in rat hepatocytes (Wang et al., 1991), Fumonisins are the first known naturally occurring inhibitors of sphingolipid biosynthesis. This may be the mechanism of some of the toxic effects of FB_1 because sphingolipids regulate cell growth, differentiation and transformation (Wang et al., 1991).

Norred et al. (1992a) and Yoo et al. (1992) confirmed that FB_1 is a potent inhibitor of sphingolipid biosynthesis in primary rat hepatocytes (IC_{50} = 0.1 μM) in the absence of a cytotoxic effect.

FB_1 also inhibits sphingolipid biosynthesis in pig kidney epithelial cells ($LCC-PK_1$) at 10-15 μM whereas concentrations >35 μM are cytotoxic. Thus the inhibition of sphingolipid biosynthesis occurs before the expression of cytotoxicity (Norred et al., 1992a; Yoo et al., 1992).

Sphingolipid biosynthesis is inhibited in ponies fed diets containing 15-44 μg/g FB_1 and changes in the sphinganine: sphingosine ratio were detected before clinical signs were seen or liver enzymes were elevated (Wang et al., 1992). Thus serum sphingolipid profiles may be an early biomarker of FB_1 exposure in horses.

Sphingolipid biosynthesis is inhibited in a dose-dependent manner in pigs fed diets containing 23 μg/g total FB_1 and FB_2 and higher (Riley et al., 1993a). Liver enzymes were elevated at a dietary level of 100 μg/g and pulmonary edema developed at 175 μg/g. Thus serum sphingolipid profiles may be an early biomarker of FB_1 exposure in pigs.

Sphingolipid biosynthesis is inhibited in a dose-dependent manner in rats fed diets containing 15 μg/g FB_1 and higher (Riley et al., 1994a). FB_1 was nephrotoxic to rats at lower dietary levels than those required for hepatotoxicity. Elevation of the sphinganine: sphingosine ratio in urine reflected the pathological changes in the kidney. Thus urinary sphingolipid profiles may be an early biomarker of FB_1 exposure in rats.

FB_1 inhibits sphingolipid biosynthesis in cultured mouse cerebellar neurons at a concentration of 25 μM (Merrill et al., 1993), and in rat hippocampal neurons in vivo and in vitro at a concentration of 2 μM (Harel & Futerman, 1993).

FB_1 Is Associated with Esophageal Cancer (EC) Risk in Humans. Sydenham et al. (1990a) published the first report on naturally occurring levels of FB_1 in corn intended for human consumption. In a sample of home-grown corn from a high incidence area of EC in Transkei, South Africa, visibly healthy kernels contained 44 000 ng/g FB_1 and *Fusarium*-infected kernels 83 000 ng/g FB_1. Thus FB_1 occurs naturally in the staple diet of people at high risk for EC in Transkei. Mean levels of FB_1 and FB_2 were significantly higher in samples of visibly healthy as well as moldy home-grown corn from a high incidence area of EC in Transkei than in corresponding samples from a low incidence area during 1985 and 1989

(Sydenham *et al.*, 1990b, 1991; Rheeder *et al.*, 1992). Some moldy corn samples intended for beer-brewing contained very high levels of FB_1 (117 520 ng/g) and FB_2 (22 960 ng/kg) (Rheeder *et al.*, 1992).

High levels of FB_1 (18 000 - 155 000 ng/g) were present in home-grown moldy corn from high incidence areas of EC in China (Chu & Li, 1994). Thus FB_1 also occurs naturally in the staple diet of people at high risk for EC in China.

Relatively high levels of FB_1 (up to 6100 ng/g) and FB_2 (up to 910 ng/g) were present in commercial corn-based human food-stuffs such as polenta in Italy (Doko & Visconti, 1993, 1994). Thus FB_1 may also occur in polenta consumed by people at high risk for EC in northern Italy.

Commercial corn-based human foodstuffs purchased in Charleston, South Carolina which is the highest incidence area of EC in the USA, contained FB_1 at levels between 105 and 1915 ng/g and FB_2 from 70 - 460 ng/g (Sydenham *et al.*, 1991). Thus FB_1 also occurs in corn-based foodstuffs intended for consumption by people at risk for EC in the USA.

Marasas (1994) emphasized that EC has not been reproduced experimentally in animals with either culture material of *F. moniliforme* or pure FB_1. Although FB_1 occurs naturally in the corn-based staple diet of several populations at high risk for EC, there is no experimental proof of a causative relationship.

This is, however, not to say that the evidence for an association between FB_1 and EC is "anecdotal" (Shelby *et al.*, 1994).

FUMONISINS: WORLD-WIDE NATURAL OCCURRENCE

Natural Occurrence : FB_1

The natural occurrence of FB_1 was first reported by Sydenham *et al.* (1990a) in moldy home-grown corn from a high incidence area of EC in Transkei, South Africa at levels of 44-83 µg/g.

Natural Occurrence: FB_2

The natural occurrence of FB_2 was first reported by Shephard *et al.* (1990) in a corn-based mixed feed sample associated with a field outbreak of LEM in South Africa at a level of 3 µg/g.

Natural Occurrence : FB_3

The natural occurrence of FB_3 was first reported by Ross *et al.* (1992) in a sample of moldy corn in the USA at a level of 64 µg/g.

World-Wide Natural Occurrence of FB_1, FB_2, FB_3

ARGENTINA: Sydenham *et al.* (1993a); Viljoen et al. (1993). AUSTRIA: Lew *et al.* (1991). BOTSWANA: Sydenham *et al.* (1993b). BRAZIL: Sydenham *et al.* (1992b). BULGARIA: Sydenham *et al.* (1993b). CHINA: Ueno *et al.* (1993); Chu & Li (1994). CANADA: Sydenham *et al.* (1991); Scott & Lawrence (1992). EGYPT: Sydenham *et al.* (1991). FRANCE: Sydenham *et al.* (1993b). GERMANY: Usleber *et al.* (1994). HUNGARY: Sydenham *et al.* (1993b). INDIA: Chatterjee & Mukkerjee (1994). ITALY: Caramelli *et al.* (1993); Doko & Visconti (1993, 1994.; Usleber et al. (1994). JAPAN: Ueno *et al.* (1993). KENYA: Sydenham *et al.* (1993b). NEPAL: Ueno *et al.* (1993). NEW ZEALAND: Mirocha *et al.* (1992b). PERU: Sydenham *et al.* (1991). POLAND: Chelkowski & Lew (1992). SOUTH AFRICA: Sydenham *et al.* (1990a,b, 1991); Rheeder *et al.* (1992); Thiel *et al.* (1992); Viljoen *et al.* (1993); Rheeder *et al.* (1994). SWITZERLAND: Pittet *et al.* (1992). UNITED KINGDOM: Scudamore & Chan (1993). U S A: Plattner *et al.* (1990); Harrison *et al.* (1990); Wilson *et al.* (1990); Ross *et al.* (1991a,b); Sydenham *et al.* (1991); Thiel *et al.* (1991a); Colvin & Harrison (1992); Eppley & Stack (1992); Haschek *et al.* (1992); Osweiler *et al.* (1992); Sydenham *et al.* (1992a); Thiel *et al.* (1992); Wilson *et al.* (1992); Binkerd *et al.* (1993); Chamberlain *et al.* (1993); Grimes *et al.* (1993); Holcomb *et al.* (1993); Hopmans & Murphy (1993); Murphy *et al.* (1993); Price *et al.* (1993); Riley *et al.* (1993); Ross *et al.* (1993); Viljoen *et al.* (1993); Maragos & Richard (1994); Motelin *et al.* (1994); Pestka *et al.* (1994); Ross (1994); Sydenham *et al.* (1994). VENEZUELA: Stack & Eppley (1992). ZIMBABWE: Sydenham *et al.* (1993b).

Natural Occurrence of FB_1, FB_2 and FB_3 in Corn and Corn-Based Foods and Feeds

Corn, Corn Screenings and Corn-Based Feeds - LEm in Horses. Plattner *et al.* (1990); Shephard *et al.* (1990); Wilson *et al.* (1990); Ross *et al.* (1991a,b); Thiel *et al.* (1991a); Sydenham *et al.* (1992a,b); Wilson *et al.* (1992); Caramelli *et al.* (1993); Ross *et al.* (1993); Ross (1994).

Corn, Corn Screenings and Corn-Based Feeds - PES in Pigs. Harrison *et al.* (1990); Ross *et al.* (1991a); Colvin & Harrison (1992); Haschek *et al.* (1992); Osweiler *et al.* (1992); Ross *et al.* (1992); Riley *et al.* (1993); Motelin *et al.* (1994); Ross (1994).

Home-Grown Corn in High Incidence Areas of EC in Humans. Sydenham *et al.* (1990a,b; 1991); Rheeder *et al.* (1992); Chu & Li (1994).

Corn Exported from the USA. Thiel *et al.* (1992); Scudamore & Chan (1993); Ueno *et al.* (1993); Viljoen *et al.* (1993); Rheeder *et al.* (1994); Sydenham *et al.* (1994).

Commercial Corn and Corn-Based Human Foods. Shephard *et al.* (1990); Sydenham *et al.* (1991); Scott & Lawrence (1992); Thiel *et al.* (1992); Stack & Eppley (1992); Pittet *et al.* (1992); Holcomb *et al.* (1993); Hopmans & Murphy (1993); Doko & Visconti (1993); Scudamore & Chan (1993); Ueno *et al.* (1993); Viljoen *et al.* (1993); Doko & Visconti (1994); Pestka *et al.* (1994); Usleber *et al.* (1994).

Highest Reported Levels of FB_1 and FB_2. See Table 1.

FUMONISINS : IMPACT

Implications of Fumonisins for Human and Animal Health

- Fumonisins Cause Outbreaks of Mycotoxicoses in Farm Animals
- Fumonisin B_1 is Carcinogenic to Rats
- Fumonisins Occur World-Wide in Corn and Corn-Based Feeds and Foods

Table 1. Highest reported levels of FB_1 and FB_2

| Corn Product | Country | Level ($\mu g/g$) | | Reference |
		FB_1	FB_2	
Corn Screenings - PES	U S A	330	48	Ross *et al.* (1991a)
Corn Screenings - LEM	U S A	160	49	Ross *et al.* (1993)
Home-grown Corn - EC	China	155	—	Chu & Li (1994)
Home-grown Corn - EC	Transkei	117	23	Rheeder *et al.* (1992)
Export Corn to Japan	U S A	4.1	10.2	Ueno *et al.* (1993)
Export Corn to South Africa	U S A	7.6	3.1	Viljoen *et al.* (1993)
Commercial Human Foods:				
"Blue Corn Meal"	U S A	6.8	—	Pestka et al. (1994)
Corn Meal and Grits	U S A	2.8	1.1	Sydenham et al. (1991)
Extruded Corn and Polenta	Italy	6.1	0.9	Doko & Visconti (1994)
Health Food	Zimbabwe	3.6	0.9	Sydenham et al. (1993b)

These and other implications have been assessed in several recent reviews (Marasas, 1993, 1994; Marasas *et al.*, 1993a,b; Nelson *et al.*, 1993, 1994; Norred, 1993; Norred & Voss, 1994; Riley *et al.*, 1993b; Ross, 1994; Thiel *et al.*, 1992).

Fumonisins Are Possibly Carcinogenic to Humans

The International Agency For Research on Cancer (IARC) evaluated the "toxins derived from *Fusarium moniliforme*" as Group 2B carcinogens, i.e. possibly carcinogenic to humans (IARC, 1993; Vainio *et al.*, 1993).

Risk Assessment of Fumonisins

According to Kuiper-Goodman (1990) there are two major components of risk assessment:

- Exposure Assessment
- Hazard Assessment

Exposure Assessment

Human exposure is calculated from estimates of the level of a mycotoxin in foodstuffs and food intake, or from direct measurements on humans (Kuiper-Goodman, 1990), and expressed as:

- PDI = Probable Daily Intake.

Exposure Assessment of Fumonisins

Thiel *et al.* (1992) calculated PDI's based on the naturally occurring levels of total fumonisins in home-grown corn in Transkei and the assumption that a 70 kg person consumes 460 g corn per day:
Person eating "healthy" corn: 14 $\mu g/kg/day$;
Person eating "moldy" corn: 440 $\mu g/kg/day$.

Hazard Assessment

Hazard is an intrinsic property of a mycotoxin with reference to toxicological effects in a specific species at a specific level of exposure (Kuiper-Goodman, 1990). Hazard is calculated from toxicological studies in experimental animals and expressed as:

NOEL = No Observed Effect Level;

TD_{50} = Dose Rate at which 50% of Animals Develop Cancer. Extrapolation is done by means of Safety Factors (NOEL: 100-1000 for toxins and 1000-5000 for carcinogens; TD_{50}: 50 000) to estimate:

- TDI = Tolerable Daily Intake.

Hazard Assessment of Fumonisins

TDI values have to be calculated with respect to carcinogenic risk to humans of fumonisins in corn and corn-based foods in order to establish tolerance levels. The mechanism of carcinogenicity of the fumonisins is important in determining the Safety Factor to calculate the TDI (Gelderblom et al., 1995, unpublished data).

FUMONISINS: MECHANISM OF ACTION

FB_1 Is a Non-Genotoxic Carcinogen

The disruption of sphingolipid metabolism is an early molecular event in the onset and progression of cell injury and the associated toxicological effects caused by FB_1 (Riley et al., 1994b). With respect to the mechanism of carcinogenicity, however, the inhibition of hepatocyte proliferation by FB_1 together with the hepatotoxicity appear to be critical determinants for cancer initiation and promotion (Riley et al., 1994b).

On the other hand Schroeder et al. (1994) reported that FB_1 disrupts sphingolipid metabolism and stimulates DNA synthesis in cultured Swiss 3T3 fibroblasts at a concentration of 10 μM. The mitogenic activity of FB_1 via the accumulation of sphingoid bases, may be the molecular mechanism of carcinogenicity.

Gelderblom et al. (1994) concluded that the hepatotoxic and hepatocarcinogenic effects of FB_1 cannot be separated and a threshold value for cancer initiation exists as a function of time. If FB_1 acts as a non-genotoxic carcinogen solely through a cytotoxic mechanism, it is likely to have a no-effect threshold.

FUMONISINS: TOLERANCE LEVELS

Tolerance levels for fumonisins in corn and corn-based feeds and foods should be realistic, scientifically sound and economically reasonable (Marasas et al., 1993a,b). These levels can only be established following comprehensive exposure and hazard assessment and on the basis of reliable PDI and TDI values.

Unrealistic Tolerance Levels

If tolerance levels are too high, human and animal health may be endangered, e.g. levels >8000 ng/g may result in LEM in horses (Marasas et al., 1993a,b).

Table 2. Fumonisins : Publications

Year	Number of publications
1988	3
1989	3
1990	20
1991	23
1992	65
1993	85
1994	62 (incomplete)

If tolerance levels are too low, the economic consequences may be disastrous, e.g. levels < 1000 ng/g may result in 34.5% of commercial corn products for human consumption in the USA being declared illegal (Thiel *et al.*, 1992).

FUMONISINS : PUBLICATIONS

During the first seven years following the isolation and characterization of the fumonisins in 1988, more than 260 papers dealing with this important new group of food-borne carcinogens have been published (Table 2). The high level of current research activity on these compounds is evident from the Symposium on Fumonisins in Food, Anaheim, California, April 4-6, 1995. It is safe to conclude that we will be hearing more about the fumonisins well into the next century.

REFERENCES

Abbas, H.K.; Boyette, C.D.; Hoagland, F.E.; Vesonder, R.F. Bioherbicidal potential of *Fusarium moniliforme* and its phytotoxin, fumonisin. *Weed Sci.* **1991**, 39, 673-677.

Abbas, H.K.; Boyette, C.D. Phytotoxicity of Fumonisin B$_1$ on weed and crop species. *Weed Technol.* **1992a**, 6, 548-552.

Abbas, H.K.; Paul R.N.; Boyette, C.D.; Duke, S.O.; Vesonder, R.F. Physiological and ultrastructural effects of fumonisin on jimsonweed leaves. *Can. J. Bot.* **1992b**, 70, 1824-1833.

Abbas, H.K.; Gelderblom, W.C.A.; Cawood, M.E.; Shier, W.T. Biological activities of fumonisins, mycotoxins from *Fusarium moniliforme*, in jimsonweed (*Datura stramonium* L.) and mammalian cell cultures. *Toxicon* **1993**, 31, 345-353.

Bezuidenhout, S.C.; Gelderblom, W.C.A.; Gorst-Allman, C.P.; Horak, R.M.; Marasas, W.F.O.; Spiteller, G.; Vleggaar, R. Structure elucidation of the fumonisins, mycotoxins from *Fusarium moniliforme*. *J. Chem. Soc. Chem. Commun.* **1988**, 743-745.

Binkerd, K.A.; Scott, D.M.; Everson, R.J.; Sullivan, J.H.; Robinson, F.R.; Fumonisin contamination of the 1991 Indiana corn crop and its effects on horses. *J. Vet. Diagn. Invest.* **1993**, 5, 653-655.

Butler, T. Notes on a feeding experiment to produce leucoencephalitis in a horse with positive results. *Amer. Vet. Rev.* **1902**, 26, 748-751.

Caramelli, M.; Dondo, A.; Cantini Cortellazzi G.; Visconti, A.; Minervini, F.; Doko, M.B.; Guarda, F. Leucoencefalomalacia nell'equino da fumonisine: prima segnalazione in Italia. *Ippologia* **1993**, 4, 49-56.

Cawood, M.E.; Gelderblom, W.C.A.; Vleggaar, R.; Behrend, Y.; Thiel, P.G.; Marasas, W.F.O. Isolation of the fumonisin mycotoxins - a quantitative approach. *J. Agric. Food Chem.* **1991**, 39, 1958-1962.

Cawood, M.E.; Gelderblom, W.C.A.; Alberts, J.F.; Snyman, SD. Interaction of ^{14}C-labelled fumonisin B mycotoxins with primary rat hepatocyte cultures. *Food Chem. Toxicol.* **1994**, 32, 627-632.

Chamberlain, W.J.; Bacon, C.W.; Norred, W.P.; Voss, K.A. Levels of fumonisin B$_1$ in corn naturally contaminated with aflatoxins. *Food Chem. Toxicol.* **1993**, 31, 995-998.

Chatterjee, D.; Mukherjee, S.K. Contamination of Indian maize with fumonisin B₁ and its effects on chicken macrophage. *Lett. Appl. Microbiol.* **1994**, 18, 251-253.

Chelkowski, J.; Lew, H. *Fusarium* species of Liseola Section - occurrence in cereals and ability to produce fumonisins. *Microbiol. Alim. Nutr.* **1992**, 10, 49-53.

Chu, F.S.; Li, G.Y. Simultaneous occurrence of fumonisin B₁ and other mycotoxins in moldy corn collected from the People's Republic of China in regions with high incidence of esophageal cancer. *Appl. Environ. Microbiol.* **1994**, 60, 847-852.

Colvin, B.M.; Harrison, L.R. Fumonisin-induced pulmonary edema and hydrothorax in swine. *Mycopathologia* **1992**, 117, 79-82.

Doko, M.B.; Visconti, A. Fumonisin contamination of maize and maize-based foods in Italy. In *Occurrence and Significance of Mycotoxins*; Scudamore, K.A., Ed.; Central Sci. Lab.: Slough, UK, **1993**; pp 49-55.

Doko, M.B.; Visconti, A. Occurrence of fumonisins B₁ and B₂ in corn and corn-based human foodstuffs in Italy. *Food Addit. Contam.* **1994**, 11, 433-439.

Dombrink-Kurtzman, M.A.; Bennett, G.A.; Richard, J.L. An optimized MTT bioassay for determination of cytotoxicity of fumonisins in turkey lymphocytes. *J. Assoc. Off. Anal. Chem. Int.* **1994**, 77, 512-516.

Gelderblom, W.C.A.; Snyman, S.D. Mutagenicity of potentially carcinogenic mycotoxins produced by *Fusarium moniliforme*. *Mycotoxin Res.* **1991**, 7, 46-52.

Gelderblom, W.C.A.; Jaskiewicz, K.; Marasas, W.F.O.; Thiel, P.G.; Horak, R.M.; Vleggaar, R.; Kriek, N.P.J. Fumonisins - novel mycotoxins with cancer-promoting activity produced by *Fusarium moniliforme*. *Appl. Environ. Microbiol.* **1988a**, 54, 1806-1811.

Gelderblom, W.C.A.; Marasas, W.F.O.; Jaskiewicz, K.; Combrinck, S.; Van Schalkwyk, D.J. Cancer promoting potential of different strains of *Fusarium moniliforme* in a short-term cancer initiation /promotion assay. *Carcinogenesis* **1988b**, 9, 1405-1409.

Gelderblom, W.C.A.; Kriek, N.P.J.; Marasas, W.F.O.; Thiel, P.G. Toxicity and carcinogenicity of the *Fusarium moniliforme* metabolite, fumonisin B₁, in rats. *Carcinogenesis* **1991**, 12, 1247-1251.

Gelderblom, W.C.A.; Marasas, W.F.O.; Thiel, P.G.; Vleggaar, R.; Cawood, M.E. Fumonisins: Isolation, chemical characterization and biological effects. *Mycopathologia* **1992a**, 117, 11-16.

Gelderblom, W.C.A.; Semple, E.; Marasas, W.F.O.; Farber, E. The cancer-initiating potential of the fumonisin B mycotoxins. *Carcinogenesis* **1992b**, 13, 433-437.

Gelderblom, W.C.A.; Cawood, M.E.; Snyman, D.; Vleggaar, R.; Marasas, W.F.O. Structure-activity relationships of fumonisins in short-term carcinogenesis and cytotoxicity assays. *Food Chem. Toxicol.* **1993**, 31: 407-414.

Gelderblom, W.C.A.; Cawood, M.E.; Snyman, S.D.; Marasas, W.F.O. Fumonisin B₁ dosimetry in relation to cancer initiation in rat liver. *Carcinogenesis* **1994**, 15, 209-214.

Gilchrist, D.G.; Ward, B.; Moussato, V.; Mirocha, C.J. Genetic and physiologic response to fumonisin and AAL-toxin by intact tissue of a higher plant. *Mycopathologica* **1992**, 117, 57-64.

Grimes, J.L.; Eleazer, T.H.; Hill, J.E. Paralysis of undetermined origin in bobwhite quail. *Avian. Dis.* **1993**, 37, 582-584.

Harel, R.; Futerman, A.H. Inhibition of sphingolipid synthesis affects axonal outgrowth in cultured hippocampal neurons. *J. Biol. Chem.* **1993**, 268, 14476-14481.

Harrison, L.R.; Colvin, B.M.; Greene, J.T.; Newman, L.E.; Cole, J.R. Pulmonary edema and hydrothorax in swine produced by fumonisin B₁, a toxic metabolite of *Fusarium moniliforme*. *J. Vet. Diagn. Invest.* **1990**, 2, 217-221.

Haschek, W.M.; Motelin, G.; Ness, D.K.; Harlin, K.S.; Hall, W.F.; Vesonder, R.F.; Peterson, R.E.; Beasley, V.R. Characterization of fumonisin toxicity in orally and intravenously dosed swine. *Mycopathologia* **1992**, 117, 83-96.

Holcomb, M.; Thompson, H.C.; Hankins, L.J. Analysis of fumonisin B₁ in rodent feed by gradient elution HPLC using precolumn derivatization with FMOC and fluorescence detection. *J. Agric. Food Chem.* **1993**, 41, 764-767.

Hopmans, E.C.; Murphy, P.A. Detection of fumonisins B₁, B₂ and B₃ and hydrolyzed fumonisin B₁ in corn-containing foods. *J. Agric. Food Chem.* **1993**, 41, 1655-1658.

IARC. *Some Naturally Occurring Substances: Food Items and Constituents, Heterocyclic Aromatic Amines and Mycotoxins*; IARC Monographs on the Evaluation of Carcinogenic Risks to Humans, Vol 56: IARC, Lyon, **1993**.

Kellerman, T.S.; Marasas, W.F.O.; Pienaar, J.G.; Naude, T.W. A mycotoxicosis of Equidae caused by *Fusarium moniliforme* Sheldon. *Onderstepoort J. Vet. Res.* **1972**, 39, 205-208.

Kellerman, T.S.; Marasas, W.F.O.; Thiel, P.G.; Gelderblom, W.C.A.; Cawood, M.E.; Coetzer, J.A.W. Leuk-oencephalomalacia in two horses induced by oral dosing of fumonisin B_1. *Onderstepoort J. Vet. Res.* **1990**, 57, 269-275.

Kriek, N.P.J.; Kellerman, T.S.; Marasas, W.F.O. A comparative study of the toxicity of *Fusarium verticillioides* (= *F. moniliforme*) to horses, primates, pigs, sheep and rats. Onderstepoort *J. Vet. Res.* **1981**, 48, 129-131.

Kuiper-Goodman, T. Uncertainties in the risk assessment of three mycotoxins: aflatoxin, ochratoxin, and zearalenone. *Can. J. Physiol. Pharmacol.* **1990**, 68, 1017-1024.

La Grenade, C.E.F.; Bean G.A. Cytotoxicity of *Fusarium moniliforme* metabolites. *Biodet. Res.* **1990**, 3, 189-195.

Lamprecht, S.C.; Marasas, W.F.O.; Alberts, J.F.; Cawood, M.E.; Gelderblom, W.C.A.; Shephard, G.S.; Thiel, P.G.; Calitz, F.J. Phytotoxicity of fumonisins and TA-toxin to corn and tomato. *Phytopathology* **1994**, 84, 383-391.

Laurent, D.; Platzer, N.; Kohler, F.; Sauviat, M.P.; Pellegrin, F. Macrofusine et micromoniline: Deux nouvelles mycotoxines isolees de Mais infeste par *Fusarium moniliforme* Sheldon. *Microbiol. Alim. Nutr.* **1989a**, 7, 9-16.

Laurent, D.; Pellegrin, F.; Kohler, F.; Costa, R.; Thevenon, J.; Lambert, C.; Huerre, M. La fumonisine B_1 dans la pathogenie de la leucoencepfalomalacie equine. Microbiol. *Alim. Nutr.* **1989b**, 7, 285-291.

Lew, H.; Adler, A.; Edinger, W. Moniliformin and the European corn borer (*Ostrinia nubilalis*). *Mycotoxin Res.* **1991**, 7, 71-76.

Maragos, C.M.; Richard, J.L. Quantitation and stability of fumonisins B_1 and B_2 in milk. *J. Assoc. Off. Anal. Chem. Int.* **1994**, 77, 1162-1167.

Marasas, W.F.O. Occurrence of *Fusarium moniliforme* and fumonisins in maize in relation to human health. *S. Afr. Med. J.* **1993**, 83, 382-383.

Marasas, W.F.O. *Fusarium*. In *Foodborne Disease Handbook*; Hui, Y.M., Gorham, J.R., Murrell, K.D., Cliver, D.O., Eds.; Marcel Dekker: New York, **1994**; Vol. 2, pp 521-573.

Marasas, W.F.O.; Kellerman, T.S.; Pienaar, J.G.; Naude, T.W. Leukoencephalomalacia: a mycotoxicosis of Equidae caused by *Fusarium moniliforme* Sheldon. *Onderstepoort J. Vet. Res.* **1976**, 43, 113-122.

Marasas, W.F.O.; Wehner, F.C.; Van Rensburg, S.J.; Van Schalkwyk D.J. . Mycoflora of corn produced in human esophageal cancer areas in Transkei, Southern Africa. *Phytopathology* **1981**, 71, 792-796.

Marasas, W.F.O.; Nelson, P.E.; Toussoun, T.A. *Toxigenic Fusarium Species: Identity and Mycotoxicology*; Pennsylvania State University Press; University Park, PA, **1984a**; pp 216-246.

Marasas, W.F.O.; Kriek, N.P.J.; Fincham, J.E.; Van Rensburg, S.J. Primary liver cancer and oesophageal basal cell hyperplasia in rats caused by *Fusarium moniliforme*. *Internat. J. Cancer* **1984b**, 34, 383-387.

Marasas, W.F.O.; Kellerman, T.S.; Gelderblom, W.C.A.; Coetzer J.A.W.; Thiel, P.G.; Van der Lugt, J.J. Leukoencephalomalacia in a horse induced by Fumonisin B_1, isolated from *Fusarium moniliforme*. *Onderstepoort J. Vet. Res.* **1988**, 55, 197-203.

Marasas, W.F.O.; Shephard, G.S.; Sydenham, E.W.; Thiel, P.G. World-wide contamination of maize with fumonisins: Foodborne carcinogens produced by *Fusarium moniliforme*. In *Cereal Science and Technology: Impact on a changing Africa*; Taylor, J.R.N., Randall, P.D., Viljoen, J.H., Eds.; CSIR, Pretoria, **1993a**, pp. 791-805.

Marasas, W.F.O.; Thiel, P.G.; Gelderblom, W.C.A; Shephard, G.S.; Sydenham, E.W.; Rheeder, J.P. Fumonisins produced by *Fusarium moniliforme* in maize: Foodborne carcinogens of Pan African importance. *African Newslett. Occup. Health Safety Suppl.* **1993b**, 2, 11-18.

Merrill, A.H.; Van Echten, G.; Wang, E.; Sandhoff, K. Fumonisin B_1 inhibits sphingosine (sphinganine) N-acyltransferase and de novo sphingolipid biosynthesis in cultured neurons *in situ*. *J. Biol. Chem.* **1993**, 268, 27299-27306.

Mirocha, C.J.; Gilchrist, D.G.; Shier, W.T.; Abbas, H.K.; Wen, Y.; Vesonder, R.F. AAL Toxins, fumonisins (biology and chemistry) and host-specificity concepts. *Mycopathologia* **1992a**, 117, 47-56.

Mirocha, C.J.; Mackintosh, C.G.; Mirza, U.A.; Xie, W.; Xu, Y.; Chen, J. Occurrence of fumonisin in forage grass in New Zealand. *Appl. Environ. Microbiol.* **1992b**, 58, 3196 - 3198.

Motelin, G.K.; Haschek, W.M.; Ness, D.K.; Hall, W.F.; Harlin, K.S.; Schaeffer, D.J.; Beasley, V.R. Temporal and dose-response features in swine fed corn screenings contaminated with fumonisin mycotoxins. *Mycopathologia* **1994**, 126, 27-40.

Murphy, P.A.; Rice, L.G.; Ross, P.F. Fumonisin B_1, B_2 and B_3 content of Iowa, Wisconsin, and Illinois corn and corn screenings. J. Agric. Food Chem. **1993**, 41, 263-266.

Nelson, P.E.; Desjardins, A.E.; Plattner, R.D. Fumonisins, mycotoxins produced by *Fusarium* species: Biology, chemistry and significance. *Ann. Rev. Phytopathol.* **1993**, 31, 233-252.

Nelson, P.E.; Dignani, M.C.; Anaissie, E.J. Taxonomy, biology and clinical aspects of *Fusarium* species. *Clin Microbiol Rev* **1994**, 7, 479-504.

Norred, W.P. Fumonisins - mycotoxins produced by *Fusarium moniliforme*. *J. Toxicol. Environ. Health* **1993**, 38, 309-328.

Norred, W.P.; Voss, K.A. Toxicity and role of fumonisins in animal diseases and human esophageal cancer. *J. Food Prot.* **1994**, 57, 522-527.

Norred WP, Bacon CW, Plattner RD, Vesonder RF. Differential cytotoxicity and mycotoxin content among isolates of *Fusarium moniliforme*. *Mycopathologia* **1991**, 115, 37-43.

Norred, W.P.; Wang, E.; Yoo, H.; Riley, R.T.; Merrill, A.H. *In vitro* toxicology of fumonisins and the mechanistic implications. *Mycopathologia* **1992a**, 117, 73-78.

Norred, W.P.; Plattner, R.D.; Vesonder, R.F.; Bacon, C.W.; Voss K.A. Effects of selected secondary metabolites of *Fusarium moniliforme* on unscheduled synthesis of DNA by rat primary hepatocytes. *Food Chem. Toxicol.* **1992b**, 30, 233-237.

Osweiler, G.D.; Ross, P.F.; Wilson, T.M.; Nelson, P.E.; Witte, S.T.; Carson, T.L.; Rice, L.G.; Nelson, H.A. Characterization of an epizootic of pulmonary edema in swine associated with fumonisin in corn screenings. *J. Vet. Diagn. Invest.* **1992**, 4, 53-59.

Park, D.L.; Rua, S.M.; Mirocha, C.J.; Abd-Alla, E.A.M.; Weng, C.Y. Mutagenic potentials of fumonisin contaminated corn following ammonia decontamination procedure. *Mycopathologia* **1992**, 117, 105-108.

Pestka, J.J.; Azcona-Olivera, J.I.; Plattner, R.D.; Minervini, F.; Doko, M.B.; Visconti, A. Comparative assessment of fumonisin in grain-based foods by ELISA, GC-MS and HPLC. *J. Food Prot.* **1994**, 57, 169-172.

Pittet, A.; Parisod, V.; Schellenberg, M. Occurrence of fumonisins B_1 and B_2 in corn-based products from the Swiss market. *J. Agric. Food. Chem.* **1992**, 40, 1352-1354.

Plattner, R.D.; Norred, W.P.; Bacon, C.W.; Voss, K.A.; Peterson, R.; Shackelford, D.D.; Weisleder, D. A method of detection of fumonisins in corn samples associated with field cases of equine leukoencephalomalacia. *Mycologia* **1990**, 82, 698-702.

Price, W.D.; Lovell, R.A.; McChesney, D.G. Naturally occurring toxins in feedstuffs: Center for Veterinary Medicine perspective. *J. Anim. Sci.* **1993**, 71, 2556-2562.

Rheeder, J.P.; Marasas, W.F.O.; Thiel, P.G.; Sydenham, E.W.; Shephard, G.S.; Van Schalkwyk, D.J. *Fusarium moniliforme* and fumonisins in corn in relation to human oesophageal cancer in Transkei. *Phytopathology* **1992**, 82, 353-357.

Rheeder, J.P.; Sydenham, E.W.; Marasas, W.F.O.; Thiel, P.G.; Shephard, G.S.; Schlechter, M.; Stockenström, S.; Cronje, D.E.; Viljoen, J.H. Ear-rot fungi and mycotoxins in South African corn of the 1989 crop exported to Taiwan. *Mycopathologia* **1994**, 127, 35-41.

Riley, R.T.; An, N-H.; Showker, J.L.; Yoo, H-S.; Norred, W.P.; Chamberlain, W.J.; Wang, E.; Merrill, A.H.; Motelin, G.; Beasley, V.R.; Haschek, W.M. Alteration of tissue and serum sphinganine to sphingosine ratio: An early biomarker of exposure to fumonisin-containing feeds in pigs. *Toxicol. Appl. Pharmacol.* **1993a**, 118, 105-112.

Riley, R.T.; Norred, W.P.; Bacon, C.W. Fungal toxins in foods: Recent concerns. *Ann. Rev. Nutr.* **1993b**, 13, 167-289.

Riley, R.T.; Hinton, D.M.; Chamberlain, W.J.; Bacon, C.W.; Wang, E; Merrill, A.H.; Voss, K.A. Dietary fumonisin B_1 induces disruption of sphingolipid metabolism in Sprague-Dawley rats: A new mechanism of nephrotoxicity. *J. Nutr.* **1994a**, 124, 594-603.

Riley, R.T.; Voss, K.A.; Yoo, H-S.; Gelderblom, W.C.A.; Merrill, A.H. Mechanism of fumonisin toxicity and carcinogenesis. *J. Food Prot.* **1994b**, 57, 638-645.

Ross, P.F. What are we going to do with this dead horse? *J. Assoc. Off. Anal. Chem. Int.* **1994**, 77, 491-494.

Ross, P.F.; Nelson, P.E.; Richard, J.L.; Osweiler, G.D.; Rice, L.G.; Plattner, R.D.; Wilson, T.M. Production of fumonisins by *Fusarium moniliforme* and *Fusarium proliferatum* isolates associated with equine leukoencephalomalacia and a pulmonary edema syndrome in swine. *Appl. Environ. Microbiol.* **1990**, 56, 3225-3226.

Ross, P.F.; Rice, L.G.; Plattner, R.D.; Osweiler, G.D.; Wilson T.M.; Owens, D.L.; Nelson, H.A.; Richard, J.L. Concentrations of fumonisin B_1 in feeds associated with animal health problems. *Mycopathologia* **1991a**, 114, 129-135.

Ross, P.F.; Rice, L.G.; Reagor, J.C.; Osweiler, G.D.; Wilson, T.M.; Nelson, H.A.; Owens, D.L.; Plattner, R.D.; Harlin, K.A.; Richard, J.L.; Colvin, B.M.; Banton, M.I. Fumonisin B_1 concentrations in feeds from 45 confirmed equine leukoencephalomalacia cases. *J. Vet. Diagn. Invest.* **1991b**, 3, 238-241.

Ross, P.F.; Rice, L.G.; Osweiler, G.D.; Nelson, P.E.; Richard, J.L.; Wilson, TM. A review and up-date of animal toxicoses associated with fumonisin-contaminated feeds and production of fumonisins by *Fusarium* isolates. *Mycopathologia* **1992**. 117: 109-114.

Ross, P.F.; Ledet, A.E.; Owens, D.L.; Rice, L.G.; Nelson, H.A.; Osweiler, G.D.; Wilson, T.M. Experimental equine leukoencephalomalacia, toxic hepatosis, and encephalopathy caused by corn naturally contaminated with fumonisins. *J. Vet. Diagn. Invest.* **1993**, 5, 69-74.

Saccardo, P.A. Fungi Italici Autographice Delinati, Fig 789. Patavii. **1881**.

Schroeder, J.J.; Crane, H.M.; Xia, J.; Liotta, D.C.; Merrill, A.H. Disruption of sphingolipid metabolism and stimulation of DNA synthesis by fumonisin B_1. A molecular mechanism for carcinogenesis associated with *Fusarium moniliforme*. *J. Biol. Chem.* **1994**, 269, 3475-3481.

Scott, P.M.; Lawrence, G.A. Liquid chromatographic determination of fumonisins with 4-fluoro-7-nitrobenzofurazan. *J. Assoc. Off. Anal. Chem. Int.* **1992**, 75, 829-834.

Scudamore, K.A.; Chan, H.K. Occurrence of fumonisin mycotoxins in maize and millet imported into the United Kingdom. In *Occurrence and Significance of Mycotoxins*; Scudamore, K.A., Eds.; Central Sci. Lab. Slough: UK, **1993**; pp 186-189.

Shelby, R.A.; White, D.G.; Bauske, E.M. Differential fumonisin production in maize hybrids. *Plant Dis.* **1994**, 78, 582-584.

Sheldon, L. A corn mold (*Fusarium moniliforme* n. sp.). *17th Ann. Rep.* Agric. Exp. Stn.: Nebraska, USA, **1904**; pp 23-32.

Shephard, G.S.; Sydenham, E.W.; Thiel, P.G.; Gelderblom, W.C.A.; Quantitative determination of fumonisins B_1 and B_2 by highperformance liquid chromatography with fluorescence detection. *J. Liq. Chromatogr.* **1990**, 13, 2077-2087.

Shier, W.T.; Abbas, H.K.; Mirocha, C.J. Toxicity of the mycotoxins fumonisins B_1 and B_2 and *Alternaria alternata* f.sp. *lycopersici* toxin (AAL) in cultured mammalian cells. *Mycopathologia* **1991**, 116, 97-104.

Stack, M.E.; Eppley, R.M. Liquid chromatographic determination of fumonisins B_1 and B_2 in corn and corn products. *J. Assoc. Off. Anal. Chem. Int.* **1992**, 75, 834-837.

Sydenham, E.W.; Gelderblom, W.C.A.; Thiel, P.G.; Marasas, W.F.O. Evidence for the natural occurrence of fumonisin B_1, a mycotoxin produced by *Fusarium moniliforme*, in corn. *J. Agric. Food Chem.* **1990a**, 38, 285-290.

Sydenham, E.W.; Thiel, P.G.; Marasas, W.F.O.; Shephard, G.S.; Van Schalkwyk, D.J.; Koch, K.R. Natural occurrence of some *Fusarium* mycotoxins in corn from low and high esophageal cancer prevalence areas of the Transkei, Southern Africa. *J. Agric. Food Chem.* **1990b**, 38, 1900-1903.

Sydenham, E.W.; Shephard, G.S.; Thiel, P.G.; Marasas, W.F.O.; Stockenström, S. Fumonisin contamination of commercial corn-based human foodstuffs. *J. Agric. Food Chem.* **1991**, 39, 2014-2018.

Sydenham, E.W.; Shephard, G.S.; Thiel, P.G. Liquid chromatographic determination of fumonisin B_1, B_2 and B_3 in foods and feeds. *J. Assoc. Off. Anal. Chem.* **1992a**, 75, 313-318.

Sydenham, E.W.; Marasas, W.F.O.; Shephard, G.S.; Thiel, P.G.; Hirooka, E.Y. Fumonisin concentrations in Brazilian feeds associated with field outbreaks of confirmed and suspected animal mycotoxicoses. *J. Agric. Food Chem.* **1992b**, 40, 994-997.

Sydenham, E.W.; Shephard, G.S.; Thiel, P.G.; Marasas, W.F.O.; Rheeder, J.P.; Peralta Sanhueza, C.E.; Gonzalez, H.H.Z.; Resnik S.L. Fumonisins in Argentinian field-trial corn. *J. Agric. Food Chem.* **1993a**, 41, 891-895.

Sydenham, E.W.; Shephard, G.S.; Gelderblom, W.C.A.; Thiel, P.G.; Marasas, W.F.O. Fumonisins: Their implications for human and animal health. In *Occurrence and Significance of Mycotoxins*; Scudamore, K.A. Ed.; Central Sci. Lab.: Slough, UK, **1993b**; pp 42-48.

Sydenham, E.W.; Van der Westhuizen, L.; Stockenström, S.; Shephard, G.S.; Thiel, P.G. Fumonisin-contaminated maize: physical treatment for the partial decontamination of bulk shipments. *Food Addit. Contam.* **1994**, 11, 25-32.

Tanaka, T.; Abbas, H.K.; Duke, S.O. Structure-dependent phytotoxicity of fumonisins and related compounds in a duckweed bioassay. *Phytochemistry* **1993**, 33, 779-785.

Thiel, P.G.; Marasas, W.F.O.; Sydenham, E.W.; Shephard, G.S.; Gelderblom, W.C.A.; Nieuwenhuis, J.J. Survey of fumonisin production by *Fusarium* species. *Appl. Environ. Microbiol.* **1991a**, 57, 1089-1093.

Thiel, P.G.; Shephard, G.S.; Sydenham, E.W.; Marasas, W.F.O.; Nelson, P.E.; Wilson, T.M. Levels of fumonisin B_1 and B_2 in feeds associated with confirmed cases of equine leukoencephalomalacia. *J. Agric. Food Chem.* **1991b**, 39, 109-111.

Thiel. P.G.; Marasas, W.F.O.; Sydenham, E.W.; Shephard, G.S.; Gelderblom, W.C.A. The implications of naturally occurring levels of fumonisins in corn for human and animal health. *Mycopathologia* **1992**, 117, 3-9.

Ueno, Y.; Aoyama, S.; Sugiura, Y.; Wang, D.S.; Lee, U.S.; Hirooka, E.Y.; Hara, S.; Karki, T.; Chen, G,; Yu, S.Z. A limited survey of fumonisins in corn and corn-based products in Asian countries. *Mycotoxin Res.* **1993**, 9, 27-34.

Usleber, E.; Straka, M.; Terplan, G. Enzyme immunoassay for fumonisin B_1 applied to corn-based food. *J. Agric. Food Chem.* **1994**, 42, 1392 - 1396.

Vainio, H.; Heseltine, E.; Wilbourn, J. Report on an IARC Working Group meeting on some naturally occurring substances. *Int. J. Cancer* **1993**, 53, 535-537.

Vesonder, R.F.; Labeda, D.P.; Peterson, R.E. Phytotoxic activity of selected water-soluble metabolites of *Fusarium* against *Lemna minor* L. (duckweed). Mycopathologia **1992a**, 118, 185-189.

Vesonder, R.F.; Petersen, R.E.; Labeda, D.; Abbas, H.K. Comparative phytotoxicity of the fumonisins, AAL-Toxin and yeast sphingolipids in *Lemna minor* L. (duckweed). *Arch. Environ. Contam. Toxicol.* **1992b**, 23, 464-467.

Viljoen, J.H.; Marasas, W.F.O.; Thiel, P.G. Fungal infection and mycotoxin contamination of commercial maize. In *Cereal Science and Technology: Impact on a Changing Africa.* Taylor, J.R.N., Randall, P.G., Viljoen, J.H., Eds; CSIR: Pretoria, SA, **1993**; pp 837-853.

Voss, K.A.; Norred, W.P.; Plattner, R.D.; Bacon, C.W. Hepatotoxicity and renal toxicity in rats of corn samples associated with field cases of equine leukoencephalomalacia. *Food Chem. Toxicol.* **1989**, 27, 89-96.

Voss, K.A.; Chamberlain, W.J.; Bacon, C.W.; Norred, W.P. A preliminary investigation on renal and hepatic toxicity in rats fed purified fumonisin B_1. *Natural Toxins* **1993**, 1, 222-228.

Wang, E.; Norred, W.P.; Bacon, C.W.; Riley, R.T.; Merrill, A.H. Inhibition of sphingolipid biosynthesis by fumonisins: implications for diseases associated with *Fusarium moniliforme. J. Biol. Chem.* **1991**, 266, 14486-14490.

Wang, E.; Ross, P.F.; Wilson, T.M.; Riley, R.T.; Merrill, A.H. Increases in serum sphingosine and sphinganine and decreases in complex sphingolipids in ponies given feed containing fumonisins, mycotoxins produced by *Fusarium moniliforme. J. Nutr.* **1992**, 122, 1706-1716.

Wilson, B.J.; Maronpot, R.R. Causative fungus agent of leucoencephalomalacia in equine animals. *Vet. Rec.* **1971**, 88, 484-486.

Wilson, T.M.; Ross, P.F.; Rice, L.G.; Osweiler, G.D.; Nelson, H.A.; Owens, D.L.; Plattner, R.D.; Reggiardo, C.; Noon, T.M.; Pickrell, J.W. Fumonisin B_1 levels associated with an epizootic of equine leukoencephalomalacia. *J. Vet. Diagn. Invest.* 1990, 2, 213-216.

Wilson, T.M.; Ross, P.F.; Owens, D.L.; Rice, L.G.; Jenkins, S.J.; Nelson, H.A. Experimental production of ELEM. A study to determine the minimum toxic dose in ponies. *Mycopathologia* **1992**, 117, 115-120.

Yoo, H-S.; Norred, W.P.; Wang, E.; Merrill, A.H.; Riley, R.T. Fumonisin inhibition of *de novo* sphingolipid biosynthesis and cytotoxicity are correlated in LLC-PK_1, cells. *Toxicol. Appl. Pharmacol.* **1992**, 114, 9-15.

2

OCCURRENCE OF FUMONISINS IN THE U.S. FOOD SUPPLY

Albert E. Pohland

U.S. Food and Drug Administration
200 C St., S.W.
Washington, DC 20204

ABSTRACT

Over the past several years a great deal of interest has been shown in assessing human exposure to the fumonisins. This interest, of course, arises as a result of the finding of fumonisins in foods and the expanding data base on toxicological effects, both acute and sub-acute. The basis for exposure assessment lies in surveys of foods as well as a knowledge of consumption patterns. An overview of such surveys, limited as they are, will be presented along with some evaluation of the methodology used.

INTRODUCTION

Over the past several years a great deal of interest has been shown in assessing human exposure to the fumonisins. This interest is reflected in the large number of research papers published since the report in 1988 describing the structure of this interesting mold metabolite, and the implication that fumonisins were in some way involved in the high incidence of human esophageal cancer in certain areas of Africa (Rheeder et al., 1992). A search of Chemical Abstracts uncovered 106 publications related to fumonisins since 1993, including 1 patent and 3 Ph.D. dissertations. This intense interest arises, of course, as a result of the finding of fumonisins in human foods, and the rapidly expanding understanding of the toxicological effects — acute, sub-acute and chronic — of these compounds. In 1993 an IARC working group concluded that there was "inadequate evidence" for the carcinogenicity of fumonisin B_1 in humans due to oral exposure, but at the same time, concluded that cultures of *F. moniliforme* did show "sufficient evidence" of carcinogenicity in experimental animals (IARC, 1993). Furthermore, the group concluded that both fumonisin-B_1 and fusarin C showed "limited evidence" of carcinogenicity in experimental animals.

It is extremely important to document the occurrence (i.e. incidence and level) of such food contaminants, since a knowledge of the contamination levels in susceptible foodstuffs, along with an estimate of consumption patterns, allows one to calculate human

Fumonisins in Food, Edited by L. Jackson *et al.*
Plenum Press, New York, 1996

exposure. Armed with an exposure estimate, and with an understanding of the known toxicological effects, one can then begin the difficult process of estimating human risk due to such contaminants in food. It is such risk assessments, coupled with economic and political realities, which form the basis for the setting of regulatory levels, the purpose of which is to limit human exposure, maximizing safety, and at the same time preserving economic health and stability.

DISCUSSION

The determination of incidence and average levels of contamination of corn or corn-based foods by fumonisins is not a simple process; numerous factors, some of which are uncontrollable, prevent the accurate determination of such quantities. For example, the natural occurence of the fumonisins is highly dependent on environmental conditions; consequently, one must measure the effects of climate, growing season, etc. In addition, good data is dependent upon: the development of statistically defensible sampling plans; the capabilities of the analytical methods used; and the degree to which quality control procedures are developed and applied. Let me discuss some of these factors briefly before going on to a discussion of the occurrence of fumonisins in the U.S. food supply.

Sampling Variability

Research on sampling has been conducted showing, that for inhomogeneously contaminated lots being sampled, the lower the contamination level, the higher the sampling error (FAO, 1993). The sampling error can be so large that it becomes the major determinant of variability. In such cases, any improvements in analytical methodology can have only a minimal effect on the interlaboratory variability (RSD_R). Each analyte/matrix combination is different with respect to the magnitude of the sampling error. There are, of course, ways to minimize this error, e.g. by taking larger samples from the lot. Subsampling errors also require careful attention, because these error components can be large depending on the success of the grinding/mixing/dividing operations. It has been generally assumed that the level of fumonisins in corn, ear to ear and kernel to kernel is relatively constant, i.e. that the fumonisins are more or less evenly distributed through the lot. If this is true, then the sampling problem becomes manageable. This may be the single biggest difference between the aflatoxins and the fumonisins; however, at this date the question of fumonisin homogeneity has not been adequately researched.

Method Variability

The only way to measure between-laboratory variability (RSD_R) is through an interlaboratory study (a method performance study). Historical evidence has clearly shown that as the contamination level is lowered, the interlaboratory coefficient of variation becomes larger (Horwitz and Albert, 1984). This generalization has been shown to be independent of matrix, analyte or method of analysis. In fact, it has been demonstrated, through the analysis of data from a large number of method performance studies that the coefficient of variation doubles as the concentration is lowered two orders of magnitude, and this principle is well described by the following equation:

$$RSD_R (\%) = 2^{(1-0.5 \, Log \, C)} = 2C^{-0.1505}$$

where concentration (C) is expressed as a decimal fraction. A plot of this equation yields the so-called "Horwitz Horn," showing the predicted precision as a function of analyte concentration (Horwitz et al., 1993). The validity of this equation has been confirmed by examining the data obtained in method performance studies on a wide variety of analytes. By making a simple calculation, one can easily show that at a concentration of 1 ppb (10^{-9}), one would expect an RSD_R of 45%.

One can evaluate the capabilities of an analytical method by comparing the RSD_R determined experimentally in an interlaboratory method performance study with that calculated using the Horwitz equation; the ratio of the two values is termed the HORRAT (Horwitz Ratio).

$$HORRAT = RSD_R(found)/RSD_R(predicted)$$

For well behaved analytical methods this ratio is close to unity. On the other hand, HORRAT values >2 indicate a method out of statistical control and unacceptable with respect to precision.

Occurrence of Fumonisins in Foods

Turning now to the "natural" occurrence of fumonisins in foods, we must keep in mind that much of the published data is not easily interpreted because: (a) in many cases the study fails to report how the sample was collected, what the sample represents in terms of a lot and whether the sampling plan was statistically developed; (b) in most cases the variability of the methodology used was unknown; and (c) in most cases positive samples were not confirmed.

In addition, one should keep in mind that: (a) Corn can be, and is, occasionally highly contaminated. In the South African epidemiology studies, levels as high as 117.5 ppm were found in moldy corn used for human consumption (Thiel et al., 1992). In the U.S., corn associated with ELEM contained up to 126 ppm FB_1; feeds implicated in outbreaks of PPE contained as much as 360 ppm FB_1 (Ross et al., 1991a). Even higher quantities of FB_1 (up to 1026 ppm) have been observed in cultures of *F. moniliforme/F. proliferatum* (Ross et al., 1992). (b) The fumonisins are produced, along with a host of other metabolites by the following mold species (Scott, 1993): *F. moniliforme*, *F. proliferatum*, *F. anthophilum*, *F. subglutinans*, *F. dlamini*, *F. napiforme*, *F. nygamai* and *Alternaria alternata f. sp. lycopersici*. (c) It has been found that *F. moniliforme* and/or *F. proliferatum* have a frequency of 90% or higher on corn, and about 90% of the isolates can be shown to produce fumonisins (Bacon and Nelson, 1994). (d)A wide variety of commodities have been analyzed for fumonisins; to date, however, these metabolites have only been reported in corn and corn based foods and feeds, with the exception of a report of fumonisins in a "black oats" feed from Brazil (Sydenham et al., 1992) and a report of the contamination of New Zealand forage grass with fumonisins (Scott, 1993). Although the fumonisins are produced readily in good yield in rice cultures of *F. moniliforme*, there are no reports of the natural occurrence of fumonisins on rice. (e) In corn the FB_1/FB_2 ratio is usually ca. 3/1, although levels of FB_2 as high as 40% of the total fumonisins have been reported (Ross et al., 1992). (f) Corn screenings have been found to contain, on average higher quantities of fumonisins by a factor of ten than does the corn from which it was derived; there does not appear to be a size associated segregation in these screenings (Mirocha et al., 1992a). (g) The fumonisins are highly polar, lipophobic compounds, extremely soluble in water and acetonitrile-water; they are also quite soluble, but unstable, in methanol (Murphy et al., 1993). (h) The fumonisins are considered to be , generally, heat stable; for example, boiling in water 30 minutes followed by drying at 60°C for 24 hours resulted in no change in FB_1 concentration. However, stability studies in corn

Table 1. Fumonisin methods

Reference	Principle	Detection Limit (µg/g)
Gelderblom et al. (1988)	LC	10
Voss et al. (1989)	TLC	
Shepherd et al. (1990)	LC/OPA	$0.05(B_1), 0.1(B_2)$
Plattner et al. (1990)	GC/MS/TFA	1
Ware et al. (1993)	LC/NDA	0.25
Ross et al. (1991b)	LC/F	1
Korfmacher et al. (1991)	FAB/MS	
Holcomb et al. (1992)	LC/FMOC	0.2
Scott and Lawrence (1992)	LC/NDB-F	0.1
Mirocha et al. (1992b)	Ion Spray-MS	
Ware et al. (1994)	Immunoassay	0.025
Stack and Eppley (1992)	LC/OPA	0.01
Murphy et al. (1993)	LC/OPA	$0.25(B_1), 0.5(B_2)$
Holcomb et al. (1993)	LC/FMOC	0.2
Holcomb et al. (1994)	LC/OPA/TBT	0.3
Scott et al. (1994)	LC/OPA	$0.003(B_1)$ in milk
Trucksess et al. (1995)	LC/OPA	0.004

meal showed that heating at 100°C and above resulted in slow loss of fumonisin (at 220°C complete loss was observed) (Scott, 1993).

Table 1 summarizes the published methods for fumonisins in foods and feeds. In general these methods reflect the analytical difficulties presented by a water soluble, contaminant having no strongly UV absorbing or fluorescing functionality. Consequently most of the methods include the formation of a suitable derivative. Clearly early methods (1988-92) were developed with a limit of detection >1 ppm in mind; today most methods are capable of much lower detection levels. No method performance studies have been reported for the bulk of these procedures; therefore the interlaboratory variability in results is unknown for these methods.

In light of these comments, then, what can we conclude with respect to the occurrence of fumonisins in U.S. foods and feeds, admitting that the data reported to date are largely deficient with respect to information on sampling plans used, the capabilities of the methodology used (e.g. detection limits), and quality control procedures. However, a considerable effort has been made to document contamination of corn by fumonisins, starting with corn-containing animal feeds. The severe animal health problems, both ELEM and PPE,

Table 2. Fumonisins in corn

Location	Year	n (+ve)	Range (ppm)	Mean (ppm)	Reference
Georgia	1991	28(24)	0-1.82	0.87	Chamberlain et al. (1993)
Indiana	1992	113(101)	0-113.4		Scott et al. (1992)
Iowa	1988	22	0-14.9	2.26	Rice and Ross (1994)
	1989	49	0-37.9	2.64	
	1990	50	0-19.1	3.28	
	1991	59	0-15.8	2.94	
	1992	80	0-1.6	0.05	
	1994	41	0-5		Ross, private communication
Pennsylvania	1992	91(91)	0-8.4	0.37	Rice and Ross (1994)
Texas	1992	234(233)	0-25		Latimer, private communication
	1993	250(248)	0->25		"

provided the impetus for this research effort. In Iowa, in a study of 119 corn screening samples, the average contamination level was found to be 20.8 ppm (Murphy et al., 1993). Feeds associated with ELEM contained up to 126 ppm fumonisin B_1; while feeds associated with PPE were found to contain as much as 330 ppm.

Several states have been actively monitoring corn produced within the state; some of these data are shown in Table 2. These data clearly indicate a high frequency of contamination, considerable year-to-year variation in contamination levels, and in some years a fairly high mean contamination level.

It is not surprising, therefore, that one would find fumonisins in corn-based foods. In Table 3 are summarized the results obtained by FDA in its 1994 Compliance Program on Fumonisins. These results were obtained by the FDA's New Orleans laboratory using the Ware procedure. Clearly corn meal is the product most frequently contaminated, and contaminated at the highest levels. The Center for Food Safety and Applied Nutrition/FDA also analyzed a variety of samples (Wood, private communication) (see Table 4), and also observed the frequent occurrence of fumonisins in corn meal; the highest level observed was 4.4 ppm. Of course, the year to year variation in results is considerable, and does not necessarily coincide with the results obtained in any particular state.

The second most frequently contaminated product was corn bread mix with the highest observed level being 2.5 ppm total fumonisins. Of the six corn muffin mix samples analyzed, 4 were positive, one containing over 1 ppm of total fumonisins.

CONCLUSIONS

It seems reasonable to conclude, at this stage of our knowledge, that corn and therefore corn containing foods and feeds, are likely to be contaminated with fumonisins from time to time, and the degree of contamination will be highly dependent upon environmental conditions. The greatest likelihood of high level contamination (i.e. >5 ppm) will be in animal feeds, particularly those containing corn crack-outs, which experience has shown, are usually found to be contaminated (high incidence and level).

It is not surprising, therefore, that corn based foods are also contaminated, the degree of contamination being dependent upon the sensitivity of the method being used. In the survey data presented, those foods in which no fumonisins was detected cannot be identified as "fumonisin free", because the numbers of samples collected and analyzed were too small,

Table 3. Fumonisins in Foods - FDA Compliance Program - 1994

Commodity	Total #	Positive	Total Fumonisins Found (ng/g)			
			0-250	251-500	501-1000	>1000
Corn bread mix	2	0				
Corn chips	6	1	1			
Corn flakes	6	0				
Corn flour	18	16	10	5	1	
Corn grits	8	4	4			
Corn meal	39	37	26	6	2	3
Corn muffin mix	3	0				
Corn pops cereal	3	0				
Corn, shelled	41	35	19	8	7	1
Corn starch	5	0				
Corn tortillas	11	3	3			
Popcorn, unpopped	17	11	10	1		

Table 4. Fumonisins in Foods - CFSAN Survey (1990-1994)

Commodity	n	+ve	Total Fumonisin Found (ng/g)			
			0-250	251-500	501-1000	>1000
Beer	22	15	12	2		1
Corn (canned)	70	28	28			
Corn (frozen)	27	9	8	1		
Corn bran cereal	7	7	4	3		
Corn bread mix	11	11	3	1	3	4
Corn cereal	3	0				
Corn chips	2	0				
Corn flakes	7	2	2			
Corn grits	13	13		9	4	
Corn hominy	1	1	1			
Corn meal	59	59	8	9	17	25
Corn muffin mix	6	4	2	1	1	
Corn pops cereal	2	0				
Corn tortillas	4	3	3			
Corn tortilla chips	2	1	1			
Fiber cereal	3	0				
Popcorn	3	3	3			
Spoon bread mix	1	1			1	

and the sensitivity of the method may not have been sufficient to detect low level contamination. Clearly, however, the data indicate substantial contamination (i.e.>250 ppb) of corn meal, corn bread mix, corn grits, and corn muffin mix. There has been some effort to examine non-corn commodities (cottonseed, peanut meal) for the presence of fumonisins, but no fumonisins were detected in the few samples that were analyzed. In addition, a feeding/transmission study has been reported which indicated no transmission to milk (Holcomb et al., 1993).

We will continue to expand our data base on the occurrence of fumonisins in foods. In doing so, emphasis must be placed on corn-based products. It is important, in reporting such data, that attention be given to carifying what the sample represents in the context of the "lot", what is meant by the "detection limit", i.e. what is meant by a "negative result", and what attempts were made at confirming the identity of the fumonisins reported.

REFERENCES

Bacon, C.W.; Nelson, P.E. Fumonisin production in corn by toxigenic strains of *Fusarium moniliforme* and Fusarium proliferatum. *J. Fd. Prot.* **1994**, 57, 514-521.

Chamberlain, W.J.; Bacon, C.W.; Norred, W.P.; Voss, K.A. Levels of fumonisin B_1 in corn naturally contaminated with aflatoxins. *Fd. Chem. Toxic.* **1993**, 31, 995-998.

Dupuy, J.; Le Bars, P.; Boudra, H.; and LeBars, J. Thermostability of fumonisin B_1, a mycotoxin from *Fusarium moniliforme*, in corn. *Appl. Environ. Microbiol.* **1993**, 59, 2864-2867.

FAO, Sampling plans for aflatoxin analysis in peanuts and corn, in *FAO Food and Nutrition Paper 55*, **1993**, Food and Agricultural Organization of the United Nations, Rome, Italy.

Gelderblom, W.C.A.; Jaskiewicz, K.; Marasas, W.F.O.; Thiel, P.G.; Horak, R.M.; Vleggaar, R.; Kriek, N.P.J. Fumonisins - novel mycotoxins with cancer-promoting activity produced by *Fusarium moniliforme*. *Appl. Environ. Microbiol.* **1988**, 54, 1806-1811

Holcomb, M.; Thompson, H.C. Analysis of fumonisin B1 in rodent feed by gradient elution HPLC using precolumn derivatization with FMOC and fluorescence detection. Abstr. 8, Int. IUPAC Symposium on Mycotoxins and Phytotoxins. Nov. 8-11, 1992, Mexico City, Mexico.

Holcomb, M.; Thompson, H.C.; Hankins, L.J. Analysis of fumonisin B1 in rodent feed by gradient elution HPLC using precolumn derivatization with FMOC and fluorescence detection. *J. Agric. Food Chem.* **1993**, 41, 764-767.

Holcomb, M.; Thompson, H.C.; Lipe, G.; Hankins, L.J. HPLC with electrochemical and fluorescence detection of the OPA/2-methyl-2-propanethiol derivative of fumonisin B_1. *J. Liquid Chromatog.* **1994**, 17, 4121-4129.

Horwitz, W.; Albert, R., Reliability of mycotoxin assays. *JAOAC,* **1984**, 67, 81.

Horwitz, W., Albert, R. and Nesheim, S., Reliability of mycotoxin assays — an update. *J.AOAC Int.*, **1993**, 76, 461-491.

IARC, Toxins derived from *Fusarium moniliforme*: Fumonisins B_1 and B_2 and fusarin C, in *IARC Monographs on the Evaluation of Carcinogenic Risks to Humans*, **1993**, Vol. 56, International Agency for Research on Cancer, Lyon, France

Korfmacher, W.A.; Chiarelli, M.P.; Lay, J.O.; Bloom, J.; Holcomb, M.; McManus, K.T. Characterization of the mycotoxin fumonisin B_1: comparison of thermospray mass spectrometry and electrospray mass spectrometry. *Rapid Commun. Mass Spectrom.* **1991**, 5, 463-468.

Mirocha, C.J.; Mackintosh, C.G.; Mirza, U.A.; Xie, W.; Xu, Y.; Chen, J. Occurrence of fumonisin in forage grass in New Zealand. *Appl. Environ. Microbiol.* **1992a**, 58, 3196-3198.

Mirocha, C.J.; Gilchrist, D.G.; Shier, W.T.; Abbas, H.K.; Wen, Y.; Vesonder, R.F. AAL toxins, fumonisins (biology and chemistry) and host-specificity concepts. *Mycopathologia* **1992b**, 117, 47-56.

Murphy, P.A., Rice, L.G.; Ross, P.F. Fumonisin B_1, B_2 and B_3 content of Iowa, Wisconsin, and Illinois corn and corn screening. *J. Agric. Food Chem.* **1993**, 41, 263-266.

Plattner, R.D.; Norred, W.P.; Bacon, C.W.; Voss, K.A.; Peterson, R.; Schackelford, D.D.; Weisleder, D.A. A method of detection of fumonisins in corn samples associated with field cases of equine leukoencephalomalacia. *Mycologia* **1990**, 82, 698-702.

Rheeder, J.P.; Marasas, W.F.O., Thiel, P.G., Sydenham, E.W., Shephard, G.S. and Van Schalwyk, D.J., *Fusarium moniliforme* and fumonisins in corn in relation to human esophageal cancer in Transkei, *Phytopathology* **1992**, 82, 353-357.

Rice, L.G.; Ross, P.F. Methods for detection and quantitation of fumonisins in corn, cereal products and animal excreta. *J. Food Protection* **1994**, 57, 536-540.

Ross, P.F.; Rice, L.G.; Plattner, R.D.; Osweiler, G.D.; Wilson, T.M.; Owens, D.L.; Nelson, H.A.; Richard, J.L. Concentrations of fumonisin B_1 in feeds associated with animal health problems. *Mycopathologia* **1991a**, 114, 129-135

Ross, P.F.; Rice, L.G.; Reagor, J.C.; Osweiler, G.D.; Wilson, T.M.; Owens, D.L.; Nelson, H.A.; Plattner, R.D.; Harlin, K.A.; Richard, J.L.; Colvin, B.M.; Banton, M.I. Fumonisin B_1 concentrations in feeds from 45 confirmed equine leukoencephalomalacia cases. *J. Vet. Diagn. Invest.* **1991b**, 3, 238-241.

Ross, P.F.; Rice, L.G.; Osweiler, G.D.; Nelson, P.E.; Richard, J.L.; Wilson, T.M. A review and update of animal toxicoses associated with fumonisin-contaminated feeds and production of fumonisins by Fusarium isolates. *Mycopathologia* **1992**, 117, 109-114.

Scott, D. Agricultural Experiment Station, W. Lafayette, Indiana, Bulletin 645, Sept. **1992**.

Scott, P.M.; Lawrence, G.A. Liquid chromatographic determination of fumonisins with 4-fluoro-7-nitrobenzofurazan. *J. AOAC Int.* **1992a**, 75, 829-834.

Scott, D. et al. Agricultural Experiment Station, W. Lafayette, Indiana, Bulletin 645, Sept., **1992**.

Scott, P.M. Fumonisins. *Int. J. Food Microbiology* **1993**, 18, 257-270.

Scott, P.M.; Delgado, T.; Prelusky, D.B.; Trenholm, H.L.; Miller, J.D. Determination of fumonisins in milk. *J. Environ. Sci. Health* **1994**, B29, 989-998.

Shephard, G.S.; Sydenham, E.W.; Thiel, P.G.; Gelderblom, W.C.A. Quantitative determination of fumonisins B_1 and B_2 by high performance liquid chromatography with fluorescence detection. *J. Liquid Chrom.* **1990**, 13, 2077-2087.

Stack, M.E.; Eppley, R.E. Liquid chromatographic determination of fumonisin B_1 and B_2 in corn and corn products. *J. AOAC Int.* **1992**, 75, 834-837.

Sydenham, E.W.; Marasas, W.F.O.; Shephard, G.S.; Thiel, P.G.; Hirooka, E.Y. Fumonisin concentration in Brazilian feeds associated with field outbreaks of confirmed and suspected animal mycotoxicoses. *J. Agric. Food Chem.* **1992**, 40, 994-997.

Thiel, P.G.; Marasas, W.F.O.; Sydenham, E.W.; Shephard, G.S.; Van Schalwyk, D.J. The implications of naturally occurring levels of fumonisins in corn from low and high esophageal cancer prevalence ares of the Transkei, southern Africa. *Mycopathologica* **1992**, 117, 3-9.

Trucksess, M.W.; Stack, M.E.; Allem, S.; Barrion, N. Immunoaffinity column coupled with liquid chromatography for determination of fumonisin B_1 in canned and frozen sweet corn. *J. AOAC Int.* **1995**, 78, 705-710,

Voss, K.A.; Norred, W.P.; Plattner, R.D.; Bacon, C.W. Hepatotoxicity and renal toxicity in rats of corn samples associated with field cases of leucoencephalomalacia. *Fd. Chem. Toxicol.* **1989**, 27, 89-96.

Ware, G.M.; Francis, O.; Kuan, S.S.; Umrigar, P.; Carman, A.; Carter, L.; Bennett, G.A. Determination of fumonisin B_1 by high performance liquid chromatography with fluorescence detection. *Anal. Lett.* **1993**, 26, 1751-1770.

Ware, G.M.; Umrigar, P.P.; Carman, A.S.; Kuan, S.S. Evaluation of Fumonitest immunoaffinity column. *Anal. Lett.* **1994**, 27, 693-715.

3

OCCURRENCE OF *FUSARIUM* AND FUMONISINS ON FOOD GRAINS AND IN FOODS

Lloyd B. Bullerman

Department of Food Science and Technology
University of Nebraska-Lincoln
Lincoln, NE 68583-0919

ABSTRACT

Fusarium moniliforme Sheldon occurs worldwide on corn intended for human and animal consumption. A closely related species *Fusarium proliferatum* also occurs frequently on corn. Yellow dent corn, white dent corn, white and yellow popcorn and sweet corn may be contaminated. Both organisms are capable of producing fumonisins, including Fumonisin B_1 (FB_1), Fumonisin B_2 (FB_2) and Fumonisin B_3 (FB_3). Fumonisins have been found in corn and corn based foods worldwide. Fumonisins may be found in sound whole kernel corn at levels at or below 1.0 µg/g. By contrast animal disease problems begin to occur at fumonisin levels above 5.0 to 10.0 µg/g. Corn based food products that have the most frequent and highest fumonisin levels, besides whole kernel corn, are corn meal, corn flour and corn grits. In the U.S., corn meal and flour have been found contaminated with FB_1 at levels from 0.5 to 2.05 µg/g, and grits from 0.14 to 0.27 µg/g. Corn flakes, corn pops, corn chips and tortilla chips have typically been found negative when tested for fumonisins. Popcorn, sweet corn and hominy corn have been found contaminated with sporadic, low levels (0.01 to 0.08 µg/g) of fumonisins. Contamination levels of corn based foods in Europe appear to be similar to slightly lower than similar products in the U.S., with the possible exception of Italy, where their corn hybrids and corn-based foods appear to be more frequently contaminated with higher levels of fumonisins.

INTRODUCTION

The fungus, *Fusarium moniliforme* Sheldon, is a soil borne plant pathogen that is found in all corn growing regions. It often produces symptomless infections of corn plants, but may infect the grain as well, and has been found worldwide on food and feed grade corn. It is not uncommon to find lots of shelled corn with 100% kernel infection (Marasas et al., 1984). The organism invades the tissues of the grain just beneath the pericarp, in the tip

Fumonisins in Food, Edited by L. Jackson *et al.*
Plenum Press, New York, 1996

region, as opposed to only surface contamination. A closely related species, *Fusarium proliferatum*, is also frequently isolated from shelled corn. These two *Fusarium* species are capable of producing the fumonisins, a group of mycotoxins which include Fumonisin B_1 (FB_1), Fumonisin B_2 (FB_2) and Fumonisin B_3 (FB_3). The fumonisins have been linked to fatal animal diseases including leukoencephalomalacia (LEM) in horses (Kellerman et al., 1990), porcine pulmonary edema (PPE) (Harrison et al., 1990) and liver cancer in rats (Gelderblom et al., 1991). Corn and fumonisins have also been associated with high incidences and increased risk of human esophageal cancer in regions of South Africa, China, Northeastern Italy and the Southeastern U.S. (Franceschi et al., 1990; Gelderblom et al., 1991; Rheeder et al., 1992). In these regions corn, in one form or another, is a dietary staple, and preliminary data indicate that corn and corn based food products from these and other regions may be contaminated with relatively high levels of fumonisins (Ross et al., 1991).

A number of studies have reported the incidences of the organisms and toxins in corn and corn-based foods. These studies can be divided into reports on a) the incidences of *F. moniliforme* and *F. proliferatum* in corn products, including yellow and white dent corn, dry milled corn fractions, popcorn and sweet corn; and b) the incidence of fumonisins in corn and corn based foods and feed.

INCIDENCE OF FUSARIUM MONILIFORME AND FUSARIUM PROLIFERATUM IN CORN

Fusarium moniliforme is one of the most prevalent fungi associated with a basic human and animal dietary staple such as corn. Recent studies reported on the *Fusarium* content of dent corn, popcorn and sweet corn (Cagampang, 1994; Katta, 1994). The data are summarized in Table 1. The dent corn included both yellow and white corn from the 1991 and 1992 crop years. The mean percent of kernels infected with *Fusarium* ranged from a

Table 1. Fusarium content of dent corn, popcorn and sweet corn (Cagampang, 1994; Katta, 1994)

Sample	# of Samples	# *Fusarium* Isolates	*Fusarium* Infected Kernels (%) High	Low	Mean
Dent corn					
Yellow Milling	12	226	28	10	18.8
NE Yellow 1991	21	275	36	3	13.1
NE Yellow 1992	63	528	35	0	8.4
NE White 1992[a]	25	906	94	1	36.2
MN Yellow 1992	10	114	38	0	11.4
Popcorn					
Microwave Yellow	41	177	15	0	4.3
Regular Yellow	13	23	7	0	1.8
Regular White	10	103	36	1	10.3
Regular Specialty	4	24	12	0	6.0
Sweet Corn					
Fresh (Cob)	24	8	2	0	0.3
Frozen (kernels)[b]	7	56	21	0	8.0

[a]Ten of the samples from one location had 100% kernel infection with *Fusarium*.
[b]Non-surface sanitized.

Table 2. Fusarium, species isolated from dent corn, popcorn and sweet corn
(Cagampang, 1994; Katta, 1994)

Sample	# Fusarium Isolates	Individual *Fusarium* Species (%)				
		Fm[a]	Fp	Fs	Fg	Others
Dent corn						
Yellow Milling	226	42.5	19.9	28.8	8.8	0
NE Yellow 1991	275	49.8	29.1	15.6	0.7	4.7
NE Yellow 1992	528	26.1	12.3	49.6	11.0	0
NE White 1992[a]	906	12.0	4.2	57.8	21.1	4.9
MN Yellow 1992	114	8.8	1.8	36.8	49.1	3.5
Popcorn						
Microwave Yellow	77	70.6	4.0	10.3	0	15.1
Regular Yellow	23	73.9	13.0	4.3	0	8.7
Regular White	103	70.9	1.9	6.8	1.0	19.4
Regular Specialty	24	91.7	8.3	0	0	0
Sweet Corn						
Fresh (Cob)	8	21.5	0	12.5	0	50.0
Frozen (kernels)[b]	56	0	0	10.7	0	87.5

[a]Fm = *F. moniliforme*; Fp = *F. proliferatum*; Fs = *F. subglutinans*; Fg = *F. graminearum*;
Others = Primarily *F. semitectum*.
[b]Non-surface sanitized.

low of 8.4% to a high of 36.2%. The highest level of infection was found in white corn from 1992. However, this mean level was raised by virtue of the fact that 10 of the samples examined were from one location and had 100% kernel infection with *Fusarium*. The high levels of contamination for the various groups of samples ranged from 28 to 94% while the low levels of contamination ranged from 0 to 10% (Cagampang, 1994).

Popcorn also had kernels infected with *Fusarium* species (Katta, 1994). Retail samples of microwave and conventionally packaged (plastic bags and jars) popcorn had high levels of contamination ranging from 7% to 36% infection of kernels. White regular packaged (non-microwave) popcorn had the highest kernel infection rate of 36%.

Fresh sweet corn on the cob had very low levels of kernel infection with *Fusarium*, while frozen sweet corn kernels had considerably higher levels of *Fusarium*. However, this may have been in part due to the fact that it was impossible to surface sanitize the frozen corn kernels, and much of the contamination may have been from cross contamination of surfaces during handling and processing.

With the yellow dent corn grown in 1991, *F. moniliforme* was the most commonly found species (Table 2). Higher levels of *F. moniliforme* were found in milling corn and yellow corn grown in 1991 than in yellow corn grown in 1992. In white and yellow corn grown in Nebraska in 1992, the predominant species found was *Fusarium subglutinans*. With yellow corn grown in 1992 and obtained from southern Minnesota, the predominant species found was *Fusarium graminearum*. This indicates that crop year and geographic location have an effect on the species of *Fusarium* that may occur in corn. In the 1992 crop year, conditions were wetter and cooler, and favored invasion by saprophytic species other than *F. moniliforme*, and did not favor *F. moniliforme*. The cool wet conditions produced less stress on the corn plants, since the plants were not suffering from either excessive heat or drought. Heat and drought stress may be factors that favor colonization of corn by *F. moniliforme* and production of fumonisins (Bacon and Williamson, 1992). Arino and

Table 3. Fusarium content of food grade corn and dry mill fractions (Cagampang, 1994).

| | | Mill Fractions | | | |
| | | Fusarium Counts (CFU/g) | | | |
Sample #	Corn % *Fusarium* Infected Kernels	Bran	Germ	Flaking Grits	Flour
10493	25	<100	4,800	<100	1600
11193	28	2000	8,900	<100	<100
11893	24	<100	6,000	<100	900
12593	21	<100	16,000	<100	100
20193	23	1000	350	<100	<100
20893	20	4000	<100	<100	1300
21593	20	<100	4,200	<100	1400
22293	13	<100	100	<100	100
30193	13	<100	300	<100	<100
30893	17	<100	4,600	<100	2700
31593	10	<100	2,900	<100	600
32293	12	64,000	9,400	<100	100
Means:	19	5,600	5,200	<100	730

Bullerman (1994) reported a higher incidence of *Fusarium* species in corn grown under dry land conditions than corn grown under irrigated conditions.

With popcorn, *F. moniliforme* was the predominant *Fusarium* sp. isolated from all samples (Katta, 1994). While the number of infected kernels was low, the percentages of *F. moniliforme* was high, ranging from 70.9 to 91.7% of the isolates. The other *Fusarium* species found in both dent corn and popcorn were primarily *Fusarium semitectum*. The sweet corn, in addition to having very low kernel infection rates, had very low to zero levels of *F. moniliforme*.

In another study (Cagampang, 1994), food grade corn and dry milled fractions were studied for *Fusarium* content (Table 3). Samples were obtained from a commercial mill on a weekly basis over a period of 12 weeks. Each weekly sample of corn and the mill fractions came from a common single lot of corn. Kernel infection rates in the corn ranged from a low of 10% to a high of 28%. Mill fractions were analyzed by selective plate count for *Fusarium* species. The fractions that had the highest counts of *Fusarium* were the germ and bran fractions, with the germ fraction having consistently higher counts. The flaking grit fraction was low in *Fusarium* count, with all samples being less than 100 CFU/g. The flour fraction was lower in *Fusarium* count than either the germ or bran fractions but consistently higher than the flaking grits. It is interesting that the germ and bran had consistently higher counts.

Table 4. Fumonisin levels associated with animal diseases and various types of corn fractions (Kellerman et al. 1990; Ross et al., 1991)

Disease/Corn Product	Fumonisin Level (μg/g)
ELEM	<1 to 126
PPE	<1 to 330
Damaged Corn	553
Cob Parts	138
Stalk Parts	54
Corn Screenings	125
Undamaged Whole Kernels	<1 to 4

Table 5. Ranges of fumonisin (FB_1, FB_2, FB_3) concentrations in corn from four different crop years (Murphy et al., 1993).

Year	n	FB_1	FB_2	FB_3
		μg of Fumonisin/g of Corn		
1988	22	ND[a]-14.9	ND-5.7	ND-2.1
1989	44	ND-37.9	ND-12.3	ND-4.0
1990	59	ND-19.1	ND-6.1	ND-2.8
1991	50	ND-5.8	ND-4.4	ND-2.3

[a]ND - None detected

This supports the suggestion that the organism localizes in the kernel tip and germ areas just beneath the pericarp. The observation that flaking grits had little or no contamination with *Fusarium* may help explain why corn flake and corn pop cereals have been consistently free of fumonisins.

Doko et al. (1995) investigated 26 inbred lines of corn grown in Italy and 72 corn hybrids grown in Croatia, Poland, Portugal, Romania, Benin and Zambia. They found that the inbred lines and hybrids from Italy, Portugal, Zambia and Benin were highly contaminated with *F. moniliforme* with 100, 100, 100 and 82% respectively contaminated. Hybrids from Croatia, Poland and Romania had very low levels of contamination.

INCIDENCE OF FUMONISINS IN CORN ASSOCIATED WITH DISEASE

Levels of Fumonisins and Animal Diseases

Fumonisins have been detected in animal feeds associated with animal diseases. In South Africa, feeds associated with leukoencephalomalacia contained from 1.3 to 27 μg/g of FB_1 and 0.1 to 12.6 μg/g of FB_2 (Kellerman et al., 1990; Marasas et al., 1979). Ross et al. (1991) reported that levels of $FB_1 \geq 10$ μg/g could be lethal to horses. The levels of fumonisins that have been associated with leukoencephalomalacia, porcine pulmonary edema, and various corn products are given in Table 4. It is thought that fumonisin levels below 5-10 μg/g will not cause leukoencephalomalacia or porcine pulmonary edema (Ross et al., 1991). Further, the highest levels of fumonisins have been found in corn products or fractions that are not associated with human food, i.e. damaged corn, cob parts, stalk parts and screenings. Most good quality whole corn kernels have fumonisin levels of 1.0 μg/g or less (Pittet et al., 1992; Pittet and Tornare, 1992; Sydenham et al., 1991).

Table 6. Fumonisin (FB_1, FB_2) levels in commercialcornmeal(Sydenham et al.,1991)

Country	Number of samples	FB_1	FB_2
		Fumonisin Levels (μg/g)	
Canada	2	ND[a]-0.05	ND
Egypt	2	1.8-3.0	0.5-0.8
Peru	4	ND-0.7	ND-0.1
South Africa	52	ND-0.5	ND-0.1
U.S.	17	ND-2.8	ND-0.9

[a]ND - None detected

Table 7. Distribution of fumonisin levels in commercial corn-based
food samples from the U.S. and South Africa (Sydenham et al., 1991)

Concentration (µg/g)	Number of samples	Percent of samples
U.S. (35 samples)		
<0.5	17	48.6
0.5-1.0	8	22.2
1.0-1.5	3	8.6
1.5-2.0	2	5.7
2.0-2.5	3	8.6
>2.5	2	5.7
South Africa (81 samples)		
<0.1	49	60.5
0.1-0.2	18	22.2
0.2-0.3	8	9.9
0.3-0.4	1	1.2
0.4-0.5	3	3.7
0.5-0.6	2	2.5

Levels of Fumonisins and Human Esophageal Cancer

Fumonisin levels in corn associated with high esophageal cancer areas of the Transkei in South Africa ranged from 10.2 µg/g in good corn to 63.2 µg/g in moldy corn and from 6.7 µg/g in good corn to 140.5 µg/g in moldy corn for the 1985 and 1989 crop years respectively (Rheeder et al., 1992). In the U.S., an area of high esophageal cancer incidence is found in and around Charleston, South Carolina. Fumonisin levels in corn meal samples from the Charleston, SC area have been found to be in the range of 0.17 to 2.4 µg/g of FB_1 and FB_2 combined (Sydenham et al., 1991). Two recent studies have reported on the fumonisin content of corn samples collected from households in Linxian County, a high incidence area of esophageal cancer in China. Chu and Li (1994) reported that high levels of FB_1 (18 to 155 µg/g; mean, 74 µg/g) were detected in samples that had heavy mold contamination, which accounted for 16 of 31 household samples. The other 15 household samples had FB_1 contents of 20 to 60 µg/g, with a mean level of 35.3 µg/g. These samples also contained low levels of aflatoxins (1.0 to 38.4 ng/g, mean 8.6 ng/g) and high levels of total type-A trichothecenes, including T-2 toxin, HT-2 toxin, iso-neosolaniol, monoacetoxy-scirpenol and others. Type-B trichothecenes were also found in half of the samples ranging from 0.47 to 5.8 µg/g. Yoshizawa et al. (1994) found lower levels of fumonisins in corn from the high esophageal cancer area. They found levels of 0.87 µg/g of FB_1 and 0.45 µg/g of

Table 8. Fumonisin (FB_1, FB_2) levels in commercial corn products purchased from supermarkets in Washington, DC in 1990 and 1991 (Stack and Eppley, 1992)

Food Product	Year	FB_1 (µg/g)	FB_2 (µg/g)
Corn Meal	1990	0.5 - 1.04	0.5 - 0.25
Corn Meal	1991	0.28 - 2.05	0.05 - 0.53
Corn Grits	1990	0.18 - 0.27	ND^a - 0.11
Corn Grits	1991	0.14 - 0.19	0.05 - 0.06
Corn Bran Cereal	1990	ND - 0.33	ND - 0.04
Corn Bran Cereal	1991	0.13 - 0.29	0.02 - 0.07

[a]ND - None Detected

Table 9. Fumonisin levels in corn-based breakfast cereals and snack products (Stack and Eppley, 1992)

Food product	Fumonisin level (µg/g)
Corn Flakes Cereal	ND[a]
Corn Pops Cereal	ND
Tortilla Chips	ND
Corn Chips	ND
Tortillas	ND - 0.15
Popcorn	0.01 - 0.06
Hominy	0.08

[a]ND - None detected

FB_2. They also reported that the incidence of fumonisins was lower, by one-half, in corn from low cancer risk areas, but that the levels of FB_1 and FB_2 from the low risk areas were about the same as from the high risk areas. They also found trichothecenes in the corn from the high risk areas. Occurrence of FB_1 with aflatoxins has also been reported in the US by Chamerlain et al. (1993), who reported the natural co-contamination of corn with aflatoxins and fumonisin B_1 in samples from the 1991 crop year in Georgia (USA), though these samples were not associated with esophageal cancer or any other disease.

INCIDENCE OF FUMONISINS IN CORN AND CORN AND CORN-BASED FOODS

Levels of Fumonisins in Whole Kernel Corn

Sydenham et al. (1993) reported on *F. moniliforme* and fumonisin levels in corn grown in field trials in Argentina. Most of the samples had total fumonisin levels in excess of 2.0 µg/g, and some were as high as 6.0 to 10.0 µg/g. Counts of *F. moniliforme* ranged from 10,000 to 700,000 CFU/g. These are relatively high levels of both the organism and the toxins for sound whole kernel corn.

Doko and Visconti (1993; 1994) reported on fumonisin contamination of corn and corn-based foods in Italy. They found that commercial corn samples and corn genotypes

Table 10. Fumonisin B_1 in corn products in Switzerland (Pittet et al., 1992; Pittet and Tornare, 1992)

Product	Incidence (Positives/Total)	Range (µg/g)	Mean or Positives[a] (µg/g)
Corn Meal	2/7	ND[b]-0.1	0.08
Corn Grits	34/55	ND-0.8	0.3
Cornflakes	1/12	ND-0.06	0.06
Sweet Corn	1/7	ND-0.07	0.07
Miscellaneous[c]	0/17	ND	—
Poultry Feed	6/22	ND-0.5	0.2

[a]Positives contained more than 0.05 µg/g.
[b]ND - None detected
[c]Miscellaneous included corn starch, popcorn, tortillas, corn cookies, and corn noodles

Table 11. Distribution of fumonisin levels in corn products in Switzerland by numbers of samples and percent of samples in each concentration range

Concentration µg/g	Number of samples	Percent of samples
<0.1	74	75.5
0.1-0.3	10	10.2
0.3-0.5	6	6.1
0.5-0.7	3	3.1
0.7-0.9	4	4.1
>0.9	1	1.0

[a]Based on 98 samples.

from breeding stations contained fumonisin levels ($FB_1 + FB_2$) ranging from 0.1 to 6.8 µg/g in commercial corn samples and 0.02 - 2.8 µg/g in corn genotypes. Of 26 corn genotypes, four had fumonisin levels above 1.0 µg/g, the rest were well below, ranging from 0.01 to 0.9 µg/g. The commercial corn samples all had quite high levels of fumonisin, with six out of seven samples having in excess of 2.0 µg/g. Doko et al. (1995), found that corn genotypes from Italy, Portugal, Zambia and Benin had fumonisin levels of 2.8, 4.4, 1.7 and 3.3 µg/g respectively. These genotypes were more consistently contaminated than genotypes from Central Europe. The authors also reported that a significant number of the samples with fumonisin levels greater than 1.0 µg/g showed no visual evidence of fungal infection. Murphy et al. (1993) studied the FB_1, FB_2 and FB_3 contents of whole corn and corn screenings from Iowa, Wisconsin and Illinois for four crop years, 1988, 1989, 1990 and 1991. Fumonisin B_1 levels ranged from 0 to 37.9 µg/g, FB_2 levels from 0 to 12.3 µg/g and FB_3 levels from 0 to 4.0 µg/g (Table 5). Levels varied with crop year, with 1989 having the highest levels. The amount of fumonisins in corn screenings was found to average about 10 times higher than the amounts in whole corn. Cagampang (1994) reported that total fumonisins ($FB_1 + FB_2 + FB_3$) in corn grown in Nebraska ranged from 0.2 to 6.5 µg/g in 1991 and 0.1 to 0.2 µg/g in 1992. The corn grown in 1992 in Nebraska had very low levels of fumonisins, as did corn grown in other parts of the midwest, such as Iowa and Illinois, though fungal contamination was high (Rice and Ross 1994). Because the 1992 crop year was cool and wet, the growth of many saprophytic fungi was favored over *F. moniliforme*, resulting in less fumonisin in the corn. Corn obtained from a commercial dry mill in 1993 contained fumonisin levels of 0.1 to 3.5 µg/g (Cagampang 1994). The highest levels of

Table 12. Fumonisin (FB_1, FB_2) levels in commercial corn foods from retail outlets in Italy (Doko and Visconti, 1993; Doko and Visconti, 1994)

Food	Fumonisin concentration (µg/g)		
	FB_1	FB_2	Total
Corn Meal	3.5	0.8	4.4
Corn Grits	3.8	0.9	4.7
Puffed Corn	0.8-6.1	0.1-0.5	0.9-6.6
Polenta	0.4-3.7	0.08-0.8	0.5-4.8
Canned Sweetcorn	0.06-0.2	ND[a]	0.06-0.2
Fresh Sweet Corn (cob)	0.8	ND	0.8
Corn Flakes Cereal	ND-0.01	ND	ND-0.01
Popcorn	ND-0.06	ND-0.02	ND-0.08
Tortilla Chips	ND-0.06	ND-0.01	ND-0.07

[a]ND - None detected

fumonisins in the mill fractions from this corn were found in the bran and germ fractions, with the least amount found in the flaking grit fraction (Cagampang 1994).

Levels of Fumonisins in Corn-Based Foods

Corn based-food products that have been examined for contamination with fumonisins in the U.S. include corn meal, corn grits, corn breakfast cereals, tortillas, tortilla chips, corn chips, popcorn and hominy corn. Sydenham et al. (1991) reported on fumonisin levels in commercial corn meal in samples from Canada, Egypt, Peru, South Africa and the U.S. (Table 6). While based on only two samples, corn meal from Egypt contained the highest levels of FB_1 and FB_2. Corn meal from the U.S. had the second highest levels of FB_1 and FB_2. Corn meal from Canada had the lowest levels. The fumonisin concentrations in corn products from the U.S. were below 1.0 µg/g in 70.8% of the samples, with 5.7% of the samples having in excess of 2.5 µg/g. By contrast corn-based food samples from South Africa had lower levels of fumonisins, since 82.7% of the samples had less than 0.2 µg/g (Sydenham et al., 1991). The distributions of fumonisin levels in commercial corn-based foodstuffs from the U.S. and South Africa are given in Table 7. All of the South African samples had fumonisin levels below 0.6 µg/g, well below 1.0 µg/g, whereas with the U.S. samples only 70.2% of the samples were below 1.0 µg/g. Just under 30% of the samples ranged from 1.0 µg/g to 2.5 µg/g, with 2.5% of the samples having more than 2.5 µg/g (Sydenham et al., 1991).

Stack and Eppley (1992) reported a survey carried out by the FDA for fumonisins in corn-based food products obtained from supermarkets in the Washington DC area (Tables 8 and 9). The highest amounts of contamination were found in corn meal and corn grits, with some also occurring in corn bran cereal. Corn flakes and corn pops cereal were negative for fumonisins, as were tortilla chips and corn chips. Very low levels of FB_1 were detected in tortillas, popcorn and hominy. Pestka et al. (1994) tested 7 grain-based foods for fumonisins and found levels ranging from 0.4 to 6.3 µg/g in corn meal, 0 to 1.2 µg/g in corn muffin mix, 0.2 µg/g in corn tortilla mix and none in corn breakfast cereals.

Corn-based food products from Europe have also been tested for fumonisins. In a study done in Switzerland by Pittet et al. (1992) it was found that the highest levels of FB_1 were found in corn meal and corn grits (Table 10). Low levels were found in corn flakes cereal and canned sweet corn. Other corn based foods including popcorn, corn starch, tortillas, corn cookies and corn noodles were negative for FB_1. Just over 75% of the samples had fumonisin levels below 0.1 µg/g, and all samples were below 1.0 µg/g (Table 11). Zoller et al. (1994) examined 104 food products from retail outlets in Switzerland for FB_1 and FB_2 and concluded that only very low levels of fumonisins were present, though 42 out of 52 samples were contaminated. Corn grits and corn flour had the highest fumonisin concentrations (0.01 to 3.4 µg/g of FB_1 and 0.005 to 0.9 µg/g of FB_2. Popcorn, corn flakes and corn snacks had lower levels of contamination. Doko and Visconti (1994) reported that in corn based foods in Italy the highest fumonisin levels were found in puffed corn, corn grits, corn meal and polenta, a type of porridge made from corn flour (Table 12). Corn grits and polenta corn flour contained rather high levels of fumonisins (FB_1 + FB_2) ranging from 0.5 µg/g to 4.7 µg/g. Canned sweet corn, cornflakes, popcorn and tortilla chips all had very low to non-detectable levels. Usleber et al. (1994) in developing an immunoassay for FB_1 tested corn-based foods in Germany for fumonisins and found low levels in popcorn, corn grits and corn semolina from Germany (<0.01 - 0.12 µg/g), but found higher levels in corn semolina imported from Italy (0.08 - 1.1 µg/g). Sanchis et al. (1994), reported the natural occurrence of FB_1 and FB_2 in corn-based

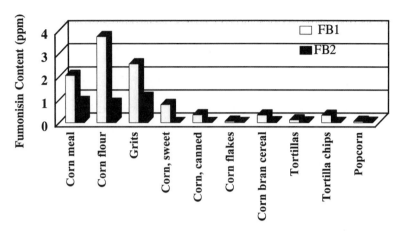

Figure 1. Ranges of (0 to highest amount) fumonisin content of corn-based foods based on data from Doko and Visconti (1994), Pittet et al., (1992), Stack and Eppley (1992), and Sydenham et al. (1992). (Graph courtesy of Dr. Lauren Jackson FDA/NCFST).

foods from Spain. Only 16% of the samples were contaminated, and the levels were very low (none detected - 0.2 µg/g).

Ueno et al. (1993) in a limited survey, studied the fumonisin content in corn and corn based products sampled in Japan, Nepal and China. They found that 8 of 9 imported corn samples in Japan contained FB_1 levels of 0.6 to 4.1 µg/g and 6 corn gluten feed samples had FB_1 levels of 0.3 to 2.4 µg/g. The samples also contained FB_2 at levels of 0.3 to 10.2 µg/g in the corn and 0.8 to 8.5 µg/g in the gluten feed. Corn grit samples contained 2.6 µg/g of FB_1 and 2.8 µg/g of FB_2. Commercial corn based food products marketed in Japan, however, contained no significant levels of fumonisins. Corn from Nepal had average levels of FB_1 of 0.6 µg/g and of FB_2 of 1.6 µg/g. FB_1 and FB_2 were also found in corn harvested in Shanghai and Beijing, China. This study further demonstrates the worldwide occurrence of fumonisins in corn and that similar background levels are found in Asian corn, as elsewhere.

The studies that have been reported on the occurrence of fumonisins in corn and corn based foods show that the highest levels of fumonisins are in the whole grain and those corn products that undergo the least or mildest forms of processing, i.e. corn meal, corn flour and corn grits, where a physical milling or grinding process is involved (Figure 1). Those corn based products that are more highly processed such as cornflakes and corn pops cereals tend to have either no detectable, or only very low levels of fumonisins. Snack foods such as corn chips and tortilla chips also tend to be negative for fumonisins while tortillas and popcorn appear to have very low levels of contamination. Canned sweet corn also appears to have low levels of contamination. Surveys of corn from around the world show that fumonisins are found in corn from all countries. Likewise, fumonisins are found in processed corn products in all parts of the world, though some regions have higher contamination levels than others. In Europe, Italy seems to have higher levels of fumonisins in both corn and processed products than other European countries. At this point these data must still be interpreted with caution because they represent the product of only a limited number of studies. Also, these studies have been limited to surveys of commercially available products. Very few studies have yet been done on the effects of various processes on the stability of fumonisins.

REFERENCES

Arino, A. A.; Bullerman, L. B. Fungal colonization of corn grown in Nebraska in relation to year, genotype and growing conditions. J. Food Prot. 1994, 57, 1084-1087.

Bacon, C. W.; Williamson, J. W. Interactions of *Fusarium moniliforme*, its metabolites and bacteria with corn. Mycopathologia 1992, 117, 65-71.

Cagampang, A. E. Incidence and effects of processing on *Fusarium moniliforme* and fumonisins in corn. M.S. Thesis. University of Nebraska, Lincoln, NE. 1994.

Chamberlain, W. J.; Bacon, C. W.; Norred, W. P.; Voss, K. A. Levels of fumonisin B_1 in corn naturally contaminated with aflatoxins. Food Chem. Toxic. 1993, 31, 995-998.

Chatterjee, D.; Mukherjee, S. K. Contamination of Indian maize with fumonisin B_1 and its effects on chicken macrophage. Lett. Appl. Microbiol. 1994, 18, 251-253.

Chu, F. S.; Li, G. Y. Simultaneous occurrence of fumonisin B_1 and other mycotoxins in moldy corn collected from the People's Republic of China in regions with high incidence of esophageal cancer. Appl. Environ. Microbiol. 1994, 60, 847-852.

Doko, M. B.; Rapior, S.; Visconti, A.; Schjth, J. E. Incidence and levels of fumonisin contamination in maize genotypes grown in Europe and Africa. J. Agric. Food Chem. 1995, 43, 429-434.

Doko, M. B.; Visconti, A. Fumonisin contamination of corn and corn based foods in Italy. UK Workshop on Occurrence and Significance of Mycotoxins. April 21-23. London. 1993.

Doko, M. B.; Visconti, A. Occurrence of fumonisins B_1 and B_2 in corn and corn-based human foodstuff in Italy. Food Add. and Contam. 1994, 11, 433-439.

Franceschi, S.; Bidoli, E.; Baron, A. E.; LaVecchia, C. Maize and the risk of cancers of the oral cavity, pharynx and esophagus in Northeastern Italy. J. Nat'l. Cancer Inst. 1990, 82, 1407-1411.

Gelderblom, W. C. A.; Kriek. N. P.; Marasas, W. F. O.; Thiel, P. G. Toxicity and carcinogenicity of the *Fusarium moniliforme* metabolite, fumonisin B_1, in rats. Carcinogenesis 1991, 12, 1247-1251.

Gelderblom, W. C. A.; Marasas, W. F. O.; Vleggaar, R.; Thiel, P. G.; Cawood, M. E. Fumonisins: Isolation, chemical characterization and biological effects. Mycopathologia 1992, 117, 11-16.

Harrison, L. R.; Colvin, B. M.; Greens, J. T.; Newman, L. E.; Cole, J. R. Pulmonary edema and hydrothorax in swine produced by fumonisin B_1, a toxic metabolite of *Fusarium moniliforme*. J. Vet. Diag. Invest. 1990, 2, 217-221.

Katta, S. K. Effect of handling and storage on popcorn quality related to mold content and expansion volume. M.S. Thesis. University of Nebraska, Lincoln, NE. 1994.

Kellerman, T. S.; Marasas, W. F. O.; Thiel, P. G.; Gelderblom, W. C. A.; Cawood, M.; Coetzer, J. A. W. Leukoencephalmomalacia in two horses induced by oral dosing of fumonisin B_1. Onderstepoort J. Vet Res. 1990, 57, 269-75.

Marasas, W. F. O.; Kriek, N. P. J.; Wiggins, V. M.; Steyn, P. S.; Towers, D. K.; Hastie, T. J. Incidence, geographic distribution and toxigenicity of *Fusarium* species in South African corn. Phytopathology 1979, 69, 1181-1185.

Marasas, W. F. O.; Nelson, P. E.; Tousson, T. A. Toxigenic *Fusarium* Species: Identity and Mycotoxicology. The Pennsylvania State University Press: University Park, PA. 1984.

Murphy, P. A.; Rice, L. G.; Ross, P. F. Fumonisin B_1, B_2 and B_3 Content of Iowa, Wisconsin and Illinois corn and corn screenings. J. Agric. Food Chem. 1993, 41, 263-266.

Pestka, J. J.; Azcona-Olivera, J. I.; Plattner, R. D.; Minervini, F.; Doko, M. B.; Visconti, A. Comparative assessment of fumonisin in grain-based foods by ELISA, GC-MS and HPLC. J. Food Prot. 1994, 57, 169-172.

Pittet, A.; Parisod, V.; Schellenberg, M. Occurrence of fumonisins B_1 and B_2 in corn-based products from the Swiss market. J. Agric. Food Chem. 1992, 40, 1352-1354.

Pittet, A.; Tornare, D. Survey of European cereals for the presence of fumonisins B_1 and B_2. Presented at the 106th Annual AOAC International Meeting. August 31 - September 2. Cincinnati, OH (Abstract). 1992.

Rheeder, J. P.; Marasas, W. F. O.; Thiel, P. G.; Sydenham, E. W.; Shephard, G. S.; van Schalkwyk, D. J. *Fusarium moniliforme* and fumonisins in corn in relation to human esophageal cancer in Transkei. Phytophathology 1992, 82, 353-357.

Ross, P. F.; Rice, L. G.; Plattner, R. D.; Osweiler, G. D.; Wilson, T. M.; Owens, D. L.; Nelson, P. A.; Richard, J. L. Concentrations of fumonisin B_1 in feeds associated with animal health problems. Mycopathologia 1991, 114, 129-135.

Sanchis, V.; Abadias, M.; Oncins, L.; Sala, N.; Vinas, I.; Canela, R. Occurrence of fumonisins B_1 and B_2 in corn-based products from the Spanish market. Appl. Environ. Microbiol. 1994, 60, 2147-2148.

Stack, M. E.; Eppley, R. M. Liquid chromatographic determination of fumonisins B$_1$ and B$_2$ in corn and corn products. J. AOAC Int'l. 1992, 75, 834-837.

Sydenham, E. W.; Shephard, G. S.; Thiel, P. G.; Marasas W. F. O.; Stockenstrom, S. Fumonisin contamination of commercial corn-based human foodstuffs. J. Agric. Food Chem. 1991, 25, 767-771.

Sydenham, E. W.; Shephard, G. S.; Thiel, P. G.; Marasas, W. F. O.; Rheeder, J. P.; Peralta Sanhueza, C. E.; Gonzalez, H. H. L.; Resnik, S. L. Fumonisins in Argentinian field-trial corn. J. Agric. Food Chem. 1993, 41, 891-895.

Ueno, Y.; Aoyama, S.; Sugiura, Y.; Wang, D. S.; Lee, U. S.; Hirooka, E. Y.; Hara, S.; Karki, T.; Chen, G; Yu, S. Z. A limited survey of fumonisins in corn and corn-based products in Asian countries. Mycotoxin Res. 1993, 9, 27-34.

Usleber, E.; Straka, M.; Terplan, G. Enzyme immunoassay for fumonisin B$_1$ applied to corn-based food. J. Agric. Food Chem. 1994, 42, 1392-1396.

Yoshizawa, T.; Yamashita, A; Luo, Y. Fumonisin occurrence in corn from high - and low - risk areas for human esophageal cancer in China. Appl. Environ. Microbiol. 1994, 60, 1626-1629.

Zoller, O.; Sager, F.; Zimmerli, B. Occurrence of fumonisins in foods. Mitt. Geb. Lebensmittel. and Hygiene 1994, 85, 81-99.

4

OCCURRENCE AND FATE OF FUMONISINS IN BEEF

J. Scott Smith and Rohan A. Thakur

Department of Animal Sciences and Industry
208 Call Hall
Kansas State University
Manhattan, Kansas 66506-1605
913-532-1219, fax:913-532-5681, e-mail:jsschem@ksu.ksu.edu

ABSTRACT

For 30 days, two groups of three steers each were fed a diet of herd-mix and alfalfa hay (control) or the diet with part of the herd-mix replaced by corn grits contaminated with 400µg/g FB_1 and 130 µg/g FB_2. Premortem examination involved liver functionality tests; serum analyses for AST, GGT, LDH, cholesterol, and total bilirubin; urinalysis; and analyses of the blood, urine, and feces for the presence of fumonisins or their metabolites. Postmortem examination involved necropsy, histopathology, and analysis of tissue for fumonisins. Results of the liver functionality test indicated some hepatobiliary compromise in the test animals. Unmetabolized FB_1 and FB_2 were detected in the feces ($\geq$80%), whereas trace amounts were detected in the urine. Postmortem analysis of the tissues showed 2070 ng/g FB_1 in the liver (SD: 1870), 97.3 ng/g FB_1 in the muscle (SD: 41), and 23.4 ng/g FB_1 in the kidney (SD: 8.7).

INTRODUCTION

Fumonisins are nongenotoxic carcinogens produced by the fungus *Fusarium moniliforme* (Figure 1). Fumonisins are responsible for the etiopathogenesis of different diseases in discrete species. Consumption of grain contaminated with *F. moniliforme* has been linked to human esophageal cancer (Marasas et al., 1981). In swine fumonisin B_1 (FB_1) causes pulmonary edema (Harrison et al., 1990), whereas in rats it is hepatocarcinogenic and nephrotoxic (Gelderblom et al., 1991). Fumonisin B_1 and B_2 have been shown to be the etiological agents of equine leukoencephalomalacia, a disease fatal to horses (Kellerman et al., 1990; Ross et al., 1994).

In all cases described above, exposure to fumonisins did induce a hepatotoxic response. The etiological mechanism of fumonisins appears to be cellular mitogenesis caused

Fumonisins in Food, Edited by L. Jackson *et al.*
Plenum Press, New York, 1996

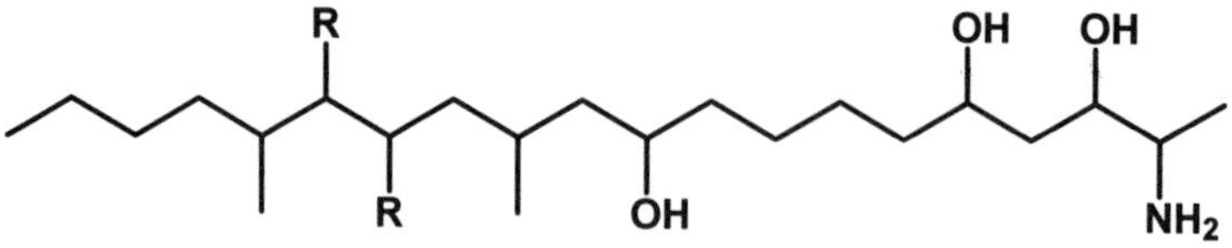

Fumonisin B1

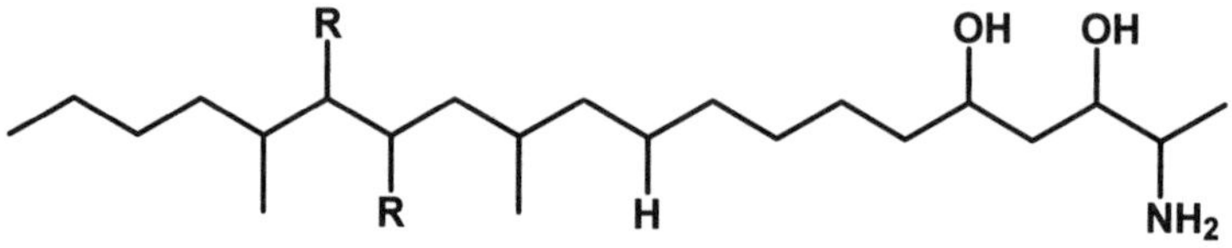

Fumonisin B2

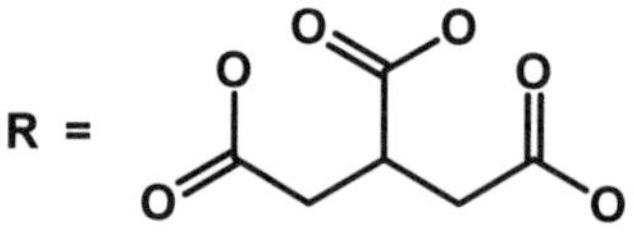

Tricarballylic acid side chain

Figure 1. Structure of fumonisin B_1 and B_2. Hydrolysis cleaves both side chains, whereas partial hydrolysis involves cleavage of a single R group.

by accumulation of sphingoid bases (Schroeder et al., 1994), via the inhibition of the de novo sphingolipid biosynthetic pathway (Riley et al., 1994).

The effect of fumonisins on ruminants is unclear. The prepeptic location of microbes and their unique physiology may alter fumonisin toxicokinetics in ruminants. The evidence for such alteration is that cattle seem to tolerate exposure to fumonisins at levels fatal to horses and swine (Osweiler et al., 1993). Cattle fed fumonisin concentrations at 148 µg/g of diet for 30 days showed significant increases in liver enzyme profiles, indicating some pathogenesis of the liver (Osweiler et al., 1993), but no signs of severe toxicosis. Similar increases in liver enzyme profiles were seen in lambs orally dosed with fumonisin containing culture material (Edrington et al., 1995). All the fumonisin-dosed lambs showed distressed primary liver and kidney functions, which ultimately resulted in death.

The ubiquitous nature of fumonisins in corn and the suspected tolerance of the bovine species to fumonisins potentially can result in a chronic exposure to fumonisins in cattle consuming corn-based diets. The absence of clinical signs of chronic exposure to fumonisins at levels fatal to other species may lead to assimilation of the fumonisins or their metabolites in tissues of cattle. A carryover effect from cattle consuming fumonisin-contaminated feed to the human diet via consumption of beef thus becomes a plausible threat to food safety.

This research answers some of these questions by dosing cattle with fumonisin-con-taminated grits. Premortem study involved blood chemistry profiling to indicate liver functionality; urinalysis; and monitoring the blood, urine, and feces for the presence of fumonisin B_1 (FB_1), FB_2, and their major hydrolysis products, the fully hydrolyzed FB_1 (HFB_1) and HFB_2 and the partially hydrolyzed FB_1 ($PHFB_1$) and $PHFB_2$. Postmortem examination involved necropsy and observation for gross lesions and histopathology, followed by analysis of the visceral organs, muscle, and fat for the presence of fumonisins and their metabolites.

MATERIALS AND METHODS

Animals

Six Holstein steers were acclimated to pens and diet for a period of 10 days. All steers were vaccinated against parainfluenza virus, infectious bovine rhinotracheitis, and bovine virus diarrhea and remained free of infectious diseases during the acclimation and experimental periods. The steers were weighed before, midway through, and at the end of the feeding study. Three steers were used as test animals, while the remaining three served as controls.

Housing and Feed

Animals were housed in individual pens at the Department of Animal Sciences and Industry's (ASI) dairy research barn. The animals had access to water ad libitum and were fed from individual feed trays. The pens were lined with rubber mats and were kept clean by scooping the excrement and washing with water three times a day. Diets were fed at 2% of the body weight and consisted of an 85:15 (w/w) blend of herd-mix and alfalfa hay. The herd-mix was prepared by ASI's feed mill facility and consisted of 49% milo; 47.65% cracked corn; 1% calcium phosphate; 0.5% trace metal salts; 0.05% vitamins A, D, E mix; and 1.5% wet molasses. The feed for the test animals was prepared by lowering the herd-mix concentration to adjust for the incorporation of the fumonisin-contaminated grits. The herd-mix was assayed for background levels of fumonisin contamination. The steers were fed twice daily for 30 days.

Preparation of Fumonisin-Contaminated Grits

Approximately 80 kg of Quaker quick-grits (Quaker, Chicago, IL) were divided into eight 10 kg batches, adjusted to 30% moisture content and allowed to stand in autoclavable plastic tubs for 8 h. At the end of 8 h, each tub was covered with aluminum foil and autoclaved for 1 h. The autoclave process was repeated after 24 h to ensure complete sterility of the grits and destruction of any surviving spores. The 80 kg of sterile grits then were spread to about 1-1½ inch thickness in disposable aluminum trays and covered with aluminum foil.

A lyophilized high fumonisin-producing strain (*F. moniliforme* M1293) was dissolved in sterile distilled water and plated onto potato dextrose agar (PDA). The PDA plates (110 x 50 mm) were incubated in the dark for 7 days at 28°C. After 7 days, the contents of each petri plate were transferred to a plastic bag to which 100 mL of sterile distilled water was added. The contents of the plastic bag were mixed using a Stomacher (Tekmar Co., Cincinnati, OH) for 30 sec to evenly distribute the culture medium in the sterile water. The contents of the bag were used as the inoculum, and 25 mL were pipetted into each of the aluminum trays containing the sterilized grits. The inoculated grits were incubated in a greenhouse with ambient temperatures of 30°C during the day and 26°C during the night. After a week, the aluminum foil covering was loosened to promote air circulation. The grits were assayed for FB_1 and FB_2 production after 7, 15, and 25 days. After 25 days, the fumonisin-contaminated grits were transferred to a 70-gallon plastic trash container and stored at 4°C.

Animal Dose

The dose (312 g containing 2614 µg/g FB_1 and 855 µg/g FB_2) was prepared every week in the laboratory by individually weighing amounts of the contaminated grits into plastic freezer bags for transfer to the animal housing facility. A week's dose of the grits was left uncovered on a tray in a chemical safety-hood for approximately 8 h (overnight). This was done in an attempt to remove any volatiles left over from the fermentation process. Prior to feeding, the dose was mixed manually with the daily ration and dumped into the feeding troughs.

In Vitro Analysis

Rumen fluid was obtained from a fistulated Hereford steer and strained immediately through three layers of cheesecloth to remove undigested feed. McDougal's artificial saliva buffer was prepared by dissolving 9.8 g $NaHCO_3$, 7.0 g $Na_2HPO_4.7H_2O$, 0.57g KCl, 0.47 g NaCl, 12 g $MgSO_4$, and 0.8 g urea in 500 mL distilled water. Before addition of buffer to the rumen fluid, 0.04 g $CaCl_2$ was added, and the buffer was acidified to pH 6.8-7.0 by bubbling CO_2 gas. Five g of ground corn was added to Erlenmeyer flasks which were flushed with CO_2 and fitted with one-way valves. The *in vitro* analysis involved two treatments, namely a 75:25 (v/v) rumen fluid+buffer mix and 100% rumen fluid added to the corn in the flasks. Both treatments were spiked with 10 µg/mL of FB_1, and the head-space was flushed with CO_2 gas to maintain anaerobic conditions. The flasks were stoppered with the one-way valves and incubated at 39°C with constant stirring. The flasks (triplicate replications at each time period) were sampled at 0, 3, 5, 7, and 9 h, and the pH change was noted to ensure fermentation.

Serum Chemistry and Urinalysis

Premortem sample collection of the blood and urine was done in the morning before the animals were fed. Blood and urine samples then were taken to the Department of Veterinary Medicine's Diagnostic Laboratory for serum chemical analyses and urinalysis on the same day. Sampling was done on days -3, 0, 7, 14, 21, and 28. The steers were bled via jugular venipuncture, and the blood was collected in sterile, preservative-free, Vacutainer™, red-top collection tubes (Becton Dickinson, Rutherford, NJ). Blood serum was analyzed for aspartate amino transferase (AST), gamma glutamyl transpeptidase (GGT), lactate dehydrogenase (LDH), bilirubin, and cholesterol. The serum chemistry was done using a Discrete Analyzer with Continuous Optical Scanning (DACOS) (Coulter Electronics, Hialeah, FL). Urine was analyzed for creatinine using the Kinetic Jaffe method and also tested for color, specific gravity, glucose, bilirubin, ketone, blood, pH, and protein using Ames Reagent strips (Miles Inc., Elkhart, IN). Presence of urinary protein was confirmed by the salicylic acid test (SSA).

Microscopic examination of the urine involved counts of erythrocytes, leukocytes, casts, epithelial cells, crystals, and bacteria.

Sample Preparation

Rumen Fluid. At the time of sampling, the flasks were opened under a continuous flow of CO_2 gas. A 1 mL sample from each flask was pipetted into a test tube containing 5 mL of acetonitrile:water (50:50, v/v). The flasks were flushed with CO_2, and the one-way valves replaced. Each test tube containing the 1 mL of rumen fluid sample and 5 mL of acetonitrile:water (50:50, v/v) was shaken for 5 min, after which it was centrifuged in a Safe-Guard bench-top centrifuge (Clay Adams Inc., New York, NY, USA) for 15 min. The resulting supernatant was subjected to a solid phase extraction (SPE) using strong anion exchange (SAX) cleanup step as described by Shephard et al. (1992a).

Blood. Blood (10 cc) was sampled for fumonisins on days -3, 0, 7, 14, 21, and 28, and collected in a purple-top Vacutainer[TM] sterile blood collection tube containing K-EDTA as an anticoagulant. The tube was centrifuged for 15 min to separate the plasma layer. This layer was transferred to a clean centrifuge tube using a Pasteur pipette, and the plasma proteins were precipitated by addition of methanol in a 1:5 (v/v) plasma:methanol ratio. The plasma was separated from its protein precipitate by centrifuging for 15 mins. The resulting supernatant was subjected to SPE-SAX cleanup (Shephard et al., 1992a).

Urine. Fresh urine samples were collected in a clean plastic cup that was held under the animal while it urinated. The urine (10 cc) was transferred to a preservative-free, sterile, red-top vacutainer[TM] collection tube. Extraction for fumonisins and their breakdown products was done by mixing the urine with equal volumes of acetonitrile:water (50:50, v/v) and applying the mixture directly to the SPE-SAX cartridges (Shephard et al., 1992a).

Feces. Feces samples were collected in the morning and evening with a gloved hand at the time of elimination and transferred to plastic freezer bags. Samples were taken on days -3, 0, 7, 14, 21, and 28. For analysis, 10 g of feces were blended at high speed with 75 mL of acetonitrile:water (50:50, v/v) for 5 min in a Waring blender (Dynamics Corp. of America, New Hartford, CT, USA). The blended contents (10 mL) then were transferred to a centrifuge tube and centrifuged for 15 mins. The resulting clear, brownish-yellow supernatant was subjected to SPE-SAX cleanup.

Meat, Fat and Visceral Organs. A total of 4 steers (3 test and 1 control), were euthanized by electrocution prior to necropsy. Immediately following necropsy sections of the liver, spleen, kidneys, gall bladder, gastrohepatic lymph nodes, pancreas, tongue, subcutaneous fat, kidney fat, longissimus dorsi, chuck, and round were placed in plastic freezer bags and frozen. For each tissue except fat, 10 g was blended in a Waring blender at high speed for 5 min with 75 mL of acetonitrile:water (50:50, v/v). The homogenized mix was filtered twice, once through a layer of cheesecloth and then through a no.4 Whatman filter paper. The blender jar and the filtered residue were re-extracted with 25 mL of acetonitrile:water (50:50, v/v). The combined filtrate was concentrated in a Buchi rotary evaporator (Brinkman Inst., Westbury, NY, USA). The concentrate then was centrifuged for 15 mins, and subjected to SPE-SAX cleanup.

Fat (10 g) was blended first in 250 mL of hexane in a Waring blender. The homogenate was filtered through a Whatman no. 4 filter paper into an extraction flask containing 50 mL acetonitrile-water (50:50, v/v). The aqueous layer was collected, concentrated, centrifuged for 15 mins, and then subjected to SPE-SAX cleanup.

Herd Mix and Grits. Fumonisin B_1 and FB_2 levels in the herd-mix and grits were assayed by blending 10 g of sample in 75 mL of acetonitrile-water (50:50, v/v) for 5 min. The blended mix (10 mL) was centrifuged for 15 mins., and 1 mL of the supernatant was applied to the SPE-SAX cartridge.

The grits also were analyzed for levels of zearalenone, T-2 toxin, deoxynivalenol (DON), and moniliformin (Sigma Chemical Co., St. Louis, MO). Standard stock solutions

(1 μg/mL each) were prepared for each of these mycotoxins in MeOH. The grits were extracted using the general mycotoxin screening method (Stahr, 1991). Briefly, 10 g of grits were blended with 50 mL of acetonitrile-water (90:10, v/v) and centrifuged for 15 mins. The supernatants were concentrated to 250 μL in a rotary evaporator and applied to a C-18 SEP-PAK column (Waters Associates, Milford, MA) for cleanup.

Analysis of Fumonisins

Chemicals. Methanol (HPLC-grade), potassium hydroxide (ACS-grade), ammonium acetate, hydrochloric acid (HCl, ACS-grade), acetonitrile (HPLC-grade), and glacial acetic acid (ACS-grade) were obtained from Fisher Scientific Co.(Fair Lawn, NJ, USA). The Fluoraldehyde^{TM} o-phthalaldehyde reagent (OPA) was obtained from Pierce (Rockford, IL, USA), and the FB_1 and FB_2 standards were obtained from Sigma Chemical Co. (St. Louis, MO, USA). Deionized water (HPLC-grade) was obtained from a Sybron-Barnstead ion exchanger.

Fumonisin B_1 stock solution was prepared by dissolving 5 mg of FB_1 standard in a 10 mL volumetric flask and brought to volume with acetonitrile-water (20:80, v/v) to give a stock solution that was nominally 500 μg/mL. The FB_2 stock solution was made up similarly. The stock solutions were used for serial dilutions. Dilutions made with acetonitrile-water (20:80, v/v) were 10 μg/mL, 5 μg/mL, 3 μg/mL, 1 μg/mL, 0.5 μg/mL, and 0.3 μg/mL. Standard curves were plotted using these dilutions and evaluated for linearity by determining the coefficient of determination (r^2).

Isolation of FB_1 and FB_2. Bond-Elut^{TM} SAX-SPE cartridges (3 mL capacity containing 500 mg sorbent) were obtained from Varian (Harbor City, CA, USA). The SAX-SPE cartridges were conditioned with 10 mL of methanol (MeOH), followed by 10 mL of MeOH-water (3:1, v/v). The supernatant (1 mL each) from each of the samples was applied to individual SAX-SPE cartridges. The cartridges were washed with 8 mL of MeOH-water (3:1, v/v), followed by 4 mL of MeOH, and the FB_1 and FB_2 (and $PHFB_1$ and $PHFB_2$, if present) were eluted by washing with 15 mL of acidified MeOH (5% glacial acetic acid in MeOH). The 15 mL acidified MeOH eluant was evaporated to dryness under vacuum in a Buchi rotary evaporator. The residue was dissolved in 5 mL of MeOH, evaporated to dryness under vacuum to ensure complete removal of acetic acid, and dissolved in acetonitrile-water (60:40, v/v). All samples were filtered through a 0.45 μm nylon syringe filter and stored in teflon-lined amber vials at - 4°C.

Preparation of HFB_1, $PHFB_1$, HFB_2, and $PHFB_2$. A 10 μg/mL stock solution of FB_1 was used to prepare the hydrolyzed fumonisin (HFB_1) and the partially hydrolyzed fumonisin ($PHFB_1$). The cleavage of the ester groups was achieved by base hydrolysis and controlling the time and temperature of the reaction. Complete hydrolysis of FB_1 was achieved by mixing 5 mL of the 10 μg/mL of stock FB_1 solution with 5 mL of 1N potassium hydroxide (KOH) and heating the mixture at 70°C for 1 h. After XAD-2 cleanup, the HFB_1 was serially diluted to achieve concentrations of 10 μg/mL, 5 μg/mL, 3 μg/mL, 1 μg/mL, 0.5 μg/mL, and 0.3 μg/mL. Partial hydrolysis of FB_1 was done by mixing 1 mL of the 10 μg/mL of stock FB_1 solution with 1 mL of 1N KOH, and reacting the solution for 10 min at room temperature. Once the desired level of hydrolysis was reached, the reactions were stopped by acidifying the solutions to pH 4.5 by the addition of 5M HCl. The HFB_2 and $PHFB_2$ were prepared similarly. Direct liquid injection thermospray mass spectrometry (TSP-MS) was used to confirm the formation of HFB_1, $PHFB_1$, HFB_2, and $PHFB_2$.

Isolation of HFB$_1$, PHFB$_1$, HFB$_2$, and PHFB$_2$. A 10 cc disposable syringe was plugged with glass wool and filled with 6 g of the XAD-2 resin (Sigma Chemicals, St. Louis, MO). A second piece of glass wool was inserted at the top to hold the resin bed in place during solvent addition. The resin was activated by washing with 10 mL of MeOH, followed by 10 mL of water. The samples were acidified to a pH of about 4.5 with 5M HCl and then added to the XAD-2 column. The column then was washed with 8 mL of water, and the hydrolyzed and partially hydrolyzed forms of FB$_1$ and FB$_2$ were eluted with 15 mL of MeOH. The eluate was evaporated to dryness under vacuum, resuspended in 1 mL of ACN-water (60:40, v/v), filtered through a 0.45 µm syringe filter, and stored in teflon-lined capped amber vials at -4°C.

Analysis of Zearalenone, T-2, DON, and Moniliformin. The C-18 SEP-PAK columns were conditioned using 6 mL of MeOH followed by 6 mL water (Stahr, 1991). The column was washed with 5 mL of water, followed by 5 mL of MeOH-water (20:80, v/v). The mycotoxins were eluted with 8 mL MeOH (100%), which was rotary evaporated under vacuum to dryness. The dried residue was resuspended in 300 µL of MeOH for analysis by HPLC and thermospray-MS as described in the next section.

High Performance Liquid Chromatography and Mass Spectrometry. A n a l y s e s were done using a Hewlett Packard (HP) Series II, 1090A HPLC (Hewlett- Packard, Palo Alto, CA, USA), fitted with a Rheodyne 7125 injector (Rheodyne Inc., Cotati, CA, USA) and having a 20 µL loop. Chromatographic separation was achieved using a double end-capped, metal-free silica, C-18 (250 mm x 4.6 mm, 5 µm), Alltech Alltima^{TM} (Alltech Associates, Deerfield, IL, USA) column equilibrated at a temperature of 40°C. Detection was performed with a HP 1046A programmable fluorescence detector set at an excitation wavelength of 229 nm, emission wavelength of 442 nm, lamp frequency of 220Hz, response time of 4000 msec., and a 418 nm filter.

The gradient mobile phase used for separation contained acetonitrile-water-acetic acid mixtures with the reservoirs set at: A (40:59:1, v/v) and B (60:39:1, v/v), pumped in a gradient of 100% A to 100%B over a period of 9 min. The flow rate was 1 mL/min. The fluorescent derivative was prepared by mixing the OPA reagent and the sample in a 2:1 ratio, and holding for exactly 1 min prior to injection into the HPLC. Mass spectral analyses were performed using an HP5989A quadrupole mass spectrometer connected to the HPLC via a HP thermospray(TSP) interface. A model 59970C processor was used for data acquisition and processing. The MS was operated in the filament-on mode. The source temperature was maintained at 225°C, and the quadrupoles were set at 100°C. The TSP was set to 98% solvent vaporization, which resulted in a TSP tip temperature of 178°C. Linear scanning within the range m/z 250-850 resulted in average rates of 0.64 scan cycles/sec. The MS was tuned using propylene glycol. The mobile phase used was an acetonitrile- 0.1M ammonium acetate mixture (50:50, pH 6.7) at a flow rate of 0.8 mL/min. All TSP- MS analyses were done using direct liquid injections.

The mycotoxin standards zearalenone, T-2, DON, and moniliformin were resolved using the Alltech Alltima^{TM} column, a mobile phase of MeOH-0.1M ammonium acetate (75:25, v/v), and detection with a UV-visible diode array detector set at 254 nm. Thermospray-MS analysis for these mycotoxins was done using a modified method of Voyksner et al. (1985).

Enzyme-Linked Immunosorbent Assay. A prototype of a direct competitive enzyme-linked immunosorbent assay (ELISA), Veratox^R (Neogen Corp., Lansing, MI) fumonisin test kit was evaluated by this laboratory for applications in biological samples. Because acetonitrile was detrimental to the antibody coated onto the microwell (Dr. Abouzied, personal communication, Neogen Corp.), 1 mL of the extracts (in acetonitrile-water, 50:50, v/v) were evaporated to dryness and resuspended in methanol-water (70:30, v/v). The enzyme-labelled toxin (conjugate), the substrate and chromophore reagent, and the stopping reagent were provided in the kit. A red color indicated a positive result, whereas a blue color indicated a negative result. The absorbance was read at a wavelength of 650 nm in an automated microwell-plate reader. The concentrations of the extracts were estimated by comparing the absorbance at 650 nm to a standard curve generated from a set of FB_1 standards in the range of 0.25-10 ng/g.

RESULTS AND DISCUSSION

In Vitro Analysis

Figure 2 shows the *in vitro* degradation of FB_1 over a period of 9 h. The viability of the fermentation was evident by the presence of head-space foam and the drop in pH during the length of the study. An average of 12.5% degradation of FB_1 in the buffered system was seen, compared to the 35% (average) degradation in the 100% rumen fluid mix at the end of the incubation period. This difference in degradation could have been related to the higher microbial activity of the 100% rumen fluid. However, it is interesting to note the lack of degradation of FB_1 by the end of 4 h in both treatments.

The samples also were analyzed for the presence of HFB_1 and $PHFB_1$, none of which could be detected. A lack of the specific signals namely, HFB_1 at 13.1 min and $PHFB_1$ at 10.3 min, in the chromatogram of the spiked rumen fluid after 9 h indicated the absence of these hydrolysis products. Thus we can assume that no hydrolysis of the ester bonds, either complete or partial, occurred *in vitro*. The complexity of the rumen fluid contents makes it difficult to explain the results obtained, although deamination and/or engulfment by protozoa could have been involved. Both essential and nonessential amino acids are rapidly deaminated (0.2-0.3 mmol/h) in the rumen (Wallace and Cotta, 1988). Because the method of detection for fumonisins and their breakdown products hinges on the presence of the primary amine group, deamination will directly affect detection. Hence, the observed degradation could be a result of deamination.

Engulfment or direct assimilation of the fumonisin molecule by protozoa could also explain the degradation. Amino acids and lipids are taken up by holotrich ciliates from their surroundings and assimilated into the protozoan cell (Williams and Coleman, 1988). Recov-

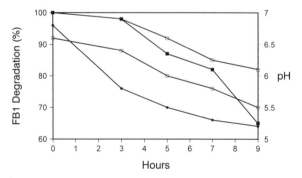

Figure 2. *In vitro* degradation of FB_1-spiked 100% rumen fluid and 75:25 rumen fluid:buffer mix; the drop in pH indicates viable fermentation. Key: ■100% rumen fluid, □ 75:25 rumen fluid:buffer, ● drop in pH for the 100% rumen fluid, and O drop in pH for the 75:25 rumen fluid:buffer mix.

ery of FB₁ thus could be affected if the molecule was engulfed and probably eluded extraction, since the sonication normally required to lyse the protozoan cell was not part of the clean up procedure. Nonetheless, the *in vitro* study indicated that FB₁ was not easily amenable to hydrolysis by the rumen fluid and between 65- 87% was intact after 9 h of incubation.

Feeding Trial-Premortem Analysis

Dose, Feed Consumption, and Weight Gain. At the end of the 25 day incubation period, the average concentration of FB₁ in the culture-containing corn grits was 2614 μg/g, whereas FB₂ was 855 μg/g. The background FB₁ concentration of the herd-mix was 0.037 μg/g, and FB₂ was below the detection limit. The test animals were each dosed at approximately 400 μg/g FB₁ (861 mg/day/steer). Because this dosage was calculated based on the concentration of FB₁ alone, the test animals were exposed to an additional 130 μg/g FB₂ (281 mg/day/steer) within the same dose, raising the total fumonisin exposure to 530 μg/g or 1142 mg/day/steer. Therefore, each of the test animals was exposed to an average of 25,830 mg of FB₁ and 8,430 mg of FB₂ during the 30-day feeding period.

The high concentration of the fumonisins in the grits, resulted in a feed-grits ratio of approximately 91:9 (w/w). Thus, the small amount of grits incorporated into the feed did not add any detectable odor to the total ration. The grits were pinkish-violet in color and did not contain any detectable (0.5 ppm) zearalenone, T-2 toxin, deoxynivalenol (DON), or moniliformin.

The test animals consumed their rations every day. However, a change in the feeding pattern was observed on days 17 through 19, when the test animals consumed approximately 70-80% of the daily ration. From day 19 to the end of the study, the test animals consumed their rations sluggishly, whereas the controls consumed the feed immediately. This decreased consumption of fumonisin-contaminated feed was similar to that observed by Osweiler et al. (1993). Beasely and Buck (1982) indicated some feed refusal associated with corn contaminated by *F. moniliforme*. However, our test animals showed no sign of feed refusal and consumed all their feed before the next feeding period. Body weights of control and test animals were not significantly different (Table 1).

Clinical Observations. The test animals appeared languid and looked visibly distressed after about 2 weeks of being fed the fumonisin-containing diet. The test animals hung their heads and generally appeared to be disinterested in the surrounding environment. Often, they would simply lie down. Body temperatures of the steers taken on days -5, -4, -3, -2, - 1, 0, 1, 2, 3, 4, 7, 14, 21, and 28 were unaffected. This result also was seen in pigs fed lethal doses of fumonisin-contaminated corn screenings (Haschek et al., 1992).

Blood Chemistry and Urinalysis. Evidence of hepatobiliary damage can be construed from the changes occurring in the serum liver function tests, AST, GGT, LDH, and cholesterol (Figures 3-6). Similar serum liver function changes were observed in horses, swine, lambs, and cattle exposed to fumonisins (Ross et al., 1993; Voss et al., 1989; Edrington et al., 1995; Osweiler et al., 1993). These significant changes suggest some hepatotoxicity at the molecular level, an effect common to all species exposed to fumonisins.

Because alkaline phosphatase assays cannot be employed in ruminants the notable increase in the GGT (Figure 4) levels is of substantial importance (Evans, 1988). Elevated GGT levels parallel the increase in serum cholesterol and may reflect fatty infiltration of the hepatocytes and bile-duct proliferation. Hepatosis characterized by cirrhosis, adenofibrosis, and bile duct proliferation was observed in rats exposed to fumonisins (Marasas et al., 1984;

Table 1. Body weights of control and test steers taken at the beginning, midway through, and at the end of the feeding period

Animals	Body weight in Kg			Change
	Day 0	Day 15	Day 30	
Control 1	296	296	310	↑
Control 2	251	248	248	↔
Control 3	277	280	285	↑
Test 1	234	216	217	↓
Test 2	210	192	198	?
Test 3	210	207	207	↔

Gelderblom et al., 1991). Tubular nephrosis of the kidney and mild hepatopathy was seen in the fumonisin-dosed lambs, which was preceded by elevated serum GGT levels (Edrington et al., 1995).

The observed increase in serum cholesterol (Figure 6) may prove to be an early indicator of fumonisin-induced hepatobiliary change (Beasely et al., 1992). A similar trend in serum cholesterol levels was observed in fumonisin-dosed lambs, calves, and swine (Edrington et al., 1995; Osweiler et al., 1993; Haschek et al., 1992). The use of free sphinganine-to-sphingosine ratio is reported to be a sensitive biomarker for fumonisin toxicity (Riley et al., 1994), but its cumbersome chemical analysis makes the measurement impractical. The use of cholesterol as an early biomarker seems to be more pragmatic, because automated equipment delivers quick results and requires minimal sample cleanup.

Except for total urinary protein, urinalysis proved to be inconclusive. A significant increase was observed in the total urinary protein after day 14 (Figure 7). Nephrotoxicity caused by fumonisins has been demonstrated in rats (Gelderblom et al., 1991). Some blood was observed in the urine of two of the test animals on day 28. Gross appearance, and microscopic examination of the urine was within normal parameters for all the animals. Urine volume over a 24-h period was not measured, and no determinations could be made in changes occurring in the kidney load and function. The specific gravity remained within normal ranges throughout the study, which indicates no significant change in urine volume in the animals.

Cattle seem to be much less susceptible to levels of fumonisins lethal to other species (Table 2). Such a resistance to fumonisins by cattle also was observed by Osweiler et al. (1993). Horses fed fumonisin at total dose levels of 8,400 mg FB_1 developed classical neurological lesions and significant hepatobiliary dysfunction (Kellerman et al., 1990). Swine exposed to a total dose of 11.3 mg FB_1 intravenously developed severe respiratory distress and died within 5 days of pulmonary edema (Harrison et al.,1990). However, higher

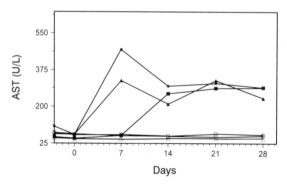

Figure 3. AST values (U/L) of control and test steers obtained from blood serum analysis during the 30-day feeding period. Key: Test 1:■ , Test 2: ●, Test 3: ▲, Control 1: O, Control 2: Δ , Control 3: □.

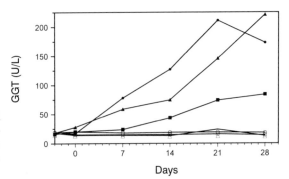

Figure 4. GGT values (U/L) of control and test steers obtained from blood serum analysis during the 30-day feeding period. Key: Test 1:■ , Test 2: ●, Test 3: ▲, Control 1: O, Control 2: Δ, Control 3: □.

total levels of oral dosage were required to achieve the same effect in swine (Haschek et al., 1992). Similar changes in the serum liver function tests were observed in this study on cattle; however, none of the test animals showed any overt signs of clinical distress.

Analysis of Blood, Urine, and Feces for Fumonisins. Only fresh samples were analyzed; therefore, the levels of fumonisins reported represent a single period of collection. The feces (Figure 8) showed the highest levels of FB_1 (127 μg/g) and FB_2 (35.1 μg/g). No evidence of HFB_1, HFB_2, $PHFB_1$, or $PHFB_2$ was found in the feces. The presence of $PHFB_1$ was detected in the feces of vervet monkeys and was attributed to hydrolytic attack occurring in the digestive tract (Shephard et al., 1994). Traces of FB_1 (18.1 μg/g) and FB_2 (9.80 μg/g) were detected in the urine of the experimental animals (Figure 8). However, the blood tested negative for the presence of fumonisins or their major metabolites.

Because the liver enzyme tests indicated a significant hepatobiliary response, the absence of fumonisins in blood was confounding. To approximate the toxicokinetics of fumonisins, a fistulated steer was used to determine the fate of a single dose of grits contaminated with fumonisins. A known amount of grits were emptied directly into the rumen to approximate a total dose of 150 μg/g FB_1 (481 mg/24 h/steer). Ensuing the intra-fistulary addition of grits, the animal was bled via jugular venipuncture at 0.5, 1, 2, 3, 4, 5, and 6 h. Only 2.67 μg/g FB_1 was detected in the blood at the end of 0.5 h. However, 1.2 μg/g of FB_1 was found in the feces by 2.2 h postdosing and the level increased to 145 μg/g FB_1 at the end of 24 h. By 24 h almost 80% (387 mg) of the dosed FB_1 had been excreted into the feces. No fumonisins were detected in the urine at any time period.

These observations lead us to believe that, in cattle, the bulk of fumonisins dosed orally are excreted unmetabolized in the feces. A similar observation was made by Shephard

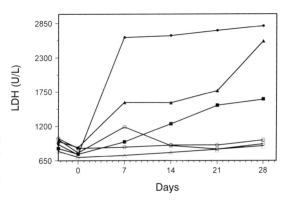

Figure 5. LDH values (U/L) of control and test steers obtained from blood serum analysis during the 30-day feeding period. Key: Test 1:■ , Test 2: ●, Test 3: ▲, Control 1: O, Control 2: Δ, Control 3: □.

Table 2. Response of various species to FB$_1$ exposure

Species	Administration route	Dose mg/kg/day	Days	Clinical Signs and Blood Chemistry Analysis
Cattle[a]	Oral	4.3	30	No overt clinical signs. Significant elevations in AST, GGT, LDH, cholesterol from day 7-30. Body temp. normal. No gross lesions.
Horse[b]	Oral	4.0	33	Severe clinical distress on days 31-33. Elevations in AST and GGT from day 21 through 30. Equine leukoencephalomalacia (ELEM) left frontal and occipital lobes.
Swine[c]	Intravenous	0.4[d]	5	Died on day 5 due to pulmonary edema (PEM), hydrothorax and pancreatic lesions.
Swine[e]	Oral	4.5	5	Severe respiratory distress, (PEM), required euthanasia. Mildly elevated AST, ALP. Prominent increase in serum cholesterol by day 15. Body temp. normal.

[a] Data obtained from current study.
[b] Data obtained from Kellerman et al., 1990.
[c] Data obtained from Harrison et al., 1990.
[d] One dose of 11.3 mg.
[e] Data obtained from Haschek et al., 1992.

et al. (1992b), when rats were orally gavaged with ^{14}C-labelled FB$_1$, and 101% radioactivity was detected in the feces. The inability of the rumen microflora to break down fumonisins was observed in the *in vitro* experiments, a surprising result considering the abundance of esterases present in the rumen. Because approximately 0.005% of the FB$_1$ was detected in the blood within 30 min, it is plausible to imagine that the site of absorption of fumonisins is the rumen wall.

This could partly explain the resistance of cattle to fumonisins. The large volume of the rumen and the secretion of copious amounts of saliva into the rumen during feeding and regurgitation cause an immediate dilution effect. Because the fumonisin molecule is not de-esterified in the rumen it retains an overall triple negative charge, making absorption through the gastro-intestinal membranes into the blood improbable. This lack of de-esteri-fication, thus assumes importance. Since the fumonisin molecule does not break down to its amino-pentol moiety, it is not converted to a partially lipophilic form conducive to absorption through the lipophilic cell walls of the gut. The trace amount of FB$_1$ that probably is absorbed through the rumen wall is eliminated rapidly through the biliary system during the first pass through the liver. The presence of 0.8 mg FB$_1$ in the feces at the end of 2.2 h is probably

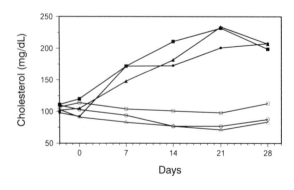

Figure 6. Cholesterol values (mg/dL) of the control and test steers obtained from blood serum analysis during the 30-day feeding period. Key: Test 1:■ , Test 2: ●, Test 3:▲, Control 1: O, Control 2: Δ, Control 3: □.

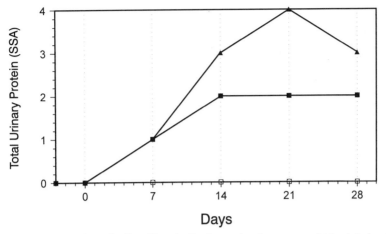

Figure 7. Total urinary protein (salicylic acid method) of control and test steers obtained during the 30-day feeding period. Key: Test 1:■ , Test 2: ●, Test 3: ▲, Control 1: O, Control 2: Δ, Control 3: □.

indicative of this hepatobiliary mechanism. At physiological pH, FB_1 will be ionized to a net pKa between 3-4, making it ideally suited for biliary excretion (Matthews, 1994). The difficulty of absorption through the highly keratinized ruminal wall and the higher rate of passage associated with a feed consisting of 85% grain, would make absorption of large amounts of fumonisins through the ruminal wall arduous. The fumonisins thus pass on through the gut unmetabolized and are excreted in the feces. Elimination of orally ingested, poorly absorbed, toxic, ionized compounds through the feces has also been demonstrated in the case of the herbicides diquat and paraquat (Matthews, 1994).

Postruminal microbial degradation of the intact fumonisins seems most unlikely, although such a degradation was reported in vervet monkeys fed fumonisins (Shephard et

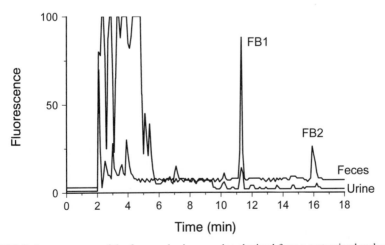

Figure 8. HPLC chromatograms of the feces and urine samples obtained from a test animal and resolved on a C-18 reversed-phase column using a mobile phase mixture of $ACN-H_2O-CH_3COOH$, set at A (40:59:1) and B (60:39:1), pumped in a gradient of 100% A to 100% B (9 min) at a flow rate of 1 mL/min. Fluorescence detector set at 229 nm excitation and 442 nm emission wavelengths.

al., 1994). The potency of the absorbed fumonisin thus appears to be significant. Riley et al. (1994) reported that $0.1\mu M$ FB_1 is enough to inhibit sphingosine N-acyl- transferase activity in rat liver microsomes by 50%. This could explain the observed elevation of the liver enzymes in response to the trace amounts of fumonisins detected in the general circulation.

Feeding Trial-Postmortem Analysis

Necropsy. No significant gross lesions were noted in any of the four animals necropsied. The rumen and reticulum, abomasum, and omasum were emptied of their contents and examined. No abnormalities were observed. The organ-to-body weight ratios of the heart, brain, lungs, liver and gall bladder, kidneys, spleen and pancreas were normal. Light microscopy of all the tissues did not reveal any abnormalities.

Analysis of Muscle and Visceral Organs for Fumonisins. Fumonisin B_1 was detected in muscle (97 ng/g, SD: 41), liver (2,100 ng/g, SD: 1,900), and kidneys (23.4 ng/g, SD: 8.7) by HPLC. Figure 9 shows the HPLC chromatograms of the muscle extract obtained from a control animal, a test animal, and a 300 ng FB_1 standard; the enlarged inset depicts the region where FB_1 elutes. Fumonisin B_1 was the only residue detected in the tissues. Fumonisins were not detected in the fat, gall bladder, spleen, pancreas, or the gastro-hepatic lymph node. The detection limit of HPLC analysis for FB_1 and FB_2 was 30 ng/mL, whereas that of the ELISA test was 1 ng/mL.

Testing of these organs using the ELISA test kit for fumonisins resulted in a strong positive response (based on the intensity of the developed red color) for the muscle, liver, and kidney, whereas the spleen and gall bladder gave a dull red color indicating only trace

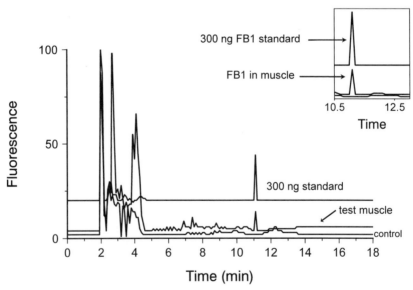

Figure 9. HPLC chromatograms of muscle obtained from a control animal, test animal, and a 300 ng FB_1 standard, indicating the presence of FB_1 residues in cattle exposed to the fumonisin-containing diet. Separation was with a C-18 reversed- phase column using a mobile phase mixture of ACN-H_2O-CH_3COOH, set at A (40:59:1) and B (60:39:1) and pumped in a gradient of 100% A to 100% B (9 min) at a flow rate of 1 mL/min. The inset depicts an enlarged portion of the chromatogram where FB_1 elutes in the samples.

Table 3. Fumonisin B_1 residues measured by HPLC in tissues from cattle fed dietary FB_1 and FB_2 for 30 days

Animal No.	Levels of $FB_1 + FB_2$ in Feed (g)	Liver		Muscle		Kidney	
		FB_1 (ng/g)	Feed/Tissue[a]	FB_1 (ng/g)	Feed/Tissue[b]	FB_1 (ng/g)	Feed/Tissue[c]
T #1	27.2 + 8.9	4,590	1600	519	1702	31	1.0×10^6
T #2	24.5 + 8.0	112	81666	145	61250	12	2.8×10^6
T #3	24.5 + 8.0	1,510	6125	640	2227	28	1.2×10^6

[a] Overall level of fumonisins in feed divided by the level in the specified tissue.
[b] Calculations based on 50% of total body weight being edible tissue.
[c] Calculations based on weight of both kidneys.

amounts of fumonisins. The fat, tongue, and gastro-hepatic lymph node gave negative responses to the ELISA test.

Table 3 shows the amounts of FB_1 residue detected in the tissues and their corresponding feed:tissue ratios. The prohibitive cost of using live large-animals in experiments and the difficulty in procuring large quantities of fumonisins limited this study to the use of three test animals. Consequently, the high standard deviations observed cannot be avoided, unless a larger pool of test animals could be used. However, the results do reflect the proclivity for residual fumonisin accumulation in tissue.

CONCLUSION

Cattle seem to tolerate high (530 µg/g fumonisins) levels of exposure to fumonisins without developing any clinical symptoms of toxicoses. Although the liver functionality tests indicated an adverse hepatobiliary response, no gross lesions were observed during necropsy. Results from the histopathological examination were not available during the writing of this manuscript; therefore, the effects at the tissue level cannot be presented.

The presence of fumonisins (97.3 ng/g) in the muscle may have been due to the continuous exposure to the very high levels of toxin in the feed. Although the level of fumonisin exposure in this study is unrealistically high relative to natural contamination, it is obvious that fumonisins are only mildly antagonistic to cattle. The high feed:tissue ratios for liver, muscle, and kidneys, indicate poor overall absorption characteristics for FB_1. While the animals tested exhibited no clinical symptoms, chronic exposure to low-to-moderate levels of fumonisins (5-25 µg/g) ultimately may damage the liver. A damaged liver may lower the first pass efficiency, resulting in a longer presence of fumonisins in the general circulation, and subsequent accumulation in the tissues. The majority of the dose was excreted as the unmetabolized parent molecule in the feces. Only trace amounts of fumonisins were detected in the blood ½ h after feeding, and in the urine indicating poor absorption. Thus, the carryover of fumonisins from cattle consuming highly contaminated feed to the human diet via consumption of beef does not appear to be a concern.

ACKNOWLEDGEMENTS

Contribution No. 95-418-B from the Kansas Agricultural Experiment Station. This work was supported by The Cooperative State Research Services, U. S. Department of Agriculture, under agreement 89-34187-4511. We are grateful to Dr. John Leslie, KSU,

Department of Plant Pathology, for allowing us the use of his greenhouse facility, and Dr. Mike Stahr, Iowa State University, for the fumonisin producing lyophilized *F. moniliforme* M1293 culture. We are indebted to Dr. T. G. Nagaraja for his adept advice throughout the study, Dr. Renee. A. H. Thakur for her microbiological expertise, and Dr. G. A. Kennedy, KSU, Department of Veterinary Medicine, for performing the necropsy and histopathological evaluation. A special thanks to Brian P. Ioerger, Neil Wallace, Lalit K. Bohra, and Basira Abdul-Karim for their invaluable help in all phases of this study.

REFERENCES

Beasely, V. R.; Buck, W. B. Feed refusal in cattle associated with *Fusarium moniliforme* in corn. Vet. Rec. 1982, 111, 393-396.

Edrington, T. S; Kamps-Holtzapple; Harvey, R. B; Kubena, L.F; Elissalde, M. H; Rottinghaus, G. E. Acute hepatic and renal toxicity in lambs dosed with fumonisin-containing culture material. 1995, 73, 508-515.

Evans, R. J. Hepatobiliary damage and dysfunction: A critical overview. In Animal clinical biochemistry-the future. Blackmore, D. J., Ed. Cambridge University Press, Cambridge, UK, 1988, pp 119-128.

Gelderblom, W. C. A.; Kriek, N. P. J.; Marasas, W. F. O.; Thiel, P. G. Toxicity and carcinogenicity of the *Fusarium moniliforme* metabolite, fumonisin B_1 in rats. Carcinogenesis. 1991, 12, 1247-1251.

Harrison, L. R.; Colvin, B. M.; Greene, J. T.; Newman, L. E.; Cole, J. R. Pulmonary edema and hydrothorax in swine produced by fumonisin B_1 a toxic metabolite of *Fusarium moniliforme*. J. Vet. Diagn. Invest. 1990, 2, 217-221.

Haschek. W. M.; Motelin, G.; Ness, D.; Harlin, K. S.; Hall, W. F.; Vesonder, R. F.; Peterson, R. E.;Beasely, V. R. Characterization of fumonisin toxicity in orally and intravenously dosed swine. Mycopathologia. 1992, 117, 83-89.

Kellerman, T. S.; Marasas, W. F. O.; Thiel, P. G.; Gelderblom, W. C. A.; Cawood, M.; Coetzer, J. A. W. Leukoencephalomalacia in two horses induced by oral dosing of fumonisin B_1. Onderstepoort J. Vet. Res. 1990, 57, 269-275.

Marasas, W. F. O.; Wehner, F. C.; Van Rensburg, S. J.; Schalkwyk, D. J. Mycoflora of corn produced in human esophageal cancer areas in Transkei, Southern Africa. Phytopathology. 1981, 71, 792-796.

Marasas, W. F. O.; Crack, N. P. J.; Fincham, J. E.; Van Rensburg, S. J. Primary liver and esophageal basal cell hyperplasia in rats caused by *Fusarium moniliforme*. Int. J. Cancer. 1984, 34, 383-387.

Matthews, H. B. Excretion and elimination of toxicants and their metabolites. In Introduction to biochemical toxicology, 2nd edition; Hodgson, E.,and Levi, P. E., Eds; Appleton and Lange, Nowalk, CT. 1994.

Osweiler, G. D.; Kehrli, M. E.; Stabel., J. R.;Thurston, J. R.; Ross, P.F.; Wilson, T. M. Effects of fumonisin-contaminated corn screenings on growth and health of feeder calves. J. Anim. Sci. 1993, 71, 459-466.

Riley, R. T.; Hinton, D. M.; Chamberlain, W. J.;Bacon, C. W.; Wang, E.;Merrill, A. H. Jr. ; Voss, K. A. Dietary fumonisin B_1 induces disruption of sphingolipid metabolism in Sprague-Dawley rats: A new mechanism of nephrotoxicity. J. Nutr. 1994, 124, 594-603.

Ross, F. P.; Rice, L. G.; Ledet, A. E.; Owens, D. L.;Nelson, H. A.;Osweiler, G. D.; Wilson, T. M. Experimental equine leukoencephalomalacia, toxic hepatosis, and encephalopathy caused by corn naturally contaminated with fumonisins. J. Vet. Diagn. Invest. 1993, 5, 69-74.

Ross, P. F.; Nelson, P. E.; Owens, D. L.; Rice, L.G.; Nelson, H. A.; Wilson, T. M. Fumonisin B_2 in cultured *Fusarium proliferatum*, M-6104, causes equine leukoencephalomalacia. J. Vet. Diagn. Invest. 1994, 6, 263-265.

Schroeder, J. J.; Crane, H. M.; Xia, J., Liotta; D.C.; Merrill, A. H. Jr. Disruption of sphingolipid metabolism and stimulation of DNA synthesis by FB_1. J.Biol. Chem. 1994, 269, 3475-3481.

Shephard, G. S.; Thiel, P. G.; Alberts, J. F.; Gelderblom, W. C. A. Fate of a single dose of the ^{14}C-labelled mycotoxin, fumonisin B_1 in rats. Toxicon.1992a, 30, 768-770.

Shephard, G. S.; Thiel, P. G.; Sydenham, E. W. Determination of fumonisin B_1 in plasma and urine by high-performance liquid chromatography. J.Chromatogr. 1992b, 574, 299-304.

Shephard, G. S.; Thiel, P. G.; Sydenham, V. R.; Alberts, J. F. Determination of the mycotoxin fumonisin B_1 and the identification of its partially hydrolyzed metabolites in the feces of non-human primates. Fd. Chem. Toxicol. 1994, 32, 23-31.

Stahr, M. H. Mycotoxin analysis. In Analytical methods in toxicology; John Wiley & Sons, NY, 1991.

Voss, K. A.; Norred, W. P.; Plattner, R. D.; Bacon, C. W. Hepatotoxicity and renal toxicity in rats of corn samples associated with field cases of equine leukoencephalomalacia. Fd. Chem. Toxic. 1989, 27, 89-96.

Voyksner, R. D.; Hagler, Jr., W. M.; Tyczkowska, K.;Haney, C. A. Thermospray high performance liquid chromatographic/mass spectrometric analysis of some fusarium mycotoxins. J. High Res. Chromatogr. & Chromatogr.Comm. 1985,8, 119-125.

Wallace, R. J.; Cotta, M. A. Metabolism of nitrogen-containing compounds. In The rumen microbial ecosystem. Hobson, P. N., Ed.; Elsevier Science Publishers Ltd., Essex, UK, 1988.

Williams, A. G.; Coleman, G. S. The rumen protozoa. In The rumen microbial ecosystem. Hobson, P. N., Ed.; Elsevier Science Publishers Ltd., Essex, UK, 1988.

5

ANALYTICAL DETERMINATION OF FUMONISINS AND OTHER METABOLITES PRODUCED BY *FUSARIUM MONILIFORME* AND RELATED SPECIES ON CORN

Ronald D. Plattner, David Weisleder, and Stephen M. Poling

Bioactive Constituents Research
National Center for Agricultural Utilization Research
Midwest Area, Agriculture Research Service
United States Department of Agriculture, Peoria, IL 61604

ABSTRACT

Fumonisins, secondary metabolites of the fungus *Fusarium moniliforme* are potent toxins that can be found in fungal contaminated corn. The detection and measurement of these toxins by HPLC with detection by an evaporative light scattering detector and by electrospray MS is reported. The light scattering detector had enough sensitivity to analyze culture materials, however, clean-up was necessary to detect fumonisins at sub-ppm levels in naturally contaminated corn extracts. The detection limit for FB_1 with the light scattering detector was in the low ng range (10-50) while the detection limit of less than 1 ng injected was observed for the electrospray detector. Several previously unreported fumonisin isomers were observed in electrospray chromatograms of culture extracts. Two of these compounds, FA_3 and FA_4 were isolated and their proposed structure confirmed by NMR experiments.

INTRODUCTION

Fumonisins, secondary metabolites of the fungus *Fusarium moniliforme* are potent liver toxins (Voss et al., 1989; Gelderblom et al., 1988) and reported to be carcinogens (Gelderblom et al., 1991). Consumption of fumonisin B_1 (FB_1), the most abundant fumonisin homolog, is known to cause diseases in animals including equine leucoencephalomalacia, a fatal disease in horses and porcine pulmonary edema (Nelson et al., 1993). *F. moniliforme* and several closely related species in the section *Liseola* are associated with ear and stalk rots in corn world-wide and are frequently found to contaminate agricultural products, especially corn (Marasas and Nelson, 1987). Fumonisins have been detected in asymptomatic apparently healthy corn in commerce at levels ranging from a few hundred parts-per-

Fumonisins in Food, Edited by L. Jackson *et al.*
Plenum Press, New York, 1996

billion to 2 parts-per-million (Sydenham et al., 1991; Pittet et al., 1992; Stack and Eppley, 1992). These findings have generated considerable effort to develop methods to detect and measure the concentrations of fumonisins in agricultural commodities and to study their toxicology.

In addition to fumonisins, strains of *F. moniliforme* and closely related species also produce several other mycotoxins and phytotoxins at high levels in laboratory culture. The toxicity of these other secondary metabolites has had much less study and their role in animal toxicoses and human exposure risk, if any, remains unclear. These components include the fusarins, potent mutagens in the Ames test; moniliformin, a potent avian toxin; the napthazarine pigment complex, a group of linear aromatics known to be phytotoxins, but untested for animal toxicity; fusaric acid, 1-carboxy-4-butene-pyridine, a known phytotoxin of unknown animal toxicity; beauvericin, a cyclic depsi-peptide, known to be a potent ionophore; and fusaproliferin, a recently reported toxic sesquiterpene. Toxicity data and reports on the concentrations of these compounds in naturally contaminated corn samples are much less available than data on fumonisins. Additional studies will be needed to determine whether they have a role in animal toxicoses.

HPLC with fluorescence detection of a suitable derivative of the free amine group (Shephard et al., 1990) is the most widely used analytical method for measuring fumonisin concentrations in food and feed matrices. This is a very good procedure to measure FB_1 and the homologs with one less hydroxyl group (FB_2 and FB_3). The major shortcoming of this technique is that it cannot detect n-acetylated fumonisin analogs which are sometimes present at lower levels in samples. A second method, hydrolysis of the esterified sidechains and gas chromatography/mass spectrometry (GC/MS) of trimethylsilyl or trifluoroacetate derivatives of fumonisin backbones, has also been reported (Plattner et al., 1990; Plattner et al., 1992). This method has been reported to give results that agree well with those obtained by fluorescence HPLC, but is also unable to measure the n-acetylated analogs. Capillary zone electrophoresis of a fluorescein isothiocynate derivative of FB_1 has also been reported (Maragos, 1995; Maragos et al., in press). This elegant and highly sensitive method uses clean-up on affinity columns and derivatization of the free amino group. Detection limits are lower than 50 parts-per-billion in corn, and good recoveries (93%) were obtained from corn spiked at 250 parts-per-billion. Another method which detects intact fumonisins without derivatization is fast atom bombardment/ mass spectrometry (FAB/MS) (Korfmacher et al., 1991). FAB is a matrix sensitive technique, so response is variable in slightly different matrices. In the presence of an easily protonated compound, another compound with a lower proton affinity may not be detected at all. For example in the presence of a few nanograms of FB_1, no signal is observed for the n-acetylated analog FA_1 (Plattner, 1995, unpublished data). Furthermore, response may not be quantitative for a single component in the presence of variable matrix components. Deuterium-labeled FB_1 has been used as an internal standard to overcome this problem for quantitative measurements (Plattner and Branham, 1994). The later method gives excellent data for FB_1; however, FB_2 and FB_3, which have the same molecular weight, are detected together in FAB/MS spectra. Detection and quantitation of underivatized fumonisins by an evaporative light scattering detector (ELSD) has been reported (Plattner, 1995,; Wilkes et al., 1995). Analysis of underivatized fumonisins by electrospray MS has also been reported (Korfmacher et al., 1991; Plattner, 1995; Musser et al., 1995). This paper provides additional details of separations of underivatized fumonisins by HPLC with detection by ELSD and electrospray, the use of stable isotope-labeled internal standards in electrospray and the identification and partial characterization of two new fumonisins in corn cultures.

MATERIALS AND METHODS

Fumonisins B_1, B_2, B_3, B_4, C_1 and C_4 were purified from cultures of isolates of *F. moniliforme* grown on cracked maize using published procedures (Desjardins et al., 1992; Nelson et al., 1993). Strains used in this study were from the culture collection at the Fusarium Research Center at the Pennsylvania State University, or from the collection of Dr. John Leslie (Kansas State University). Extracts of culture materials were made as reported previously (Nelson et al., 1993).

HPLC experiments were performed on a Spectra-Physics 8700 liquid chromatograph. Either a 3 μm x 3 cm C-18 column (column 1-Perkin Elmer, Norwalk, CT) with a mobile phase flow rate of 0.5 ml/min or a 4.5 cm x 250 mm 5-μm base deactivated C-8 column (column 2-YMC, Wilmington, NC) with mobile phase flow rate of 0.2 ml/min was used. The Varex MK III evaporative light-scattering detector (Alltech Associates, Deerfield, IL) was used as a detector (tube temperature of 75°C) with column 1. The solvent consisted of a binary gradient containing water (solvent A) and acetonitrile with 0.1% trifluoroacetic acid (TFA) (solvent B). Typically with column 1, all major fumonisins were well resolved with a gradient from 35% solvent B to 50% solvent B in 15 min. The HPLC was equipped with a 20-μL loop injector. Injections of 1-10 μL of aqueous samples of fumonisins (nominal concentrations in the range of 0.1-50 μg/ml) were made and the resulting detector response was measured with a Spectra-Physics integrator.

Alternatively the HPLC was coupled to a Finnigan-MAT TSQ 700 mass spectrometer via the Finnigan electrospray interface (Finnigan MAT, San Jose, CA). For electrospray experiments, several elution solvents were tried. These included mixtures of methanol/water either unamended or amended with either 0.1-1% acetic acid, 0.1% trifluoroacetic acid, 0.1N ammonium acetate, or 0.1N ammonium formate and acetonitrile/water with 0.1% trifluoroacetic acid. The eluting solvent was delivered using a binary gradient delivery system. Best results were obtained with solvent A as 50/50/1 methanol/water/acetic acid and solvent B as 100/1 methanol/acetic acid. With column 1, good separations were obtained with a gradient from about 30% solvent A to 100 % solvent B over 15-20 min, while on column 2, best separations were obtained with a gradient from 80% solvent A to 50/50 solvent A/solvent B in 20 min with a final hold of 10 min. Satisfactory results were also achieved with solvent A being 50/50 methanol/water with 0.1 N ammonium acetate or ammonium formate and solvent B of methanol with 0.1 N ammonium acetate or ammonium formate. Solvents containing acetonitrile or trifluoroacetic acid were less useful because the efficiency of electrospray in the presence of these solvents was poor. The total HPLC eluent was coupled into the detector. Flow rates from 0.1-0.5 ml/min gave satisfactory electrospray results. Mass spectra were obtained by scanning from 350-950 daltons in 1-3 sec (full scans) or by selected ion monitoring.

RESULTS AND DISCUSSION

As is the case with derivatized fumonisins, underivatized fumonisins do not elute satisfactorily from C-8 or C-18 bonded silica HPLC columns without the presence of either acids or buffers in the solvent system. Either methanol or acetonitrile can be used with water to achieve separations. However, acetonitrile is a much stronger solvent than methanol (e.g., 80/20/1 methanol/water/acetic acid and 30/70/1 acetonitrile/water/acetic acid give similar separations). With the ELSD, the presence of salts degrades the signal-to-noise ratio; and in the worst case, results in plugging problems. Consequently, it is recommended that salts be kept to a minimum. We obtained best results eluting with solvent A being water and solvent

B being acetonitrile with 0.1% trifluoroacetic acid. Good separation of major naturally occurring fumonisin homologs and types were achieved with the C-18 column using a gradient from 35% solvent B to 50% solvent B in 15 min. Detectable signals were observed with injection of as little as 10 ng of FB_1, FB_2, or FB_3. However, the response of fumonisins was not linear with concentration. A good fit between amount injected and response was obtained using a second order equation. The detection limit for underivatized fumonisins with this detector was not nearly as low fluorescence detection limits for OPA derivatives, but the detection limit of the detector was not the limiting factor in the method. Satisfactory chromatograms were obtained from injections of 1 μl of extracts of corn cultures (0.2 g corn/ml acetonitrile:water extraction solvent). Good agreement was obtained for the fumonisin concentrations in culture extracts by this method compared to fluorescence detection of OPA derivatives. The detection limit for FB_1 with this injection volume was about 50 ppm. While this is adequate for the analysis of culture materials where fumonisin levels are often greater than 1000 ppm, additional clean-up will be needed to achieve 100-200 fold fumonisin concentration to achieve the sensitivity required for practical detection of fumonisins in naturally contaminated extracts.

Electrospray MS is an ideal technique to measure and detect fumonisins. Both positive and negative ion signals are observed when FB_1 is analyzed by electrospray. In the positive ion mode, the protonated molecule (m/z 722) is the base peak, and little fragmentation is observed. Occasionally small amounts of sodiated or potassiated molecules are seen as well as some doubly charged protonated salt ions. In the negative ions observed, the carboxylate anion (M-H)⁻ at m/z 720 is the base peak. Smaller amounts of doubly charged anions are also observed. The relative sensitivity of positive ions to negative ions is a function of compound and the pH of the solvent used. For FB_1, in the presence of water/methanol and acetic acid, the positive ions give about a 10 fold greater response than do the negative ions. At approximately pH 7 in the presence of water/methanol and ammonium acetate, the yield of positive ions and negative ions is more nearly equal. At pH 4, FA_1 response of (M-H)⁻ is greater than the response of MH⁺. This is probably because the carboxylate anion is more easily formed. In FB_1, the free amine is available for a proton atom but not in FA_1. Sodiated and potassiated molecular ions are also much more abundant in FA_1 than in FB_1. At the higher pH values, negative ions predominate and almost no positive ions are detected for FA_1. Hydrolyzed FB_1 which has no carboxyl groups forms abundant positive ions but no negative ions are observed. In HPLC coupled electrospray experiments with pure fumonisins, methanol as the organic modifier gave much better sensitivity and higher ion currents than acetonitrile containing solvent systems.

Electrospray sensitivity and column capacity and selectivity are all greatly affected by pH and ion strength in the eluting solvent. Injections of greater than about 1 μg of fumonisins results in column overload under acidic conditions (pH <4), and peaks begin to broaden and elute earlier; the total response with larger injected amounts becomes non-linear. At higher pH values the capacity of the column increases to about 10 μg before these effects are observed. However, the total yield of positive ions at higher pH values is lower so that overall detection sensitivity is not improved by the increased column capacity. Generally fumonisins elute significantly earlier in the gradient (at higher water concentrations) in the presence of 0.1 N buffers (ammonium acetate or formate) than when the modifier is 0.1 or 1 % acetic acid. Large differences in selectivity are then seen among the various fumonisin homologs and types. The B series fumonisins have the greatest change in selectivity, while the half-hydrolyzed analogs ($HHFB_x$) and A series fumonisins are less affected. The fully hydrolyzed B (HFB_x) series fumonisins are hardly affected by the change in ion strength in the solvent. For example, on the C18 column eluting with a gradient from 50/50/1 to 80/20/1 methanol/water/acetic acid, the two $HHFB_1$ isomers co-elute slightly before FB_1, followed closely by HFB_1. When eluted with a gradient from 50/50 to 80/20 methanol/water with 0.1

N ammonium acetate the elution order is FB_1, followed by the two completely separated $HHFB_1$ isomers and then by HFB_1.

We previously reported that the use of stable isotope-labeled internal standards gave results similar to those observed in FAB/MS when samples were directly injected into the detector without chromatography (Plattner, 1995, in press), but with large injections (>0.5 µg) some fumonisin is retained on the column. This bound fumonisin from the previous run elutes with the next injection. This carryover can result in inaccurate isotope ratios or too large a response for samples injected subsequently to a large sample. With HPLC separation, injections of as little as 0.5 ng of FB_1 can be detected; and good linearity of response for both FB_1 (m/z 722) and its labeled analog with deuterium labels on the two methyl groups (m/z 728) were obtained across the range from 5 to 900 ng injected. When equal amounts of FB_1 and our labeled standard are injected, the ratio of the area for FB_1 to the six deuterium-labeled analog (m/z 722/728) observed was 1.12. This is in close agreement of the expected ratio of 1.1 since the deuterium labeled standard was only 90% pure. (It contains 9% three deuterium atom labeled and 1% FB_1.) With approximately 100-ng mixtures of FB_1 and labeled FB_1 across the ratio range of 0.2 to 5, a linear relationship was observed between the area ratios and the amounts ratios. The coefficient of variation of the response factor calculated from 24 injections (3 replicate injections of 8 samples across the previously mentioned range) was about 5%. This data indicates that the labeled FB_1 should provide a suitable internal standard for isotope dilution MS measurements using HPLC separation with electrospray detection provided that the total level of fumonisins injected into the HPLC is closely monitored to avoid results being corrupted by carryover (Plattner, 1995, in press) from previously injected samples with high amounts of FB_1. Preliminary experiments indicate good agreement between quantitation by this method and other methods.

The analysis by HPLC with electrospray detection of culture material extracts from *F. moniliforme* and closely related species indicates the presence of many fumonisin homologs and analogs. Generally, similar patterns are observed. FB_1 is generally the most abundant form with lesser amounts of FB_2 and/or FB_3, have been reported (Nelson et al., 1993). One advantage of electrospray and ELSD detection is that the A series fumonisins, which do not react to form the fluorescent derivatives because they lack a free amine group, are clearly seen. This series of analogs always occurs at much lower levels than the B series. They are seen in culture extracts that have never been exposed to acetic acid and thus are not likely to be artifacts of extraction or clean-up. While they are unusual, strains which produce distinctly different patterns of fumonisins have been reported. Several strains of *F. moniliforme* were isolated from maize in Nepal which produce no detectable fumonisins (Desjardins et al., 1992). Isolates of *F. proliferatum* which produce no FB_1 but make either FB_2 or FB_3 have been reported (Nelson et al., 1993). We recently identified two strains with unusual fumonisin ratios during a screen of over 240 fertile mating population A strains of *F. moniliforme*. The HPLC chromatograms of these two strains are shown in Fig. 1 along with a strain which produces a normal pattern of fumonisins. One of these unusual strains (KSU 819) (Fig. 1c) accumulates no FB_1 or FB_2, but has high levels of FB_3, FB_4 and FC_4. This strain accumulates a component that elutes near FB_1. This component shows signals at m/z 748 and 770 in the positive ions and m/z 746 in the negative ions. These signals are consistent with the expected response of FA_2, the 10-deoxy analog of FA_1 (Ross et al., 1990) but would be expected to be FA_3, the 5-deoxy analog of FA_1, which has not yet been reported, because the culture produces FB_3 and no FB_2.

The unknown component was isolated from culture extract on 10-g of C-18 bonded silica in a 35-ml plastic syringe barrel (Waters Associates, Milford, MA), eluted with water/acetonitrile mixtures, then passed through 10 g of SCX bonded silica (Varian Associates). Full details of this rapid clean-up which can produce mg quantities of purified $FB_{1,2,3}$ and 4 and FA_1, the new compound, will be reported separately. The proton NMR spectrum of

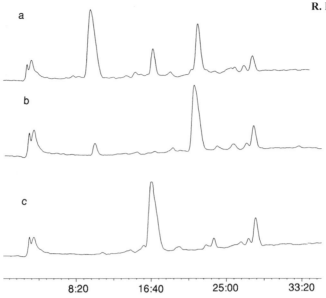

Figure 1. Electrospray MS chromatograms of *Fusarium moniliforme* culture extracts. Trace a is M-3125; trace b is K-816; trace c is K-819. One microliter (0.2 ml equivalent of culture) was injected onto a C-8 column and eluted with methanol/water/acetic acid (gradient from 60/40/1 to 75/25/1 in 20 min hold at final) at a flow rate of 0.2 ml/min. Approximate retention times were: fumonisin B_1 - 11 min: B_3 - 17 min; B_2 - 23 min; C_4 - 28 min; B_4 - 29 min.

the new component gives important clues to its structure. As expected, the signal from the protons on C-1 was observed as a doublet at 1.12 ppm compared with 1.26 ppm for the doublet in the free amine FB_3. The signal for the proton at C-2 was shifted from 3.1 to 3.8 as in the proton NMR of FA_1 (Bezuidenhout et al., 1988). The signal for the protons on the C-14 and C-15, two carbon atoms with esters (TCA), were at the same locations as in the spectrum of FB_3 (data not shown). The protons on the two carbons with free hydroxyl groups were observed as multiplets at 3.47 and 3.67 ppm compared to 3.61 for C-10 and 3.46 for C-3 in the spectrum of FB_3. The multiplet at 3.47 was coupled to the signal from the C-2 proton at 3.8 showing that it must be from a hydroxyl group at C-3. Thus, the structure of this new n-acetylated analog is FB_3.

A component was also observed which eluted slightly after FB_3 and which had signals at m/z 732 and 764. This is consistent with the expected signals for the n-acetylated analog of FB_4. This component eluted from the C-18 cartridge after FB_3 in the fraction that contained FB_4 and FC_4. The proton NMR spectrum of that fraction containing a mixture of three components showed signals for C-1, C-2, and C-3 for FB_4; and the signals for C-2 and C-3 for FC_4, all seen at the expected locations and in the expected ratios. Signals expected for the acetate methyl protons and the protons on C-1 and C-2 from a n-acetylated analog of FB_4 are also seen in the proper ratios. Hydrogen-hydrogen correlation experiments verified the connectivity of the C-1 and C-2 proton signals. Thus, this component is likely to be FA_4. Current studies are underway to purify the FA_4 and completely establish the structures of the two new n-acetylated fumonisins. The other unusual strain, KSU 817, made little or no FB_1 or FB_3 but accumulated high levels of FB_2, FB_4 and FC_4 (Figure 1b). In addition, chromatograms signals from components that elute just before FB_2 and FB_3 are observed. They are 14 dalton less (m/z 692) than the signal from FB_2, as would be expected for FC_2 and FC_3. They are present in very low concentration and have not yet been isolated nor their structures confirmed. It is interesting to note that most producing cultures contain significant amounts of FB_4, usually about 10-20% the level of FB_2. Measurement of this compound by any other method has not been reported. No strains have yet been found that produce FB_4 without either high levels of FB_2 or FB_3. In all cultures examined to date, the level of FC_4, while always quite low, is much greater than the level of FC_1. It is unlikely that the C series fumonisins are precursors or metabolites of the B series fumonisins but rather probably are formed because the enzyme that couples the amino acid and ketide precursor is not completely specific for alanine. It is not known if the A series compounds are precursors or metabolites of the B series or whether $FB_{2,3 \text{ or } 4}$ are precursors of FB_1. The hypothesis that the unusual phenotypes of KSU-817 and KSU-819 arise from mutations in the C-5 and C-10 hydroxylation enzymes is presently under test using genetic crosses of these naturally occurring variants with each other and with fumonisin non-producing strains in our laboratory and will be reported separately.

In summary, ELSD and electrospray detection of fumonisins are useful in the identification and quantitation of fumonisins in cultures and could be successfully applied to quantitation in naturally contaminated samples. Excellent sensitivity and selectivity are obtained and new fumonisin analogs and homologs have been discovered and can be measured.

ACKNOWLEDGMENTS

We thank Dr. Filmore Meredith and Dr. Steven Musser for helpful discussions during the course of this work.

REFERENCES

Bezuidenhout, S. C.; Gelderblom, W. C. A.; Gorst-Allman, C. P.; Horak, R. M.; Marasas, W. F. O.; Spiteller, G.; Vleggaar, R. Structure elucidation of the fumonisins, mycotoxins from *Fusarium moniliforme. J. Chem. Soc., Chem. Commun.* **1988**, 743-745.

Desjardins, A. E.; Plattner, R. D.; Shackelford, D. D.; Leslie, J. F.; Nelson, P. E. Heritability of fumonisin B_1 production in *Gibberella fujikuroi* mating population A. *Appl. Environ. Microbiol.* **1992**, *58*, 2799-2805.

Gelderblom, W. C. A.; Jaskiewicz, K.; Marasas, W. F. O.; Thiel, P. G.; Horak, R. M.; Vleggaar, R.; Kriek, N. P. J. Fumonisins — Novel mycotoxins with cancer-promoting activity produced by *Fusarium moniliforme. Appl. Environ. Microbiol.* **1988**, *54*, 1806-1811.

Gelderblom, W. C. A.; Kriek, N. P. J.; Marasas, W. F. O.; Thiel, P. G. Toxicity and carcinogenicity of the *Fusarium moniliforme* metabolite fumonisin B_1 in rats. *Carcinogenesis* **1991**, *12*, 1247-1251.

Korfmacher, W. A.; Chiarelli, M. P.; Laly, J. O.; Blom, J.; Holcomb, M.; McManus, K. T. Characterization of the mycotoxin fumonisins B_1: Comparison of thermospray, fast-atom bombardment and electrospray mass spectrometry, *Rapid Commun. Mass Spec.* **1991**, *5*, 463-468.

Maragos, C. M. Capillary zone electrophoresis and HPLC for the analysis of fluorescein isothiocyanate labeled fumonisin B_1, *J. Agric. Food Chem.* **1995**, *43*, 390-394.

Maragos, C. M.; Bennett, G. A.; Richard J. L. Analysis of fumonisin B_1 in corn by capillary electrophoresis. American Chemical Society National Meeting, Anaheim, CA, April 2-7, 1995.

Marasas, W. F. O.; Nelson, P. E. Mycotoxicology: Introduction to the mycology, plant pathology, chemistry, toxicology and pathology of naturally occurring mycotoxicoses in animals and man, Pennsylvania State Univ. Press: University Park, PA, 1987.

Musser, S. M.; Gay M. L.; Eppley, R. M.; Chiarelli, M. P.; Lay, J. O. Identification of fumonisin metabolites in bulk preparations of fumonisin B_1, Proc. 42nd ASMS Conference on Mass Spectrometry, 29 May-3 June 1994, Chicago IL.

Nelson, P. E.; Desjardins A. E.; Plattner, R. D. Fumonisins, mycotoxins produced by *Fusarium* species: Biology, Chemistry and Significance. *Annu. Rev. Phytopathol.* **1993**, *31*, 233-252.

Pittet, A.; Parisod, V.; Schellenberg, M. Occurrence of fumonisins B_1 and B_2 in corn based products from the Swiss market. *J. Agric. Food Chem.* **1992**, *40*, 1352-1354.

Plattner, R. D.; Norred, W. P.; Bacon, C. W.; Voss, K. A.; Peterson, R. A method of detection of fumonisins in corn samples associated with field cases of equine leukoencephalomalacia, *Mycologic* **1990**, *82*, 698-702.

Plattner, R. D.; Weisleder, D.; Shackelford, D. D.; Peterson, R.; Powell, R. G. A new fumonisins from solid cultures of *Fusarium moniliforme*. Mycopathologia **1992**, *117*, 23-28.

Plattner, R. D.; Branham, B. E. Labeled fumonisins: Production and use of fumonisin B_1 containing stable isotopes. *JAOAC Int.* **1994**, *77*, 525-532.

Plattner, R. D. Detection of fumonisins produced in *Fusarium moniliforme* cultures by HPLC with electrospray MS and evaporative light scattering detectors. Natural Toxins in press, 1995.

Ross, P. F.; Nelson, P. E.; Richard, J. L.; Osweiler, G. D.; Rice, L. G.; Plattner, R.D.; Wilson, T.W. Production of fumonisins by *Fusarium moniliforme* and *Fusarium proliferatum* isolates associated with equine leukoencephalomalacia and a pulmonary edema syndrome in swine, Appl. Environ. Microbiol. **1990**, *56*, 3225-3226.

Shephard, G. S.; Sydenham, E. W.; Thiel, P. G.; Gelderblom, W. C. A. Quantitative determination of fumonisins B_1 and B_2 by high pressure liquid chromatography with fluorescence detection. *J. Liquid Chromatogr.* **1990**, *13*, 2077-2087.

Stack, M. E.; Eppley, R. M. Liquid chromatographic determination of fumonisins B_1 and B_2 in corn and corn products, *J. Assoc. Off. Anal. Chem.* **1992**, *75*, 834-837.

Sydenham, E. W.; Shephard, G. S.; Thiel, P. G.; Maracas, W. F. O.; Stochestrom, S. Fumonisin contamination of commercial corn-based human foodstuffs, *J. Agric. Food Chem.* **1991**, *39*, 2014-2018.

Voss, K. A.; Norred, W. P.; Plattner, R. D.; Bacon, C. W.; Porter, J. K. Hepatoxicity in rats of aqueous extracts of *Fusarium moniliforme* strain MRC 826 corn cultures, *Toxicologist* **1989**,*9*, 258.

Wilkes, J. G.; Sutherland, J. B.; Churchwell, M. I.; Williams, A. J. Determination of fumonisins B_1,B_2,B_3 and B_4 by high-performance liquid chromatography with evaporative light scattering detection. *J. Chromatogr.* **1995**, *695*(2):319-323.

QUANTITATION AND IDENTIFICATION OF FUMONISINS BY LIQUID CHROMATOGRAPHY/MASS SPECTROMETRY

Steven M. Musser

Instrumentation and Biophysics Branch
Center for Food Safety and Applied Nutrition
U.S. Food and Drug Administration
Washington, DC 20204

ABSTRACT

A method was evaluated for the quantitation and identification of fumonisins by on-line liquid chromatography/mass spectrometry (LC/MS) with electrospray ionization. A linear response in the full-scan mode with positive ion detection was obtained for fumonisin B_1 (FB_1) over the range of 5-5000 ng injected on-column. Purified FB_1, FB_2, and half-hydrolyzed FB_1 showed equimolar responses. Fully hydrolyzed FB_1 did not show a similar molar response profile and produced a signal which was approximately 2 times that obtained for an equal quantity of FB_1. Most known fumonisins and preparative by-products such as methyl esters were chromatographically resolved and identified by MS by using an acetonitrile gradient and positive ion detection. Negative ion electrospray was used to differentiate fumonisin amides from esters on the basis of differences in response. Response factors for FB_1 and the acetyl amide of FB_1 in the negative ion mode at pH 4.5 were approximately 1:3, respectively.

INTRODUCTION

Fumonisins are sphingosine-like compounds produced by several species of the fungus *Fusarium* (Figure 1). Among the most common and highest-producing fumonisin species are *F. moniliforme* and *F. proliferatum,* both of which are frequently found on corn (Ross et al., 1990). Seven different fumonisins have been characterized (Bezuidenhout et al., 1988; Branham and Plattner, 1993; Cawood et al., 1991; Plattner et al., 1992). The fumonisins identified as FB_1, FB_2, and FB_3 are the most abundant forms found in nature. Interest in the characterization and identification of fumonisins in the food supply stems

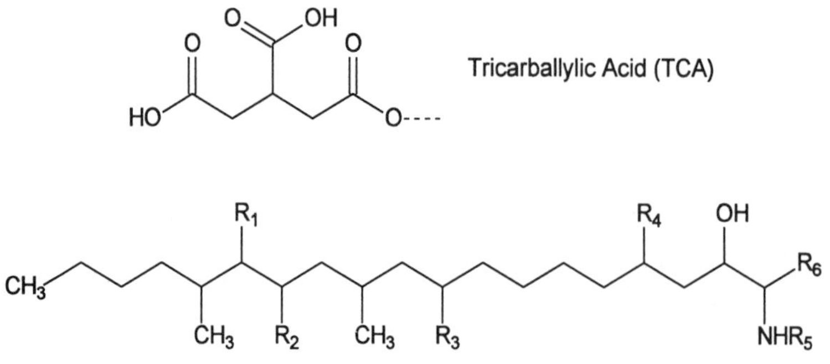

Figure 1. Structures of fumonisins.

	R_1	R_2	R_3	R_4	R_5	R_6	m.w.
FB$_1$	TCA	TCA	OH	OH	H	CH$_3$	721
FB$_2$	TCA	TCA	H	OH	H	CH$_3$	705
FB$_3$	TCA	TCA	OH	H	H	CH$_3$	705
FB$_4$	TCA	TCA	H	H	H	CH$_3$	689
FA$_1$	TCA	TCA	OH	OH	COCH$_3$	CH$_3$	763
FA$_2$	TCA	TCA	H	OH	COCH$_3$	CH$_3$	747
FA$_3$	TCA	TCA	OH	H	COCH$_3$	CH$_3$	747
FC$_1$	TCA	TCA	OH	OH	H	H	707
HHFB$_{1a}$	TCA	OH	OH	OH	H	CH$_3$	563
HHFB$_{1b}$	OH	TCA	OH	OH	H	CH$_3$	563
AP$_1$	OH	OH	OH	OH	H	CH$_3$	405

from toxicological evidence that fumonisins induce a wide range of adverse effects in animals. Among the observed toxicological endpoints are equine leucoencephalomalacia (Kellerman et al., 1990), porcine pulmonary edema (Colvin and Harrison, 1992), and hepatic cancer in rats (Gelderblom et al., 1993). Because of the toxic effects of fumonisins on animals and the obvious implications for human toxicity, reliable analytical methods for the quantitation and identification of fumonisins in food products are desirable. Because fumonisins are known contaminants of corn-based food products (Hopmans and Murphy, 1993), accurate methods for confirmation of fumonisin identity are needed.

The analysis of food products for fumonisins and the identification of new fumonisins from culture material have proven to be difficult analytical problems. Current analytical methodology generally requires derivatization with a chromophore for analysis by liquid chromatography (LC) (Scott and Lawrence, 1994; Stack and Eppley, 1992) or hydrolysis followed by derivatization for analysis by gas chromatography/mass spectrometry (GC/MS) (Plattner et al., 1990). Although several LC/MS methods have been described for the separation and quantitation of fumonisins, each suffers from one or more problems including thermal decomposition, matrix interferences, and incomplete derivatization that have prevented their routine use. A solution to these problems can be found in the recently developed atmospheric pressure ionization technique known as electrospray (Bruins et al., 1987). This soft ionization technique provides practical on-line LC/MS analysis that can be performed with conventional LC systems and does not require derivatization or cause thermal decomposition of fumonisins (Korfmacher et al., 1991; Thakur and Smith, 1994; Young and Lafontaine, 1993). Although fumonisins have been determined by electrospray LC/MS, a comprehensive investigation of factors affecting the ruggedness of the method have not been investigated (Doerge et al., 1994). This paper describes the effectiveness of electrospray LC/MS as a practical technique for evaluating the purity of fumonisin preparations and for identifying and quantitating fumonisins other than FB_1 that may be present in culture extracts.

EXPERIMENTAL PROCEDURES

Isolation of Analytes and Preparation of Analytical Standards

An analytical standard of FB_1 was obtained from Sigma Chemical Company (St. Louis, MO). Solid cultures of *Fusarium proliferatum* M-1597 (provided by P.E. Nelson of the Fusarium Research Center, Pennsylvania State University) were grown on solid corn. Levels of production of FB_1 were approximately 3-5 mg/g solid culture material. Approximately 500 g of solid culture material was extracted with 2 liters of a 75:25 MeOH:H_2O solution and filtered though a Buchner funnel (5 mL of this solution was freeze-dried, and the solid was reserved for LC/MS analysis). The eluant was diluted with H_2O to a final MeOH concentration of 30%. The diluted solution was pumped at 20 mL/min onto a Waters Corp. BONDPAK C-18 preparative chromatography column (40 X 400 mm, 20 μm, 125 A). The column was washed with H_2O for 80 min and then washed with 50:50 H_2O:MeOH for 20 min, followed by elution of the fraction containing FB_1 and FB_1 amides with 30:70 H_2O:MeOH. The methanol was removed under vacuum and the aqueous portion of the eluate was applied to a Rainin Corp. Dynamax C-18 column (41.4 x 250 mm, 8 μm, 60 A) at a flow rate of 50 mL/min. The column was washed with a solution of 25:75 acetonitrile (ACN):H_2O for 50 min, and the fraction containing fumonisins and structurally related amides was eluted with 50:50 ACN:H_2O. The ACN was removed under vacuum, and the aqueous solution was freeze-dried. Fractionation of the individual fumonisins remaining in the freeze-dried residue was performed on a YMC Inc. (Greensboro, NC) C-8 basic column (4.6 x 250 mm) with a mobile phase flow rate of 1 mL/min. An acetonitrile gradient containing 0.1% acetic acid (pH 3) beginning at 20:80 ACN:H_2O and changing linearly to 45:55 ACN:H_2O in 35 min was used to separate the fumonisins. The collected fractions were freeze-dried and yielded FB_1, FB_2, half-hydrolyzed FB_1, and N-acetyl FB_1 at purities greater than 97% by LC/MS. Fully hydrolyzed FB_1 was prepared from FB_1 by hydrolysis in 0.1 N NaOH and rechromatography of the reaction mixture on the analytical column as described above.

LC/MS Analysis

A Hewlett-Packard (Palo Alto, CA) Model 1050 LC pump was used to provide linear gradients and a constant flow rate of 200 μL/min. All chromatography was performed on a YMC Inc. J-sphere ODS-L80 LC column (2 x 250 mm). Chromatographic elution for positive ion electrospray analysis started with 25% ACN for 5 min, followed by a linear gradient to 40% ACN in 30 min; a buffer concentration of 40 mM formic acid was maintained throughout elution. Under these conditions fully hydrolyzed FB_1 elutes at 12.0 min, the

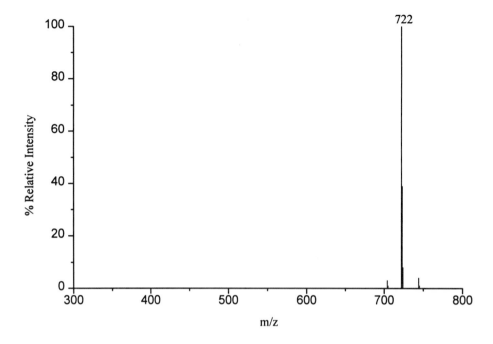

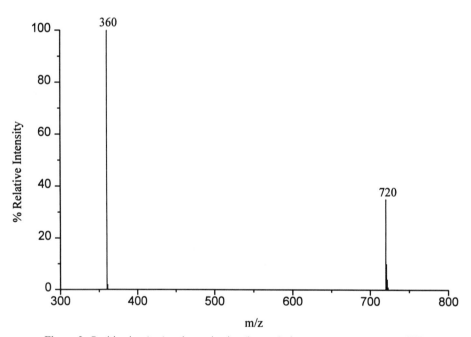

Figure 2. Positive ion (top) and negative ion (bottom) electrospray mass spectra of FB$_1$.

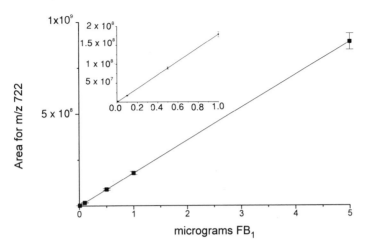

Figure 3. Response for the integrated area of protonated FB_1. Each value represents the mean of three determinations.

half-hydrolyzed FB_1 isomers elute at 18.5 and 19.7 min, FB_1 elutes at 24.8 min, and FB_2 elutes at 33.4 min. Chromatographic elution for negative ion analysis used the same gradient, but 0.1% acetic acid and 5 mM ammonium acetate (pH 4.5) were substituted for the formic acid buffer. A Finnigan Model TSQ-7000 triple-quadrupole mass spectrometer with the standard Finnigan electrospray ion source was used for MS. Nitrogen was used as a nebulizing gas and the capillary temperature was 250°C. The instrument was scanned over the range of 300-800 amu at 1 s/scan. The entire 200 μL/min column effluent was directed into the ion source.

RESULTS AND DISCUSSION

Positive ion electrospray ionization produced protonated molecules $[MH]^+$ of all fumonisins tested and did not produce any fragmentation (Figure 2, top). Although it is possible to induce fragmentation in the skimmer region of the ion source, this procedure was not performed because it might have led to incorrect identification of fumonisins during analysis of mixtures and culture extracts. Negative ion analysis of fumonisins produced spectra in which the base peak was the doubly charged molecular anion and the intensity of the singly charged molecular anion was approximately 25-50% of that of the base peak (Figure 2, bottom).

Multiple charging of the analyte is generally a useful characteristic of electrospray, although in this case it makes analysis and quantitation difficult, because the relative amounts of singly and doubly charged ions are different for each species. For this reason and because the response of fumonisin amides is greater than that of FB_1 in the negative ion mode, negative ion analysis was used only to confirm the presence or absence of amides in the extract.

Full-scan analysis in the positive ion mode of various concentrations of FB_1 showed a linear response over the range of 5-5000 ng for the integrated area of the protonated molecule (m/z 722) (Figure 3).

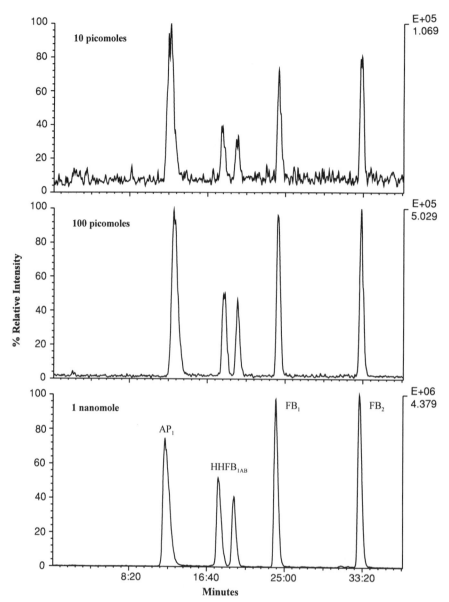

Figure 4. Relative responses for equimolar mixtures of AP_1, $HHFB_1$, FB_1, and FB_2.

When amounts greater than 5000 ng were injected onto the LC column, the response for FB_1 tended to flatten, indicating that the ionization process was saturated. Figure 4 shows the separation and total ion response of an equimolar mixture of AP_1, half-hydrolyzed FB_1, FB_1, and FB_2.

The molar responses of AP_1 and FB_1 under these conditions varied from 1.2:1 to 2:1, respectively, with the ratio increasing at low (<100 pmoles) concentrations. Integrated areas for all other fumonisins except AP_1 were equal, indicating that the response factors for fumonisins which retain at least one tricarballylic acid are equal in the positive ion mode.

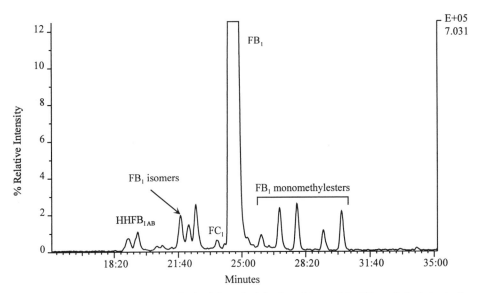

Figure 5. Total ion chromatogram obtained from the injection of 1 μg of the FB₁ analytical standard.

In addition, the response for individual fumonisins does not change as LC analytical time increases, as is the case for LC methods using *o*-phthalaldehyde as a derivatization reagent (Holcolm et al., 1994). These findings indicate that an accurate determination of the percentage of individual fumonisin congeners present in a mixture can be made relative to the most abundant fumonisin, usually FB₁. Although this approach cannot be used to determine the actual amount of each congener present without knowing the amount of FB₁,

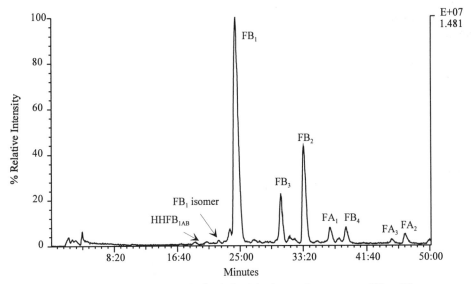

Figure 6. Total ion chromatogram obtained after injecting a culture extract of *F. proliferatum*.

it is useful for determining purity and calculating the relative amounts of fumonisin metabolites in different fumonisin cultures.

The principal fumonisins, FB_1, FB_2, and FB_3, were chromatographically resolved by using a wide range of C-18 columns and mobile phase systems. However, separation of half-hydrolyzed FB_1, methyl esters, and minor fumonisin metabolites from the major components was more difficult. LC columns packed with either C-8 or low carbon load (<10%) C-18, base-deactivated phases produced the best separations. In addition, use of acetonitrile rather than methanol as the organic phase produced better peak shape and resolution. Several mobile phase modifiers other than formic acid were used in developing an LC method for the separation of fumonisin congeners; they included 0.1% acetic acid, 0.05% trifluoroacetic acid (TFA), and ammonium formate. Although TFA produced the best resolution and peak shape for fumonisin mixtures, it induced a very high needle current, resulting in unstable electrospray conditions, and therefore could not be used. Of the remaining modifiers, formic acid gave the best combination of chromatographic resolution and stable electrospray currents. Formic acid could not be used in the negative ion mode, because the pH (<2) of the mobile phase caused the carboxylate groups on the tricarballyllic acid side chains to exist primarily in a neutral form rather than as anions in solution, thus inducing an unacceptable decrease in sensitivity. The optimum conditions for negative ion analysis were found to be 0.1% acetic acid and 2-5 mM ammonium acetate; the final pH of the solution was 4.5. Although a decrease in analyte resolution was observed at this pH, the order of elution was not changed. Higher mobile phase pH values resulted in undesirable changes in peak retention time, with individual analytes changing their order of elution. Although the sensitivity of negative ion analysis could be increased by using higher pH values, it would not be possible to directly compare an analysis made in the positive ion mode with one made in the negative ion mode because of large shifts in retention time.

Although the chromatographic resolution of impurities present in fumonisin preparations is not a requirement for quantitation by LC/MS, since the known impurities could be determined by integration of the molecular ion traces, it would be very difficult to identify new metabolites and unknown impurities unless the components were resolved into separate chromatographic peaks. Electrospray LC/MS analyses of an analytical standard of FB_1 (Figure 5) and an *F. proliferatum* culture extract (Figure 6) illustrate this point.

The total ion chromatogram (TIC) of the FB_1 analytical standard clearly shows the presence of two isomers of FB_1 at a level of ≈1% each. Poor chromatographic conditions would cause the isomers to coelute with FB_1, thus masking their presence. Examination of the TIC of the culture extract shows the presence of several fumonisin isomers in addition to FB_{1-3}, which might have been overlooked or assumed to be adducts had they not been separated into discrete peaks. Another obvious difference between the analytical standard and the culture extract is the absence of methyl esters in the extract. Although the amides were present in the culture extract as well as in the analytical standard, the methyl esters were present only in the purified FB_1 preparation. This would indicate that while the amides of fumonisins are naturally occurring, methyl esters are an artifact of the purification method. Comparison of the ion chromatograms for FB_1 and its N-acetyl amide present in the culture extract in both the positive and negative ionization modes illustrates the difference in response and allows the identification of the amide (Figure 7).

CONCLUSION

Electrospray LC/MS allows the separation and identification of fumonisins in both analytical standards and fungal culture extracts without derivatization or pretreatment. In

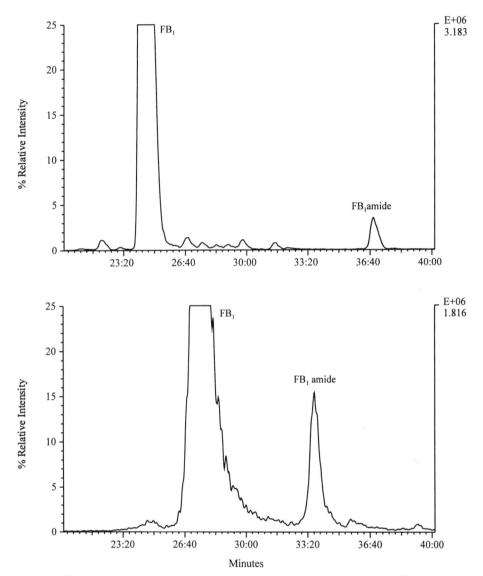

Figure 7. Positive ion (top) and negative ion (bottom) selected ion chromatograms for FB₁ and FA₁ in *F. proliferatum* culture extract.

addition, the total amount of fumonisins present in a solution can be estimated relative to the amount of the most abundant fumonisin isomer present. This method allows differentiation of naturally occurring fumonisin metabolites from those produced as a result of bulk workup procedures, because the finished product can be compared directly with the original culture extract. Comparison of positive ion chromatograms with negative ion chromatograms can confirm the presence of amides and easily differentiate them from esters, thus providing positive identification of both species.

REFERENCES

Bezuidenhout, S.C.; Gelderblom, W.C.A.; Gorst-Allan, C.P.; Horak, R.M.; Marasas, W.F.O.; Spiteller, G.; Vleggaar, R. Structure elucidation of the fumonisins, mycotoxins from *Fusarium moniliforme*. *J. Chem. Soc. Chem. Commun.* **1988**, 743-745.

Branham, B.E.; Plattner, R.D. Isolation and characterization of a new fumonisin from liquid cultures of *Fusarium moniliforme*. *J. Nat. Prod.* **1993**, *56*, 1630-1633.

Bruins, A.P.; Covey, T.R.; Henion, J.D. Ion spray interface for combined liquid chromatography/atmospheric pressure ionization mass spectrometry. *Anal. Chem.* **1987**, *59*, 2642-2646.

Cawood, M.E.; Gelderblom, W.C.A.; Vleggaar, R.; Behrend, Y.; Thiel, P.G.; Marasas, W.F.O. Isolation of the fumonisin mycotoxins: A quantitative approach. *J. Agric. Food Chem.* **1991**, *39*, 1958-1962.

Colvin, B.M.; Harrison, L.R. Fumonisin induced pulmonary edema and hydrothorax in swine. *Mycopathologia.* **1992**, *117*, 79-82.

Doerge, D.R., Howard, P.C.; Bajic, S.; Preece, S. Determination of fumonisins using on-line liquid chromatography coupled to electrospray mass spectrometry. *Rapid Commun. Mass Spectrom.* **1994**, *8*, 603-606.

Gelderblom, W.C.A.; Cawood, M.E.; Snyman, S.D.; Vleggaar, R.; Marasas, W.F.O. Structure-activity relationships of fumonisins in short-term carcinogensis and cytotoxicity assays. *Food Chem. Toxicol.* **1993**, *31*, 407-414.

Holcolm, M.; Thompson, H.C., Jr.; Lipe, G.; Hankins, L.J. HPLC with electrochemical and fluorescence detection of the OPA/2-methyl-2-propanethiol derivative of fumonisin B_1. *J. Liq. Chromatogr.* **1994**, *17*, 4121-4129.

Hopmans, E.C.; Murphy, P.A. Detection of fumonisins B_1, B_2, and B_3 and hydrolyzed fumonisin FB_1 in corn-containing foods. *J. Agric. Food Chem.* **1993**, *41*, 1655-1658.

Kellerman, T.S.; Marasas, W.F.O.; Thiel, P.G.; Gelderblom, W.C.A.; Cawood, M.E.; Coetzer, J.A.W. Leukoencephalomalacia in two horses induced by oral dosing of FB_1. *Onderstepoort J. Vet. Res.* **1990**, *57*, 269-275.

Korfmacher, W.A.; Chiarelli, M.P.; Lay, J.O., Jr.; Bloom, J.; Holcolb, M.; McManus, K.T. Characterization of the mycotoxin fumonisin B_1: Comparison of thermospray, fast-atom bombardment and electrospray mass spectrometry. *Rapid Commun. Mass Spectrom.* **1991**, *5*, 463-468.

Plattner, R.D.; Norred, W.P.; Bacon, C.W.; Voss, K.A., Peterson, R.; Shackelford, D.D.; Weisleder, D. A method of detection of fumonisins in corn samples associated with field cases of equine leukoencephalomalacia. *Mycologia.* **1990**, *82*, 698-702.

Plattner, R.D.; Weisleder, D.; Shackelford, D.D.; Peterson, R.; Powell, R.G. A new fumonisin from solid cultures of *Fusarium moniliforme*. *Mycopathologia.* **1992**, *117*, 23-28.

Ross, P.F.; Nelson, P.E.; Richard, J.L.; Osweiler, G.D.; Rice, L.G.; Plattner, R.D.; Wilson, T.M. Production of fumonisins by *Fusarium moniliforme* and *Fusarium proliferatum* isolates associated with equine leukoencephalomalacia and pulmonary edema in swine. *Appl. Environ. Microbiol.* **1990**, *56*, 3225-3226.

Scott, P.M.; Lawrence, G.A. Stability and problems in recovery of fumonisins added to corn-based foods. *J. AOAC Int.* **1994**, *77*, 541-545.

Stack, M.E.; Eppley, R.M. Liquid chromatographic determination of fumonisins B_1 and B_2 in corn and corn products. *J. AOAC Int.* **1992,** *75*, 834-836.

Thakur, R.A.; Smith, J.S. Analysis of fumonisin B_1 by negative-ion thermospray mass spectrometry. *Rapid Commun. Mass Spectrom.* **1994**, *8*, 82-88.

Young, J.C.; Lafontaine, P. Detection and characterization of fumonisin mycotoxins as their methyl esters by liquid chromatography/particle beam mass spectrometry. *Rapid Commun. Mass Spectrom.* **1993**, *7*, 352-359.

7

NMR STRUCTURAL STUDIES OF FUMONISIN B$_1$ AND RELATED COMPOUNDS FROM *FUSARIUM MONILIFORME*

B. A. Blackwell, O. E. Edwards[1], A. Fruchier[2], J. W. ApSimon[1], and
J. D. Miller

Mycotoxin Research Group
Plant Research Centre
Agriculture Canada
Ottawa, Canada K1A 0C6
[1] Ottawa-Carleton Chemistry Institute
Carleton University
Ottawa, Canada K1S 5B6
[2] Ecole Normale Superieure de Chimie
8 Rue de l'École Normale
34053 Montpellier, France

ABSTRACT

Fumonisin B$_1$ (FB$_1$) is the primary mycotoxin produced by *Fusarium moniliforme* and appears to be responsible for the varied toxigenic effects associated with ingestion of this mold, particularly that of the inhibition of sphingolipid biosynthesis. Understanding the structure and biosynthesis of fumonisins is a key factor in determining structure/activity relationships. To this end, Nuclear Magnetic Resonance (NMR) methods have been used to identify various derivatives of FB$_1$, both naturally occurring and synthetic. With accurate chemical shift assignments, NMR may be used to determine the level of impurities in toxicological grade FB$_1$ preparations. Specifically enriched FB$_1$ was prepared from *F. moniliforme* cultures using ^{13}C-enriched acetate as well as several ^{13}C-enriched amino acids. ^{13}C NMR analysis indicates that the biosynthesis of fumonisins involves the addition of methionine-derived methyl functions, glutamate-derived tricarballylic ester functions and alanine to an 18 carbon hydrocarbon backbone that is likely polyketide in origin. With the goal of obtaining a crystalline compound for the determination of absolute configuration, several derivatives of FB$_1$ have been prepared, and NMR analysis used to determine the relative and absolute configuration of the 10 stereocenters present in this molecule.

Fumonisins in Food, Edited by L. Jackson *et al.*
Plenum Press, New York, 1996

INTRODUCTION

The pathogenic nature of certain species of fungi to plants has historically been a significant agricultural problem since these fungi often produce metabolites that can induce toxic effects upon ingestion. In recent years, the species *Fusarium moniliforme* has been the focus of much research attention since this fungus and the metabolites that it produces are an area of growing concern for both producers and consumers of corn-based products for both animal and human food. The occurrence of *F. moniliforme* is widespread and often symptomless. The toxicities associated with consumption of contaminated corn by animals are highly varied and the actual number of fumonisins, fumonisin-like compounds and precursors that occur in contaminated corn is not known.

F. moniliforme produces the mycotoxins moniliformin, fusarins, and fumonisins, but it is the fumonisins that are currently receiving the most toxicological and chemical interest. Fumonisin was first characterized by Benzuidenhout et al. (1988), who determined the structure of fumonisin B_1, the most prevalent of the fumonisins in naturally contaminated corn. Fumonisins B_2, B_3, and B_4 as well as their hydrolysis products (the "HB" series) have been characterized (Plattner et al., 1992; Powell and Plattner, 1995), as well as the "A" series (Benzuidenhout et al., 1988) and the "C" series (Branham and Plattner, 1993b). The structures of some of these fumonisins and their derivatives are shown in Figure 1.

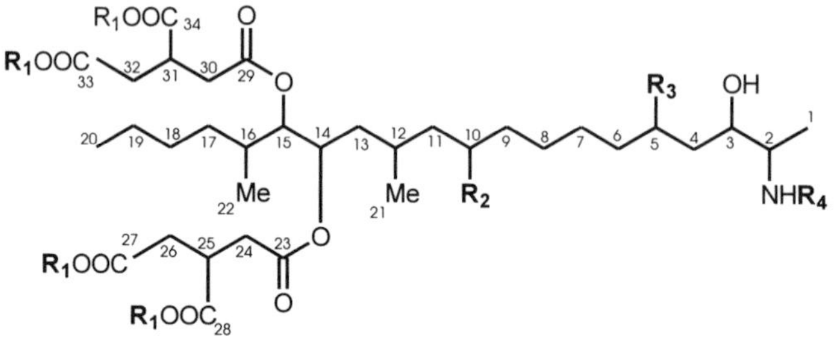

	R_1	R_2	R_3	R_4
Fumonisin B_1	H	OH	OH	H
Tetramethyl FB_1	CH_3	OH	OH	H
Fumonisin B_2	H	H	OH	H
Fumonisin B_3	H	OH	H	H
Fumonisin A_1	H	OH	OH	$COCH_3$
Fumonisin C_1	H	C-1, H replaces CH_3		
Hydrolyzed FB_1	C-14, C-15 are OH			

Figure 1. Structure of some fumonisins.

This active area of research has resulted in a high volume of literature on fumonisins including several recent reviews covering the subjects of taxonomy (Nelson et al., 1994), biology (Nelson et al., 1993), analysis and toxicology (Norred, 1993; Riley et al., 1993) as well as chemistry and biosynthesis (Powell and Plattner, 1995).

The NMR spectra of fumonisin was first characterized by Bezuidenhout et al. (1988). In polar solvents, FB_1 exists as a zwitterion and undergoes strong interactions with metal cations, leading to a highly unresolved ^1H NMR spectrum (Laurent et al., 1990). The chemical shifts are also strongly pH dependent (Plattner et al., 1992). In methanol (CD_3OD) however, the spectrum is much better resolved although there is a slight risk of the formation of methyl esters, and $^1H/^{13}C$ correlation experiments can be used to locate the chemical shifts of the protons in the remaining unresolved methylene region. 2D NMR methods and higher magnetic fields have been used to obtain unambiguous chemical shift assignments for the major fumonisins (Savard and Blackwell, 1994; Powell and Plattner, 1995). With this information, NMR can be used to monitor for minor contaminants in FB_1 preparations and to study both the biosynthetic pathways and the stereochemistry of these compounds.

NMR ANALYSIS OF PURE FUMONISIN B₁

Since the health implications of low levels of fumonisin in human foods is unknown it is essential to provide highly pure FB_1 for studies leading to the establishment of safe levels for foods and to provide analytical standards. Quantities of pure FB_1 as well as ^{14}C-enriched FB_1 are also required to understand the effects on farm animals (Prelusky et al., 1994).

The high hydrophilicity of fumonisins makes them very difficult to isolate, since unlike most known mycotoxins, fumonisins are not soluble in organic solvents. The only satisfactory extraction methods to date require a combination of water with either methanol or acetonitrile. Even then, many polar impurities including salts, sugars and peptides are co-extracted. The similarity in structure of all the fumonisins also makes it very difficult to purify them from one another. The molecular weights of FB_1 and FB_2 differ by only 16 mass units and FB_2 and FB_3 are isomers. Although NMR is not as sensitive as other analytical tools, it is the analysis of choice in combination with mass spectrometry (MS) since derivatization is not required and signal strength is directly proportional to the relative amount of a particular compound present in the sample. In contrast, standard GC methods of quantitation introduce an error factor due to derivatization.

The production, isolation and purification of fumonisins from 10 liter cultures of *F. moniliforme sheldon* has been previously described (Miller et al., 1994; Blackwell et al., 1994). The use of a liquid medium greatly reduces the number of contaminants present, however it also changes the nature of those found. For example, more salts are present in liquid culture. Careful attention to pH, oxygenation, and innoculum conditions has increased yields to over 500 mg FB_1 per liter of medium.

The ^1H NMR spectra of two batches of FB_1 purified by the method of Miller et al. (1994) are shown in Figure 2A. The sensitivity of proton NMR permits a better chance of seeing lower levels of impurities, but this is offset by the lack of resolution. If impurities are very similar in structure or lie in the methylene region of the spectrum (2-3 ppm), they are difficult to detect. The ^1H spectrum can be used to determine the presence of acetylated or methylated products, giving rise to sharp resonances at 2 and 3.4 ppm, that can be easily integrated to determine the relative amounts of these compounds. In these two batches of FB_1 there are less than 1% methylated or acetylated products. The sharp resonance at 2 ppm

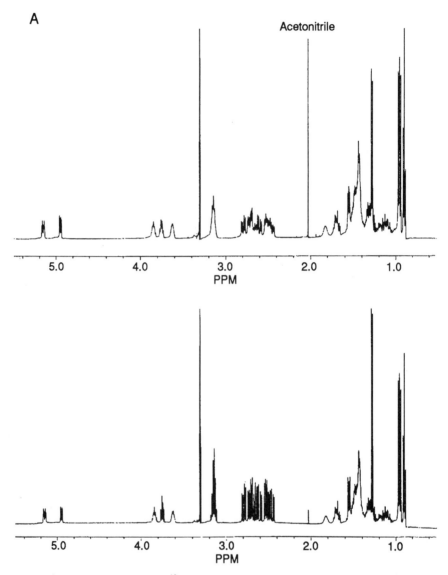

Figure 2. 500 MHz ^{1}H NMR (A) and 125 MHz ^{13}C NMR (B) spectra of two batches of purified FB$_1$. Samples are approximately 15 mg/ml in methanol-d$_4$. The resonances of the major impurities are identified.

is due to residual acetonitrile used in the purification process and amounts to 0.76% and 0.05% in the two samples.

Although not as sensitive, ^{13}C NMR can detect small changes in structure since the resonances due to each carbon in the molecule are discreet. The ^{13}C NMR spectra of the same two batches (Figure 2B) were acquired under conditions designed to result in very little intensity reduction due to relaxation effects. The spectra were accumulated until a signal to noise ratio of approximately 1000:1 was attained, in order to see impurities on the order of 1% or less. The spectra also show negligible presence of acetylated or methylated derivatives, but do show the presence of small amounts of fumonisin C$_1$ (FC$_1$)

B

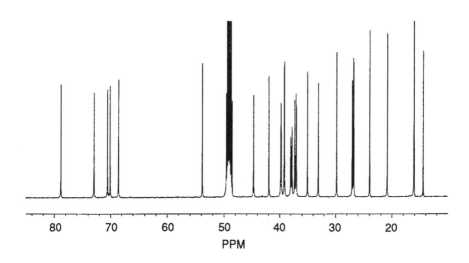

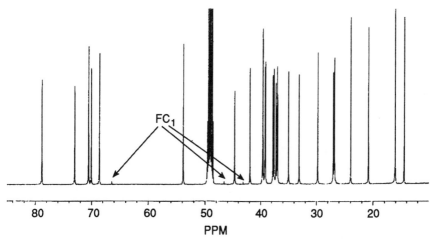

Figure 2. Continued.

in both batches. The resonances of C-1 to C-5 of FC₁ are well resolved from those of FB₁ (Figure 3), due to the lack of a methyl group at C-1 (see Figure 1). The relative amounts of FC₁ to FB₁ were determined by comparing the peak heights of three resolved resonances of FC₁ with their equivalent resonances in FB₁ (to minimize intensity effects due to differences in relaxation times or nOe's) and averaged, giving calculated impurity levels of 1.4% (Figure 3A) and 0.4% (Figure 3B) in the two batches tested. By this technique, organic contaminants down to 0.5% can be detected, provided their resonances are resolved from those of FB₁.

The NMR resonances of the tricarballylic ester functions can be uniquely identified and the chemical shifts of the esters at C-14 and at C-15 are consistent and slightly different

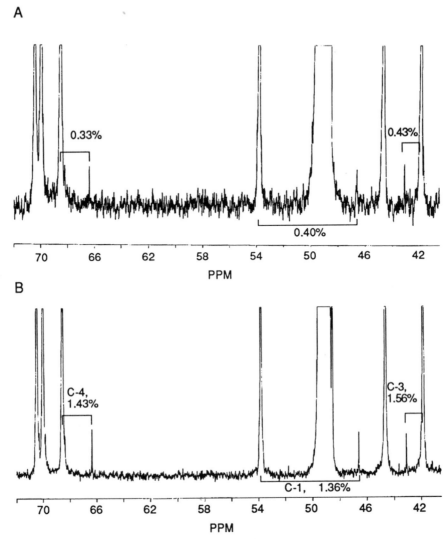

Figure 3. Expansion of the 40-70 ppm region of the two ^{13}C spectra of Figure 2B, showing the resonance intensity of those carbons of FC$_1$ which are resolved. The spectra are a result of 25,000 scans with a 70° pulse and 2.8 sec recycle time. A and B correspond to the upper and lower spectra of Figure 2B respectively.

due the different molecular environnments. However, the ^{13}C resonances, particularly those of the carbonyls, are very sensitive to the presence of trace cations and will shift and broaden proportionally to the quantity of cation present. Figure 4 shows that the appearance of the spectrum can be a qualitative indicator of the quantity of cations (particularly sodium) present. The effect is greater for the free acid functions (C-27, 28, 33, 34) than for the esterified carbonyls (C-23 and C-29). The adjacent methylenes (C-26 and C-32) are more affected than C-24 and C-30. Inorganic analysis of these two samples revealed that the sample of Figure 4A contains 659 ppm sodium (total inorganics 1330 ppm), while that of Figure 4B contained 2315 ppm sodium (3386 total inorganics), correlating well with the appearance of the spectrum.

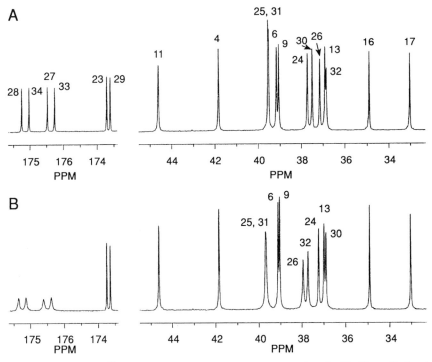

Figure 4. Comparison of the ^{13}C NMR spectra of the two batches of FB₁ (described in Figure 2) showing the effect of the presence of cations on the linewidth and chemical shift of the resonances of the tricarballylic ester groups.

BIOSYNTHESIS OF FB₁

Fumonisins are structurally similar to sphingosine, the base backbone of sphingolipids, prompting the idea that they may be biosynthetically related (Abbas and Shier, 1992; Plattner and Shackelford, 1992). The biosynthesis of sphingosine proceeds via the condensation of palmitoyl CoA with serine. Past studies have shown the advantage of incorporating specifically enriched ^{13}C-acetate into fungal cultures at the appropriate time of biosynthesis of a particular metabolite as a preliminary step to the preparation of radiolabelled (^{14}C) compounds for toxicological research (Miller and Blackwell, 1986). The ^{13}C studies permit the determination of the most cost effective method to manufacture highly enriched ^{14}C-FB₁ labelled in multiple locations for tracer analysis. An earlier study has shown that the ^{13}C-acetate is incorporated into the fumonisin backbone in a manner more suggestive of polyketide biosynthesis rather than lipid biosynthesis, but the analyses were performed on impure FB₁ (Blackwell et al., 1994).

Specifically enriched ^{13}C-acetate (^{13}C-1 and ^{13}C-2) as well as ^{13}C-3 *L*-alanine, ^{13}C-5 *L*-glutamic acid, ^{13}C-3 *L*-serine and ^{13}C-CH₃ *L*-methionine were incorporated into 50 ml cultures of *F. moniliforme* as previously described (Blackwell et al., 1994). The precursors were added as small aliquots spread over a 24 hour period after the log phase growth of the culture and were designed to coordinate with the onset of rapid fumonisin production. The addition of labelled acetate resulted in some reduction in the yield of fumonisin, likely due to a pH effect. FB₁ was purified from the fungal extracts in a similar manner to that used for the 10 liter cultures (Miller et al., 1994), only on a smaller scale.

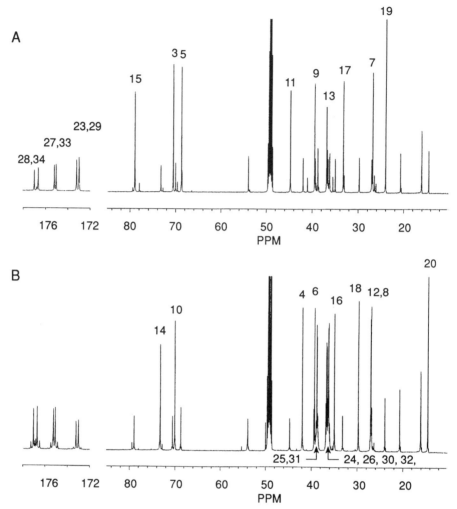

Figure 5. 125 MHz ^{13}C NMR spectra of purified FB$_1$ enriched in A) ^{13}C-1 Acetate and B) ^{13}C-2 Acetate.

The ^{13}C NMR spectra of FB$_1$ enriched with the specifically labelled acetate is shown in Figure 5. Those positions receiving enrichment are indicated. ^{13}C enrichment is evenly distributed down the hydrocarbon backbone from C-3 to C-20 (by approximately a factor of 4 to 5). The methyl group of acetate contributed to enrichment in the even numbered carbons, while the carbonyl position of acetate enriched the odd numbered carbons. Note that no enrichment is observed at C-1 at 16.0 ppm or C-2 at 53.7 ppm which form the amino terminal of fumonisin from either precursor. Also no enrichment from acetate is seen at C-21 (20.8 ppm) and C-22 (also at 16.0 ppm), indicating that the acetate pool is not available for methylation. Enrichment in the side chains is less than in the backbone of the fumonisin (approximately a factor of 2), and is unevenly distributed - the two carbons at the esterified end receiving somewhat less of the label than the other carbons - for both of the tricarballylic groups. This suggests that a 4-carbon unit is formed first, likely from the the Kreb's acid cycle and a third acetate unit is added later. The precursor to the esterification step must therefore be unsymmetrical in nature, since symmetry would induce an even labelling pattern.

The pattern of acetate incorporation is consistent with the head-to-tail pattern from the condensation of acetyl CoA units expected from both lipid and polyketide biosynthesis. However, the pattern alone cannot distinguish between the two pathways. The fact that substantial enrichment has occurred after initial lipid has been formed in these cultures which are not carbon compromised favors the polyketide pathway. There is also no label incorporated into the cellular lipid of the fungal hyphae. In addition, the hydroxyl functions at C-3, C-5 and C-15 are in the correct position for oxidation from polyketide carbonyl groups and the polyketide model is more flexible in terms of patterns of substitution. The large number of functionalities (7) would be difficult to evoke from stearic or oleic CoA.

The addition of relatively small quantities (0.2 mg\ml of culture) of ^{13}C-enriched amino acids reduced fumonisin production dramatically, with methionine and serine having the most effect and glutamic acid having the least, an observation that has also been made by others (Branham and Plattner, 1993a). This could be due to the addition of small quantities of available nitrogen from these precursors lifting the nitrogen limitation that is required to produce fumonisin (Miller et al., 1994).

The label from C-3 of alanine was incorporated into C-1 of fumonisin (by a factor of 4), in agreement with previous studies (Branham and Plattner, 1993a). However, alanine was also catabolized as evidenced by the labelling of the even positions of the backbone and the ester functions, although to a lesser extent (x2). ^{13}C-3 serine showed poor incorporation, resulting in some increase in the intensities of the resonances due to C-1 and the backbone carbons, but also the C-21 and C-22 methyl groups. Why these methyl groups were enriched by serine but not alanine, is unclear but may be indicative of the greater extent of catabolism of serine and the more direct incorporation of alanine into fumonisin. Under no conditions was C-2 enriched, indicating that "scrambling" was minimal. The S-CH₃ group of methionine was incorporated uniquely and efficiently (x20) into the C-21 and C-22 methyl functions, as predicted by previous studies using deuterium enrichment and mass spectrometric analysis (Plattner and Shackelford, 1992). Plattner and Branham (1994) have used this efficient incorporation of methionine to produce stable, highly enriched deuterated FB₁ (FB₁-d₆) to be used as an analytical standard for accurate quantitation of fumonisins by GC\MS or FAB\MS techniques. These results are also consistent with the ease of interconversion of alanine and serine and their catabolism into acetyl CoA via pyruvate, while methionine is exclusively utilized for protein synthesis and methylation. ^{13}C-5 glutamic acid was the least efficiently incorporated into FB₁, but showed a unique enrichment of the secondary carboxyl functions of both side chains at C-28 and C-34. When combined with the data from acetate enrichment, this indicates that the precursor to esterification involves a condensation between α-keto glutarate (the deaminated precursor of glutamate) and a second acetyl CoA unit.

The results of the labelling studies are summarized in Figure 6. The condensation of alanine with an 18-carbon polyketide chain confirms previous studies on AAL toxin, where alanine was shown to be directly incorporated into C-1 and C-2 (Shier et al., 1991). The array of functionalities is consistent with the carbonyl derivation of the hydroxyl groups at C-3,5 and 15, subsequent hydroxylation at C-10 and C-14, methionine-derived methylation at C-12 and C-16, followed by esterification by a precursor yet to be defined. Less incorporation of acetate into the side chains suggests that this precursor is formed either earlier or later than the fumonisin backbone, and that esterification occurs after the C-14 and C-15 hydroxyl groups are formed. This is supported by a recent report of a related metabolite with a carbonyl function at C-15 (S. Musser, private communication).

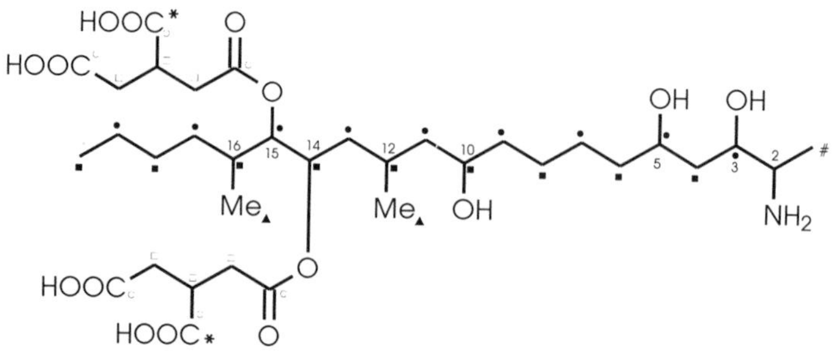

Figure 6. Summary of the pattern of incorporation of ^{13}C-enriched precursors into FB$_1$. The sites of incorporation as designated as (■) 2-^{13}C acetate, (●) 1-^{13}C acetate, (#) 2-^{13}C alanine, (▲) S-CH$_3$ methionine and (*) 5-^{13}C glutamic acid.

STEREOCHEMISTRY OF FB$_1$

The structural similarity between sphingosine and fumonisin has prompted interest in its mode of action in addition to its biosynthesis as discussed above. Several of the fumonisins (FB$_1$, FB$_2$ and HB$_1$) have been shown to be potent inhibitors of sphingolipid biosynthesis (Wang et al., 1991) and along with FB$_3$ and AAL toxin, to act specifically as an inhibitor of ceramide synthetase (Merrill et al., 1993a). The detailed role of fumonisin in this process has recently been reviewed (Riley et al., 1994a,b). It has been specifically shown that this disruption of sphingolipid metabolism plays an important early role in fumonisin induced disease (Riley et al., 1994b).

Examination of the structure of FB$_1$ shows that the molecule contains 10 stereocenters, providing a possible 1024 different stereochemical structures (ApSimon et al., 1994a). The presence of a single unique high resolution ^{1}H NMR spectrum for FB$_1$ derived from different species of *Fusarium* and isolated by different methodologies suggests that only one of these 1024 structures is adopted in solution. Since fumonisin does not crystallize, several groups have been involved in determining the stereochemistry of the backbone through the synthesis of derivatives and NMR analysis. To date, all have come to the same conclusions using different derivatization schemes. The determination of this specific configuration should assist in the understanding of the molecular interaction involved in the toxic response.

With the hope of obtaining a crystalline compound for analysis, several derivatives containing "bulky" functions and semi-rigid units were prepared. All of the derivatives failed to provide useful crystals, but they did yield the relative stereochemistry for portions of the FB$_1$ backbone. The synthetic strategy is summarized in Figure 7. Generally the derivatives were formed from tetramethyl FB$_1$ (FB$_1$(CH$_3$)$_4$) (see Figure 1) so that organic solvents could be used in the purification steps and so as to not involve the tricarballylic ester functions. FB$_1$(CH$_3$)$_4$ was converted to the 2,3-carbamate (**2**) using triphosgene and triethylamine in benzene. The 3,5-carbonate (**4**) was prepared using phosgene in pyridine after protecting the amide function with N-p-bromobenzoate (**3**). These two compounds were used to determine the relative stereochemistry of the C-1 to C-5 fragment of FB$_1$.

The relative stereochemistry of the C-10 to C-16 moiety was determined by comparing the NMR parameters of FB$_1$ to those of the 10,14-cyclic ether derivative of N-p-bromobenzoate FB$_1$ aminopentaol (**6**). This compound was formed by hydrolysis with KOH in ethanol of the 3,5-carbonate-N-p-bromobenzoate analogue of **4** in which the hydroxyl at C-10 had been replaced by chlorine (**5**). Details of the synthetic procedures are given in ApSimon et al. (1994b) and Blackwell et al. (1995).

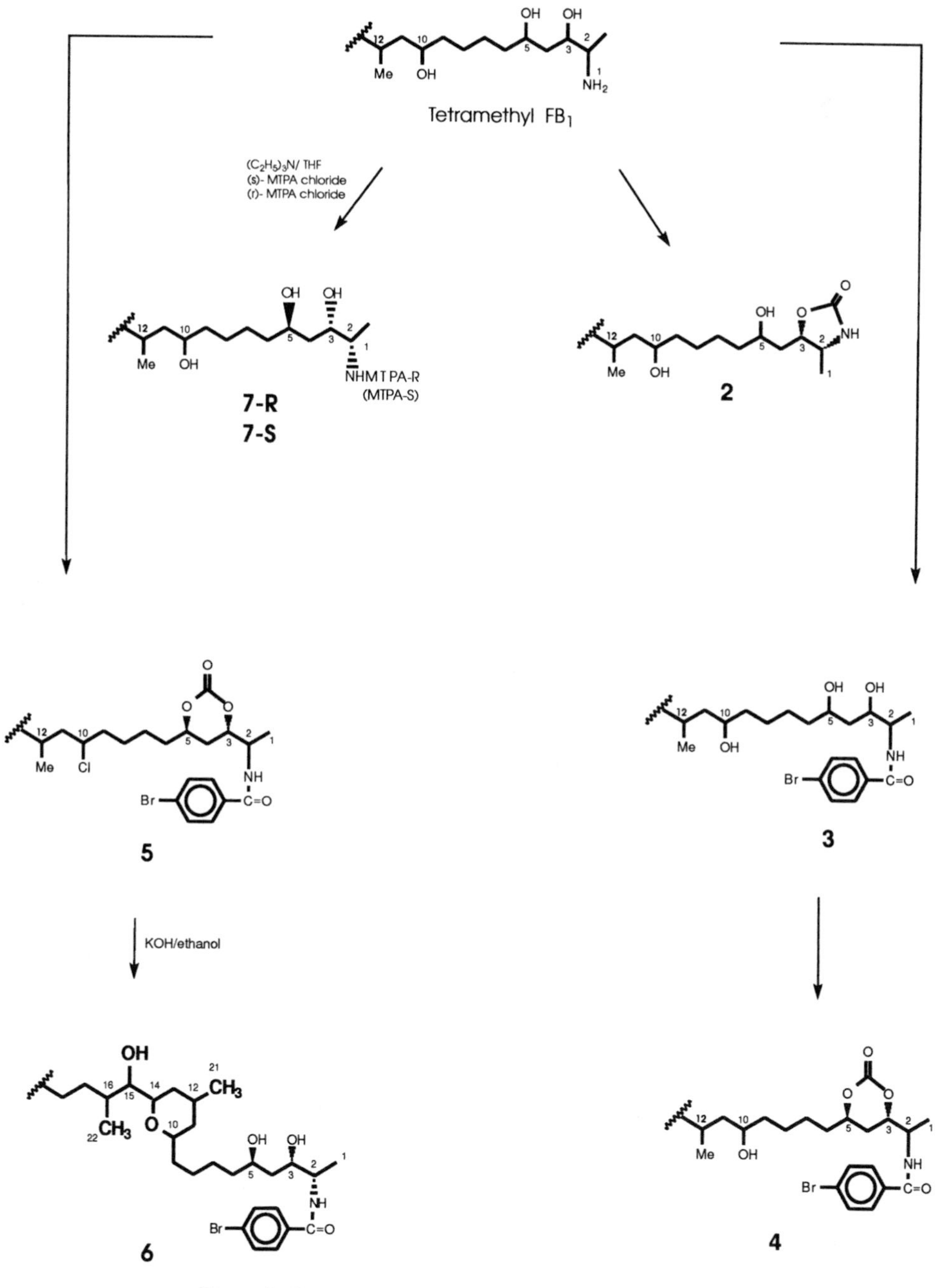

Figure 7. Summary of the synthetic scheme of FB₁ derivatives.

Selected NMR data for compounds **2** to **6** as well as the parent compounds FB₁ and FB₁(CH₃)₄ are given in Tables 1, 2 and 3. Detailed NMR assignments for all compounds were made using a combination of ¹H\¹H, ¹H\¹³C correlation, long range inverse detected ¹H\¹³C (HMBC), and nOe (NOESY) 2D NMR experiments. Chemical shifts varied little

Table 1. Selected ^1H chemical shifts in $CDCl_3$ of fumonisin derivatives (PPM from TMS)

Position	FB$_1$	(CH$_3$)$_4$FB$_1$	2	3	4	5	6
NH	—	—	5.21	6.63	6.36	—	6.45
H-1	1.27	1.06	1.26	1.30	1.43	1.50	1.29
H-2	3.14	2.80	3.57	4.17	4.49	4.90	4.16
H-3	3.74	3.48	4.38	3.98	4.55	5.23	3.97
H-4a	1.55	1.43	1.63	1.57	1.95	2.10	—
H-4b	1.55	1.63	1.74	1.73	2.13	2.10	—
H-5	3.84	3.84	3.88	3.92	4.48	4.56	3.91
H-10	3.62	3.56	3.60	3.61	3.55	3.94	3.34
H-11	1.15;1.45	1.02;1.38	1.03;1.44	1.03;1.46	1.00;1.40	1.00;1.62	0.85;1.55
H-12	1.81	1.80	1.80	1.81	1.85	1.73	1.76
H-13	1.55;1.70	1.25,1.68	1.31;1.69	1.33;1.69	1.30;1.59	1.52;1.72	1.20;1.92
H-14	5.16	5.15	5.17	5.19	5.18	5.13	3.74
H-15	4.94	4.88	4.90	4.91	4.90	4.89	3.78
H-16	1.70	1.59	1.62	1.63	1.60	1.62	1.76
CH$_3$-21	0.96	0.92	0.94	0.96	0.95	0.95	0.90
CH$_3$-22	0.94	0.88	0.89	0.91	0.91	0.92	0.97

*FB$_1$ in CD_3OD

from those of the parent compounds except at the sites of substitution. The presence of the N-p-bromobenzoate ester induced a large (1.4 ppm) downfield shift in the H-2 resonance of **3**, whereas the formation of the carbamate ring induced a 0.8-0.9 ppm downfield shift. The introduction of the 6-membered ring in **4** caused only a 0.6 ppm shift in the relevant proton peaks. The ^{13}C resonances follow a similar pattern. The formation of the carbonate (**4**) induces a 12 ppm upfield shift at C-4 and smaller downfield shifts at C-3 and C-5. The NMR spectra of **5** were virtually the same as those of **4**, except for the resonances of C-10. The presence of chlorine induces an upfield shift of 10 ppm for the C-10 resonance and a downfield shift of 0.4 ppm for the H-10 resonance. The structure of **6** was established by comparison to the spectra of the N-p-bromobenzoyl aminopentaol (not shown). Large differences in the coupling constants, specifically $J_{13,14}$, $J_{14,15}$, and $J_{15,16}$ were observed due to the presence of the 6-membered ring. An HMBC spectrum established the relationships between the vicinal carbons and protons along the backbone and confirmed the location of the ring by showing correlations between H-10 and C-14 and between H-14 and C-10 through the common ether linkage.

Table 2. $J_{H,H}$ coupling constants of fumonisin derivatives (Hz)

Position	FB$_1$	(CH$_3$)$_4$FB$_1$	2	3	4	5	6
NH, H-2	—	—	—	8.7	9.0	—	8.3
H-1,H-2	6.7	6.5	6.2	6.8	7.0	6.9	6.8
H-2,H-3	6.8	—	6.5	3.1	2.2	10.2	3.1
H-3,H-4a	3.2	—	3.1	3.1	3.8	5.4	3.2
H-3,H-4b	9.6	—	9.7	9.4	10.3	7.3	9.4
H-4 (AB)	—	—	—	14.6	14.6	14.4	—
H-4a,H-5	—	—	—	7.5	3.4	5.8	3.2
H-4b,H-5	—	—	—	3.5	5.7	5.0	9.0
H-13,H-14	2.4,10.8	2.4,11.5	2.7,11.5	2.3,11.3	2.5,11.5	6.2,6.2	1.7,5.5
H-14,H-15	3.7	3.2	3.4	3.4	3.4	2.9	9.4
H-15,H-16	8.1	8.8	8.7	8.7	8.7	8.7	2.6
H-12,H-21	6.7	6.6	6.6	6.6	6.6	6.5	6.5
H-16,H-22	6.8	6.8	6.8	6.8	6.8	6.8	7.0

Table 3. Selected ^{13}C NMR assignments in CDCl$_3$ of fumonisin derivatives (ppm from TMS)

Position	FB$_1$	(CH$_3$)$_4$FB$_1$	2	3	4	5	6
1	16.06	20.10	20.15	18.30	17.89	14.56	18.35
2	53.76	51.23	53.93	50.19	47.52	56.78	50.11
3	70.43	73.21	81.29	71.63	78.09	76.36	71.83
4	41.83	39.75	41.58	40.25	28.49	29.19	40.25
5	68.48	68.49	67.73	69.26	77.38	76.36	69.86
6	39.01	37.56	38.00	37.18	34.56	34.75	37.20
9	39.01	38.74	38.62	38.56	38.52	37.75	36.81
10	69.98	68.82	68.80	68.87	68.72	61.44	71.15
11	44.61	43.11	43.02	43.07	43.06	45.04	40.00
12	26.96	25.20	25.11	25.18	25.07	27.66	25.00
13	36.92	35.37	35.41	35.44	35.55	34.12	33.25
14	72.80	71.26	71.10	71.22	71.06	72.12	72.73
15	78.82	77.82	77.83	77.86	77.83	77.82	73.68
16	34.92	33.70	33.69	33.72	33.70	33.75	33.39
CH$_3$-21	20.83	20.39	20.43	20.47	20.40	20.70	22.53
CH$_3$-21	16.04	15.39	15.41	15.44	15.43	15.44	16.99

*FB$_1$ in CD$_3$OD

Comparison of the coupling constants for **2** and **4** to those of model compounds 2-oxo-1,3-dioxanes shows that the C-3 and the C-5 hydroxyl functions are *trans* and that the C-2 amino and C-3 hydroxyl functions are *cis* in these cyclic derivatives (see Figure 8A). The J$_{2,4}$ value of 10.4 Hz in **4** is somewhat larger than that observed in FB$_1$ (~9Hz) due to the fixed configuration of the ring. The J$_{2,3}$ of 6.5 Hz in **2** indicates a dihedral angle of either 110° or 60° between H-2 and H-3. The 110° orientation is determined by the observation of an nOe effect between CH$_3$-1 and H-3. The observation of an nOe effect between H-2 and H-3 in the spectra of both FB$_1$(CH$_3$)$_4$ and compound **6** confirms the *threo* disposition in the parent FB$_1$. A parallel study also confirmed these results (Poch et al., 1994).

Analysis of nOe data defines the relative configuration of the 6-membered ring of **6** as shown in Figure 8B. Enhancements were observed between H-10, H-12 and H-15, but not between H-10 and H-14, thus indicating these protons to be *trans*. Independent analysis of nOe effects observed in the spectra of FB$_1$ yields the relative configuration of the C-12 to C-16 moiety also shown (Figure 8B). Enhancements are observed between H-14 and H-16 as well as H-12 and between H-15 and CH$_3$-22. This indicates that the parent compound is relatively rigid in solution and accounts for the distinctive ^{1}H and ^{13}C shifts observed for the two tricarballylic esters. The relative configuration between C-10 and the C-12 to C-16 moiety observed in **6** is the same as that of the parent FB$_1$.

The relative configuration of two portions of the molecule thus defined, the absolute stereochemistry at some one position must be determined, in order to define the absolute stereochemistry at each assymetric center. The incorporation of *L*-alanine biosynthetically does not define the absolute stereochemistry of the C-1 to C-3 moiety since the assymetric center may be lost during decarboxylation. For example, *L*-serine is also the precursor of sphinganine which possesses the 2*S*, 3*R* (opposite) stereochemistry. Therefore, Mosher's method (Ohtani et al., 1991) was applied directly to FB$_1$ to form the α-methoxy-α-tri-fluoromethylphenylacetyl (MPTA) amide derivatives (see Figure 7) in a method similar to that of Hoye et al. (1994), except that in this case positions C-3 and C-5 were underivatized. FB$_1$(CH$_3$)$_4$ was treated with MPTA chloride in THF for 30 min and the amides purified on silica using 10% methanol in chloroform. The *S*-amide was prepared using the *R*-MPTA chloride, while the *R*-amide was prepared using the *S*-MPTA chloride. The resultant amides

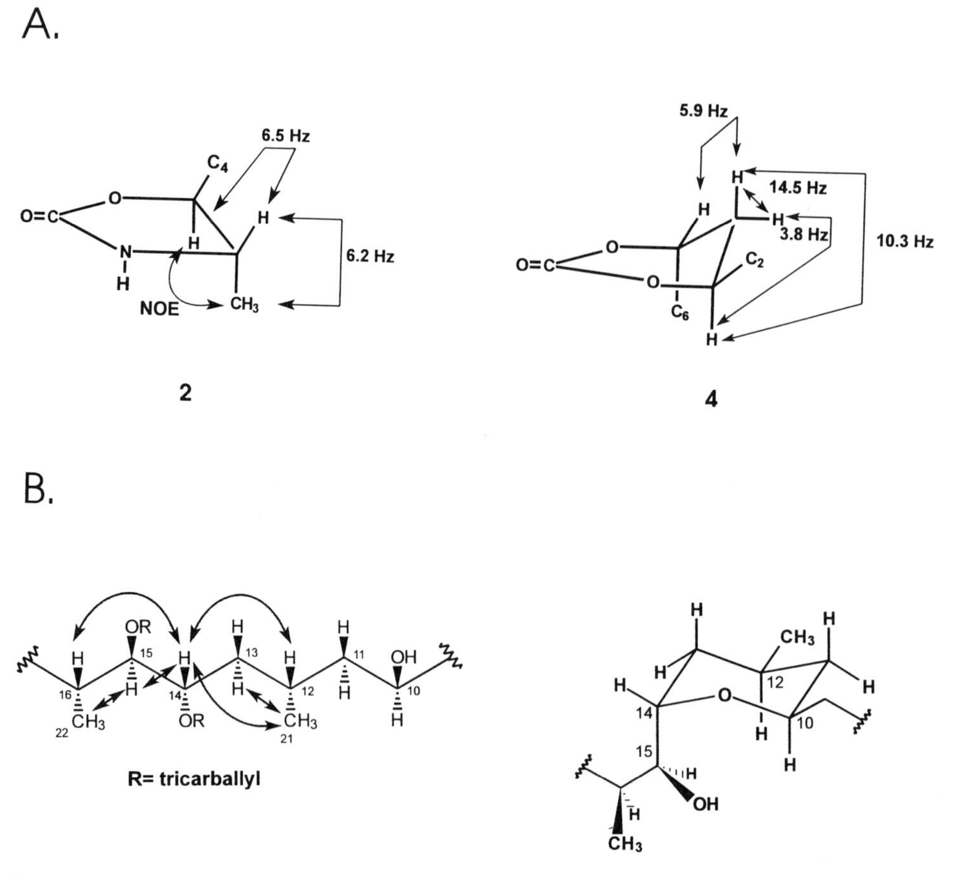

A.

2

B.

R= tricarballyl

FB₁

6

Figure 8. The configuration, $J_{H,H}$ values and nOe's observed in A) the 2,3-carbamate (**2**) and 3,5-N-p-bromobenzoate (**4**) derivatives of FB₁(CH₃)₄ and B) FB₁ and the 10,14-cyclic ether of FB₁ aminopentaol (**6**).

were analysed by NMR and selected chemical shifts of the two amides are shown in Table 4.

Comparison of the magnitude and sign of the differences in the proton chemical shifts between the two amides to theoretical ones (Ohtani et al., 1991) determined that the absolute configuration at C-2 was **S**, identical to that of the naturally occuring amino acids. Thus *L*-alanine is incorporated with retention of configuration into FB₁.

The resultant absolute configuration of FB₁ is shown in Figure 9. The relative configuration between C-10 and C-5 has been determined by Hoye et al. (1994). This also agrees with the studies of Harmange et al. (1994) on FB₂ and shows that the fumonisins have the same absolute configuration as the AAL toxins as determined by Boyle et al. (1994). The configuration of the side chains has been designated as **S** (Shier et al., 1995), but this has recently been challenged to be **R** in a study on FB₂ by Boyle and Kishi (1995). This is independently being investigated in our laboratory by using a different methodology. All of the NMR data indicates that FB₁ adopts a linear configuration in aqueous solution, rather than a globular one in spite of the possibility of hydrogen binding between the ester functions and the hydrophilic end (C-1 to C-5).

Table 4. Selected Chemical Shifts of the **R** and the **S** Mosher Amides of $FB_1(CH_3)_4$

	Position (δ, ppm for TMS)				
	1	2	3	4	5
¹H Chemical Shift					
(R)-MTPA	1.22	3.97	3.84	—	3.77
(S)-MTPA	1.18	3.98	3.87	—	3.89
Δδ ($δ_s$-$δ_r$)	-0.04	+0.01	+0.03	—	+0.12
¹³C Chemical Shift					
(R)-MTPA	20.40	49.78	71.14	40.10	68.99
(S)-MTPA	20.39	49.72	71.04	39.98	69.22
Δδ	-0.01	-0.04	-0.10	-0.12	+0.23

Comparison of the configuration of FB_1 to that of other *Fusarium* metabolites including sphingosine shows that the fumonisins and AAL toxins have the same configuration at the amino terminus (2*S*,3*S*), but that AAL toxin has the opposite configuration for the hydroxyl function analogous to C-5 of FB_1 (5*R* for FB_1 and 4*S* for AAL). Thus FB_1 has the opposite configuration to that of the related compounds sphingofungins (2*S*, 3*R*) which has been determined by VanMiddlesworth et al. (1992) to be the same as sphingosine. Both AAL toxin and fumonisin therefore are shown to have the opposite absolute configuration at the C-1 to C-4 positions to sphinganine (or sphingosine), which could be related to the fact that FB_1 has been shown to bind more tightly to sphinganine N-acyl transferase (ceramide synthase) than sphinganine (Merrill et al., 1993b). A notable structural difference is the lack of hydroxyl on C-1 of FB_1 which may prevent it from undergoing the metabolic pathways of sphingosine. The fact that both AAL toxins and fumonisins have the opposite configuration at this terminus to not only sphingosines, but also to the analogs sphingofungins may be related to their respective roles in inhibiting different steps in the biosynthesis of sphingolipids (Van Middlesworth et al., 1992).

These studies show that NMR is a valuable analytical tool for studying fumonisin. The structural similarity of this class of compounds to those produced by a wide variety of organisms indicates that there is a need for further analytical and biochemical investigations into these metabolites, especially in determining the structures of minor metabolites. As this knowledge improves, progress will be made in the determination of what levels of fumonisin should be considered acceptable for human or animal consumption.

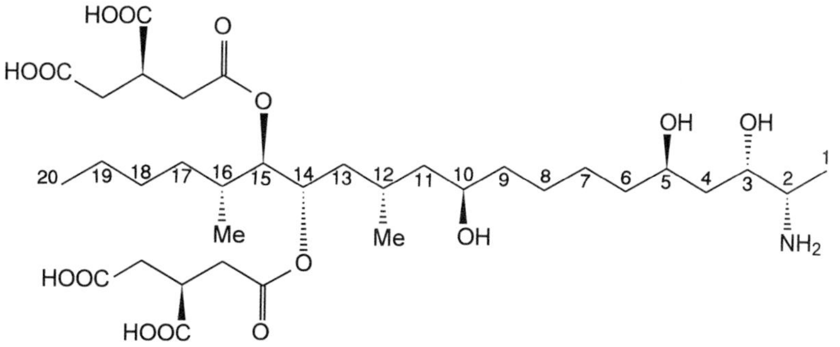

Figure 9. Absolute configuration of FB_1.

ACKNOWLEDGEMENTS

This constitutes PRC no. 1598. The authors would like to thank John Nikiforuk for the NMR spectra, David Fielder for assistance with the drawings, Marc Savard for the purification of FB_1 and Tim Bender for assistance with the biosynthetic experiments.

REFERENCES

Abbas, H. K.; Shier, W. T. Evaluation of biosynthetic precursors for the production of radiolabelled fumonisin B_1 by *Fusarium moniliforme* on rice medium. 106th Annu. Assoc. Off. Anal. Chem. Meeting. Cinncinnati, 1992; p. 234 (Abstr.)

ApSimon, J.W.; Blackwell, B.A.; Edwards, O.E.; Fruchier, A.; Miller, J.D.; Savard, M.E.; Young, J.C. The chemistry of fumonisins and related compounds. Fumonisins from *Fusarium moniliforme*, chemistry, structure and biosynthesis. *Pure and Appl. Chem.* **1994a**, *66*, 2315-2318.

ApSimon, J.W.; Blackwell, B.A.; Edwards, O.E.; Fruchier, A. Relative configuration of C-1 to C-5 fragment of fumonisin B_1. *Tetrahedron Letts.* **1994b**, *35*, 7703-7706.

Bezuidenhout, S.C.; Gelderblom, W.C.A.; Gorst-Allman, C.P.; Horak, R.M.; Marasas, W.F.O.; Spiteller, G.; Vleggaar, R. Structure elucidation of fumonisins, mycotoxins from *Fusarium moniliforme*. *J. Chem. Soc. Chem. Commun.* **1988**, 743-745.

Blackwell, B.A.; Miller, J.D.; Savard, M.E. Production of carbon 14-labelled fumonisin in liquid culture. *J. A.O.A.C. International* **1994**, *77*, 506-511.

Blackwell, B.A.; Edwards, O.E.; ApSimon J.W.; Fruchier, A. Relative configuration of the C-10 to C-16 fragment of fumonisin B_1. *Tetrahedron Letts.* **1995**, *36*, 1973-1976.

Boyle, C.D.; Harmange, J.C.; Kishi, Y. Novel structure elucidation of AAL toxin T_A backbone. *J. Amer. Chem. Soc.* **1994**, *116*, 4995-4996.

Boyle, C.D.; Kishi, Y. Absolute configuration at the tricarballylic acid moieties of fumonisin B_2. *Tetrahedron Letts.* **1995**, *36*, 4579-4582.

Branham, B.E.; Plattner, R.D. Alanine is a precursor in the biosynthesis of fumonisin B_1 by *Fusarium moniliforme*. *Mycopathologia* **1993a**, *124*, 99-104.

Branham, B.E.; Plattner, R.D. Isolation and characterization of a new fumonisin from liquid cultures of *Fusarium moniliforme*. *J. Nat. Prod.* **1993b**, *56*, 1630-1633.

Harmange, J.C.; Boyle, C.D.; Kishi, Y. Relative and absolute stereochemistry of the fumonisin B_2 backbone. *Tetrahedron Letts.* **1994**, *35*, 6819-6822.

Hoye, T.R.; Jimenez, J.I.; Shier, W.J. Relative and absolute configuration of the fumonisin B_1 backbone. *J. Amer. Chem. Soc.* **1994**, *116*, 9409-9410.

Laurent, D.; Lanson, M.; Goasdoue, N.; Kohler, F.; Pellegrin, F.; Platzer, N. Étude en RMN ^{1}H and ^{13}C de la macrofusine, toxine isolée des maïs infesté par *Fusarium moniliforme sheld*. *Analusis* **1990**, *18*, 172-179.

Merrill, A.H. Jr.; Wang, E.; Gilchrist, D.G.; Riley, R.T. Fumonisins and other inhibitors of de novo sphingolipid biosynthesis. *Adv. Lipid Res.* **1993a**, *26*, 215-234.

Merrill, A.H. Jr.; vanEchten, G.; Wang, E.; Sandhoff, K. Fumonisin B_1 inhibits sphingosine (sphinganine) N-acyltransferase and de novo sphingolipid biosynthesis in cultured neurons in situ. *J. Biol. Chem.* **1993b**, *268*, 27299-27306.

Miller, J.D.; Blackwell, B.A. Biosynthesis of 3-acetyldeoxynivalenol and other metabolites by *Fusarium culmorum* HLX1503 in a stirred jar fermentor. *Canadian J. Bot.* **1986**, *64*, 1-5.

Miller, J.D.; Savard, M.E.; Rapior, S. Production and purification of fumonisins from a stirred jar fermenter. *Natural Toxins* **1994**, *2*, 354-359.

Nelson, P.E.; Desjardins, A.E.; Plattner, R.D. Fumonisins, mycotoxins produced by *Fusarium* species - biology, chemistry and significance. *Annu. Rev. Phytopathol.* **1993**, *31*, 233-252.

Nelson, P.E.; Dignani, M.E.; Anaissie, E.J. Taxonomy, biology and clinical aspects of *Fusarium* species. *Clinical Microbiol. Rev.* **1994**, *7(4)*, 479-504.

Norred, W.P. Fumonisins, mycotoxins produced by *Fusarium moniliforme*. *J. Toxic and Environ. Health* **1993**, *38*, 309-328.

Ohtani, I.; Kusumi, T; Kashna, Y.; Kakisawa, H. High field FTNMR application of Mosher's method. The absolute configurations of marine terpenoids. *J. Amer. Chem. Soc.* **1991**, *113*, 4092-4098.

Plattner, R.D.; Shackelford, D.D. Biosynthesis of labelled fumonisins in lipid cultures of *Fusarium monili-forme. Mycopathologia* **1992**, *117*, 17-22.

Plattner, R.D.; Weislander, D.; Shackelford, D.D.; Peterson, R.; Powel, R.G. A new fumonisin from solid cultures of *Fusarium moniliforme. Mycopathologia* **1992**, *117*, 23-28.

Plattner, R.D.; Branham, B.E. Labelled Fumonisins: Production and use of fumonisin B₁ containing stable isotopes. *J. A.O.A.C. International* **1994**, *77*, 525-532.

Poch, G.K.; Powell, R.G.; Plattner, R.D.; Weisleder, D. Relative stereochemistry of fumonisin B₁ at C-2 and C-3. *Tetrahedron Letts.* **1994**, *35*, 7707-7710.

Powell, R.G.; Plattner, R.D. Fumonisins, In *Alkaloids, Chemical and Biological Perspectives*, Vol. 9, S. William Pelletier (Ed.); Pergamon Press: NY, **1995**, In Press.

Prelusky, D.B.; Trenholm, H.L.; Savard, M.E. Pharmacokinetic fate of 14C-labelled fumonisin B₁ in swine. *Natural Toxins* **1994**, *2*, 73-80.

Riley, R.T.; Wang, E.;Merrill, A.H. Jr. Liquid chromatographic determination of sphinganine and sphingosine: Use of the free sphinganine to sphingosine ratio as a biomarker for consumption of fumonisins. *J. A.O.A.C. International* **1994a**, *77*, 533-540.

Riley, R.T.; Voss, K.A.; Yool, H.S.; Gelderblom, W.C.A.; Merrill, A.F., Jr. Mechanism of fumonisin toxicity and carcinogenesis. *J. Food Protection* **1994b**, *57*, 528-535.

Riley, R.T.; Norred, W.P.; Bacon, C.N. Fungal toxins in food - recent concerns. *Ann. Rev. Nutrition* **1993**, *13*, 167-189.

Savard, M.E.; Blackwell, B.A. Spectral characteristics of secondary metabolites from *Fusarium* fungi, In *Mycotoxins in Grain, Compounds Other Than Aflatoxin*, J.D. Miller and H.L. Trenholm (Eds.); Eagan Press: St. Paul, MN, **1994**, pp. 59-257.

Shier, W.J.; Abbas, H.K.; Mirocha, C.J. Toxicity of the mycotoxins fumonisins B₁ and B₂ and *Alternaria alternata* f. sp. lycopersici toxin (AAL) in cultured mammalian cells. *Mycopathologia* **1991**, *116*, 97-104.

Shier, W.J.; Abbas, H.K.; Badria, F.A. Complete structures of the sphingosine analog mycotoxins fumonisin B₁ and AAL toxin T_A: Absolute configuration of the side chains. *Tetrahedron Letts.* **1995**, *36*, 1571-1574.

Van Middlesworth, F.; Dufresne, C.; Wincott, F.F.; Mosley, R.P.; Wilson, K.F. Determination of the relative and absolute stereochemistry of sphingofungins A, B, C, D. *Tetrahedron Letts.* **1992**, *33*, 297-300.

Wang, E.; Norred, W.P.; Bacon, C.W.; Riley, R.T.; Merrill, A.H., Jr. Inhibition of sphingolipid biosynthesis by fumonisins, implications for diseases associated with *Fusarium moniliforme. J. Biol Chem.* **1991**, *266*, 14486-14490.

8

DETERMINATION OF UNDERIVATIZED FUMONISIN B$_1$ AND RELATED COMPOUNDS BY HPLC

Jon G. Wilkes*, Mona I. Churchwell, Stanley M. Billedeau,
David L. Vollmer, Dietrich A. Volmer, Harold C. Thompson, Jr. , and
Jackson O. Lay, Jr.

USFDA/NCTR
Jefferson, AR 72079

ABSTRACT

A method is presented for determining the purity of the mycotoxin fumonisin B$_1$ (FB$_1$) by high performance liquid chromatography (HPLC) with evaporative light scattering detection (ELSD). The ELSD is a universal HPLC detector that exhibits a non-linear relationship between analyte amount and the resulting response. A log-log plot of ELSD response with the mass of FB$_1$ injected was used as a calibration curve for determining the quantities of both FB$_1$ and also individual impurities present in samples. Assumptions related to the uniformity of ELSD response for different but related compounds and other issues implied in this use of ELSD data were examined.

One potential error produced by use of this method for purity analysis comes from the ELSD's decreased sensitivity for low-concentration analytes. Because analytes become more dilute the longer they remain on a chromatographic column, this sensitivity discrimination can be related to the retention times at which they appear. The ELSD response for FB$_1$ at retention time 15.5 minutes was used to construct a general purpose calibration curve. Whenever a peak appeared at any time other than 15.5 minutes, the discrimination effect was corrected using a an empirically determined weighting factor and a proportion calculated from the retention time difference compared to 15.5 minutes.

Purities for two fumonisin samples were calculated using both the ELSD method described above and an electrospray/mass spectrometric method. The quantitative assumptions underlying each method were discussed in order to understand and reconcile differences between the two sets of purity values obtained.

---

* corresponding author

Fumonisins in Food, Edited by L. Jackson *et al.*
Plenum Press, New York, 1996

INTRODUCTION

Fumonisin mycotoxins are produced by the fungus *Fusarium moniliforme* which grows under certain conditions in corn and other grains. Fumonisins are believed to be the agents responsible for a variety of toxic (and sometimes severe) effects in many species and they have been suggested as a cause of esophageal cancer in humans who eat *Fusarium*-infected grain (Nelson, 1993; Norred, 1993; Scott, 1993; Kriek, 1981). Determination of fumonisin purity is an important element in the validation of chronic, low dose toxicity studies currently being initiated at the National Center for Toxicological Research. These studies include examination of the effects in rats and mice of fumonisin B_1 (FB_1).

Ideally toxicological studies involve tests with substances of a purity exceeding 98% and the impurities must be identified and quantified to the extent possible. In the case of FB_1, this criterion presented significant production, purification, and analytical challenges. The usual way of producing quantities of the mycotoxin is to culture a microbe, *Fusarium moniliforme*, which produces the analyte (often in a mixture including other mycotoxins) and then extract the toxin from the culture matrix and separate it from related compounds. Separation of the toxin from its mixture concomitants can be a laborious process (Alberts, 1990), and one which is aggravated if the common laboratory chromatographic detectors do not respond to the toxin. Also, obtaining a valid measurement of the relative amounts of disparate components is especially difficult in the absence of analytical standards.

The fumonisins are relatively high molecular weight species (FB_1, M.W. = 721 Da) lacking a UV chromophore. **Figure 1** shows the structure for FB_1. The fact that the molecule contains four carboxylic acid moieties, with their associated polarity and hydrogen bonding, helps explain FB_1's very low volatility. It is possible to derivatize FB_1 at the primary amine with fluorescamine, for example, so as to allow sensitive HPLC detection by fluorescence methods (Holcomb, 1993). But expected impurities of FB_1 include species which will not react with fluorescamine and, as such, would remain invisible to a fluorescence detector. Since neither the fluorescamine method nor the fluorescence detector is universal, they should not be used for a purity analysis.

Below we present work with HPLC and an evaporative light scattering detector (ELSD) for determining fumonisin purity. The ELSD is sensitive to underivatized FB_1 and appears to be equally suited for detection of the other low volatility components present in partially purified *F. moniliforme* culture extracts. A method of using LC/ELSD for purity analysis is described. As a comparison and partial validation of the ELSD purity method, we also present some HPLC data with detection by electrospray mass spectrometry (LC/ES/MS).

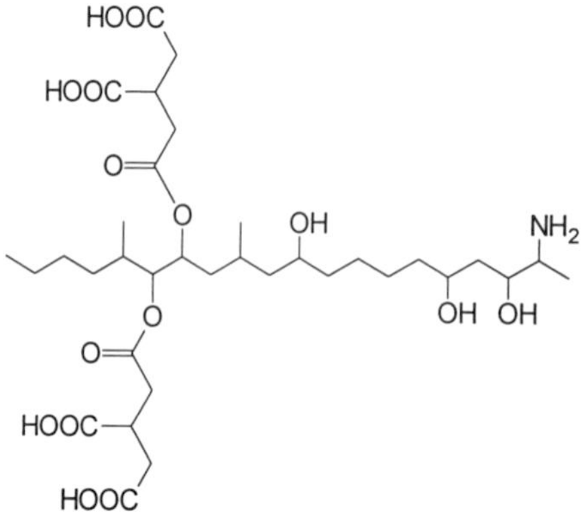

Figure 1. Molecular structure for fumonisin B_1 (FB_1). The molecule contains no strong UV chromophore. HPLC is the chromatographic technique of choice for separation and purification of FB_1.

EXPERIMENTAL

Samples studied were partially purified and well purified *F. moniliforme* culture extracts containing high percentages of FB₁ with several contaminants consisting primarily of fumonisin-like compounds and other compounds chemically dissimilar from the fumonisins. For purposes of purity analysis each sample was dissolved to a concentration of 1 to 3 mg/mL in 18 megohm deionized water from a Milli-Q water purification system (Millipore Products, Bedford, MA) and stored in plastic centrifuge tubes. (An unknown contaminant appeared in the FB₁ ELSD chromatograms after overnight solution storage in Pyrex sample containers.) Depending on the inherent sensitivity of the detector used, from 4 to 20 μL of each solution were injected.

Chromatographic separations were achieved using reversed-phase, mobile phase composition gradients. Mobile phase mixtures used with two different models of ELSD, described below, were constituted from acetonitrile (HPLC Reagent grade from Baker, 0.1 ppm residue after evaporation), 18 megohm deionized water, and 0.025% trifluoroacetic acid (TFA). The LC/ELSD systems also used a YMCBasic base deactivated C-8 column (YMC, Wilmington, NC), 4.6 mm i.d.×150 mm, with 5μm particles, equilibrated at 40° C. The mobile phase flow rate was 1 mL/min.

For the LC/ES/MS analyses similar mobile phase mixtures were used. These were constituted from Fisher Optima grade acetonitrile (0.8 ppm residue after evaporation), the same water and 0.025% formic acid (FA) buffer. An Ultracarb 31% carbon loaded ODS HPLC column (Phenomenex, Torrance, CA), 3.0 mm i.d.×150 mm, was substituted for the YMCBasic. (The mass spectrometer electrospray ion source suffered greatly reduced sensitivity for all analytes when TFA was used as a mobile phase buffer. In order to obtain good chromatographic integrity without the TFA, substitution of the Ultracarb column with FA was necessary.) The mobile phase flow rate used with the smaller diameter column was 0.6 mL/min followed by a 1:4.5 post-column split in which 110 μL/min was introduced into the electrospray ion source of a model TSQ7000 triple quadrupole tandem mass spectrometer (Finnigan-MAT, San Jose, CA).

Instrumental Setup. Two ELSD systems were used: a Varex Mk III (Alltech Assoc., Deerfield, IL) and a SEDEX 55 (Richard Scientific, Novato, CA). While the principle of detection used by the designs is similar, there are significant technical differences between the two which result in somewhat different experimental setups and result in different analytical strengths and weaknesses.

The Varex Mk III was used with a 50:50 mobile phase split. The HPLC pump was a Spectrasystem P4000 (Thermo Separations Products, Fairmont, CA). Typical operating parameters were varied in several experiments over these ranges: detector temperature, 100 to 110° C; nitrogen nebulizing gas flow, 1.9 to 2.3 standard liters per minute. Output signal was processed using a Chromatopac C-R3A integrator (Shimadzu, Kyoto, Japan).

The SEDEX 55 did not require a post-column mobile phase split. In its design an *aerosol* split is made within the instrument so that the majority of the chromatographic effluent (up to 99%) is diverted from the detector. The fact that so little solvent is introduced to the detection chamber permits much cooler operation of the device than is the case for the Varex Mk III. The SEDEX 55 was operated at 40° C using a nitrogen nebulizing gas at a pressure of 2 bar. The HPLC pump was an SP8800, tertiary, with low pressure mixing (Spectra-Physics, San Jose, CA). The same integrator was used.

The TSQ7000 was operated in a pneumatically assisted electrospray mode using 60 psi nitrogen. With the electron multiplier set at 1400 volts, full scan mass spectra were acquired over a range from *m/z* 300 to 800. The mass scan range actually used for detection and quantitation of fumonisin and impurities included the sum of two segments: from *m/z*

560 to 570 and from 685 to 760. The HPLC pump was a Constametric 4100MS, quartenary, with low pressure mixing (Thermo Separations Products, Fairmont, CA).

Experimental Procedures. Several samples were chromatographically separated and detected using both ELSD and ES/MS. A series of dilutions of one FB_1 sample were made spanning over 2.5 orders of magnitude variation in concentration. These were injected into the Varex and Sedex ELSD systems. Log-log plots of fumonisin peak area as a function of injected mass were made. Analogous log-log plots were made over a much smaller dynamic range to show the variation of four impurity peaks' responses with a nominal amount injected. Since the actual amount of each impurity was not known with certainty, it was assumed that each caused the same response per unit mass as FB_1 itself. This kind of assumption is not usually valid for chemically dissimilar compounds in most common HPLC detectors. The ELSD is sometimes referred to as a "mass" detector because its response is, in most cases, proportional to the mass of analyte present regardless of the analyte's identity. Nevertheless, for purposes of a purity analysis, it was decided that the mass detector assumption did require validation. The slope of the impurity log-log plots was compared to that of FB_1 to see whether these values varied between components. If so, this would imply that these substances differed with respect to transport efficiency between the nebulizer and the light scattering cell.

To test the assumption that a given mass of an impurity scattered light as efficiently as the same amount of fumonisin, two weighed impurity samples were analyzed and the systems' sensitivity for them was compared to that of FB_1.

Calculations. The non-linearity of ELSD response necessarily implied a second order discrimination effect which would compromise the use of a calibration curve based on fumonisin response for determining an impurity which appeared at a significantly different retention time than fumonisin or even fumonisin itself, if chromatographic conditions were changed. Early chromatographic peaks are always sharper and elute as a more concentrated band in the detector than peaks from the same amount of material would be if they appeared later in the separation. The commercially available models of ELSD give a significantly reduced response per unit mass for most compounds at low concentrations. To compensate for this phenomenon, a weighted correction factor was developed:

The plot of log(response [peak area]) versus log(fumonisin mass) from the Varex Mk III for an isocratic separation was algebraically inverted to give a function which could be used as a calibration curve. The use of logarithmic calibration curves or polynomial response approximations for quantitation of light reflected in thin layer chromatography is a standard quantitative technique (Poole, 1984). From the observed peak area a nominal impurity mass was calculated. Each peak's true mass (m_{true}) was calculated from this calibration curve mass (m_{cal}) using the weighted proportion, $(T_R - k \, \Delta T_R)/T_R$, thus:

$$m_{true} = [(T_R - k \, \Delta T_R)/T_R] \, m_{cal}$$

The proportion was defined using the retention time difference (ΔT_R) between each peak's T_R and 15.5 minute (fumonisin retention time for data used in the original log-log response plot).

The weighting factor (k) was empirically determined by slightly increasing the percent organic contribution of the isocratic separation so that, when an identical injection of fumonisin was made, it eluted at about 13.5 minutes, two minutes earlier. An apparently larger mass of FB_1, m_{cal}, was calculated using the calibration curve. Using this m_{cal} and the true mass, m_{true}, 2 min for ΔT_R and 15.5 min for T_R, the proportional correction equation was solved for k.

To confirm the validity and general utility of both the correction factor concept and the determined magnitude of k, the percent organic in the mobile phase was reduced so that the fumonisin appeared about ten minutes later, at 25 minutes. The value of k determined from the early peak was used in the correction factor expression to compensate for the reduced signal of this now very late peak. The calculated, corrected peak mass was then compared to the known true mass.

RESULTS AND DISCUSSION

ELSD and LC/ES/MS Detection. Figure 2 is a truncated Varex LC/ELSD chromatogram from a 4 L injection of a "dirty" FB_1 sample dissolved in water at 3.03 g/ L. Figure 3 shows a truncated LC/ES/MS reconstructed ion chromatogram of the same sample. The latter is obviously more sensitive than the former, but there is a reassuring congruity between the two chromatograms with respect to the number and approximate relative size of peaks.

Comparison of the Varex and Sedex ELSDS for FB_1 Analysis. Two chromatograms of the same, fairly pure sample of FB_1 are shown as **Figure 4** and **Figure 5**. The former was obtained from 10 μL of a 3.09 μg/μL solution in water detected by the Varex after the 50:50 post-column mobile phase split. The latter was obtained from 5 μL of the 3.09 μg/μL solution detected by the Sedex without a split. The Varex has a more stable baseline than the Sedex. The reduced peak resolution observed in the Sedex chromatogram is attributable to degradation of the HPLC column between the particular analyses compared rather than to any excess dead volume in the detector. The Sedex, with significantly greater sensitivity, is able to distinguish as many as seven small impurity peaks eluting after FB1, none of which could be detected at this level by the Varex unit.

Generally, the Varex unit surpassed the Sedex model with respect to reproducibility and stability of response. The Sedex excelled in matters of analytical sensitivity. Because it operates at much lower temperature, the Sedex had several orders of magnitude greater sensitivity than the Varex for semi-volatile analytes. (Data not shown.) This did not seem to affect fumonisin purity analysis using the Varex because, in this instance, the contaminants detected were non-volatile. Still, even for the nonvolatile fumonisins, the Sedex was noticeably more sensitive than the Varex.

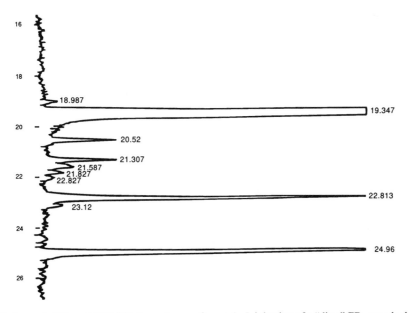

Figure 2. Truncated Varex LC/ELSD chromatogram from a 4 μL injection of a "dirty" FB_1 sample dissolved in water at 3.03 μg/μL.

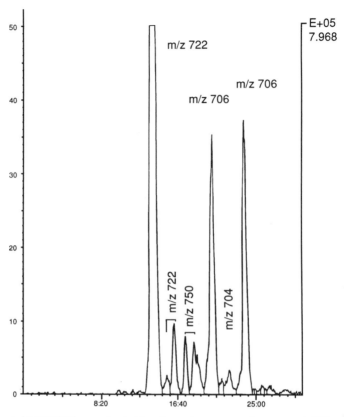

Figure 3. Truncated LC/ES/MS reconstructed ion chromatogram, over the range m/z 563 to 566 plus 700 to 760, of the same "dirty" FB_1 sample. Other experimental details in text. Considering the dissimilarity in equipment and chromatographic conditions used in the separations shown in Figures 2 and 3, there is a striking qualitative correspondence between the traces.

ELSD Quantitation of Impurities. In the Varex Mk III, the slope of the log-log plot of FB_1 response as a function of mass was 1.46. This confirmed the high degree of non-linearity of fumonisin response in the ELSD. Response lies somewhere between linear (log-log slope = 1) and quadratic (log-log slope = 2). There was no significant difference in slope of the log-log plot between the Varex and the Sedex ELSDs. **Figure 6** shows the log-log plot obtained from the Varex MK III. Each concentration was injected three to five times. Linear regression of the base ten log-log data yielded the function

$$\log(\text{area}) = 1.46 \log(\text{mass}) + 0.487$$

with a coefficient of correlation value, 0.99267. (See *caveat* below which discusses the significance of these numbers.) This equation was solved for log(mass) and used as a calibration curve for determining the nominal amount of each impurity from its peak area:

$$\log(\text{mass}) = 0.684 \log(\text{area}) - 0.334$$

Response for the fumonisin and two impurities that had been collected and weighed was found to be identical, within experimental error. This suggested that the ELSD for these

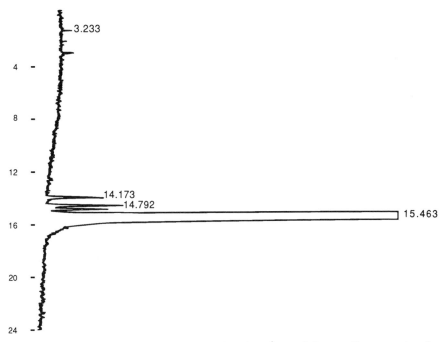

Figure 4. 10 µL injection of a relatively pure FB₁ sample, using a reversed phase gradient separation, detected after a 50:50 mobile phase split on the Varex Mk III ELSD.

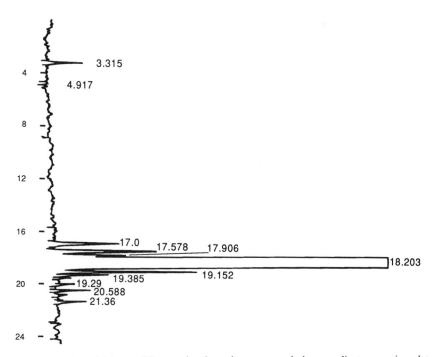

Figure 5. 5 µL injection of the same FB₁ sample, also using a reversed phase gradient separation, detected without a mobile phase split on the Sedex 55 ELSD.

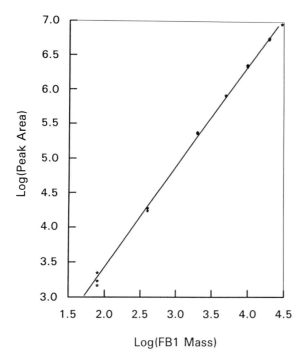

Figure 6. Log-log plot of detector response (peak area) versus FB_1 mass injected from the Varex Mk III. Linear regression of data, three to five replicates at each concentration, gave **log(area) = 1.46 log(mass) + 0.487** with a coefficient of correlation value of 0.99267.

compounds acted as a true mass detector. In other words, no difference in these impurities' efficiency at scattering light from their particles in aerosol was observed.

The log-log slopes observed over a limited range for four impurities were 1.63, 1.65, 1.96, and 1.41, respectively. Of these the only one which is believed to vary significantly from the 1.46 value for FB_1 is 1.96. Closer examination of the chromatograms showed that the third impurity eluted immediately prior to FB_1 and was not completely resolved from it in the higher concentration injections. The leading edge of the huge fumonisin peak was integrated with impurity number three at higher concentrations and this exaggerated the non-linearity of the response curve. Thus, there was no demonstrable difference in log-log slopes between any of these fumonisin impurities. This implied that there were no differences between fumonisin and these impurities with respect to analyte loss mechanisms (such as electrostatic phenomena [Wilkes, 1995a,b] or volatility). This further validated the use of the ELSD for the fumonisin purity determination.

One *caveat* should be included. The mathematical transformations involved in using log-log calibration curves can be deceptive. The transformation of the original data points to their logarithmic expressions *appears* to decrease variation between replicates. It almost always produces a calibration curve which shows excellent linearity and extremely small variations between replicates. Since the calibration uses the inverse transformation to quantitate unknowns, small errors are magnified. Therefore, rounding off of figures should never be done in the logarithmic form, for either the data or the coefficients of the calibration curve. In this paper we have presented the regression obtained for log-log response curves to a few significant digits only for purposes of qualitative comparison. In actual use, these figures were extended several more digits to eliminate the possibility of propagating any rounding error. Also, for purposes of assessing the experimental error in measurements between replicates, one should always work with the original data values rather than with their logarithmic transforms.

Comparison of Purity by LC/ELSD and LC/ES/MS. Table 1 shows the overall purity values obtained by the LC/ELSD method and by LC/ES/MS for two different fumonisin samples, one very dirty sample and one fairly pure. Each purity value was obtained by subtracting the value for the impurities from one hundred percent. The values for the impurities were derived from their portion of the total signal. Qualitatively, results of the two methods agreed. (Agreement in the sense that the two methods ranked half a dozen samples of differing purity in the same order. Comparative data for only two of these are shown in Table 1.) Quantitatively, they differed substantially, with the numerical purity values obtained by the two methods converging for clean samples: when assessing purity of very clean samples, ones from lots which might be actually used in the toxicological study, the purity values almost agreed.

There are several reasons for discrepancies between these two measurements. First, the ELSD measures mass and determines a mass % purity while the EC/MS counts molecules and its data yields a mole % purity. Unfortunately, this difference does not entirely explain the discrepancies. When these factors are taken into account by using the molecular weights of the known major impurities to transform the ELSD purities from mass to mole percent, the differences are only somewhat reconciled. A significantly different contribution to the mass or mole percent purity occurs only for impurities which differ greatly in molecular weight from FB₁ (721 Da). This would not explain the magnitude of the discrepancies observed, particularly for the "dirty" fumonisin sample, in which the major contaminants appear by LC/EC/MS to have molecular weights in the range of 700 to 750 Da.

Second, in order to obtain sufficient sensitivity for minor components by LC/ELSD, very concentrated injections of sample are necessary. It is intuitively obvious that at some as yet undetermined point there must occur a decreased efficiency for scattering of light from aerosols containing large residual particles because of the greater probability that some of the particles are shielded from the light source in the shadow of other large particles. This effect would reduce response for the most concentrated component (here FB₁) and decrease the apparent sample purity. Based on the relative values in Table 1, this is not an explanation for the present discrepancies.

Another potential problem with the ELSD method would occur if any of the analytes absorbed rather than scattered light ("black" particles) or if they fluoresced. For the known impurities in the cleaner fumonisin samples, neither of these phenomena are believed to be a problem. "Dirty" fumonisin samples are slightly colored and, for such samples, the absorption effect might explain a systematic ELSD discrimination against some impurities and, consequently, an apparent greater total purity value by ELSD than by ES/MS.

There are also good reasons to question the validity of purity assays based solely on electrospray mass spectrometry data. Determination of impurity contributions by LC/ES/MS, in the absence of analytical standards for individual calibration, requires the assumption that all compounds present be ionized with equal efficiency. While

Table 1. % FB₁ purity by LC/ELSD or LC/ES/MS

A "dirty" sample	
ELSD (mass %)	ES/MS (mole %)
73.2 ± 0.9	52.2 ± 1.2

A cleaner sample	
ELSD (mass %)	ES/MS (mole %)
97.3 ± 0.8	94.0 ± 0.4

arguments can be made that this assumption may be true for structurally similar compounds, it is well known that this assumption is certainly not true for dissimilar compounds.

The procedure for determining purity by LC/ES/MS is straightforward, especially in comparison to the calculations and corrections used for the LC/ELSD. The instrumental software is set up to produce a table of component percentages based on peak area. In these LC/EC/MS experiments, in order to gain sufficient sensitivity to observe the less concentrated impurities, the mass range scanned was limited to the two aforementioned, relatively narrow windows, a subset of the available spectrometric range. The assumption was implicit that (1) no impurities exist in the sample with masses outside the two windows *and* (2) no significant fragmentation occurred which would lead to ions appearing outside the windows. In either event, significant ions would not be counted in the purity calculation. Validity of the measurement presupposed that the *only* significant product of the electrospray process was a protonated molecule or some other pseudomolecular ion. This assumption was partially validated by looking for, and finding little evidence of, other analyte fragments in the lower mass range. It is probably not possible to determine whether even smaller ions are produced in an electrospray source by fragmentation of large analytes, because such small fragments would appear in the region of intense solvent ion background. But if such a phenomenon did exist, it would compromise attempts to use LC/ES/MS for purity analysis.

FUTURE WORK

Because of the uncertainties discussed above, work is ongoing to develop alternative methods of purity analysis for the fumonisin mycotoxins and similar compounds. Two methods being considered at present are LC with flame ionization detection (Wilkes , 1991) and LC with detection using a chemical reaction interface mass spectrometer (Teffera, 1994). Both of these systems should give a linear, universal response. Either would not only prevent the clumsiness associated with log-log calibration curves, proportional correction factors, and mass to mole percent conversions, as in this LC/ELSD method, but also resolve discrimination questions associated with the use of electrospray (or most other) mass spectrometric techniques for analytical quantitation of HPLC-separated components.

It should be stressed that quantitative methods for fumonisin purity analysis based on either LC/ELSD or LC/ES/MS data have proved extremely useful and, we believe, a very significant improvement on earlier approaches. The instrumentation is also commercially available in well designed packages. These instruments and methods will continue to be used in our laboratories on a routine basis even while we explore other possibilities.

CONCLUSIONS

LC/ELSD seems to have reasonable integrity for purity analysis. Because of the possibility that some impurities exist which are not chromatographically resolved, its values for overall sample purity measurements should be considered an upper bound relative to the true purity of the sample. The Varex Mk III gives a particularly reliable quantitative analysis for fumonisin. Using the conditions described in this study, the Sedex 55 appears to have significantly greater sensitivity for trace components, especially if they are marginally volatile, than the Varex. However, for these analyses, it produced a slightly less stable baseline.

The LC/ES/MS is excellent for identifying and, at least qualitatively, determining impurities in fumonisin. This is particularly true for relatively pure compounds having impurities that are similar to FB_1.

ACKNOWLEDGEMENTS

The authors are indebted to Dr. Steve Musser of the FDA's Center for Food Safety and Applied Nutrition, Washington, D. C., who exhaustively collected two FB_1 impurities from multiple analytical-scale separations and kindly provided them to assist our efforts at evaluating the ELSD quantitation method.

REFERENCES

Alberts, J. F.; Gelderblom, W. C. A.; Thiel, P. G.; Marassas, W. F. O.; van Schalkwyk, D. J.; Behrend, Y. Effects of temperature and incubation period on production of fumonisin B_1 by *Fusarium moniliforme*. *Appl. Environ. Microbiol.*, 56, **1990**, 1729-1733.

Holcomb, M.; Thompson, H. C.; Hankins, L. J. Analysis of fumonisin B_1 in rodent feed by gradient elution HPLC using precolumn derivatization with FMOC and fluorescence detection. *J. Agric. Food Chem.* 41, **1993**, 764.

Kriek, N. P. J.; Kellerman, T. S.; Marasas, W. F. O. A comparative study of the toxicity of *Fusarium verticillioides* (= *F. moniliforme*) to horses, primates, pigs, sheep, and rats. *Onderstepoort J. Vet. Res.* 48, **1981**, 129.

Nelson, P. E.; Desjardins, A. E.; Plattner, R. D. Fumonisins, mycotoxins produced by *Fusarium* species - biology, chemistry, and significance. *Annu. Rev. Phytopathol.* 31, **1993**, 233.

Norred, W. P. Fumonisins - mycotoxins produced by *Fusarium moniliforme* [Review]. *J. Toxicol. Environ. Health,* 38, **1993**, 309.

Poole, C. F.; Schuette, S. A. Quantitative evaluation of thin layer chromatograms, in *Contemporary Practice of Chromatography*, Chapter 9, section 10; Elsevier, Amsterdam, **1984**, 657.

Scott, P. M. Fumonisins" [Review]. *Int. J. Food Microbiol.* 18, **1993**, 257.

Teffera, Y.; Abramson, F. P. Application of HPLC/CRIMS for the analysis of conjugated metabolites: a demonstration using deuterated acetaminophen. *Proceedings of the 42nd ASMS Conference on Mass Spectrometry and Allied Topics*, Chicago, Illinois, May 29-June 3, **1994**, 616.

Wilkes, J. G. Diffusion based solvent removal interface for coupling liquid and supercritical fluid chromatography to gas chromatographic detectors. *Ph.D. Dissertation*, University of Houston, December, **1991**, 252 - 270.

Wilkes, J. G.; Zarrin, F.; Lay, Jr., J. O.; Vestal, M. L. Particle size distribution is not the major factor explaining variable analyte transmission efficiency in liquid chromatography/particle beam/mass spectrometry. *Rapid Comm. Mass Spectrom.*, 9, **1995a**, 133-137.

Wilkes, J. G.; Freeman, J. P.; Heinze, T. M.; Lay, Jr., J. O.; Vestal, M. L. AC corona discharge aerosol-neutralization device adapted to liquid chromatography/particle beam/mass spectrometry, *Rapid Comm. Mass Spectrom.*, 9, **1995b**, 138-142.

9

ANALYSIS OF FUMONISIN B$_1$ IN CORN BY CAPILLARY ELECTROPHORESIS

Chris M. Maragos, Glenn A. Bennett, and John L. Richard

Mycotoxin Research Unit
National Center for Agricultural Utilization Research
USDA/ARS, 1815 N. University St.
Peoria, IL 61604

ABSTRACT

Intact fumonisins contain two tricarballylic acid groups and can therefore acquire a net negative charge. The anionic nature of the fumonisins is the basis behind the widely used method for cleanup of corn with strong anion exchange (SAX) columns. This property also enables the fumonisins to be separated by electrophoretic techniques which, until now, have not been applied to the analysis of fumonisins in corn. Fumonisin B$_1$, extracted from corn with 80/20 (v/v) methanol/water and isolated with a commercially available affinity column, was derivatized with fluorescein isothiocyanate for analysis by capillary zone electrophoresis with laser-induced fluorescence detection (CZE-LIF). Recoveries from corn fortified with 0.25 to 5.0 ppm FB$_1$ averaged 89% (range 71 to 102%). As little as 0.05 ppm FB$_1$ could be detected in corn. For corn naturally contaminated with FB$_1$, the CZE-LIF method compared favorably to established SAX/HPLC and C$_{18}$/HPLC methods. Capillary electrophoresis can be used for quantitation of FB$_1$ in corn, with minimal use of organic solvents and provides an additional tool for confirming fumonisin contamination.

INTRODUCTION

Fumonisins have been associated with several diseases in animals, including leukoencephalomalacia in horses (Wilson et al., 1992), pulmonary edema in pigs (PPE; Colvin and Harrison, 1992), and hepatocarcinoma in rats (Gelderblom et al., 1988). The hazard that fumonisins pose to humans has not been determined. The *Fusarium* species that produce fumonisins are common parasites and saprophytes of corn. Swine feeds suspected in causing PPE have been found to contain 20 to 360 ppm fumonisin B$_1$ (FB$_1$), while problem feeds for horses contained 8 to 117 ppm (Ross et al., 1992). The widespread use of corn in feed for pigs, combined with the potential for PPE and the potential for residual carry-over of fumonisin into liver and kidney after slaughter (Prelusky et al., personal communication)

Fumonisins in Food, Edited by L. Jackson *et al.*
Plenum Press, New York, 1996

have lead to the development of a variety of analytical methods for the fumonisins in corn, including HPLC (Shephard et al. 1992; Stack and Eppley, 1992; Bennett and Richard, 1994; Sydenham et al., 1990; Scott and Lawrence, 1992; Holcomb et al., 1993), TLC (Rottinghaus et al., 1992), and ELISA (Azcona-Olivera, 1992; Fukuda et al., 1994; Usleber et al., 1994).

Because the fumonisins lack a strong chromophore, the majority of analytical methods rely upon derivatization with fluorescent labels such as: o-phthaldialdehyde (OPA; Shephard et al. 1992; Stack and Eppley, 1992), naphthalene dicarboxaldehyde (NDA; Bennett and Richard, 1994), fluorescamine (Sydenham et al., 1990; Rottinghaus et al., 1992), 4-fluoro-7-nitrobenzofurazan (Scott and Lawrence, 1992), fluorescein isothiocyanate (Maragos, 1995), and (9-fluorenylmethyl) chloroformate (Holcomb et al., 1993).

Intact fumonisins, having two tricarballylic acid groups, can be separated from components of dissimilar charge using electrophoretic techniques. Application of capillary electrophoresis to mycotoxins has been limited to the analysis of fumonisin B_1 in horse serum (Maragos, 1995). The present report describes the first application of capillary zone electrophoresis (CZE) to the analysis of fumonisins in corn. Fumonisins were isolated from corn using commercially available affinity columns, derivatized with fluorescein isothiocyanate, and analyzed by CZE with laser-induced fluorescence (LIF) detection. The CZE-LIF method was compared to two established methods that use strong anion exchange columns or C_{18} columns for cleanup and HPLC analysis of naturally contaminated corn.

MATERIALS AND METHODS

Materials

Fumonisin B_1 was purchased from Sigma Chemical Co. (St. Louis, MO, lot #121H0066). Fluorescein-5-isothiocyanate (FITC) and naphthalene-2,3-dicarboxaldehyde (NDA) were purchased from Molecular Probes, Inc. (Eugene, OR). Affinity columns for isolation of fumonisin (FumoniTest, lot #369) were purchased from Vicam (Watertown, MA). Strong anion exchange (SAX) columns (12 cc, 500 mg sorbent mass) and C_{18} columns (12 cc, 500 mg sorbent mass) were obtained from Varian (Sunnyvale, CA). For fortification studies yellow dent corn, in 50 lb bags, was purchased from Garver Feeds (Decatur, IL) and analyzed by HPLC to establish that the background level of FB_1 was less than 0.2 ppm before spiking. Deionized water (Nanopure II, Sybron/Barnstead) was used throughout, while methanol and acetonitrile were both HPLC grade.

Sample Preparation

Corn retained upon a No. 5 sieve (4.00 mm opening, Fisher Scientific Co.) was ground for 30 seconds in a model M-2 Stein mill (Seedburo Eg. Co., Chicago, IL). Twenty five-gram portions were weighed and fortified from an FB_1 stock solution in acetonitrile/water (1/1). Samples were extracted with 100 mL of 80/20 (v/v) methanol/water as described previously (Bennett and Richard, 1994). The methanol/water extract (equivalent to 0.25 g corn/mL) was diluted 1:7.5 with deionized water. The affinity column was washed with 3 mL of phosphate buffered saline (PBS; 10 mM phosphate with 0.85% NaCl, pH 7.4) and 6.0 mL of diluted extract, equivalent to 0.20 g of corn, was loaded at a flow rate of 1.5 mL/min. The column was washed with 5.0 mL phosphate buffer (10 mM, pH 7.5) and 5.0 mL deionized water. Fumonisins were eluted with 1.6 mL methanol into 2 mL amber vials, dried under a gentle stream of nitrogen at 60°C, and stored at -20°C until analysis. For determination of recoveries, FB_1 standards in 1/1 (v/v) acetonitrile/water were added to 2 mL vials, 1.6 mL methanol was added, then standards were dried under nitrogen and derivatized as described for corn samples.

Derivatization with FITC and NDA

For reaction with fluorescein isothiocyanate (FITC), dried samples or standards were solubilized in 120 µL of a freshly prepared solution of dimethylsulfoxide (DMSO) and borate buffer (3 volumes DMSO/ 1 volume borate buffer, 100 mM, pH 9.50). Thirty µL of freshly prepared FITC in DMSO (3 mM) was added and mixed. Vials were sealed and kept at room temperature (23°C) for 2.5 h then stored at -20°C and analyzed within 8 h by CZE. For analysis by reverse-phase HPLC samples were derivatized with naphthalene-2,3-dicarbox-aldehyde (NDA) and quantitated as described previously (Bennett and Richard, 1994).

Capillary Zone Electrophoresis

Samples derivatized with FITC were quantitated using a Beckman P/ACE System 5000 capillary electrophoresis unit, with Argon-ion laser induced fluorescence (LIF) detection (Laser module 488). The capillary (50 cm x 75 µm I.D) was housed in a cartridge configured for LIF detection (Beckman #360604) and was maintained at 35°C. Derivatized samples (150 µL) were brought to 4.70 mL with electrophoresis buffer (60 mM borate, pH 9.50). Before each sample was injected, the capillary was rinsed 5 min with electrophoresis buffer at 0.5 psi. Sample was injected under 0.5 psi (hydrodynamic injection) for 5.0 seconds (equivalent to a volume of 30 nL). After immersion in electrophoresis buffer, voltage was applied and the current maintained at 250 µA. At the conclusion of the run, the capillary was rinsed once with 0.5 N NaOH (2.0 min, 0.5 psi) and once with deionized water (2.0 min, 0.5 psi). Fumonisin B₁ standards were injected at 5-6 different levels within the range of 0 to 5.0 pg. FB₁ content of samples was determined by comparison to the fluorescence data of the standards fit with a sigmoidal curve (TableCurve, Jandel Scientific).

RESULTS AND DISCUSSION

Derivatization with FITC and NDA

Previous research has lead to the widespread use of a number of fluorescent markers for fumonisin analysis. Chief among these have been o-phthaldialdehyde (OPA; Shephard et al. 1992; Stack and Eppley, 1992), and naphthalene dicarboxaldehyde (NDA; Bennett and Richard, 1994), although fluorescamine (Sydenham et al., 1990; Rottinghaus et al., 1992), 4-fluoro-7-nitrobenzofurazan (Scott and Lawrence, 1992) and (9-fluoroenylmethyl) chloroformate (Holcomb et al., 1993) have also been used. For HPLC analyses we currently use NDA for derivatization because of the relative stability of the product. In general, fluorescein isothiocyanate (FITC) derivatives are less desirable for labeling small molecular weight compounds due to the presence of substantial reagent peaks, and the extended reaction times required for adequate derivatization. However, commercial CE instrumentation with laser-induced fluorescence is currently limited to the argon-ion laser, which is ideally suited to fluorescein derivatives. For this reason the FITC derivatives of fumonisins were prepared for analysis by CZE-LIF (Maragos, 1995). Derivatization of up to 783 ng of purified FB₁ (equivalent to injection of 5.0 pg) was reproducible, and yielded a standard curve that was sigmoidal over this range (Fig. 1).

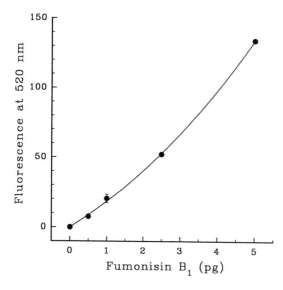

Figure 1. Standard curve of FB$_1$ derivatized with FITC. Fumonisin B$_1$ over the range of 0-783 ng was derivatized with FITC as described in the text, and amounts up to 5.0 pg were analyzed by CZE-LIF. The curve is a sigmoidal fit of the average of four trials. Error bars represent ± 1 S.D. from the mean, r^2=0.9995.

Analysis of Corn

Corn samples were extracted with 80/20 (v/v) methanol/water and the fumonisins were isolated using a commercially available affinity column. Extracts were then derivatized with FITC and analyzed by CZE-LIF (Fig. 2). After correcting for the fumonisin content of the unfortified corn (0.18 ppm), recoveries of added FB$_1$ over the range of 0.2 to 5.0 µg/g ranged from 71% to 102% (Table I). Quantitation below 0.2 ppm was quite variable, with coefficients of variation of 9% to 30% (Tables I & II). However, the variability dropped when FB$_1$ was present at levels greater than 0.2 ppm, with a CV of 7.7% at 1 ppm.

The limited capacity of the affinity columns for fumonisins, 1.2 µg (Ware et al., 1994), limits the upper and lower ranges of the method. For example, a corn sample having 5.0 ppm FB$_1$ would have 1.0 µg FB$_1$ loaded onto the affinity column. Above 5 ppm the recovery of FB$_1$ from corn can be expected to diminish rapidly. This effect may be responsible for the decrease in recovery observed between 2.0 ppm and 5.0 ppm (91% and 71% respectively), and is further compounded by the presence of FB$_2$, which will also compete for binding to the affinity column. For highly contaminated samples this limitation can be overcome by loading less extract onto the column, or increasing the sample dilution.

The capacity of the affinity columns also influenced the sensitivity of the method. Analysis of samples containing less than 0.2 ppm could be enhanced by increasing the amount of sample loaded onto the column (0.5 g equivalents of corn instead of 0.2 g). Figure

Table I. Recovery of Fumonisin B$_1$ from fortified corn

FB$_1$ Added (µg/g corn)	FB$_1$ Found (µg/g corn)	Recovery of Added FB$_1$[b]
0	0.18 ± 0.06	—
0.25	0.41 ± 0.05	92.8%
1.00	1.20 ± 0.09	102.2%
2.00	1.99 ± 0.10	90.7%
5.00	3.76 ± 0.12	71.7%

[a] Samples of corn were fortified with FB$_1$ in quadruplicate at each of the levels indicated (total, 20 samples).

[b] Recovery after subtraction of FB$_1$ in the unspiked sample.

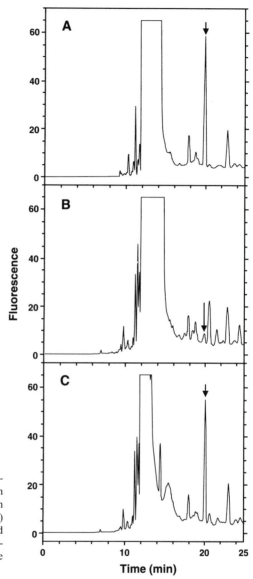

Figure 2. Electropherograms of FITC derivatized FB_1. The arrows indicate the FB_1 migration time. (A) 2.5 pg purified FB_1, (B) unfortified corn containing 0.18 ppm FB_1, (C) the same corn as (B) fortified with 2.0 ppm FB_1. The volume injected was 30 nL, CZE electrophoresis buffer was composed of 60 mM sodium borate, pH 9.50. The current was maintained at 250μA.

3 shows the electropherogram of a corn sample containing 133 ppb FB_1, for which 0.5 g equivalents was loaded onto the affinity column. This sample had the lowest FB_1 content of any we have analyzed (by HPLC or CZE), because of this the sensitivity of the method could not be measured, but was estimated to be at least 50 ppb based upon the response of this 133 ppb sample.

While the sensitivity was enhanced by loading a greater amount of sample, we do not routinely use this modification because it reduced the concentration level at which sample dilution was required for highly contaminated samples. With 0.5 g equivalents of corn loaded onto the affinity column the upper applicable limit would be reduced to roughly 2 ppm. We have chosen to optimize our method so that as little as 0.2 and as much as 5.0 μg FB_1/g could be quantitated without changing the sample dilution. Samples that gave a response near 5 ppm were diluted and reanalyzed if it was necessary to quantitate above 5 ppm.

Table II. Comparison of three methods for FB_1 in naturally incurred corn

Sample	SAX/HPLC[a]	C_{18}/HPLC	Affinity/CZE-LIF
White Corn	0.35 ± 0.08	ND	0.27 ± 0.08
White Corn	53.5 ± 1.7	48.1 ± 3.7	47.7 ± 3.8
Yellow Corn	1.3 ± 1.7^{b}	< 0.5	0.13 ± 0.01
Yellow Corn	4.4 ± 0.6	4.4 ± 0.2	2.9 ± 0.1
Yellow Corn	8.0 ± 0.1	NA	9.6 ± 1.9
Yellow Corn[c]	NA[d]	NA	3.9 ± 0.3

[a] All corn samples were extracted with 80/20 (v/v) methanol/water. FB_1 was isolated using either strong anion exchange (SAX), C_{18}, or affinity columns. Quantitation was either by HPLC or capillary electrophoresis (CZE-LIF).

[b] Actual analyses were: 0, 0.6, 3.3 µg/g.

[c] The sample was also analyzed by HPLC after affinity column cleanup and NDA derivatization (4.2 ± 0.2 µg/g)

[d] NA, not analyzed.

Comparison to HPLC with Naturally Incurred Samples

The method reported here, affinity column cleanup with CZE-LIF detection, was compared to two established methods for fumonisin analysis that use either SAX or C_{18} cleanup of corn extracts followed by HPLC analysis of the NDA derivative of FB_1 (Bennett and Richard, 1994). For five samples of white and yellow corn naturally contaminated with 0.13 to 50 ppm FB_1 the two HPLC methods agreed well for the three samples having more than 4 ppm FB_1, but were substantially different for the samples containing less than this

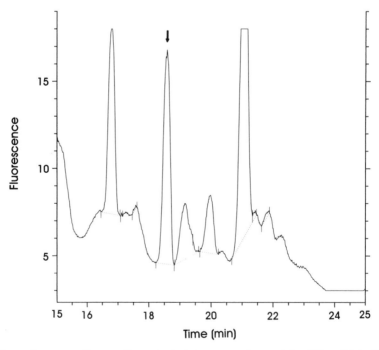

Figure 3. Electropherogram of a naturally contaminated corn sample containing 133 ng FB_1/g. The amount of corn loaded onto the affinity column was equivalent to 0.5 g.

amount. The Affinity/CZE-LIF method yielded estimates slightly lower than those obtained by SAX/HPLC for three of the five samples, and dramatically less for one of the samples of yellow corn (1.3 ± 1.8 by SAX/HPLC and 0.13 ± 0.01 ppm by Affinity/CZE-LIF). The SAX/HPLC estimates for this sample were extremely variable (range 0 to 3.3 ppm) and the C_{18}/HPLC method, like the Affinity/CZE-LIF method, indicated that the amount of FB_1 present was low. For further comparison, additional samples of naturally contaminated corn are currently being analyzed by the HPLC and CZE-LIF methods.

CONCLUSIONS

Fumonisins can be separated and reliably quantitated by capillary electrophoresis with LIF detection. The method described here is comparable in sensitivity to established HPLC methods for corn. The advantages of CZE include the use of much smaller sample volumes and the generation of substantially smaller volumes of hazardous waste, while the major disadvantage of this method is the long derivatization time (3 h). While CZE-LIF has not yet reached the stage where it can rival HPLC for ease of use, the method holds considerable promise for further improvements in sensitivity and analysis times.

REFERENCES

Azcona-Olivera, J.I.; Abouzied, M.M.; Plattner, R.D.; Pestka, J.J. Production of monoclonal antibodies to the mycotoxins fumonisins B_1, B_2, and B_3. *J. Agric. Food Chem.* **1992**, 40, 531-534.

Bennett, G.A.; Richard, J.L. Liquid chromatographic method for analysis of the naphthalene dicarboxaldehyde derivative of fumonisins. *J. AOAC Int.* **1994**, 77, 501-506.

Colvin, B.M.; Harrison, L.R. Fumonisin-induced pulmonary edema and hydrothorax in swine. *Mycopathologia* **1992**, 117, 79-82.

Fukuda, S.; Nagahara, A.; Kikuchi, M.; Kumagai, S. Preparation and characterization of anti-fumonisin monoclonal antibodies. *Biosci. Biotech. Biochem.* **1994**, 58, 765-767.

Gelderblom, W.C.A.; Jaskiewicz, K.; Marasas, W.F.O.; Thiel, P.G.; Horak, R.M.; Vleggaar, R.; Kriek, N.P.J. Fumonisins-novel mycotoxins with cancer-promoting activity produced by Fusarium moniliforme. *Appl. Environ. Microbiol.* **1988**, 54, 1806-1811.

Holcomb, M.; Thompson, H.C.; Hankins, L.J. Analysis of fumonisin B_1 in rodent feed by gradient elution HPLC using precolumn derivatization with FMOC and fluorescence detection. *J. Agric. Food Chem.* **1993**, 41, 764-767.

Maragos, C.M. Capillary zone electrophoresis and HPLC for the analysis of fluorescein isothiocyanate labeled fumonisin B_1. *J. Agric. Food Chem.* **1995**, 43, 390-394.

Prelusky, D.B.; Trenholm, H.L.; Vudathala, D.; Miller, J.D.; Savard, M.E.; Scott, P.M.; Delgado, T. Residues in Meat, Milk, and Eggs. Personal Communication.

Ross, P.F.; Rice, L.G.; Osweiler, G.D.; Nelson, P.E.; Richard, J.L.; Wilson, T.M. A review and update of animal toxicoses associated with fumonisin-contaminated feeds and production of fumonisins by Fusarium isolates. *Mycopathologia*, **1992**, 117, 109-114.

Rottinghaus, G.E.; Coatney, C.E.; Minor, H.C. A rapid, sensitive thin layer chromatographic procedure for the detection of fumonisin B_1 and B_2. *J. Vet. Diagn. Invest.* **1992**, 4, 326-329.

Scott, P.M.; Lawrence, G.A. Liquid chromatographic determination of fumonisins with 4-fluoro-7-nitroben-zofurazan. *J. AOAC Intl.* **1992**, 75, 829-834.

Stack, M.E.; Eppley, R.M. Liquid chromatographic determination of fumonisins B_1 and B_2 in corn and corn products. *J. AOAC Intl.* **1992**, 75, 834-837.

Shephard, G.S.; Thiel, P.G.; Sydenham, E.W. Determination of fumonisin B_1 in plasma and urine by high-performance liquid chromatography. *J. Chromatography* **1992**, 574, 299-304.

Sydenham, E.W.; Gelderblom, W.C.A.; Thiel, P.G.; Marasas, W.F.O. Evidence for the natural occurrence of fumonisin B_1, a mycotoxin produced by *Fusarium moniliforme*, in corn. *J. Agric. Food Chem.* **1990**, 38, 285-290.

Usleber, E.; Straka, M.; Terplan, G. Enzyme immunoassay for fumonisin B$_1$ applied to corn-based food. *J. Agric. Food Chem.* **1994**, 42, 1392-1396.

Ware G.M.; Umrigar, P.P.; Carman, A.S. Jr.; Kuan, S.S. Evaluation of fumonitest immunoaffinity columns. *Anal. Lett.* **1994**, 27, 693-715.

Wilson, T.M.; Ross, P.F.; Owens, D.L.; Rice, L.G.; Green, S.A.; Jenkins, S.J.; Nelson, H.A. Experimental reproduction of ELEM. *Mycopathologia* **1992**, 117, 115-120.

10

ISOLATION AND PURIFICATION OF FUMONISIN B_1 AND B_2 FROM RICE CULTURE[*]

F. I. Meredith[1], C. W. Bacon[1], W. P. Norred[1] and R. D. Plattner[2]

[1] Toxicology and Mycotoxin Research Unit
R. B. Russell Agriculture Research Center USDA/ARS
P. O. Box 5677
Athens, GA 30604
[2] National Center for Agricultural Utilization Research
USDA/ARS
1815 N University St.
Peoria, IL 61604

ABSTRACT

Procedures are presented for growing *Fusarium moniliforme* MRC 826 on rice, separation of fumonisin B_1 (FB_1) from fumonisin B_2 (FB_2), purification of FB_1 and preliminary procedures for purification of FB_2. The mycotoxins were extracted from rice culture material (RCM) with acetonitrile-water (1:1), filtered, and the acetonitrile removed on a rotary evaporator. Preparative reverse phase liquid chromatography (LC) was used to isolate and partially purify FB_1 and FB_2 from the extract. The extract was applied to a C_{18} reverse phase cartridge. FB_1 and FB_2 were eluted from the cartridge by a gradient of water-acetonitrile at a flow rate of 30 mL/min. A second preparative LC procedure using 0.5% pyridine-water and two CN cartridges was used to purify FB_1.

The FB_2 fraction was concentrated on a rotary evaporator to remove the acetonitrile. Acetonitrile was added back in sufficient quantity to redissolve the crystalline material in the fraction. An aliquot of the FB_2 fraction was added to a centrifugal spinning silicic acid TLC plate. The centrifugal TLC plate was washed at 3 mL/min with a linear gradient of (A) chloroform-acetone(4:3) and (B) methanol-acetone (1:1) to elute the FB_2. Gradient starting conditions were 10% methanol and ending conditions were 50% methanol. This preliminary study using the centrifugal spinning TLC showed the procedure to have the potential to be useful for purification of FB_2.

---

[*] The mention of firm names or trade products does not imply the endorsement by the U. S. Department of Agriculture over firms or similar products not mentioned.

Fumonisins in Food, Edited by L. Jackson *et al.*
Plenum Press, New York, 1996

INTRODUCTION

The mycotoxins fumonisin B_1 (FB_1) and fumonisin B_2 (FB_2), isolated from *Fusarium moniliforme* Sheldon, have been the subject of extensive studies due to recent investigations that implicate FB_1 as a possible factor in the etiology of esophageal cancer, which occurs at very high rates in the Transkei region of South Africa and in other isolated regions of the world (Riley et al., 1993; Rheeder et al., 1992; Marasas et al., 1988a; Gelderblom et al., 1988; Bezuidenhout et al., 1988). Equine leucoencephalomalacia (ELEM), a fatal disease of horses in which the white and gray matter of the brain develop liquefactive lesions has been found to occur after feeding either purified fumonisin FB_1 or culture material that contained fumonisins (Kellerman et al., 1990; Marasas et al., 1988b; Plattner et al., 1990; Ross et al., 1994; Ross et al., 1991). Fumonisin B_1 and culture material containing the fumonisins were toxic to broiler chicks, ducklings and turkey poults (Javed et al., 1993; Brydon et al., 1987). Swine dosed with fumonisin B_1 or fumonisin containing culture material developed porcine pulmonary edema (PPE) (Harrison et al., 1990; Kriek et al., 1981). Fumonisin B_1, FB_2 and the structurally related AAL toxin, two esters of 1,2,3-propanetricarboxylic acid esterified to the backbone of 1-amino-11,15-dimethylheptadeca-2,4,5,13,14-pentol (Bottini and Gilchrist, 1981; Bottini et al., 1981) are cytotoxic to cultured mammalian liver cells (Shier et al., 1991), rat liver cells and dog kidney cells (Mirocha et al., 1992). Studies with rats fed purified FB_1 and naturally contaminated corn showed that kidney and liver cells were affected. Only very low levels of these toxins were required to produce cell changes (Voss et al., 1989; Voss et al., 1993).

The mode of action of these mycotoxins is not fully understood. The role of the fumonisins in animal diseases and the possible effect on humans has created a need for developing methods capable of obtaining large amounts of highly purified mycotoxins for toxicology studies. A preparative procedure was developed for obtaining FB_1 using Amberlite, silica gel, and C_{18} reverse phase chromatography (Cawood et al., 1991). Total recovery of FB_1 by this procedure was 80% and purity was $\geq$ 90%, however, the recovery values reported did not give the amount of FB_1 recovered at 90% or above. This paper presents procedures for separation of FB_1 and FB_2, purification of FB_1 by preparative LC and a preliminary investigation on the purification of FB_2 using centrifugal spinning TLC.

MATERIAL AND METHODS

Rice Culture

The fungus *F. moniliforme* MRC 826 was cultured for 7 days on potato dextrose agar, which was used as an inoculum source. To each 2.8 L wide mouth Fernbach flask, 300 g of Uncle Ben's Converted Rice™, and 300 mL of distilled water were added. The flasks were held at room temperature for 12 hr then autoclaved for 30 min. Each flask was cooled to room temperature, inoculated with 5 mL of an aqueous suspension of conidia (10^9/mL), stoppered loosely with a cotton plug and incubated at 26°C for 28-35 days in the dark. For the first 10 days of the incubation period, each flask was shaken once daily.

Extraction

Extraction of the rice culture material (RCM) was carried out in the ratio of 100 g to 500 mL of acetonitrile-water (1:1 vol/vol). The RCM was stirred occasionally during the extraction and after 4 hrs was vacuum filtered through Whatman #4 paper. The filtered solids were resuspended in fresh solvent (1:1) and allowed to stand over night. The RCM was filtered as described above, the extracts were combined, and the volume was reduced approximately 60% with a 10 L vacuum rotary evaporator at a water bath temperature of 38°C. The final acetonitrile concentration of the extract was approximately 18-20%. The concentrated extract was stored in brown 4 L jugs at 2°C until preparative LC separations were conducted.

HPLC with Fluorescence Detection

Fumonisin B_1 and FB_2 concentrations were determined using a Hewlett Packard 1090 HPLC equipped with a Rainin C_{18} reverse phase 10 cm x 4.6 mm column containing 3 μm particle size packing material and a Rainin guard column. Components from the column were eluted isocratically with methanol (75%) and water (25% containing 1% orthophosphoric acid). Components were derivatized using a preinjection program in which 10 μL of o-phthalaldehyde (OPA)(Fluoraldehyde, Pierce Chemical Co.), 6 μL of sample, and an additional 10 μL of OPA were drawn into the sample loop. The OPA, sample and OPA were allowed to mix in the capillary tube and after 1 min were injected onto the column. Derivatives were detected with a Hewlett Packard 1046A fluorescence detector at 304 nm excitation and 440 nm emission.

HPLC with Light Scattering Detection

FB_2 was determined on a HPLC system equipped with an evaporative light scattering detector (ELSD) (Alltech Associates). This detector is a mass detector that has a detection limit of 50 ng of FB_2. The pump system was a Waters 6000 pump modified with glass pump heads (Bodman Industries) operated isocratically with a solvent containing 75% methanol and 25% water. The pH of the solvent was adjusted to 2.43 by the addition of trifluoroacetic acid. The column was a Rainin reverse phase Microsorb C_{18}, 10 cm x 4.6 mm with 3 μm particle size packing material.

FB₁ and FB₂ Standards

An analytical standard of FB_1 was prepared as described by Plattner and Branham (1994). The FB_2 standard was obtained from Sigma Chemical Co.

Preparative LC of FB₁ and FB₂

The preparative LC system was a Waters Delta 4000 equipped with a gradient elution controller and gradient mixer. A Waters Bondapak PrepPak 500/C_{18} reverse phase cartridge (47 mm x 300 mm) was pre-conditioned with 4 void volumes (1 void volume = 501 mL) of acetonitrile and 6 void volumes of water. The sample containing 2800 mg FB_1 and 1000 mg FB_2 was diluted 1:2 with water (3 L of extract with 6 L of water for a total of 9 L) and pumped onto the cartridge at 50 mL/min. The extract was a deep red to black color before the dilution and turned a cloudy light red-brown color upon the addition of the water.

After the sample was pumped onto the cartridge, water was pumped through the cartridge for 10 min at a rate of 30 mL/min. The water-acetonitrile step-wise gradient was then initiated with a flow rate of 30 mL/min. The gradient was completed after 230 min and a solvent concentration of 50% water-50% acetonitrile.

Sixteen fractions were collected and analyzed for FB_1 and FB_2. Three fractions were found to contain FB_1. Acetonitrile was removed from each fraction under vacuum on the rotary evaporator in preparation for additional purification. Two fractions containing FB_2

were also reduced in volume by rotary evaporation to approximately 20% acetonitrile. The two FB$_2$ fractions were stored at 2°C for purification at a later date.

Purification of FB$_1$

The FB$_1$ sample (fraction 12) was purified by isocratic preparative LC using a Waters Prep LC/System 500A equipped with two Waters PrepPak 47 mm x 300 mm Bondapak CN cartridges connected in series. Flow rate of the water-0.5% pyridine elution solvent was 50 mL/min for 100 min. Immediately before the sample was applied to the cartridges, the cartridges were conditioned with 2 L of methanol, 1 L of acetonitrile and 8 L of 0.5% pyridine-water. Collection of the fractions was started as soon as the sample was introduced to the cartridges. FB$_1$ was determined by HPLC. Fractions that contained FB$_1$ were freeze dried and the weights of recovered material determined. Purity was determined by HPLC/fluorescence detection and by mass spectrometry.

Centrifugal TLC

FB$_2$ was purified by centrifugal TLC with a Chromatotron (Harrison Scientific). A round TLC plate was coated 2mm thick with TLC grade silica gel (Merck 7749 with gypsum binder). The silicic acid coated plate was air dried, and the sorbent was activated by heating for 12 hr at 70°C.

The FB$_2$ sample was placed on a rotary evaporator and the acetonitrile was removed. Acetonitrile was added dropwise with shaking, back to the aqueous FB$_2$ solution until all the colloidal material was dissolved. Approximately 1.5 mL of the concentrated FB$_2$ solution (240 mg) was slowly and carefully applied under nitrogen to the spinning absorbent material on the plate. This resulted in a small circular colored band (5mm wide) of material being deposited on the inner circumference of the silicic acid sorbent.

Studies with various solvent systems and silicic acid TLC plates were used in developing the solvent system required for the separation of the individual constituents. Two Altex 110 pumps, Altex 420 gradient controller and Altex solvent mixer were used to pump solvent onto the silicic acid plate for elution. The solvent was made into 2 parts (A) chloroform-acetone (4:3) and (B) methanol-acetone (1:1) applied as a linear gradient at 3 mL/min for 180 min with the starting gradient at 90% (A) and 10% (B) and the ending gradient at 50% (A) and 50% (B). Six mL fractions were collected from the Chromatotron with a fraction collector. The fractions were dried under vacuum with a Savant Speedvac, then dissolved in 2 ml of acetonitrile-water (1:1). Every fifth fraction was analyzed by HPLC (fluorescent detection) to determine FB$_2$. Fractions containing FB$_2$ were reanalyzed by HPLC connected to a ELSD.

Mass Spectrometry

Fast atom bombardment (FAB/MS) spectra were obtained on a Finnigan TSQ-700 mass spectrometer equipped with an Ion Tech fast-atom gun. Procedures for preparation of the deuterium-labeled FB$_1$ and the method for determining purity are presented in the report of Plattner and Branham (1994).

RESULTS AND DISCUSSION

The use of rice as a substrate for cultures of *F. moniliforme* gave cleaner extracts with less small fine particles than cultures grown on corn, which reduced problems with the check

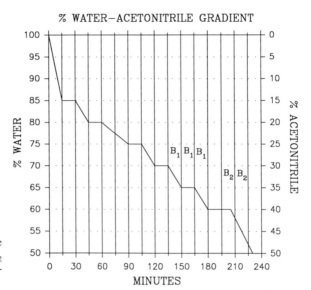

Figure 1. Percent water-acetonitrile gradient used for eluting FB_1 and FB_2 from preparative C_{18} reverse phase cartridge.

valves on the preparative pumps. The extract also filtered faster and appeared to contain less pigments than extracts from cultures grown on corn. The reduced concentration of contaminating red pigments made purification of FB_1 and FB_2 easier and resulted in higher purity toxins. The yield of FB_1 from rice was similar to yields obtained from corn (Cawood et al., 1991).

Sixteen fractions were collected within 230 min from the C_{18} cartridge with a gradient of 100% water to 50% water-acetonitrile (Figure 1).

The concentration of FB_1 applied to the C_{18} cartridge was approximately 2800 mg. As determined by HPLC, the FB_1 levels in fractions 10, 11, and 12 were 137 mg, 1275 mg, and 1277 mg, respectively. The amount of FB_2 in fractions 14 and 15 were 472 mg and 498 mg, respectively. The combined quantity of FB_1 from fractions 10, 11, and 12 was 2689 mg for a recovery of 96%. The total amount of FB_2 from fractions 14 and 15 was 970 mg for 97% recovery. Unidentified red colored constituents were found in all the five fractions containing fumonisin.

Fraction 12 containing FB_1 was purified further with the preparative LC system. The fraction was applied to the pre-conditioned cartridges and simultaneously with the sample introduction a timer was started. Total elution time was 100 mins. The fractions collected, time, volume, concentration of the fractions and the percent purity of the fractions collected are presented in Table 1.

A total of 1018 mg of FB_1 was recovered (79.7% recovery). Purity among the fractions ranged from 41 to 95.5%. Of the FB_1 recovered in the second LC procedure, 841 mg had a purity of 90% or better. This represents a recovery of 66% of the FB_1 in the starting extract. The FB_1 that was less than 90% pure were combined with other low purity FB_1 fractions for further purification by chromatography.

A method for confirming the presence of FB_1 is FAB/MS (Figure 2). This method produces little fragmentation and showed abundant protonated molecules at m/Z 722 (Plattner and Branham, 1994).

Purity was determined by mixing equal amounts of the isolated FB_1 and deuterated FB_1 and determining the spectra by FAB/MS (Plattner and Branham, 1994). Ratio of the

Table 1. Preparative isocratic HPLC purification of FB$_1$ (1277 mg) using two cyano cartridges and water-0.5% pyridine. The fractions collected, time, volume, concentration of the fraction, and percent purity of each fraction are presented

Fraction	Minutes	mL	mg FB$_1$	% Purity HPLC	% Purity FAB/MS
1	0-20	1000	a—	—	—
2	20-25	250	71	95	93
3	25-30	250	165	95	—
4	30-35	250	217	93	—
5	35-40	250	178	92	101
6	40-45	250	121	92	—
7	45-50	250	89	91	104
8	50-55	250	65	88	—
9	55-60	250	59	73	—
10	60-70	500	41	50	40
11	70-80	500	12	41	—
12	80-100	1000	—	—	—
TOTAL			1018		

a— Not determined

signal from FB$_1$ to that of labeled FB$_1$ gave a calculated purity of 93% (Plattner and Branham, 1994).

A TLC plate demonstrating the effectiveness of the chloroform, methanol and acetone (4:3:3) solvent system that was developed for the centrifugal TLC system is shown in Figure

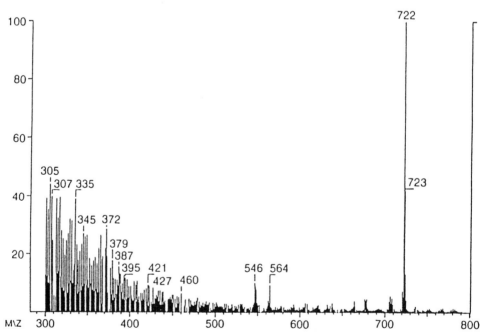

Figure 2. FAB/MS of fraction 5 of FB$_1$ separated by preparative LC using CN cartridges. The major signal m/z 722 is fumonisin B$_1$.

Figure 3. Silicic acid TLC plate showing the separation of FB$_2$ chloroform-methanol-acetone (4:3:3).

FB$_2$ Sample
STD

3. Good separation of the FB$_2$ on the silicic acid TLC plate was obtained using this solvent system. This solvent system reduces the chance that FB$_2$ will be degraded, as organic acids are not used.

The HPLC chromatogram of the partially purified extract of LC fraction 15 (Table 1) before it was applied to the centrifugal TLC plate is presented in Figure 4 A. The quantity of FB$_2$, which is the major peak in the chromatogram, applied to the centrifugal TLC plate was approximately 240 mg.

The eluting solvent in the first 15 fractions from the centrifugal TLC plate was red in color but all of the remaining fractions that were collected were colorless. TLC fraction 20 was the first fraction to contain FB$_2$. FB$_2$ levels increased rapidly in each fraction until TLC fraction 25 at which point the concentration remained about the same until TLC fraction 40. After TLC fraction 40 the level of FB$_2$ rapidly decreased.

HPLC analysis with fluorescent detection of fraction 35 shows FB$_2$ as the major peak present (Figure 4. B). The chromatogram of TLC fraction 35 shows big decreases in the levels of contaminates compared to the starting material (fraction 15). The chromatogram of TLC fraction 35 is representative of the fractions from 25 to 42 in that FB$_2$ is the major constituent. The HPLC/ELSD chromatogram of TLC fraction 35 is presented in Figure 5. FB$_2$ is the major component in the chromatogram. The determination of FB$_2$ as the major peak by two independent detection systems indicates that it is properly identified and the calculated quantity is significant (12.8 mg). The FB$_2$ peak accounted for 80% of the total area counts in the HPLC/fluorescence chromatogram and 90% of the area in the HPLC/ELSD chromatogram.

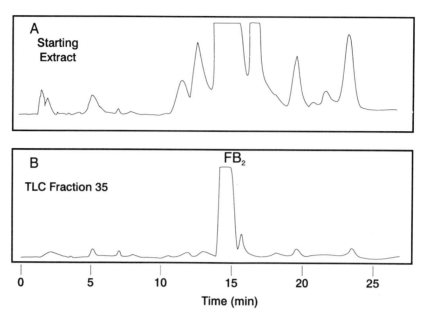

Figure 4. A. HPLC chromatogram with fluorescent detection of FB_2 present in the partially purified extract that was applied to the centrifugal TLC plate. B. HPLC chromatogram with fluorescence detection of FB_2 in TLC fraction 35 after preparative centrifugal TLC.

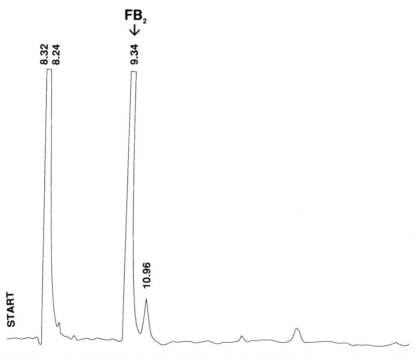

Figure 5. FB_2 determined in TLC fraction 35 by HPLC equipped with ELSD after preparative centrifugal TLC.

The quantity of FB_2 applied to the TLC plate was approximately 240 mg and the recovered FB_2 from all fractions combined was 218 mg, resulting in a recovery of 91%. The FB_2 obtained after centrifugal TLC was greatly improved in purity. Subsequent rechromatography of the FB_2 on the TLC system may allow for additional purification. Further refinement of the solvent system should allow for the complete separation of contaminates from FB_2. The preparative centrifugal TLC system is a rapid method that requires small quantities of solvent for the partial purification of FB_2.

SUMMARY

Rice and corn, inoculated *with F. moniliform,* produce similar amounts of FB_1 and FB_2. The advantage of RCM over corn is that rice produces cleaner extracts with reduced particulate matter and pigments. This makes the purification of FB_1 and FB_2 easier. Advantages of a water-acetonitrile gradient over a methanol-water gradient for separation of FB_1 and FB_2 on the C_{18} reverse phase are:

- 1) complete separation of FB_1 and FB_2,
- 2) reduced solvent requirements,
- 3) rapid chromatography and
- 4) less chance for degradation of FB_1 and FB_2.

The use of CN cartridges eluted isocratically with water-0.5% pyridine enables high purity (93% or better) FB_1 to be obtained. Preliminary investigation of preparative centrifugal TLC for purification of FB_2 shows good promise and may be the method of choice to obtain FB_2 in good quantity and high purity. The centrifugal TLC procedure is fast and requires small amounts of solvents.

ACKNOWLEDGMENTS

We wish to thank L. K. Tate and R. M. Bennett for technical assistance in this research.

REFERENCES

Bezuidenhout, S. C.; Gelderblom, W. C. A.; Grost-Allman, C. P.; Horak, R. M.; Marasas, W. F. O.; Spiteller, G.; Vleggaar, R. Structure elucidation of the fumonisins, mycotoxins from *Fusarium moniliforme. J. Chem. Soc., Chem. Commun.* **1988**, 743-745.

Bottini, A. T., Gilchrist, D. G. Phytotoxins. I. A 1-aminodimethylheptadecapentol from *Alternaria alternata* f.sp. *lycopersicion. Tetrahedron Lett.* **1981**, 22, 2719-2722.

Bottini, A. T.; Bowen, J. R.; Gilchrist, D. G. Phytotoxins. II. Characterization of a phyto-toxic fraction from *Alternaria alternata* f. sp. *lycopersici. Tetrahedron Letters,* **1981**, 22, 2723-2726.

Brydon, W. L.; Love, R. J.; Burgress, L. W. Feeding grain contaminated with *Fusarium graminearum* and *Fusarium moniliforme* to pigs and chickens. *Aust. Vet J.* **1987**, 64, 225-226.

Cawood, M. E.; Gelderblom, W. C. A.; Vleggaar, R.; Behrend, Y.; Thiel, P. G.; Marasas, W. F. O. Isolation of the fumonisins mycotoxins: A quantitative approach. *J. Agric. Food Chem.* **1991**, 39, 1958-1962.

Gelderblom, W. C. A.; Jaskiewicz, K.; Marasas, W. F. O.; Thiel, P. G.; Vleggaar, R.; Kriek, N. P. J. Fumonisins-Novel mycotoxins with cancer-promoting activity produced by *Fusarium moniliforme. Applied Environ. Micro.* **1988**, 54, 1806-1811.

Harrison, L. R.; Colvin, B. M.; Greene, J. T.; Newman, L. E.; Cole, J. R. Pulmonary edema and hydrothorax in swine produced by fumonisin B₁, a toxic metabolite of *Fusarium moniliforme. J. Vet. Diag. Invest.* **1990**, 2, 217-221.

Javed, T.; Bennett, J. L.; Richard, J. L.; Dombrink-Kurtzman, M.A.; Cote, L. M.; Buck, W. B. Mortality in broiler chicks on feed amended with *Fusarium proliferatum* culture material or with purified fumonisin B₁ and moniliformin. *Mycopathologia,* **1993**, 123, 171-184.

Kellerman, T. S.; Marasas, W. F. O.; Thiel, P. G.; Gelderblom, W. C. A.; Cawood, M.; Coetzer, J. A. W. Leucoencephalomalacia in two horses induced by oral dosing of fumonisin B₁. *Onderstrpoort J. Vet. Res.* **1990**, 57, 269-275.

Kriek, N. P. J.; Kellerman, T. S.; Marasas, W. F. O. A comparative study of the toxicity of *Fusarium verticillioides* (=*F. moniliforme*) to horses, primates, pigs, sheep, and rats. *Onderstepoort J. Vet. Res.* **1981**, 48, 129-131.

Marasas, W. F. O.; Jaskiewicz K.; Venter, F. S.; Van Schalkwyk, D. J. *Fusarium moniliforme* contamination of maze in oesophageal cancer areas in Transkei. *S. Afr. Med. J.* **1988a**, 74, 110-114.

Marasas, W. F. O.; Kellerman, T. S.; Gelderblom, W. C. A.; Coetzer, J. A. W.; Thiel, P. G.; Vanderlugt, J. J. Leukoencephalomalacia in horses induced by fumonisin B₁ isolated from *Fusarium moniliforme.* *Onderstepoort J. Vet. Res.* **1988b**, 55, 197-203.

Mirocha, C. J., Gilchrist, D. G., Shier, W. T., Abbas, H. K., Wen, Y., Vesponder, R. F. AAL toxin, fumonisins (biology and chemistry) and host-specificity concepts. *Mycopathologia,* **1992**, 117, 47-56.

Plattner, R. D.; Norred, W. P.; Bacon, C. W. Voss, K. A.; Peterson, R.; Shackelford, D. D. A method of detection of fumonisin in corn samples associated with field cases of equine leucoencephalomalacia. *Mycologia,* **1990**, 82, 698-702.

Plattner, R. D.; Branham B. E. Labeled fumonisins: Production and use of fumonisin B₁ containing stable isotopes. *AOAC Intern.* **1994**, 77, 525-532.

Rheeder, J. P.; Marasas, W. F. O.; Thiel, P. G.; Sydenham, E. W.; Shephard, G. S.; Van Schalkwyk, D. J. *Fusarium moniliforme* and fumonisin in corn in relation to human esophageal cancer in Transkei. *Phytopathology,* **1992**, 82, 353-357.

Riley, R. T.; Norred, W. P.; Bacon, C. W. Fungal toxins in foods: Recent concerns. *Annu. Rev. Nutr.* **1993**, 13, 167-189.

Ross, P. F.; Rice, L. G.; Reagor, J. C.; Osweiler, G. D.; Wilson, T. M.; Owen, D. L.; Plattner, R. D.; Harlin, K. A.; Richard, J. L.; Colvin, B. M.; Banton, M. I. Fumonisin B₁ concentrations in feeds from 45 confirmed equine leucoencephalomalacia cases.. *Vet. Diag. Invest.* **1991**, 35, 1551-1555.

Ross, P. F.; Nelson, P. E.; Owen, D. L.; Rice, L. G.; Nelson, H. A.; Wilson, T. M. Fumonisin B₂ in cultured *Fusarium proliferatum* M-6104, causes equine leucoencephalomalacia. *J. Vet. Diagn. Invest.* **1994**, 6, 263-265.

Shier, W. T.; Abbas, H. K.; Mirocha, C. J. Toxicity of the mycotoxins fumonisin B₁, and B₂ and *Alternaria alternata* f. sp. *lycopersici* toxin (AAL) in cultured mammalian cells. *Mycopathologia,* **1991**, 116, 97-104.

Voss, K. A.; Norred, W. P.; Plattner, R. D.; Bacon, C. W.; Porter, J. K. Hepatotoxicity in rats of aqueous extracts of *Fusarium moniliforme* strain MRC 826 corn cultures. *Toxicologist,* **1989**, 9, 225.

Voss, K. A.; Chamberland, W. J.; Bacon, C. W.; Norred, W. P. A preliminary investigation of renal and hepatic toxicity in rats fed purified fumonisin B₁. *Nat. Toxins,* **1993**, 1, 222-228.

11

IMMUNOCHEMICAL METHODS FOR FUMONISINS

Fun S. Chu

Department of Food Microbiology and Toxicology
University of Wisconsin
Madison, WI 53706

ABSTRACT

Both monoclonal and polyclonal antibodies have been raised against fumonisins (Fm) and the hydrolyzed toxin in several laboratories. They have been used in several immunoassays and can detect 5 to 5000 ng of FmB_1/mL in a "clean" standard solution. However, sample matrices still give severe interferences. Thus, without a cleanup treatment, most immunoassays can only be used as a screening tool for Fm in corn/feed in the 0.5 to 10 ppm range. The antibodies have also been used to generate anti-idiotype antibodies, to use as immunohistological reagents and to make affinity columns that are being used as a cleanup tool. In the present symposium, approaches used to generate antibodies, antibody specificity, sensitivity of various immunoassays and problems encountered in the immunoassays will be reviewed. Approaches used to overcome such problems and methods to improve the sensitivity and specificity of immunoassays of Fm are discussed.

INTRODUCTION

Fumonisins (Fm) are a group of toxic metabolites produced primarily by *Fusarium moniliforme,* one of the most common fungi colonizing corn throughout the world. Fumonisin B_1 (FmB_1, diester of propane-1, 2, 3-tricarboxylic acid of 2 amino -12, 16-dimethyl-3, 5, 10, 14, 15-pentahydroxyicosane), the most common naturally occurring Fm in this group of mycotoxins, has been found to be a potent cancer promoter as well as an etiological toxic agent responsible for equine leukoencephalomalacia and for porcine pulmonary edema (Gelderblom et al., 1988; Marasas, 1986; Marasas et al., 1984; Norred, 1993; Riley and Richard, 1992; Thiel et al., 1992). Preliminary reports on the carcinogenicity of FmB_1 in rats and recent reports on the worldwide occurrence of high concentrations of this group of mycotoxins in foods and feeds, generally at the ppm level, have prompted intensive research on Fm in the last few years (Chu and Li, 1994; Gelderblom et al., 1991, 1994; Murphy et al., 1993; Nelson et al., 1993; Norred, 1993; Norred and Voss, 1994; Riley et al., 1994; Ross et al., 1991; Thiel et al., 1992). The impact of Fm on human and animal health can be seen from various discussions in the present symposium and different reports in this book.

In view of their potential hazard to human and animal health (Norred, 1993; Riley et al., 1993), extensive research has been conducted to develop more efficient methods for Fm determination. Several HPLC (Bennett and Richard, 1994; Holcomb et al., 1993; Murphy et al., 1993; Scott and Lawrence, 1992; Shephard et al., 1990; Stack and Eppley, 1992; Sydenham et al., 1992; Thiel et al., 1993; Ware et al., 1993) and TLC (Rottinghaus et al., 1992) methods with good sensitivity and accuracy have been developed. However, these methods generally require extensive sample cleanup and precolumn derivatization. Because Fm contains an amino group and two carboxylic acid groups, several fluorogenic reagents such as o-phthaldialdehyde, fluorescamine, naphthalene dicarboxaldehyde-potassium cyanide, 4-fluoro-7-nitrobenzofurazan and (9-fluorenylmethyl) chloroformate (FMOC) were used in the derivatization. The sensitivity of these chemical methods is generally in the range of 50 ppb to 1 ppm. With the increased use of immunochemical methods for the detection of mycotoxins and other low molecular weight contaminants and naturally occurring toxins in foods and feeds (Chu, 1992, 1994a, 1994b, 1995; Morgan and Lee, 1990; Pestka, 1987; 1994), specific antibodies against FmB_1 have been made available; more versatile immunochemical methods for Fms have been established (Azcona-Olivera et al., 1992a,b; Fukuda et al., 1994; Shelby et al., 1994; Usleber et al., 1994). In the present report, the current status of the immunochemical methods for the detection of fumonisins is reviewed.

PREPARATION OF FUMONISIN-PROTEIN/ENZYME CONJUGATES

Like most other mycotoxins, fumonisins are low molecular weight secondary fungal metabolites that are not immunogenic. Thus, FmB_1 must first be conjugated to a protein/polypeptide carrier before subsequent use in immunization (Chu, 1994a, 1994b; 1995). Since fumonisin contains two reactive carboxylic and hydroxyl groups as well as a primary amine group, the most logical position for conjugation of this hapten to the protein carrier is through the amine group at the C-2 position of the backbone chain. Thus, one of the most common methods used in the preparation of immunogens for antibody production involved cross-linking of the amine group of FmB_1 to a carrier protein that was activated with glutaraldehyde. The linkage was then stabilized with sodium borohydride (Avrameas and Ternyck, 1969). Proteins, including bovine serum albumin (BSA), cholera toxin (CT), and keyhole limpet hemocyanin (KLH) and ovalbumin (OVA) have been used (Azcona-Olivera et al., 1992a,b; Fukuda et al., 1994; Shelby and Kelley, 1992; Usleber et al., 1994). This approach has also been used for the preparation of an immunogen that was used for the production of antibodies against the hydrolyzed FmB_1 ($HFmB_1$) by cross-linking $HFmB_1$ to CT (Maragos and Miklasz, 1995; Maragos et al., 1995). However, the water soluble carbodiimide (WSC) method, an approach that has been commonly used to couple many haptens/mycotoxins to proteins (Chu, 1986, 1994a, 1994b, 1995), has also been tested for conjugation of FmB_1 to protein carriers (Yu and Chu, 1995). Although the exact position in the FmB_1 molecule involved in the coupling reaction is not known, conjugates prepared by this approach have been found to be good immunogens for generating both monoclonal (mAb) and polyclonal (pAb) antibodies against FmB_1 (Yu and Chu, 1995).

To alleviate the problem that the antibodies may have non-specific cross-reaction with the residues in the approximate linking-bridge region, methods or carrier proteins different from those used in the preparation of conjugates for immunization are often used (Chu, 1995). Thus, several other approaches were used for conjugation of FmB_1 to protein/enzymes in various immunoassays. For example, Fukuda et al. (1994) conjugated FmB_1 to BSA via the formation of m-maleimidobenzoyl-N-hydroxysuccinimide ester (MBS). The FmB_1-BSA was then used as the immunogen coated to the microtiter plate in an indirect ELISA where the mAb generated from the FmB_1-OVA via the glutaraldehyde method was used. Conjugation of FmB_1 to a marker enzyme such as horseradish peroxidase (HRP) was generally done by the periodate method of Nakane and Kawaoi (1974).

PRODUCTION AND CHARACTERIZATION OF ANTIBODIES AGAINST FUMONISINS

Both polyclonal (pAb) and monoclonal (mAb) antibody techniques have been used for the production of antibodies against fumonisins (Azcona-Olivera et al., 1992a,b; Fukuda et al., 1994; Maragos and Miklasz, 1995; Maragos et al., 1995; Shelby and Kelley, 1992; Usleber et al., 1994; Yu and Chu, 1995). Whereas rabbits were commonly used in most cases for the production of antibodies against mycotoxins including FmB_1 (Chu, 1995; Usleber et al., 1994; Yu and Chu, 1995), pAb produced in the ascites fluid of BALB/c mice by the method of Kurpisz et al. (1988) were also useful for immunoassay of Fms (Azcona-Olivera et al., 1992a). Because of the adjuvant nature of CT, Azcona-Olivera et al. (1992a, 1992b) coupled FmB_1 to this protein, which was then directly injected to the BALB/c mice for the antibody production without use of adjuvant. Different FmB_1-protein conjugates have been used for the production of antibodies against Fms; properties of these antibodies are summarized in Table 1.

Several conclusions could be drawn from these data: (1) most FmB_1 antibodies cross-reacted with Fms B_2 and B_3 to some degree, but not with the hydrolyzed fumonisins, tricarballylic acid, sphingosine and sphinganine; (2) the antibodies cross-reacted weakly with AAL toxin (Yu and Chu, 1995); (3) the affinity of fumonisin antibodies with FmB_1 appears to be much lower than the antibodies for most other mycotoxins such as aflatoxins, trichothecenes and ochratoxins (Chu, 1994) with their respective homologous mycotoxins; thus, the sensitivity of most immunoassays for FmB_1 detection was relatively low; and (4) the affinity of the polyclonal antibodies for FmB_1 obtained by Usleber et al. (1994) appears to be much higher than those from other laboratories; it is not known whether the pAb obtained from rabbit immunized with FmB_1-KLH indeed had high apparent affinity to FmB_1 or whether it might be attributed to the use of a more effective toxin-enzyme marker in the assay (Usleber et al., 1994).

Monoclonal antibody against the hydrolyzed FmB_1 has recently been obtained by Maragos et al. (1995) when the hydrolyzed FmB_1 conjugated to CT was used as the immunogen. The antibody belongs to IgG1, κ light chain and has good sensitivity for detection of the hydrolyzed Fms (ID_{50} of 36 ng $HFmB_1$/mL). The relative cross-reactivity of this mAb for $HFmB_1$, $HFmB_2$, $HFmB_3$, and $HFmB_4$ was found to be 100, 11, 21, and 2.1, respectively. The antibody did not cross-react with the unhydrolyzed Fms, sphingosine, sphinganine and the tricarballylic acid (TCA). Most recently, antibodies against AAL-toxin were obtained by Szurdoki et al. (1995). The AAL-toxin, which has a chemical structure similar to FmB_1, is a host specific toxin produced by *Alternaria alternata* f sp. *lycopersici* (Gilchrist and Grogan, 1976). These antibodies did not cross-react with fumonisin derivatives.

Data from cross-reactivity of various fumonisin antibodies with different fumonisins, suggest that the C-11 to C-20 regions of Fm and the hydrolyzed Fm molecule play an important role in generating the specificity of the antibodies when a toxin-conjugate prepared through the C-2 amine group was used as the immunogen. The observation that mAb

Table 1. Cross-reactivity (ID_{50}) of various antibodies against fumonisins [a,b]

Immunogen	ID_{50} values (ng of FmB_1/mL)	Relative cross-reactivity (%) FmB_2	FmB_3	References[c]
Polyclonal antibodies				
FmB1-CT (m, cl)[b]	260	87	40	1
FmB1-KLH (r, cl)	0.62 (d-c)	24	55	2
FmB1-KLH (r,wsc)	0.45 (d-c)	62	1.8	3
Monoclonal antibodies				
FmB1-OVA (2 clones, cl)	255-239	520-555	nd	4
FmB1-OVA (2 clones, cl)	65-72	130-150	nd	4
FmB1-KLH (cl)	90	100	nd	4
FmB1-CT (7 clones, cl)	630 (d-c)	35	27.4	5
FmB1-CT (P2A5-3-F3, cl)	140	93	nd	6
FmB1-BSA (wsc)	99	49	nd	3

[a]Except otherwise noted, most were determined by a competitive indirect ELISA. The cross-reactivity is expressed as ID_{50}, i.e. the concentrations of free FmB_1 causing 50% inhibition of binding of antibody with the solid phase FmB_1-protein conjugate. The relative cross-reactivities of various Fms with the antibody was calculated by dividing the ID_{50} of FmB_1 by the respective Fm and multiplying by 100 (analyzed under the same conditions).

[b]Abbreviation used: m, mouse; r, rabbit; cl, cross-linking method; WSC, water soluble carbodiimide method; d-c, direct competitive ELISA.

[c]References: (1) Azcona-Olivera et al., 1992a; (2) Usleber et al., 1994; (3) Yu and Chu (1995); (4) Fukuda et al., 1994; (5)Azcona-Olivera et al., 1992b; (6) Chu et al., 1995.

obtained from various hybridoma cell lines shows varied degrees of cross-reactivity with different fumonisins and has low cross-reactivity with AAL-toxin suggests that the flexibility of the backbone structure as well as the size and position of the side chain interaction modulate antibody specificity. The similarity in cross-reactivity of the antibodies generated by a conjugate prepared by the WSC method with those prepared by the cross-linking method suggests that it is likely that the linking position of the conjugate prepared by the WSC method was also via the amine group in the FmB_1 molecule.

ANTI-IDIOTYPE AND ANTI-ANTI-IDIOTYPE ANTIBODIES AGAINST FUMONISINS

The development of immunochemical methods for mycotoxin detection has led to a great demand for specific antibodies and related immunochemical reagents for the assay. An alternate approach to preparing immunochemical reagents is through generating anti-idio-type (anti-ID) antibodies (Nisonoff, 1991). Anti-idiotype antibodies (Ab2) for large mole-cules have been well-developed and have been applied to clinical diagnosis and immunotherapy (Kennedy et al., 1987, Nisonoff, 1991). Recent success in generating Ab2 against a number of small molecular weight haptens, including mycotoxins such as T-2 toxin (Chanh et al., 1989, 1990, 1992) and aflatoxin (Hsu and Chu, 1994), insecticides, herbicides (Spinks et al., 1993), hormones (Khole and Hegde, 1992;), and phycotoxins (Shestowsky et al., 1992, 1993), prompted our interest in generating anti-ID antibodies for FmB_1. A mAb generated from the hybridoma cell line P2A5-3-F3 that was obtained from mice immunized with FmB_1–CT conjugate was chosen as the idiotype antibody (mAb1) for generating Ab2.

Anti-idiotype antibodies for FmB_1 were demonstrated in rabbits after immunization with the affinity purified P2A5-3-F3-mAb (Chu et al., 1995). Ab2 bound specifically to FmB_1–mAb1. Indirect competitive ELISA revealed that the binding of FmB_1–mAb1 to FmB_1–OVA was inhibited by Ab2. Ab2 could also be used as FmB_1–OVA surrogate in the ELISA for FmB_1. In the FmB_1–OVA-based ELISA, the concentrations causing 50% inhibition (ID_{50}) of binding of mAb1 to FmB1-OVA by FmB_1 and FmB_2 were found to be 0.14 and 0.15 µg/mL, respectively. In the Ab2-based ELISA, the ID_{50} values of binding of mAb1 to Ab2 by FmB_1 and FmB_2 were found to be 0.46 and 0.61 µg/mL, respectively. Using the affinity-purified Ab2 Fab fragment as an immunogen, polyclonal anti-Ab2 antibodies (pAb3) were generated in BALB/c mice (Kurpisz et al., 1988). The pAb3 was found to have characteristics similar to those of original mAb1. Most recently, a hybridoma cell line that generates mAb3 was obtained in our laboratory (Yu and Chu, 1995). The ID_{50} values of binding of mAb1 (original mAb), pAb3 and mAb3 to FmB_1–OVA by FmB_1 and FmB_2 are compared in Table 2. It is apparent that the anti-anti-idiotype antibodies have similar characteristics as the original mAb.

The availability of anti-idiotype and anti-anti-idiotype antibodies for fumonisin and other mycotoxins have provided a new generation of immunochemical reagents, which could be used for both therapeutic and analytical purposes. Ab2, a surrogate of the toxin, could be used as the immunogen in generating antibodies for the toxin (thus, a vaccine) and could be used in the immunoassay. Ab3 could be used as an antibody for toxin detection and could also serve as a surrogate of the receptor for the toxin.

APPLICATION OF IMMUNOASSAY FOR FUMONISIN DETECTION

A. Competitive Enzyme-Linked Immunosorbent Assays (ELISA)

With the availability of antibodies against FmB_1, both direct (d-c) and indirect competitive (id-c) ELISA have been established for the analysis of FmB_1 in corn and corn-based foods and feed. In d-c-ELISA, FmB_1-conjugated to horseradish peroxidase was most commonly used as the marker enzyme. In contrast, FmB1 conjugated to a protein-carrier different from the immunogen or a conjugate prepared by a method different from the one used in the preparation of conjugate for immunization, was coated to the wells of the microtiter plate in id-c-ELISA. Thus, FmB_1-OVA was used most commonly as the coating

Table 2. Comparison of the concentrations of free FmB_1 causing 50% inhibition of binding (ID_{50}, ng/mL) of anti-FmB_1 (pAb1 or mAb1) and anti-anti-id (pAb3, mAb3)

Antibody	ID_{50} Concentrations (ng/mL)	
Type	FmB_1	FmB_2
FmB_1-mAb1[a]	140	150
FmB_1-pAb3[b]	190	260
FmB_1-mAb3[c]	75	95

[a]FmB_1-mAb1 was mAb against FmB_1-CTconjugate (provided by Dr. C. M. Maragos of USDA, ref. Chu et al., 1995).

[b]FmB_1-pAb3 (anti-anti-idiotype antibody) was generated from the ascites fluid in BALB/c mice after immunizing with affinity purified anti-idiotype antibodies (FmB_1-mAb2) from rabbits that had been immunized with FmB_1-mAb1.

[c] FmB_1-mAb3 (monoclonal anti-anti-idiotype antibody) was generated from BALB/c mice after immunizing with affinity purified anti-idiotype antibodies from rabbits (FmB_1-pAb2) that had been immunized with FmB_1-mAb (Yu and Chu, 1995).

Table 3. Sensitivity of ELISA for Fumonisin B_1 and hydrolyzed Fumonisin B_1[a]

Extraction solvent[b]	Foods/Feeds	Standard range (ng/mL)	Detection limits (ng/g or ng/mL)	References[c]
Fumonisin B_1				
A, mAb-d-c[b]	Corn	5-5000	200	1
B, pAb-d-c	Corn	0.1-5.0	10	2
C, mAb-id-c	Corn	200-2000	1000	3
B, mAb-id-c	Corn (clp)	20-1000	200	4
	Corn (clp)	50-5000	200	4
	Milk	100-5000	100	5
Hydrolyzed FmB_1				
D, mAb-d-c	Corn (clp)	2-300	5-10	6

[a]Abbreviation used: A-E, represents solvent system used; mAb, monoclonal antibody; pAb, polyclonal antibodies; d-c, direct competitive ELISA; id-c, indirect competitive ELISA; clp, cleanup.

[b]Extraction solvent used: A, methanol/water (3/1); B, acetonitrile/water (1/1); C, phosphate buffer + 5% Tween; D, methanol/phosphate buffer (8/2).

[c]References: (1) Azcona-Olivera et al., 1992a,1992b; (2) Usleber et al., 1994; (3) Shelby et al., 1994; (4) Chu et al., 1995; (5) Maragos and Richard, 1994; (6) Maragos and Miklasz, 1995; Maragos et al., 1995.

antigen. Polyclonal anti-idiotype antibodies (pAb2) have also been found to be useful immunogens for coating to the solid-phase in id-c-ELISA. A good correlation (r = 0.86, p < 0.0001) between the data obtained from the FmB_1–OVA-based and the Ab2-based ELISA for the analysis of FmB_1 in corn was found (Chu et al., 1995).

The sensitivity of ELISA for the determination of Fms in corn and corn-based foods and feed is summarized in Table 3. In general, reliable data could be obtained from ELISA at FmB_1 levels in food above 1 ppm. However, using a pAb-based d-c-ELISA, Usleber et al. (1994) were able to detect as low as 18 and 21 ppb of FmB_1 in a popcorn and a corn grit sample, respectively. Although cleanup was generally not necessary for the immunoassay of Fm in corn and corn-based foods at high levels of contamination, both accuracy and sensitivity of ELISA of fumonisin were improved when the sample extracts were subjected to a solid-phase column cleanup such as a SAX column (Usleber et al., 1994) or a C-18 Sep-Pak reversed-phase column (Chu, et al., 1995; Maragos and Miklasz, 1995; Maragos et al., 1995) before ELISA. The effect of organic solvent on a mAb-based d-c-ELISA of $HFmB_1$ in corn was studied by Maragos and Miklasz (1995). These investigators found that the amount of solvent in the assay system greatly affected ELISA sensitivity; the effect was more predominate in the acetonitrile-based system than in the methanol-based system. Thus, more dilution is necessary when the acetonitrile-based solvent is used as the extraction solvent.

Attempts to correlate the ELISA data with other chemical methods have been made in various laboratories, but the results were inconsistent. Whereas some reports showed good correlations, other investigators have found a wide discrepancy. The Fm levels obtained by ELISA were generally higher than those obtained by chemical methods (Dreher and Usleber, 1995; Usleber et al., 1994; Shelby et al., 1994; Pestka et al., 1994; Trucksess, 1995). For example, Shelby et al. (1994) found that ELISA detected more samples at levels between 1 to 10 ppm than the TLC methods among 322 samples tested. Treatment of samples with a SAX column cleanup gave lower ELISA values than without treatment, but those data were still higher than the HPLC results (Usleber et al., 1994). A comparative assessment of ELISA of FmB_1 in grain-based foods with GC-MS and HPLC was made by Pestka et al. (1994). Among 71 samples tested, 11 samples, all corn-based foods, were positive for Fm by ELISA, which had a detection limit of 200 ng of FmB_1/g. The 11 positive samples and randomly selected 9 negative samples were further analyzed by chemical methods. Whereas excellent correlation was found between the data obtained from HPLC and GC-MS ($r = 0.946$, $p < 0.01$) by these investigators, the correlations between the HPLC and ELISA ($r = 0.512$, $p < 0.05$) and GC-MS with ELISA ($r = 0.478$, $p < 0.05$) were poor. The ELISA data were always considerably higher than the chemical methods. In one sample, FmB_1 level obtained from the ELISA was 30 times higher than levels obtained by chemical methods (Pestka et al., 1994). Similar results were obtained when the fumonisin-producing *Fusarium* cultures were tested (Tejada-Simon et al., 1995). Although these inconsistencies were generally attributed to the possible cross-reaction of the FmB_1 antibodies with other Fms, with Fms that might be conjugated to other compounds in the biological system, or with other structurally related compounds, the low analytical recovery in the chemical methods was also considered as one of the possible factors.

B. Immunoaffinity Chromatography

With the availability of antibodies against various mycotoxins, immunoaffinity (IAF) columns were made by conjugating the antibodies to a solid-phase matrix. These columns are then used either in a screening test or as a cleanup column for subsequent chemical analysis (Chu, 1994a, 1994b, 1995). In general, sample extracts diluted in phosphate buffer are applied to the column. After washing to remove the unbound materials, a specific mycotoxin is then eluted from the column with the appropriate solvent system and the eluant subjected to other chemical analyses. For mycotoxins with native fluorescence such as aflatoxin, ochratoxin and zearalenone, the toxin level in the eluate could be directly determined fluorometrically or be determined after a derivatization step, which enhances the fluorescence. For fumonisin screening, it is necessary to introduce a fluorophore to the materials eluted from the IAF column. A commercial kit (FumoniTest, Vicam) for screening fumonisin in corn at a level of 1 ppm or above has been produced utilizing this principle.

The IAF column technique has gained wide application as a cleanup tool for a number of mycotoxins in recent years (Chu, 1994, 1995). Immunoaffinity columns for a number of mycotoxins are also commercially available, including fumonisin (Jordan, et al., 1994; Trucksess, 1995; Ware et al., 1994). Ware et al. (1994) evaluated a commercially available IAF column for fumonisin (FumoniTest, Vicam) that had a total capacity of 1.2 μg fumonisin and an equal affinity for FmB_1 and FmB_2. A sample size of 0.5 g of corn containing a level of less than 2 ppm fumonisin, preferably 1 ppm, was recommended. Using this column as a cleanup tool for the extracts obtained from corn samples containing fumonisins and followed by naphthalene-2,3-dicarboxaldehyde derivatization and subsequent HPLC analysis, these investigators could detect levels of fumonisins as low as 10 and 4 ppb of FmB_1 and FmB_2, respectively in the corn. Good analytical recovery (85-87%) was achieved for FmB_1 and FmB_2 added to the corn at level of 1 ppm. Other studies also found that the fumonisin IAF column was an effective cleanup tool for subsequent HPLC analysis (Hansen et al., 1994; Trucksess, 1995). Jordan et al. (1994) have also incorporated the IAF column into an automated system for quantifying fumonisin.

C. Other Immunohistochemical Methods

Instead of coating antibodies to the ELISA plate, Abouzied and Pestka (1994) immobilized different monoclonal antibodies against aflatoxin B_1, fumonisin B_1 and zearale-neone, as multiple lines on a nitrocellulose strip; respective mycotoxin-peroxidase conjugates were used as the testing markers. In the assay, free mycotoxin and the mycotoxin-peroxidase compete for the binding site of the antibody immobilized on the solid-phase nitrocellulose. After reacting with substrate, the color intensity is measured; a video- and computer-assisted system estimates the toxin levels. These investigators were able to screen samples for all three mycotoxins simultaneously with detection limits of 0.5, 500, and 3 ng/mL for aflatoxin B_1, FmB_1 and zearalenone, respectively. Monoclonal antibody against FmB_1 has also been used as an immunohistological reagent to localize the fumonisins in the mycelium of *Fusarium* cultures. However, fumonisin levels estimated from the intensity of immunochemical stain were considerably higher than those obtained from by other chemical studies (Tejada-Simon et al., 1995).

CONCLUDING REMARKS

Both monoclonal and polyclonal antibodies against fumonisin B_1 have been made available for various immunochemical studies on fumonisins in the last few years. Most antibodies have good cross-reactivity to fumonisins B_2 and B_3, but not with other derivatives. Various immunoassays have also been established for the analysis of fumonisin in grain, especially corn and corn-based foods. With the exception of the antibodies obtained from one laboratory, most antibodies appeared to have low affinity to fumonisin; thus, the sensitivity of most immunoassays for fumonisin was low with a detection limit in the range of 0.5 to 1 ppm. Data obtained from comparative studies on the efficacy of immunoassays with other chemical methods were inconsistent. In most cases, higher estimates were obtained from immunoassays. Thus, immunoassay of fumonisin has only been recommended as a screening method. The fumonisin antibodies have also been conjugated to a solid-phase gel and then successfully used as an affinity tool for chemical analysis. ELISA kits for screening of fumonisin and immunoaffinity columns for cleanup purposes are also commercially available (Chu, 1995; Pestka, 1994).

While the exact reasons for the higher results from ELISA methods relative to other methods are not known, it has been suggested that the antibodies may cross-react with other fumonisin derivatives, i.e. other Fm and conjugates of Fm, and sample matrices; analytical loss in the chemical methods could also be one of the reasons. The low antibody affinity is considered by this reviewer as one of the major factors contributing to such problems. Thus, future efforts should be directed at generating high affinity antibodies against this group of mycotoxins. Polyclonal antibodies with higher affinity to both FmB_1 and FmB_2 have recently been obtained in our laboratory (Yu and Chu, 1995) and others (Abouzied et al., 1995; Dreher and Usleber, 1995). At present, improvement of the ELISA sensitivity could be achieved by incorporating a cleanup and concentrating step into the ELISA protocol and by using a more efficient marker enzyme preparation in the assays. Better labelling techniques, including fluorescent-labelled antibodies/fumonisins, should also be tested to evaluate the possibility of achieving a more sensitive and rapid method for the toxin detection (Chu, 1995). New approaches to the production of antibodies, including using anti-idiotype and anti-idiotype antibodies (Chu et al., 1995) and structural modulation (Elissalde et al., 1995) have been tested or proposed; there is optimism that better and higher affinity antibodies for fumonisin could be obtained in the near future. Several laboratories, including our lab, have initiated work in cloning these antibodies. It will not be too long before we understand how these antibodies are reacted with these mycotoxins; thus, a new generation of antibodies could be made available through point mutation.

ACKNOWLEDGEMENTS

This work was supported by Grant NC-129 from the College of Agricultural and Life Sciences, University of Wisconsin-Madison and funds from food industries. We thank C. M.

Maragos, R. M. Eppley and M. W. Trucksess for providing purified fumonisins B_1 and B_2, C. J. Mirocha for providing AAL-toxin for our work, and Carole Ayres and Barbara Cochrane for their help in preparing the manuscript.

REFERENCES

Abouzied, M. M.; Askegard, S.; Bird, C.; Miller, B. Fumonisins Veratox® : A rapid quantitative ELISA for determination of fumonisin in food and feed. In *Fumonisins in Food*; L. Jackson, J. DeVries, L. Bullerman, eds.; Plenum Publishing Co.: New York, 1995 (in press) or 209th meeting of the American Chemical Society (Abstr. of fumonisin symposium), ACS, Washington, D. C. 1995.

Abouzied, M. M.; Pestka, J. J. Simultaneous screening of fumonisin B_1, aflatoxin B_1, and zearalenone by line immunoblot: a computer-assisted multianalyte assay system. *J. AOAC Intern.* **1994**, *77*, 495-501.

Avrameas, S.; Ternynck, T. The cross-linking of proteins with glutaraldehyde and its use for the preparation of immunoadsorbents. *Immunochemistry*, **1969**, *56*, 1729-1733.

Azcona-Olivera, J. I.; Abouzied, M. M.; Plattner, R. D.; Norred, W. P.; Pestka, J. J. Generation of antibodies reactive with fumonisins B_1, B_2, and B_3 by using cholera toxin as the carrier-adjuvant. *Appl. Environ. Microbiol.* **1992a**, *58*, 169-173.

Azcona-Olivera, J. I.; Abouzied, M. M.; Plattner, R. D.; Pestka, J. J. Production of monoclonal antibodies to the mycotoxins fumonisins B_1, B_2, and B_3. *J. Agric. Food Chem.* **1992b**, *40*, 531-534.

Bennett, G. A.; Richard, J. L. Liquid chromatographic method of analysis for the naphthalene dicarboxaldehyde derivative of fumonisins. *J. AOAC Intern.* **1994**, *77*, 501-506.

Chanh, T. C.; Huot, R. I.; Schick, M. R.; Hewetson, J. F. Anti-idiotypic antibodies against a monoclonal antibody specific for the trichothecene mycotoxin T-2. *Toxicol. Appl. Pharmacol.* **1989**, *100*, 201-207.

Chanh, T. C.; Rappocciolo, G.; Hewetson, J. F. Monoclonal anti-idiotype induces protection against the cytotoxicity of the trichothecene mycotoxin T-2. *J. Immunol.* **1990**, *144*, 4721-4728.

Chanh, T. C.; Kennedy, R. C.; Hewetson, J. F. Anti-idiotype vaccines in toxicology. *Int. J. Clin. Lab. Res.* **1992**, *22*, 28-35.

Chu, F. S. Immunoassay for mycotoxins. In *Modern Methods in the Analysis and Structural Elucidation of Mycotoxins*; Cole, R. J., Ed.; Academic Press: New York, 1986; pp 207-237.

Chu, F. S. Immunoassay for mycotoxins, current state of the art, commercial and epidemiological applications. *Vet. Hum. Toxicol.* **1990**, *32* (Suppl.), 42-50.

Chu, F. S. Development and use of immunoassay in detection of the ecologically important mycotoxins. In *Handbook of Applied Mycology; Mycotoxins*; Bhatnagar, D., Lillehoj, E. B., Arora, D. K., Eds.; Dekker: New York, 1992; Vol. V, pp 87-136.

Chu, F. S. Development of antibodies against aflatoxins. In *The Toxicology of Aflatoxins—Human Health, Veterinary and Agricultural Significance*; Eaton, D. L. and Groopman, J. D., Eds; Academic Press: New York, 1994a; Chapter 21, pp. 451-481.

Chu, F. S. Mycotoxin analysis: Immunochemical techniques. In *Foodborne Disease Handbook*; Hui, Y. H., Gorham, J. R., Murrell, K. D. and Cliver, D. O., Eds.; Marcel Dekker: New York, 1994b; Vol. 2, pp 631-668.

Chu, F. S. Recent studies on immunoassays for mycotoxins. In *Residue Analysis in Food Safety: Applications of Immunoassay Methods*; Beier, R. C. and Stanker, L. H., Eds. ACS symposium series book or 209th meeting of the American Chemical Society (Abstr.), ACS, Washington, D. C. 1995 (In press).

Chu, F. S.; Li, G. Y. Simultaneous occurrence of fumonisin B_1 and other mycotoxins in moldy corn collected from the People's Republic of China in regions high in esophageal cancer. *Appl. Environ. Microbiol.* **1994**, *60*, 847-852.

Chu, F. S.; Huang, X.; Maragos, C. M. Production and characterization of anti-idiotype and anti-anti-idiotype antibodies against fumonisin B_1. *J. Agric. Food Chem.* **1995**, 43, 261-267.

Dreher, R. M.; Usleber, E. Comparison study of a fumonisin EIA and HPLC. In *Residue Analysis in Food Safety: Applications of Immunoassay Methods*; Beier, R. C. and Stanker, L. H., Eds. ACS symposium series book or 209th meeting of the American Chemical Society (Abstr.), ACS, Washington, D. C. 1995 (In press).

Elissalde, M. H.; Kamps-Holtzapple, C.; Beier, R. C.; Plattner, R. D.; Rowe, L. D.; Stanker, L. H. Molecular modeling of fumonisin B_{1-3} and hydrolyzed backbone of fumonisin B_1: approaches to the development of higher affinity monoclonal-based ELISA for these mycotoxins. In *Residue Analysis in Food Safety: Applications of Immunoassay Methods*; Beier, R. C. and Stanker, L. H., Eds. ACS symposium series book or 209th meeting of the American Chemical Society (Abstr.), ACS, Washington, D. C. 1995 (In press).

Fukuda, S.; Nagahara, A.; Kikuchi, M.; Kumagai, S. Preparation and characterization of anti-fumonisin monoclonal antibodies. *Biosci. Biotechnol. Biochem.* **1994**, *58*, 765-767.

Gelderblom, W. C. A.; Jaskiewicz, K.; Marasas, W. F. O.; Thiel, P. G.; Horak, R.M.; Vleggaar, R.; Kriek, N. P. J. Fumonisins–novel mycotoxins with cancer-promoting activity produced by *Fusarium moniliforme*. *Appl. Environ. Microbiol.* **1988**, *54*, 1806-1811.

Gelderblom, W. C. A.; Kriek, N. P. J.; Marasas, W. F. O.; Thiel, P. G. Toxicity and carcinogenicity of the *Fusarium moniliforme* metabolite, fumonisin B_1, in rats. *Carcinogenesis* **1991**, *12*, 1247-1251.

Gelderblom, W. C. A.; Cawood, M. E.; Synman, S. D.; Marasas, W. F. O.; Fumonisin B_1 dosimetry in relation to cancer initiation in rat liver. *Carcinogenesis* **1994**, *15*, 209-214.

Holcomb, M.; Thompson, H. C.; Hankins, L. J. Analysis of fumonisin B_1 in rodent feed by gradient elution HPLC using precolumn derivatization with FMOC and fluorescence detection. *J. Agric. Food Chem.* **1993**, *41*, 764-767.

Hsu, K.-H.; Chu, F. S. Production and characterization of anti-idiotype and anti-idiotype antibodies from a monoclonal antibody against aflatoxin. *J. Agric. Food Chem.* **1994**, *42*, 2353–2359.

Kennedy, R. C.; Zhou, E. M.; Lanford, R. E.; Chanh, T.C.; Bona, C. A. Possible role of anti-idiotypic antibodies in the induction of tumor immunity. *J. Clin. Invest.* **1987**, *80*, 1217-1224.

Khole, V.; Hegde, U. Monoclonal anti-idiotypic antibody to progesterone with internal image properties. *Reprod. Fertil. Dev.* **1992**, *4*, 223-230.

Kurpisz, M.; Gupta, S. K.; Fulgham, D. L.; Alexander, N. J. Production of large amounts of mouse polyclonal antisera. *J. Immunol. Methods.* **1988**, *115*, 195-198.

Jordan, L.; Hansen, T. L.; Zabe, N. A. Automated mycotoxin analysis. *American Lab.* **1994**, *March*, 18-24.

Maragos, C. M.; Richard, J. L. Quantitation and stability of fumonisins B-1 and B-2 in milk. *J. AOAC Int.* **1994**, *77*, 1162-1167.

Maragos, C. M.; Miklasz, S. D. Monoclonal antibody-based competitive ELISAs for the hydrolysis product of fumonisin B_1 (HFB$_1$). In *Residue Analysis in Food Safety: Applications of Immunoassay Methods*; Beier, R. C. and Stanker, L. H., Eds. ACS symposium series book or 209th meeting of the American Chemical Society (Abstr.), ACS, Washington, D. C. 1995 (In press).

Maragos, C. M.; Plattner, R. D.; Miklasz, S. D. Determination of hydrolyzed fumonisin B_1 (HFB$_1$) in corn by competitive direct enzyme linked immunosorbent assay. *Food Add. Contam.* **1995** (In press).

Marasas, W. F. O. *Fusarium moniliforme:* A mycotoxicological miasma. In *Mycotoxins and Phycotoxins*; Steyn, P. S., and Vleggaar, R., Eds.; Elsevier: Amsterdam, 1986; pp 19-28.

Marasas, W. F. O.; Kriek, N. P. J.; Fincham, J. E.; van Rensburg, S. J. Primary liver cancer and oesophageal basal cell hyperplasia in rats caused by *Fusarium moniliforme. Int. J. Cancer* **1984**, *34*, 383-387.

Morgan, M. R. A.; Lee, H. A. Mycotoxins and natural food toxicants. In *Development and Application of Immunoassay for Food Analysis*; Rittenburg, J. H., Ed.; Elsevier: New York, 1990; pp 143-170.

Murphy, P. A.; Rice, L. G.; Ross, P. F. Fumonisins-B_1, B_2, and B_3 content of Iowa, Wisconsin, and Illinois corn and corn screenings. *J. Agric. Food Chem.* **1993**, *41*, 263-266.

Nakane, P. K.; Kawaoi, A. Peroxidase-labeled antibody a new method of conjugation. *J. Histochem. & Cytochem.* **1974**, 22, 1084-1091.

Nelson, P. E.; Desjardins, A. E.; Plattner, R. D. Fumonisins, mycotoxins production by *Fusarium* species: biology, chemistry and significance. *Ann. Rev. Phytopathol.* **1993**, *31*, 233-252.

Nisonoff, A. Idiotypes: Concepts and applications. *J. Immunol.* **1991**, *147*, 2429-2438.

Norred, W. P. Fumonisins–mycotoxins produced by *Fusarium moniliforme. J. Toxicol. Environ. Health* **1993**, *38*, 309-328.

Norred, W. P.; Voss, K. A. Toxicity and role of fumonisins in animal diseases and human esophageal cancer. *J. Food Prot.* **1994**, *57*, 522-527.

Pestka, J. J. Enhanced surveillance of foodborne mycotoxins by immunochemical assay. *J. Assoc. Off. Anal. Chem.* **1988**, *71*, 1075-1081.

Pestka, J. J. Application of immunology to the analysis and toxicity assessment of mycotoxins. *Food & Agric. Immunol.* **1994**, *6*, 219-234.

Pestka, J. J.; Azcona-Olivera, J. I.; Plattner, R. D.; Minervini, F.; Doko, B.; Visconti, A. Comparative assessment of fumonisin in grain-based foods by ELISA, GC-MS, and HPLC. *J. Food Prot.* **1994**, *57*, 167-172.

Riley, R. T.; Richard, J. L., Eds. Fumonisins: A current perspective and view to the future. *Mycopathologia* **1992**, *117*, 1-124.

Riley, R. T.; Norred, W. P.; Bacon, C. W. Fungal toxins in foods: recent concerns. *Ann. Rev. Nutr.* **1993**, *13*, 167-189.

Riley, R. T.; Voss, K. A., Yoo, H. S.; Gelderblom, W. C. A.; Merrill, A. H. Mechanism of fumonisin toxicity and carcinogenesis. *J. Food Prot.* **1994**, *57*, 528-535.

Ross, P. F.; Nelson, P. E.; Richard, J. L.; Osweiler, G. D.; Rice, L. G.; Plattner, R. D.; Wilson, T. M. Production of fumonisins by *Fusarium moniliforme* and *F. proliferatum* isolates associated with equine leukoencephalomalacia and a pulmonary edema syndrome in swine. *Appl. Environ. Microbiol.* **1990**, *56*, 3225-3226.

Ross, P. F.; Rice, L. G.; Plattner, R. D.; Osweiler, G. D.; Wilson, T. M.; Owens, D. L.; Nelson, H. A.; Richard, J. L. Concentrations of fumonisin B_1 in feeds associated with animal health problems. *Mycopathologia* **1991**, *114*, 129-132.

Rottinghaus, G. E.; Coatney, C. E.; Minor, H. C. A rapid, sensitive thin layer chromatography procedure for the detection of fumonisin B_1 and B_2. *J. Vet. Diagn. Invest.* **1992**, *4*, 326-329.

Scott, P. M.; Lawrence, G. A. Liquid chromatographic determination of fumonisins with 4-fluoro-7-nitrobenzofurazan. *J. AOAC Int.* **1992**, *75*, 5829-5834.

Shelby, R. A.; Kelley, V. C. Detection of fumonisins in corn by thin-layer immunoassay. *Phytopathology* **1992**, 82, 500 (Abstr.).

Shelby, R. A.; Rottinghaus, G. E.; Minor, H. C. Comparison of thin-layer chromatography and competitive immunoassay methods for detecting fumonisin on maize. *J. Agric. Food Chem.* **1994**, 42, 2064-2067.

Shephard, G. S.; Sydenham, E. W.; Thiel, P. G.; Gelderblom, W. C. A. Quantitative determination of fumonisins B_1 and B_2 by high-performance liquid chromatography with fluorescence detection. *J. Liquid Chromatogr.* **1990**, *13*, 2077-2087.

Shestowsky, W.S.; Quilliam, M. A.; Sikorska, H. M. An idiotypic-anti-idiotypic competitive immunoassay for quantitation of okadaic acid. *Toxicon* **1992**, *30*, 1441-1448.

Shestowsky, W. S.; Holmes, C. F. B.; Hu, T.; Marr, J.; Wright, J. L. C.; Chin, J.; Sikorska, H. M. An anti-okadaic acid-anti-idiotypic antibody bearing an internal image of okadaic acid inhibits protein phosphatase PP1 and PP2A catalytic activity. *Biochem. Biophys. Res. Commun.* **1993**, *192*, 302-310.

Spinks, C. A.; Wang, B.; Mills, E. N. C.; Morgan, M. R. A. Production and characterization of monoclonal anti-idiotype antibody mimics for the pyrethroid insecticides and herbicide paraquat. *Food Agric. Immunol.* **1993**, *5*, 13-25.

Stack, M. E.; Eppley, R. M. Liquid chromatographic determination of fumonisin-B_1 and fumonisin-B_2 in corn and corn products. *J. AOAC Int.* **1992**, *75*, 5834-5837.

Sydenham, E. W.; Shephard, G. S.; Thiel, P. G. Liquid chromatographic determination of fumonisins B_1, B_2, and B_3 in foods and feeds. *J. AOAC Int.* **1992**, *75*, 313-318.

Szurdoki, F.; Trousdale, E.; Gee, S. J.; Gilchrist, D.; Hammock, B. D. Development of immunoassays for AAL toxins. In *Residue Analysis in Food Safety: Applications of Immunoassay Methods*; Beier, R. C. and Stanker, L. H., Eds. ACS symposium series book or 209th meeting of the American Chemical Society (Abstr.), ACS, Washington, D. C. 1995 (In press).

Tejada-Simon, M. V.; Marovatsanga, L. T.; Pestka. J. J. Comparative detection of fumonisin by HPLC, ELISA, and immunocytochemical localization in *Fusarium* cultures. *J. Food Prot.* **1995** (In press).

Thiel, P. G.; Marasas, W. F. O.; Sydenham, E. W.; Shephard, G. S.; Gelderblom, W. C. A. The implication of naturally occurring levels of fumonisins in corn for human and animals health. *Mycopathologia* **1992**, *119*, 3-9.

Thiel, P. G.; Sydenham, E. W.; Shephard, G. S.; Vanschalkwyk, D. J. Study of the reproducibility characteristics of a liquid chromatographic method for the determination of fumonisins B-1 and B-2 in corn—IUPAC collaborative study. *J. AOAC Int.* **1993**, *76*, 361-366.

Trucksess, M. W. Comparison of immunochemical and liquid chromatography methods for determination of fumonisin B_1 in corn. In *Residue Analysis in Food Safety: Applications of Immunoassay Methods*; Beier, R. C. and Stanker, L. H., Eds. ACS symposium series book or 209th meeting of the American Chemical Society (Abstr.), ACS, Washington, D. C. 1995 (In press).

Usleber, E.; Straka, M.; Terplan, G. Enzyme immunoassay for fumonisin B_1 applied to corn-based food. *J. Agric. Food Chem.* **1994**, *42*, 1392-1396.

Ware, G. M.; Francis, O.; Kuan, S. S.; Umrigar, P.; Carman, A.; Carter, L.; Bennett, G. A. Determination of fumonisin B_1 in corn by high performance liquid chromatography with fluorescence detection. *Anal. Lett.* **1993**, *26*, 1751-1770.

Ware, G. M.; Umrigar, P. P.; Carman, A. S.; Kuan, S. S. Evaluation of Fumonitest immunoaffinity columns. *Anal. Lett.* **1994**, *27*, 693-715.

Yu, F. Y.; Chu, F. S. Production and characterization of monoclonal anti-anti-idiotype antibodies against fumonisin B_1. *(in preparation) 1995.*

12

FUMONISINS VERATOX®

A New Rapid Quantitative ELISA for Determination of Fumonisin in Food and Feed

Mohamed M. Abouzied, Scott D. Askegard, Charles B. Bird, and Brinton M. Miller

Neogen Corporation
Lansing MI 48912

ABSTRACT

Polyclonal antibodies against fumonisin B_1 were produced by immunizing sheep with fumonisin B_1-keyhole limpet hemocyanin as an immunogen. A quantitative competitive enzyme-linked immunosorbent assay was developed whereby free fumonisins or sample extract containing fumonisins and enzyme-labelled fumonisin competed for binding to the solid phase-bound antibodies. The color intensity of wells, formed by substrate reaction with the enzyme, was inversely related to FB_1 concentration. Detection limits for the assay were 0.1 ng/mL fumonisin B_1 and concentrations of fumonisins B_1, B_2, and B_3 required for 50% binding inhibition were 5.5, 23 and 18 ng/mL, respectively. For food and feed analyses, samples were extracted with 70% methanol and dilutions of the extracts were used directly for ELISA. ELISA results were compared to HPLC analyses by a reference laboratory and the correlation (r value) between ELISA and HPLC was 0.967. The assay may be used to quantitate fumonisins in food and feed samples within 30 minutes.

INTRODUCTION

Fumonisins are a group of mycotoxins (Figure 1) that are produced by *Fusarium moniliforme* (Thiel et al., 1991), a worldwide common fungus in corn, and by some other related species (Chen et al, 1992; Nelson et al., 1992). The most abundant of these, fumonisin B_1 (FB_1), has been associated with equine leukoencephalomalacia (Kellerman et al., 1990; Marasas et al., 1988; Wilson et al, 1990) and porcine pulmonary edema syndrome (Harrison et al., 1990). It causes hepatic cancer and exhibits cancer-promoting activity in rats (Gelderblom et al., 1988). Epidemiological evidence also indicates a possible correlation between the fumonisins and human esophageal cancer (Sydenham et al., 1990b). Due to the natural occurrence of FB_1 and its analogues in corn (Sydenham et al., 1990a,b; Thiel et al., 1991;

Fumonisins in Food, Edited by L. Jackson *et al.*
Plenum Press, New York, 1996

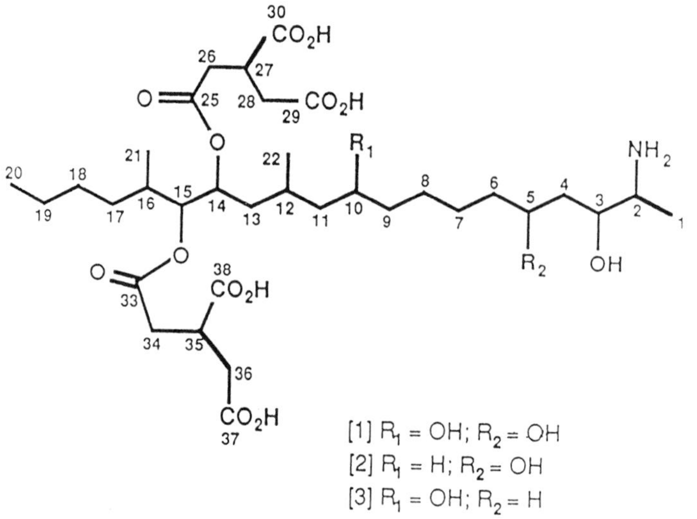

Figure 1. Structure of fumonisins: [1] fumonisin B_1, [2] fumonisin B_2, [3] fumonisin B_3.

Wilson et al., 1990) and their wide range of toxicological effects, it is essential to monitor human and animal exposure via food and feed.

Incidences of mycotoxins in plant and animal foods continue to be a serious issue for United States agriculture. The Food and Agriculture Organization (FAO) estimates that at least 25% of the world's crops are affected by mycotoxins (Mannon and Johnson, 1985). Samples of corn-based food and feed from United States, South Africa and parts of South America have been reported to have FB_1, and FB_2 (Thiel et al., 1992).

The methods used for quantitation of fumonisins in different commodities are laborious and require extensive extraction, clean up and derivatization procedures. These methods include liquid chromatography (Bennet and Richard, 1994), gas chromatography (GC) (Jackson and Bennet, 1990; Sydenham et al., 1990a), thin layer chromatography (TLC) (Gelderblom et al., 1988; Jackson and Bennet, 1990; Sydenham et al., 1990a; Wilson et al., 1990), liquid secondary ion mass spectroscopy (LSIMS) (Bezuidenhout et al., 1988; Plattner et al., 1990; Voss et al., 1989), high performance liquid chromatography (HPLC) (Alberts et al., 1990; Gelderblom et al., 1988; Shephard et al., 1990; Sydenham et al., 1990a; Sydenham et al., 1992; Wilson et al., 1990), and gas chromatography-mass spectroscopy (GC-MS) (Jackson and Bennett, 1990; Plattner et al., 1990; Voss et al., 1989; Wilson et al., 1990).

Immunoassays are an alternative analytical method for the detection of fumonisins in food and feed samples. Immunochemical methods, as screening tools, are rapidly gaining acceptance as an option for residue and other environmental analyses. They offer the advantages of being rapid, reliable, simple, relatively inexpensive, and are a field adaptable alternative to conventional chromatographic and colorimetric methods.

Monoclonal antibodies (MAb) for FB_1 were produced (Azcona-Olivera et al., 1992) but the MAb-based ELISA gave higher values when compared to HPLC (Pestka et al., 1994). That phenomenon might be a result of higher cross reactivity of the monoclonal antibodies to fumonisin related compounds that are not detected by HPLC. Currently, the food and agriculture industry is using commercial ELISA kits for the detection and quantitation of mycotoxins and other residues (Pestka et al., 1995).

We report here the development of a sensitive, quantitative polyclonal antibody-based immunoassay for fumonisins in agricultural commodities.

MATERIALS AND METHODS

Fumonisin B_1 (FB_1) was purchased from Sigma Chemical Co. (St. Louis, MO) and fumonisin B_3 (FB_3) and fumonisin B_2 (FB_2) were obtained from PROMEC, Tygerberg, South Africa. All inorganic chemicals and organic solvents were reagent grade or better.

Fumonisin FB_1 Conjugates

FB_1 is a low molecular weight compound. To be immunogenic FB_1 has to be coupled to a large molecule, usually a protein. Different proteins can be used as carriers e.g., bovine serum albumin (BSA), ovalbumin (OA) and keyhole limpet hemocyanin (KLH). For the present work, fumonisin B_1 was conjugated to KLH using the method of Avrameas and Ternynck (1969). FB_1 was also conjugated to horseradish peroxidase for use in the direct competitive ELISA following the method of Nakane and Kawaoi (1974) and to ovalbumin (FB_1-OA) (Avrameas and Ternynck, 1969) to use it as the solid phase antigen in the indirect ELISA.

Animal Immunization

A wide range of animals can be used for the production of polyclonal antibodies. The most commonly used laboratory animals are rabbits, mice, rats, hamster, and guinea pigs (Harlo and Lane, 1988). Sheep, goats, donkeys and horses are frequently used for commercial production of polyclonal antibodies. The polyclonal antibodies used here were produced in sheep. Each sheep was injected subcutaneously with an emulsion consisting of 0.5 to 1.0 mg of FB_1-KLH dissolved in 0.5 to 1.0 mL of saline solution (0.85%) and an equal volume of Freund's complete adjuvant. Booster injections were given every 4-6 weeks with the same volume of immunogen except that incomplete Freund's adjuvant was used. One week to ten days after each boost, blood was collected from the animals to determine antibody titer and specificity.

Indirect ELISA

For antisera titration, wells of polystyrene microtiter plates (Immunolon-2-Re-movawells®, Dynatech Laboratories, Alexandria, VA) were coated overnight (4°C) with 100 μL of FB_1-OA (5 μg/mL) in 0.1M sodium carbonate-bicarbonate buffer, pH 9.6. Plates were washed 4 times with 300 μL of deionized water. Wells were blocked for 30 min at 37°C with 300 μL of 1% polyvinyl alcohol (PVA) (w/v) in PBS and then washed three times with deionized water. Next, 50 μL of serially diluted serum were added to each well and incubated for 30 min at 37°C. Unbound antibody was removed by washing four times with water, and 100 μL of donkey anti-sheep IgG peroxidase conjugate (diluted 1:5000) was added to each well. Plates were incubated for 30 min at 37°C, then washed eight times with water. Bound peroxidase was determined with K-Blue Substrate® (Neogen Corp., Lansing, MI). Absorbance was read at 650 nm using a Bio-Kinetic Reader® EL-312e (Bio-Tek Instruments, Inc., Winooski, VT). The serum titer was arbitrarily designated as the maximum dilution that yielded at least twice the absorbance of the same dilution of non-immune control serum.

Sample Preparation

Fifty g of ground corn were blended with 250 mL of 70% (v/v) methanol/water solution for 2 minutes in a high-speed blender. The extract was filtered through Whatman #1 filter paper. Samples were diluted 1:80 by mixing 100 μL of sample filtrate with 7.9 mL of 10% (v/v) methanol-water.

Competitive Indirect ELISA (CI-ELISA)

CI-ELISA was used to assess the presence of specific FB_1 antibodies in sheep sera. The assay was essentially identical to the indirect ELISA described above, except that, after blocking and washing, 50 μL of FB_1 dissolved in PBS were simultaneously incubated with 50 μL of the appropriate dilution of serum.

Competitive Direct ELISA (CD-ELISA)

The Veratox® Fumonisin Kit is based on a competitive direct ELISA (CD-ELISA), whereby free fumonisins or sample extract containing fumonisins and enzyme-labeled fumonisin compete for binding to the solid phase-bound antibodies. The color intensity of wells, formed by substrate reaction with enzyme, is inversely related to FB_1 concentration.

Microtiter wells were coated with sheep anti-fumonisin polyclonal antibodies in 0.1M carbonate buffer (pH 9.6) overnight at 40°C. Wells were washed with deionized water. Nonspecific binding was minimized by blocking the unbound sites of microtiter wells with 300 μL of 1% (w/v) PVA in PBS for 30 min at 37°C. After washing with deionized water, strips were dried and packed in foil pouches with desiccant, sealed and stored at 4°C until use.

Fumonisin standards or diluted sample extracts (100 μL) were added to each well of an untreated microtiter strip (mixing wells). To each mixing well, 100 μL of fumonisin B_1-horseradish peroxidase (FB_1-HRP) conjugate were added. Using a multi-channel pipettor, wells contents were mixed by pipetting the liquid up and down in the tips. One hundred μL were transferred to the antibody-coated wells and incubated for 15 min at room temperature. The wells were washed five times using a wash bottle or running stream of water. Thereafter, to each well, 100 μL K-Blue Substrate were added and incubated for 15 min. To stop the reaction, 100 μl of stopping reagent were added to each well (Red Stop®, Neogen Corp., Lansing, MI). Absorbance was read in a microwell reader using a 650 nm filter.

Spiking Study

Fumonisin-free ground corn samples (5 g) in 50 mL centrifuge tubes were spiked with known amounts of fumonisin B_1 dissolved in 2 mL methanol to give final concentrations of 0.5, 1, 2, 3, 5 and 10 ppm of FB_1. The methanol was allowed to evaporate overnight in a chemical fume hood. Fumonisin B_1 was extracted from the spiked samples by adding 25 mL of 70% (v/v) methanol-water to each tube, followed by shaking for 20 min at 200 rpm (S/P Rotator, Baxter Scientific Products). Extracts were filtered through Whatman # 1 filter paper. Extracts were diluted 1:80 in 10% (v/v) methanol-water and analyzed by ELISA.

HPLC Analyses

For the correlation study between ELISA and HPLC analyses, samples (25g) were extracted with 70% (v/v) methanol-water (250 mL) by blending for 2 min at high speed. Crude extracts were centrifuged, filtered and used for ELISA. For HPLC, filtrate was purified

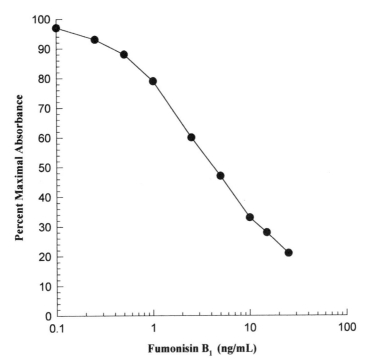

Figure 2. Competitive direct ELISA for fumonisin B₁ using polyclonal antibodies produced by sheep immunized with fumonisin B₁-KLH.

with a strong anion-exchange (SAX) column (Sydenham et al., 1992). Fumonisins were eluted with 1% (v/v) acetic acid in methanol. Eluates were evaporated to dryness under nitrogen then dissolved in 70% (v/v) methanol-water prior to HPLC analysis (Sydenham et al., 1992).

RESULTS AND DISCUSSION

Small molecules such as fumonisins will not elicit an immune response unless they are bound to larger, immunogenic molecules. Cholera toxin (Azcona-Olivera et al., 1992) and KLH (Usleber et al., 1994) were used previously to produce fumonisin antibodies in mice and rabbits. In this study, sheep injected with FB₁-KLH, successfully produced antibodies specific for FB₁ eight weeks after initial injection as determined by indirect competitive ELISA. Antibodies titers increased (>1:50,000) with repeat injections as determined by indirect ELISA. Bleeds that showed the highest affinity towards FB₁ by direct competitive ELISA were used to produce the Veratox® test.

The Veratox® fumonisin kit is a quantitative direct competitive enzyme-linked immunosorbent assay (ELISA) in a microtiter plate format whereby free fumonisins or sample extract-containing fumonisins and enzyme-labeled fumonisin (FB₁-HRP) compete for binding to the solid phase-bound antibodies. Color is inversely proportional to the amount of fumonisins in the sample. Thus, absence of blue color development indicates higher levels of fumonisins in the sample. The maximum color absorbance compared with that of the

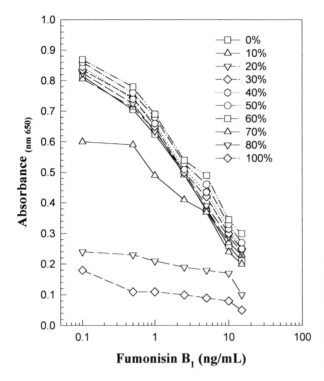

Figure 3. Reactivity of fumonisin B₁ polyclonal antibodies towards fumonisin B₂ (FB₂) and fumonisin B₃ (FB₃) as determined by competitive direct ELISA. Concentration of fumonisins required to inhibit 50% antibody binding were 5.5, 23, and 18 ng/mL for FB₁, FB₂, and FB₃, respectively.

negative control (0 ng FB_1/ml) indicates absence of fumonisins as shown in the assay standard curve (Figure 2).

The limit of detection of FB_1 in 10 % (v/v) methanol-water solution was <0.1 ng/mL and the concentrations of FB_1, FB_2 and FB_3 required for a 50% inhibition of antibody binding in a competitive ELISA were 5.5, 23 and 18 ng/mL, respectively (Figure 3). Mean cross-reactivities (n=6) for FB_2 and FB_3 relative to FB_1 were 24 and 30%, respectively. Cross-reactivity is particularly desirable since it enables simultaneous detection of all three fumonisins.

The effect of methanol, the most common solvent for mycotoxin(s) extractions, on ELISA performance was examined. FB_1 was dissolved in different concentrations of methanol and incubated with FB_1-HRP in anti-fumonisin polyclonal antibodies-coated microtiter wells. As shown in Figure 4, up to 35% final concentrations of methanol (70% methanol of FB_1 standard or sample extract) can be incorporated in the microtiter wells with the enzyme conjugate without any significant loss of ELISA performance.

The high sensitivity of the assay used here required sample extracts to be diluted to put FB_1 concentrations in the range of the standard curve (0 to 15 ng/mL). The assay was balanced so that FB_1 standards were prepared in 10% methanol-water and sample extracts were diluted 1:80 in 10% methanol-water. Dilution factors were calculated and the standard curve values were expressed as ppm so that actual concentration of fumonisins in samples could be calculated directly.

The recovery of FB_1 from corn samples spiked with known amounts of FB_1 and determined by the competitive direct ELISA is shown in Table 1. Overall recovery varied from 74 to 91%.

Reproducibility of the ELISA kit was based on a set of FB_1 standards that were evaluated on different plates on the same day (intra-assay) and on different days (inter-assay).

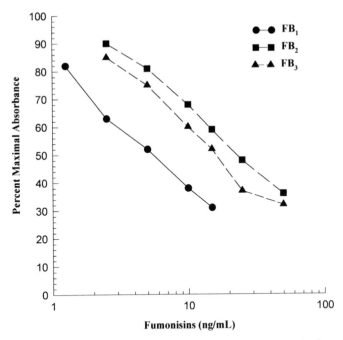

Figure 4. Effect of methanol concentrations on competitive direct ELISA for fumonisin B₁.

The intra-assay coefficient of variation (% CV) was 3.8% and the inter-assay coefficient of variation was 7.6%.

Correlation data between ELISA and HPLC were generated using 20 naturally contaminated samples (Table 2; Figure 5). An excellent correlation of 0.969 (p <0.001) was obtained for FB$_1$. The correlations for FB$_2$ and FB$_3$ were 0.865 and 0.832, respectively. The correlation of FB$_1$ and FB$_2$ combined was 0.966 and the three combined fumonisins (FB$_1$ + FB$_2$ + FB$_3$) was 0.967.

Prior to our report on this new assay, a monoclonal antibody-based competitive ELISA for fumonisins had been developed (Azcona-Olivera et al., 1992). Concentration of fumonisins in naturally contaminated samples analyzed by the monoclonal antibody-based ELISA were higher than those obtained by HPLC or GC-MS for the same samples (Pestka et al., 1994). These higher values with the MAb-based ELISA may be the result of cross-reactivity of the monoclonal antibody with fumonisin related compounds that exist in

Table 1. Recovery of fumonisin B$_1$ from corn spiked with different concentrations of FB$_1$

FB$_1$ added ppm	FB$_1$ recovered[a] ppm	Recovery %
0.5	0.423 ± 0.02 (6)	84
1	0.744 ± 0.08 (6)	74
2	1.569 ± 0.16 (6)	78
3	2.250 ± 0.31 (6)	75
5	3.700 ± 0.38 (6)	74
10	9.146 ± 0.97 (6)	91

[a]Mean amount determined: μg/g ± SD (number of determinations).

Table 2. Fumonisins in naturally contaminated samples
determined by ELISA and HPLC

Samples	ELISA μg/g	HPLC μg/g			
		FB$_1$	FB$_2$	FB$_3$	Total
1	0.2	< 0.05	< 0.05	< 0.05	< 0.05
2	0.2	< 0.05	< 0.05	< 0.05	< 0.05
3	0.1	0.10	< 0.05	< 0.05	0.10
4	0.3	0.16	< 0.05	< 0.05	0.16
5	0.3	0.15	0.07	< 0.05	0.22
6	0.5	0.27	< 0.05	< 0.05	0.27
7	0.3	0.22	0.15	< 0.05	0.37
8	1.0	0.56	0.13	0.07	0.76
9	1.1	0.62	0.22	< 0.05	0.84
10	1.3	0.62	0.35	< 0.05	1.02
11	2.0	1.01	0.23	0.07	1.31
12	1.7	1.13	0.46	0.08	1.67
13	1.7	1.05	0.31	0.07	1.43
14	2.5	1.28	0.44	0.10	1.82
15	2.6	1.40	0.44	0.10	1.94
16	3.2	1.54	0.48	0.17	2.19
17	2.8	1.23	0.86	0.42	2.51
18	3.8	1.70	0.67	0.16	2.53
19	2.5	1.62	1.05	0.20	2.87
20	5.6	4.10	0.99	0.35	5.44

the naturally contaminated sample and which could not be detected by HPLC. Fumonisins-free corn samples spiked with different concentrations of fumonisins (FB$_1$, FB$_2$, FB$_3$) and analyzed by the MAb-based ELISA assay gave results that correlated well to those obtained by HPLC (data not shown here). This suggests that naturally incurred samples contain compounds that cross react with the monoclonal antibody. This problem indicated the need

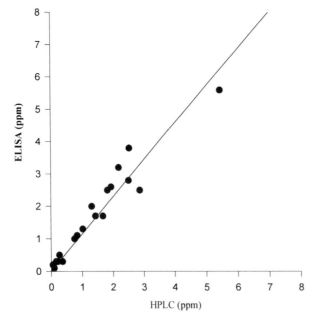

Figure 5. Correlation of results obtained by analyzing 20 naturally contaminated samples for fumonisin content by ELISA with results obtained by HPLC. Two samples were negative by both methods.

for a fumonisin immunoassay which correlates with the best known reference test (i.e., HPLC) using naturally contaminated samples.

The assay we report here is a 30 min assay, having an excellent correlation with HPLC for fumonisins in naturally incurred corn-based samples. The potential applications of this immunoassay include analyses of fumonisins in feeds, foods and clinical samples. In addition, our polyclonal antibodies to fumonisins may be utilized for preparation of immunoaffinity chromatography columns to be applied to sample preparation (concentration and clean up) for other analytical procedures, such as, HPLC and GC-MS.

ACKNOWLEDGEMENTS

The authors would like to thank Eric W. Sydenham (PROMEC, South Africa) for conducting the correlation study between ELISA and HPLC.

REFERENCES

Alberts, J.F.; Gelderblom, W.C.A.; Thiel, P.G.; Marasas, W.F.O.; Van Schalkwyk, D.J.; Behrend, Y. Effects of temperature and incubation period on production of fumonisin B$_1$ by *Fusarium moniliforme.* Appl. Environ. Microbiol. **1990**, 56, 1729-1733.

Avrameas, S.; Ternynck, T. The cross-linking of proteins with glutaraldehyde and its use for the preparation of immunoadsorbents. Immunochemistry. **1969**, 6, 53-66.

Azcona-Olivera, J.I.; Abouzied, M.M.; Plattner, R.D.; Pestka, J.J. Production of monoclonal antibodies to the mycotoxins fumonisins B$_1$, B$_2$, and B$_3$. J. Agric. Food Chem. **1992**, 40, 531-534.

Bennett, G.A.; Richard, J.L. Liquid chromatographic method for analysis of the naphthalene dicarboxyaldehyde derivative of fumonisins. J. AOAC International. **1994**, 77, 501-505.

Bezuidenhout, S.C.; Gelderblom, W.C.A.; Gorst-Allman, C.P.; Horak, R.M.; Marasas, W.F.O.; Spiteller, G.; Vegglaar, R. Structure elucidation of the fumonisins, mycotoxins from *Fusarium moniliforme.* J. Chem. Soc. Chem. Commun. **1988**, 743-745.

Chen, J. ; Mirocha, C.J.; W. Xie, W.; Hogge, L.; Olson, D. Production of the mycotoxin fumonisin B$_1$ by *Alternaria alternata* f. sp. *lycopersici*. Appl. Environ. Microbiol. **1992**, 58, 3928-3931.

Gelderblom, W.C.A.; Jaskiewicz, K.; Marasas, W.F.O.; Thiel, P.G.; Horak, R.M.; Vegglaar, R.; Kriek, N.P.J. Fumonisin-Novel mycotoxins with cancer-promoting activity produced by *Fusarium moniliforme.* Appl. Environ. Microbiol. **1988**, 54, 1806-1811.

Harlo E.; Lane. D. Antibodies: A Laboratory Manual. Cold Spring Harbor Laboratory, New York, 1988.

Harrison, L.R.; Colvin, B.M.; Greene, J.T.; Newman, L.E.; Cole, J.R. Pulmonary edema and hydrothorax in swine produced by fumonisin B$_1$, a toxic metabolite of *Fusarium moniliforme.* J. Vet. Diagn. Invest. **1990**, 2, 217-221.

Jackson, M.A.; Bennett, G.A. Production of fumonisin B$_1$ by *Fusarium moniliforme* NRRL 13616 in submerged culture Appl. Environ. Microbiol. **1990**, 56, 2296-2298.

Kellerman, T.S.; Marasas, W.F.O.; Thiel, P.G.; Gelderblom, W.C.A.; Cawood, M.; Coetzer, J.A.W. Leucoencephalomalacia in two horses induced by oral dosing of fumonisin B$_1$. Onderstepoort. J. Vet. Res. **1990**, 57, 269-275.

Mannon, J.; Johnson, E. Fungi down on the farm. Scientist. **1985**, 105(1446), 12-16.

Marasas, W.F.O.; Kellerman, T.S.; Gelderblom, W.C.A.; Coetzer, J.A.W.; Thiel, P.G.; Van der Lugt, J.J. Leukoencephalomalacia in a horse induced by fumonisin B$_1$ isolated from *Fusarium moniliforme.* Onderstepoort J. Vet. Res. **1988**, 55, 197-203.

Nakane, P.K.; Kawaoi, A. Peroxidase-labeled antibody: a new method of conjugation. J. Histochem. Cytochem. **1974**, 22, 1048-1091.

Nelson, P.E.; Plattner, R.D.; Shackelford, D.D.; Desjardins, A.E. Production of fumonisins by *Fusarium moniliforme* strains from various substrates and geographic areas Appl. Environ. Microbiol. **1992**, 58, 984-989.

Pestka, J.J.; Azcona-Olivera, J.I.; Plattner, R.D.; Minervini, F.; Doko, M.B.; Visconti. A. Comparative assessment of fumonisin in grain-based foods be ELISA, GC-MS and HPLC. J. Food Protec. **1994**, 57, 169-172.

Pestka, J.J.; Abouzied, M.M.; Sutikno. Immunological assays for mycotoxin detection. Food Technol. **1995**, 49(2), 120-128.

Plattner, R.D.; Norred, W.P.; Bacon, C.W.; Voss, K.A.; Peterson, R.; Shackelford, D.D.; Weisleder, D. A method of detection of fumonisins in corn samples associated with field cases of equine leukoencephalomalacia Mycologia. **1990**, 82, 698-702.

Shephard, G.S.; Sydenham, E.W.; Thiel, P.G.; Gelderblom, W.C.A. Quantitative determination of fumonisin B_1 and B_2 by high performance liquid chromatography with fluorescence detection J. Liq. Chromatogr. **1990**, 13, 2077-2087.

Sydenham, E.W.; Gelderblom, W.C.A.; Thiel, P.G.; Marasas, W.F.O. Evidence for the natural occurrence of fumonisin B_1, a mycotoxin produced by *Fusarium moniliforme*, in corn. J. Agric. Food Chem. **1990a**, 38, 285-290.

Sydenham, E.W.; Thiel, P.G.; Marasas, W.F.O.; Shephard, G.S.; Van Schalkwyk, D.J.; K.R. Koch. Natural occurrence of some *Fusarium* mycotoxins in corn from low and high esophageal cancer prevalence areas of the Transkei, Southern Africa J. Agric. Food Chem. **1990b**, 38, 1900-1903.

Sydenham, E.W.; Shephard, G.S.; Thiel, P.G. Liquid chromatographic determination of fumonisins B_1, B_2, B_3 in foods and feeds J. Assoc. Off. Anal. Chem. Int. **1992**, 75, 313-318.

Thiel., P.G.; Marasas, W.F.O.; Sydenham, E.W.; Shephard, G.S.; Gelderblom, W.C.A.; Nieuwenhuis, J. Survey of fumonisin production by *Fusarium species* Appl. Environ, Microbiol. **1991**, 57, 1089-1093.

Thiel, P.G.; Marasas, W.F.O.; Sydenham, E.W.; Shephard, G.S.; Gelderblom, W.C.A. The implications of naturally occurring levels of fumonisins in corn for human and animal health Mycopathalogia. **1992**, 117, 3-9.

Usleber, E.; Straka, M.; Terplan, G. Enzyme immunoassay for fumonisin B_1 applied to corn-based food J. Agric. Fd Chem. **1994**, 42, 1392-1396.

Voss, K.A.; Plattner, R.D.; Bacon, C.W.; Norred, W.P. Comparative studies of hepatotoxicity and fumonisin B_1 and B_2 content of water and chloroform/methanol extracts of *Fusarium moniliforme* strain MRC 826 culture material. Mycopathologia. **1989**, 112, 81-92

Wilson, T.M.; Ross, P.F.; Rice, L.G.; Osweiler, G.D.; Nelson, H.A.; Owens, D.L.; Plattner, R.D.; Reggiardo, C.; Noom, T.H.; Pickrell, J.W. Fumonisin B_1 levels associated with a epizootic of equine leukoencephalomalacia J. Vet. Diagn. Invest. **1990**, 2, 213-216.

13

THE RELIABILITY AND SIGNIFICANCE OF ANALYTICAL DATA ON THE NATURAL OCCURRENCE OF FUMONISINS IN FOOD

P.G. Thiel, E.W. Sydenham, and G.S. Shephard

Programme on Mycotoxins and Experimental Carcinogenesis
Medical Research Council
PO Box 19070, Tygerberg 7505
South Africa

ABSTRACT

Several methods are presently used to identify and quantify fumonisins in foods and feeds. HPLC procedures on derivatized fumonisins with fluorescence detection are most commonly used. The validity and significance of reported fumonisin levels depend on several factors such as the specificity, detection limit, accuracy and reproducibility of the analytical method as well as on the sampling procedure used, and the integrity and purity of the analytical standards. The importance of these factors is discussed and the results of two international collaborative studies are presented on the determination of fumonisins in corn by a reversed-phase HPLC method on o-phthaldialdehyde (OPA) derivatized fumonisins using fluorescence detection.

INTRODUCTION

Interest in the levels of fumonisins in foods and feeds has increased continuously since their structures were first reported by Bezuidenhout *et al.* (1988). The attention given to the fumonisins is at present rivalling that given to the aflatoxins, the mycotoxins which have historically received by far the most attention of the numerous mycotoxins described to date. This focus on the fumonisins has arisen for several reasons. The fumonisins have been shown to be directly responsible for the animal diseases, leukoencephalomalacia (LEM) in horses (Kellerman *et al.*,1990) and pulmonary edema syndrome (PES) in pigs (Harrison *et al.*, 1990). Fumonisin B_1 (FB_1) has subsequently also been shown to be hepatocarcinogenic in rats (Gelderblom *et al.*, 1991). The most important justification for the attention given to the fumonisins is, however, the fact that they occur worldwide at significant concentrations in corn, a major dietary staple food for humans and animals in many countries.

Fumonisins in Food, Edited by L. Jackson *et al.*
Plenum Press, New York, 1996

In developed countries corn is used primarily for animal feeding, while it forms the major part of the human diet in many communities in the developing world.

Studies on the fumonisins have been directed towards investigating the toxic and carcinogenic effects (Gelderblom *et al.*,1991; Voss *et al.*, 1993), pharmacokinetic investigations (Shephard *et al.*, 1994), mechanism of action studies (Wang *et al.*, 1991; Riley *et al.*, 1993) and detoxification procedures (Norred *et al.*,1991; Sydenham *et al.*,1994) but considerable attention has also been given to the development of analytical procedures for the detection of fumonisins. The methods employed for the detection of fumonisins in corn, foods and feeds include thin-layer chromatography (Gelderblom *et al.*,1988; Rottinghaus *et al.*, 1992; Stockenström *et al.*, 1994), gas chromatography (Plattner *et al.*, 1990, 1992; Sydenham *et al.* 1990) and mass spectrometry (Holcomb *et al.*, 1993a; Korfmacher *et al.*, 1991; Plattner *et al.*, 1990; Plattner and Branham, 1994). The most commonly used procedures employ high-performance liquid chromatographic separation of derivatized fumonisins, as underivatized fumonisins do not fluoresce or absorb UV or visible light (Gelderblom *et al.*, 1988; Shephard *et al.*, 1990; Sydenham *et al.*,1990; Bennett and Richard, 1994; Holcomb *et al.*, 1993b; Scott and Lawrence, 1992; Ross *et al.*, 1991; Miller *et al.*, 1993).

Factors which are important in selecting a suitable analytical procedure include the specificity of the procedure, the concentration ranges to be detected and the detection limit of the procedure. The significance of analytical data on fumonisin levels in foods and feeds depend on the analytical method employed, the representativeness of the samples analyzed, the integrity of the analytical standards used and the accuracy and reproducibility of the procedure. The importance of these factors in evaluating the significance of data on fumonisin levels in foods and feeds is discussed.

THE SPECIFICITY OF ANALYTICAL PROCEDURES FOR FUMONISINS

Toxicological evidence indicates that the biological activity of the fumonisins depends on a free amino group in the molecule, as in the fumonisin B toxins, and that the fumonisin A toxins, which are the N-acetyl derivatives, lack biological activity (Gelderblom *et al.*,1993). Furthermore if one assumes that all fumonisin B toxins have similar toxic potencies, then it will be necessary to determine all fumonisin B analogues in a sample in order to assess its potential toxicity. It has however already been established that in practice only FB_1, FB_2 and FB_3 occur in significant concentrations in corn and in foods and feeds.

Methods which determine only FB_1 or FB_1 and FB_2 would therefore be underestimating the potential toxic hazard of a commodity contaminated by fumonisins. The first report on the involvement of fumonisins in pulmonary edema in pigs (Harrison *et al.*, 1990) quantified only the FB_1 level in incriminated feeds, thereby overestimating the toxic potency of FB_1 as significant quantities of both FB_2 and FB_3 were probably also present. The same is true for the first reports on the detection of fumonisins in feeds associated with LEM (Shephard *et al.*,1990; Plattner *et al.*,1990; Thiel *et al.*, 1991) where only FB_1 and FB_2 were quantified. There are however indications that FB_3 is not involved in LEM in horses (Ross *et al.*, 1994) making it unnecesary to quantify FB_3 when testing feeds for horses. Because of the lack of availability of FB_3 standards most reported data on fumonisin levels in foods and feeds refer only to the FB_1 and FB_2 content.

Several immunochemical techniques have been developed for the detection of fumonisins (Azcona-Olivera *et al.*,1992a,b; Usleber *et al.*,1994; Shelby *et al.*, 1994; Maragos and Richard, 1994; Ware *et al.*, 1994). These methods employed either mono- or polyclonal antibodies and used these antibodies in either ELISA assays or for extract purification using immunoaffinity columns. Both mono- and polyclonal antibodies, prepared against FB_1, showed specificity towards other fumonisins but the degree of cross-reactivity differed. Immunochemical techniques would therefore be responsive to all the fumonisins and it is claimed that they are therefore good screening techniques. Because the cross-reactivities to FB_2 and FB_3 are often very low, an immunochemical technique could underestimate the total concentration of fumonisins in a commodity. In practice, however, it has been found that ELISA assays register higher values on contaminated corn than recorded by HPLC techniques (Pestka et al.,1994; Usleber et al., 1994). This would imply that the antibodies lack specificity for the known fumonisins and might react with other compounds in the corn extracts.

In general it can be concluded that immunochemical techniques can be used for screening purposes but that more specific techniques, such as HPLC procedures, should be used when it is necessary to quantitatively determine the potential toxic hazard of a specific food or feed.

DETECTION LIMIT AND CONCENTRATION RANGE

Toxicological investigations showed that feeds containing 50,000 ng/g FB_1 were both hepatotoxic and hepatocarcinogenic to rats (Gelderblom et al., 1991). Feeds containing 15,000 to 50,000 ng/g FB_1 were nephrotoxic to rats (Voss et al., 1993) while Harrison et al. (1993) described pulmonary edema in pigs consuming feeds containing 155,000 ng/g FB_1. Leukoencephalomalacia (LEM) in horses was associated with feeds containing 8,000 ng/g FB_1 (Wilson et al., 1992). The following limits will possibly be set in the future for the total fumonisin B toxins in animal feeds: 50,000 ng/g for cattle and poultry, 10,000 ng/g for swine and 5,000 ng/g for horses. It is at present impossible to say what tolerance level will be decided upon for fumonisin B toxins in human food. From an economic point of view, however, it is highly unlikely that such a level will be far below 1,000 ng/g. According to Thiel et al. (1992) in a small survey in the United States, 34.5% of commercial corn products for human consumption contained more than 1,000 ng/g FB_1 and FB_2.

With the exception of the determination of fumonisins by HPLC with UV detection of maleyl derivatives (detection limit 10,000 ng/g; Sydenham et al., 1992), all the methods presently available should be able to detect fumonisins in foods and feeds at levels well below 1,000 ng/g. HPLC methods using fluorescence detection have much lower detection limits - 50 ng/g in the case of OPA derivatization (Shephard et al., 1990) and 100 ng/g for derivatization with NBD-F (Scott et al., 1992). The detection limit for a TLC method was reported to be 100 ng/g (Rottinghaus et al., 1992), for a LC-MS method as 50 ng/g (Doerge et al., 1994) and 500 ng/g for an ELISA procedure (Azcona-Olivera et al., 1992b).

Several methods are therefore presently available which could cover the concentration range of 100 to 10,000 ng/g fumonisins. All these methods should perform even better at higher concentrations as extracts can merely be diluted with less interference from matrix components.

SAMPLING

It is well known that when samples of an agricultural commodity are analysed to assess potential contamination with a mycotoxin, the largest contribution to the variability of the results is derived from the variability in sampling and not from the inherent variability of the analytical procedure. This is definitely also the case for fumonisin contamination of corn. There are however indications that the variation in concentration of fumonisins within one batch of corn is normally far less than the variations in aflatoxin concentrations seen in batches of nuts. Fumonisins seem to be more evenly distributed between all the kernels of a batch of corn, while in the case of aflatoxins one contaminated nut in a batch of nuts could often render the batch unacceptable for consumption. Although sampling seems to be less of a problem than with aflatoxins, the contribution of sampling error to the total error should not be underestimated.

ANALYTICAL STANDARDS

Pure analytical standards are absolutely essential to obtain accurate results and to be able to compare the results originating from two different laboratories. At present, fumonisin standards are obtained from different sources and it is most likely that these standards vary in composition. The fumonisins have an affinity for water and the moisture content of different preparations can differ considerably. It has not been possible to crystallize any of the fumonisins and there is no direct and absolute method available whereby the purity of a fumonisin standard can be quantitatively determined. At present the purity of fumonisin preparations are judged by indirect techniques, for instance, by checking if contaminants can be observed by chromatographic techniques or by nuclear magnetic resonance (NMR) or mass spectrometry (MS) as well as by analysis for inorganic contaminants. It is unlikely that attempts to crystallize any of the fumonisins will be successful. A more promising approach would be to attempt to crystallize a fumonisin salt.

ACCURACY AND REPRODUCIBILITY

It is important to accumulate accurate and reproducible data on the natural occurrence of fumonisins in human food to aid in the establishment of realistic tolerance levels for fumonisins. A method which is suitable to acquire such data will eventually also be used to monitor fumonisins in food for control purposes and to monitor animal feeds to determine their suitability for use. Immunochemical methods should prove to be suitable for screening purposes but lack specificity. Of the methods presently available, liquid chromatographic procedures seem to be the most suitable for universal use as their specificity, sensitivity and accuracy should be acceptable, while the instrumentation needed is more readily available than that needed for procedures based on mass spectrometry.

Two international collaborative studies on fumonisin methods have been performed to date. In the first study (Thiel et al.,1993), performed by 11 laboratories from 6 countries, the reproducibility characteristics of the method described by Shephard et al. (1990) were investigated using naturally contaminated corn. This method was described for the determination of FB_1 and FB_2 by reversed-phase HPLC of the OPA derivatives with fluorescence detection after cleanup of extracts on strong anion exchange solid phase extraction media. In the second study, performed by 12 laboratories from 10 countries, both the accuracy and the reproducibility of the method of Sydenham et al. (1992), were investigated using corn "spiked" with FB_1, FB_2 and FB_3. This method is basically similar to the method of Shephard et al. (1990) but includes the determination of FB_3 and was improved from the suggestions received in response to the first collaborative study. Both of these studies were carried out as projects of the Commission on Food Chemistry of the Applied Chemistry Division of the International Union of Pure and Applied Chemistry (IUPAC), while the second study was also an accredited study of the Association of Official Analytical Chemists International (AOAC Int.).

In the second study mean recoveries of FB_1 spiked into non-contaminated corn ranged from 81.1% to 84.2% at a spiking range of 500 to 8,000 ng/g, recoveries of FB_2 ranged from 75.9% to 81.9% at a spiking range of 200 to 3,200 ng/g and recoveries of FB_3 ranged from 75.8% to 86.8% at a spiking range of 100 to 1,600 ng/g. It must be emphasized that these recoveries refer to fumonisins which were added to corn meal. It is difficult to determine the recovery of fumonisins from naturally contaminated corn meal. Scott and Lawrence (1994) made a thorough study of the recovery of fumonisins added to corn-based foods and showed that recoveries depended on the extraction solvent and that methanol:water was a poor extraction solvent for corn bran flour.

In the second collaborative study the relative standard deviations for within-laboratory repeatability (RSD_r) ranged from 5.8% to 13.2% for FB_1, from 7.2% to 17.5% for FB_2 and from 8.0% to 17.2% for FB_3. Relative standard deviations for between-laboratory reproducibility (RSD_R) varied from 13.9% to 22.2% for FB_1, from 15.8% to 26.7% for FB_2 and from 19.5% to 24.9% for FB_3. To further assist in determining the acceptability of the reproducibility of the procedure, the so-called HORRAT values were calculated (Horwitz and Albert, 1991; Horwitz et al.,1993). These HORRAT ratios are considered to be "*the best single index of acceptability of method performance*" while ratios "*above 2 are considered unacceptable*". All the HORRAT values for the second collaborative study were within the acceptable range (i.e. less than 2) ranging from 0.75 to 1.73 for FB_1, from 0.77 to 1.72 for FB_2 and from 0.83 to 1.34 for FB_3. These values were generally lower than those reported by Thiel et al. (1993) where HORRAT values ranged from 1.0 to 1.7 for FB_1 and from 1.4 to 2.2 for FB_2. The study by Thiel et al. (1993) was, however, done on naturally contaminated material compared to the "spiked" samples analyzed in the second collaborative study, and it was to be expected that the variation of data on naturally contaminated material would be higher.

CONCLUSIONS

When interpreting quantitative data on the natural occurrence of fumonisins in foods and feeds, it is important to establish which fumonisins were tested for and the purity of the available standards employed. Selection of a method to be used universally for fumonisin analysis will depend on the availability of instrumentation, the detection limit of the method as well as the reproducibility and accuracy of the procedure. Determination of FB_1, FB_2 and FB_3 in corn by reversed-phase HPLC of OPA derivatized fumonisins using fluorescence detection, was studied collaboratively. The results of two international collaborative studies indicated that both the accuracy and the reproducibility of this method are suitable for the method to be considered for official status at analyte concentrations of up to 8,000 ng/g FB_1, 3,200 ng/g FB_2 and 1,600 ng/g FB_3 in corn.

REFERENCES

Azcona-Olivera, J.I.; Abouzied, M.M.; Plattner, R.D.; Norred, W.P.; Pestka, J.J. Generation of antibodies reactive to fumonisins B_1, B_2, and B_3 by using cholera toxin as the carrier-adjuvant. *Appl. Environ. Microbiol.* **1992a**, *58*, 169-173.

Azcona-Olivera, J.I.; Abouzied, M.M.; Plattner, R.D.; Pestka, J.J. Production of monoclonal antibodies to the mycotoxins fumonisins B_1, B_2, and B_3. *J. Agric. Fd Chem.* **1992b**, *40*, 531-534.

Bennet, G.A.; Richard, J.L. Liquid chromatographic method for analyses of the naphthalene dicarboxaldehyde derivative of fumonisins. *J. Assoc. Off. Anal. Chem. Int.* **1994**, *77*, 501-505.

Bezuidenhout, S.C.; Gelderblom, W.C.A.; Gorst-Allman, C.P.; Horak, R.M.; Marasas, W.F.O.; Spiteller, G.; Vleggaar, R. Structure elucidation of the fumonisins, mycotoxins from *Fusarium moniliforme. J. Chem. Soc. Chem. Commun.* **1988**, 743-745.

Doerge, D.R.; Bajic, S.; Preece, S.W.; Howard, P.C. Determination of fumonisins using on-line liquid chromatography coupled to electrospray mass spectrometry. *Rapid Commun. Mass Spectrom.* **1994**, *8*, 603-606.

Gelderblom, W.C.A.; Jaskiewicz, K.; Marasas, W.F.O.; Thiel, P.G.; Horak, R.M.; Vleggaar, R.; Kriek, N.P.J. Fumonisins - novel mycotoxins with cancer-promoting activity produced by *Fusarium moniliforme. Appl. Environ. Microbiol.* **1988**, *54*, 1806-1811.

Gelderblom, W.C.A.; Kriek, N.P.J.; Marasas, W.F.O.; Thiel, P.G. Toxicity and carcinogenicity of the *Fusarium moniliforme* metabolite, fumonisin B$_1$, in rats. *Carcinogenesis* **1991**, *12*, 1247-1251.

Gelderblom, W.C.A.; Cawood, M.E.; Snyman, D.; Vleggaar, R.; Marasas, W.F.O. Structure-activity relationships of fumonisins in short-term carcinogenesis and cytotoxicity assays. *Fd. Chem. Toxicol.* **1993**, 31, 407-414.

Harrison, L.R.; Colvin, B.M.; Greene, J.T.; Newman, L.E.; Cole, J.R. Pulmonary edema and hydrothorax in swine produced by fumonisin B$_1$, a toxic metabolite of *Fusarium moniliforme. J. Vet. Diagn. Invest.* **1990**, *2*, 217-221.

Holcomb, M.; Sutherland, J.B.; Chiarelli, M.P.; Korfmacher, W.A.; Thompson, H.C.Jr.; Lay, J.O.Jr.; Hankins, L.J.; Cerniglia, C.E. HPLC and FAB mass spectrometry analysis of fumonisins B$_1$ and B$_2$ produced by *Fusarium moniliforme* on food substrates. *J. Agric. Fd Chem.* **1993a**, *41*, 357-360.

Holcomb, M.; Thompson, H.C.Jr.; Hankins, L.J. Analysis of fumonisin B$_1$ in rodent feed by gradient elution HPLC using precolumn derivatization with FMOC and fluorescence detection. *J. Agric. Fd Chem.* **1993b**, *41*, 764-767.

Horwitz, W.; Albert, R. Performance characteristics of methods of analysis used for regulatory purposes. Part II. Pesticide Formulations. *J. Assoc. Off. Anal Chem.* **1991**, *74*, 718-744.

Horwitz, W.; Albert, R.; Nesheim, S. Reliability of mycotoxin assays - An update. *J. Assoc. Off Anal. Chem.* **1993**, *76*,461-491.

Kellerman, T.S.; Marasas, W.F.O.; Thiel, P.G.; Gelderblom, W.C.A.; Cawood, M.E.; Coetzer, J.A.W. Leukoencephalomalacia in two horses induced by oral dosing of fumonisin B$_1$. *Onderstepoort J. Vet. Res.* **1990**, *57*, 269-275.

Korfmacher, W.A.; Chiarelli, M.P.; Lay, J.O.Jr.; Bloom,J.; Holcomb, M.; McManus, K.T. Characterization of the mycotoxin fumonisin B$_1$: comparison of thermospray, fast-atom bombardment and electrospray mass spectrometry. *Rapid Commun. Mass Spectrom.* **1991**, *5*, 463-468.

Maragos, C.M.; Richard, J.L. Quantification and stability of fumonisins B$_1$ and B$_2$ in milk. *J. Assoc. Off. Anal. Chem. Int.* **1994**, *77*, 1162-1167.

Miller, J.D.; Savard, M.E.; Sibilia, A.; Rapior, S.; Hocking, A.D.; Pitt, J.I. Production of fumonisins and fusarins by *Fusarium moniliforme* from Southeast Asia. *Mycologia* **1993**, *85*, 385-391.

Norred, W.P.; Voss, K.A.; Bacon, C.W.; Riley, R.T. Effectiveness of ammonia treatment in detoxification of fumonisin-contaminated corn. *Fd Chem. Toxicol.* **1991**, *29*, 815-819.

Pestka, J.J.; Azcona-Olivera, J.I.; Plattner, R.D.; Minervini, F.; Doko, M.B.; Visconti, A. Comparative assessment of fumonisin in grain-based foods by ELISA, GC-MS, and HPLC. *J. Fd Prot.* **1994**, *57*, 169-172.

Plattner, R.D.; Norred, W.P.; Bacon, C.W.; Voss, K.A.; Peterson, R.; Shackelford, D.D.; Weisleder, D. A method of detection of fumonisins in corn samples associated with field cases of equine leukoencephalomalacia. *Mycologia* **1990**, *82*, 698-702.

Plattner, R.D.; Weisleder, D.; Shackelford, D.D.; Peterson, R.; Powell, R.G. A new fumonisin from solid cultures of *Fusarium moniliforme. Mycopathologia* **1992**, *117*, 23-28.

Plattner, R.D.; Branham, B.E. Labelled fumonisins: Production and use of fumonisin B$_1$ containing stable isotopes. *J. Assoc. Off. Anal. Chem. Int.* **1994**, *77*, 525-532.

Riley, R.T.; An, N-H.; Showker, J.L.; Yoo, H-S.; Norred, W.P.; Chamberlain, W.J.; Wang, E.; Merrill, A.H.; Motelin, G.; Beasley, V.R.; Haschek, W.M. Alteration of tissue and serum sphinganine to sphingosine ratio: An early biomarker of exposure to fumonisin-containing feeds in pigs. *Toxicol. Appl. Pharmacol.* **1993**, *118*, 105-112.

Ross, P.F.; Rice, L.G.; Plattner, R.D.; Osweiler, G.D.; Wilson T.M.; Owens, D.L.; Nelson, H.A.; Richard, J.L. Concentrations of fumonisin B$_1$ in feeds associated with animal health problems. *Mycopathologia* **1991**, *114*, 129-135.

Ross, P.F.; Nelson, P.E.; Owens, D.L.; Rice, L.G.; Nelson, H.A.; Wilson, T.M. Fumonisin B$_2$ in cultured *Fusarium proliferatum*, M-6104, causes equine leukoencephalomalacia. *J. Vet. Diagn. Invest.* **1994**, *6*, 263-265.

Rottinghaus, G.E.; Coatney, C.E.; Minor, H.C. A rapid, sensitive thin layer chromatography procedure for the detection of fumonisin B_1 and B_2. *J. Vet. Diagn. Invest.* **1992**, *4*, 326-329.

Scott, P.M.; Lawrence, G.A. Liquid chromatographic determination of fumonisins with 4-fluoro-7-nitrobenzofurazan. *J. Assoc. Off. Anal. Chem. Int.* **1992**, *75*, 829-834.

Scott, P.M.; Lawrence, G.A. Stability and problems in recovery of fumonisins added to corn-based foods. *J. Assoc. Off. Anal. Chem. Int.* **1994**, *77*, 541-545.

Shelby, R.A.; Rottinghaus, G.E.; Minor, H.C. Comparison of thin-layer chromatography and competitive immunoassay methods for detecting fumonisin on maize. *J. Agric. Fd Chem.* **1994**, *42*, 2064-2067.

Shephard, G.S.; Sydenham, E.W.; Thiel, P.G.; Gelderblom, W.C.A.; Quantitative determination of fumonisins B_1 and B_2 by highperformance liquid chromatography with fluorescence detection. *J. Liq. Chromatogr.* **1990**, *13*, 2077-2087.

Shephard, G.S.; Thiel, P.G.; Sydenham, E.W.; Alberts, J.F.; Cawood, M.E. Distribution and excretion of a single dose of the mycotoxin fumonisin B_1 in a non-human primate. *Toxicon* **1994**, *32*, 735-741.

Stockenström, S.; Sydenham, E.W.; Thiel, P.G. Determination of fumonisins in corn: evaluation of two purification procedures. *Mycotoxin Res.* **1994**, *10*, 9-14.

Sydenham, E.W.; Gelderblom, W.C.A.; Thiel, P.G.; Marasas, W.F.O. Evidence for the natural occurrence of fumonisin B_1, a mycotoxin produced by *Fusarium moniliforme* in corn. *J. Agric. Fd Chem.* **1990**, *38*, 285-290.

Sydenham, E.W.; Shephard, G.S.; Thiel, P.G. Liquid chromatographic determination of fumonisin B_1, B_2 and B_3 in foods and feeds. *J. Assoc. Off. Anal. Chem.* **1992**, *75*, 313-318.

Sydenham, E.W.; Van der Westhuizen, L.; Stockenström, S.; Shephard, G.S.; Thiel, P.G. Fumonisin-contaminated maize: physical treatment for the partial decontamination of bulk shipments. *Food Addit. Contam.* **1994**, *11*, 25-32.

Thiel, P.G.; Shephard, G.S.; Sydenham, E.W.; Marasas, W.F.O.; Nelson, P.E.; Wilson, T.M. Levels of fumonisin B_1 and B_2 in feeds associated with confirmed cases of equine leukoencepha lomalacia. *J. Agric. Fd. Chem.* **1991**, *39*, 109-111.

Thiel. P.G.; Marasas, W.F.O.; Sydenham, E.W.; Shephard, G.S.; Gelderblom, W.C.A. The implications of naturally occurring levels of fumonisins in corn for human and animal health. *Mycopathologia* **1992**, *117*, 3-9.

Thiel, P.G.; Sydenham, E.W.; Shephard, G.S.; van Schalkwyk, D.J. Study of the reproducibility characteristics of a liquid chromatographic method for the determination of fumonisins B_1 and B_2 in corn: IUPAC collaborative study. *J. Assoc. Off. Anal. Chem. Int.* **1993**, *76*, 361-366.

Usleber, E.; Straka, M.; Terplan, G. Enzyme immunoassay for fumonisin B_1 applied to corn-based food. *J. Agric. Fd Chem.* **1994**, *42*, 1392-1396.

Voss, K.A.; Chamberlain, W.J.; Bacon, C.W.; Norred, W.P. A preliminary investigation on renal and hepatic toxicity in rats fed purified fumonisin B_1. *Natural Toxins* **1993**, *1*, 222-228.

Wang, E.; Norred, W.P.; Bacon, C.W.; Riley, R.T.; Merrill, A.H. Inhibition of sphingolipid biosynthesis by fumonisins: implications for diseases associated with *Fusarium moniliforme*. *J. Biol. Chem.* **1991**, *266*, 14486-14490.

Ware, G.M.; Umrigar, P.P.; Carman, A.S.Jr.; Kuan, S.S. Evaluation of fumonitest immunoaffinity columns. *Anal. Lett.* **1994**, *27*, 693-715.

Wilson, T.M.; Ross, P.F.; Owens, D.L.; Rice, L.G.; Jenkins, S.J.; Nelson, H.A. Experimental production of ELEM. A study to determine the minimum toxic dose in ponies. *Mycopathologia* **1992**, *117, 115-120.*

14

INTRODUCTORY BIOLOGY OF *FUSARIUM MONILIFORME*

John F. Leslie

Department of Plant Pathology
4002 Throckmorton Plant Sciences Center
Kansas State University
Manhattan, Kansas 66506-5502

ABSTRACT

Fusarium moniliforme is a name that has been applied to any of six biological species (or mating populations) that share the teleomorph (sexual stage) *Gibberella fujikuroi*. Two of these six biological species, termed "A" and "D", are known to produce fumonisin mycotoxins. Strains from the "A" biological species grow as endophytes on maize and often comprise 90+% of the *Fusarium* isolates recovered from healthy maize seed. It is possible to distinguish all six biological species using sexual fertility and isozymes. Other attributes, such as morphological characters and sequences from the ribosomal DNA internally transcribed spacer (rDNA-ITS) region, can be used to identify some, but not all, of the biological species. Within a biological species, genetic variability and population structure can be assessed with anonymous RFLPs and tests of vegetative compatibility. The "A" biological species is genetically diverse, and the sexual cycle appears to be important in the life cycle of field populations of this organism in the United States.

INTRODUCTION

Gibberella fujikuroi (Sawada) Ito in Ito and K. Kimura is the teleomorph for many of the species of conidial anamorphs in *Fusarium* section *Liseola* including *Fusarium moniliforme* Sheldon, *Fusarium subglutinans* (Wollenweber and Reinking) Nelson, Toussoun and Marasas, and *Fusarium proliferatum* (Matsushima) Nirenberg [Nelson *et al.*, 1983]. *Fusarium* spp. in this group are distributed worldwide on a variety of hosts including: asparagus [Elmer and Ferrandino, 1992], bananas [Shaw *et al.*, 1993a], figs [Caldis, 1927], maize [Leslie *et al.*, 1990], mango [Varma *et al.*, 1974], pine [Correll *et al.*, 1992], pineapple [Rohrbach and Pfeiffer, 1976], rice [Sun and Snyder, 1981], sorghum [Leslie *et al.*, 1990], stone fruits [Subbarao and Michailides, 1993], and sugarcane [Martin *et al.*, 1989]. Although these fungi are widespread in native and cultivated plants, they are most commonly

Fumonisins in Food, Edited by L. Jackson *et al.*
Plenum Press, New York, 1996

associated with stalk and root rots of maize and sorghum [Frederiksen, 1986; Shurtleff, 1980], and economic losses attributable to these fungi in these crops alone are in the hundreds of millions of dollars per year.

In addition to their capabilities as plant pathogens, strains from this group of fungi are capable of producing a variety of secondary metabolites. Members of this group have been confirmed as producers of the fumonisin and moniliformin mycotoxins [Leslie *et al.*, 1992b; Marasas *et al.*, 1984; Thiel *et al.*, 1991], and have been used commercially for the synthesis of gibberellic acid [Phinney and West, 1960]. They also have been credited with producing a variety of other secondary metabolites including: deoxynivalenol, diacetoxy-scirpenol, fusaric acid, fusarins, fusariocins, T-2 toxin, and zearalenone [Marasas *et al.*, 1984], but not all of these reports have been substantiated. Difficulties with the identification of the fungal strains have made a tricky situation even worse.

In spite of both their economic significance and their broad distribution, the demographic organization and constraints on populations of these fungi are not well understood. For example, the anamorphic entity described by Snyder and Hansen [1945] as *F. moniliforme* has been subdivided into at least three different morphological species (*F. moniliforme*, *F. proliferatum*, and *F. subglutinans*) and at least six biological species (designated by the letters A - F). The ensuing nomenclatural difficulties have resulted in massive confusion for those who work with these organisms and for those who try to mitigate the losses they inflict on agriculture. As more researchers begin to work with this organism because of its ability to synthesize fumonisins, the need to work with carefully defined standardized strains and the development of a common taxonomic nomenclature will be essential. At the most recent International *Fusarium* workshop (July, 1993), the *G. fujikuroi* species complex was targeted for a realignment of the nomenclature to bring it into accord with what is known of the biology. Thus changes are coming in the nomenclature of these organisms that should simplify discussions of the fumonisin toxins.

In my laboratory, we have been studying the basic biology of this group of fungi since 1984, with the goals of making it a readily manipulable laboratory system and of understanding the population biology of the organism sufficiently well so that we could devise new methods to reduce the losses attributable to this organism. The objectives of this paper are: (1) to describe the divisions of this species-complex into different biological species (or mating populations), (2) to describe some of the characteristics of these biological species and how these biological species can be distinguished from one another, and (3) to highlight some of the progress we have made in understanding the field biology of the organism.

BIOLOGICAL SPECIES IN *GIBBERELLA FUJIKUROI*

The use of mating as a diagnostic character is one means of skirting the difficulties that arise when distinct biological entities are morphologically very similar to one another. The presence of mating populations is consistent with the evolutionary concept of a species consisting of all of the members of a population that are capable of interbreeding with one another and producing fertile offspring. Mating has, therefore, been a very useful tool in defining the biological species in the *G. fujikuroi* species complex. These distinctions can be very important even if sexual recombination is rare under field conditions since they can provide useful *in vitro* diagnostics, a means to assess the maximum level of genetic exchange that can occur, and a framework in which the possibilities of sympatric speciation can be evaluated. Once the biological species have been delineated, then the affinities of strains that are infertile can be determined based on asexual characters that are conserved within a mating population. The proportions of hermaphrodites and female-sterile strains, and the relative frequencies of strains with different mating types can be used to determine the relative importance of sexual and asexual reproduction in the life history of the organism [Leslie and Klein, 1995].

Three mating populations (or biological species), termed A, B and C were described by Hsieh et al. [Hsieh _et al._, 1977], and a fourth, D, was added by Kathariou [1981] but never formally described. Kuhlman [1982] described all four mating populations as varieties with overlapping characters. Over the past decade we have amassed a collection in excess of 6,000 field isolates from this section and have identified two additional mating populations - E and F [Klittich and Leslie, 1992; Leslie, 1991], one of which, F, has been formally described [Klittich and Leslie, 1992]. A seventh mating population, G, with a _Fusarium nygamai_ anamorph, has been reported by others [Klaasen and Nelson, 1993], but has not yet been formally described. Since there are still entities within this section that have not been associated with an identified mating population, additional mating populations probably remain to be identified. A summary of the known mating populations and the anamorphs with which they have been associated is given in Table 1.

Nomenclatural confusion in the literature has two primary causes. First, there are disagreements over which name is proper for use by the experts who have described the species. Such disagreements include the use of both _F. moniliforme_ and _F. verticillioides_ to describe the same organism. Problems such as this are relatively easy to solve once a suitable resolution of priority issues is reached. Of greater concern are instances in which the best available morphological characters are insufficient to distinguish the biological species. These entities are commonly referred to as sibling species. The best described example of sibling species in this group is with the A and F mating populations, which share the _F. moniliforme_ anamorph but which are genetically isolated from one another. The B and E

Table 1. Biological species (mating populations) of
Gibberella fujikuroi and their corresponding
Fusarium anamorphs

Mating Population	_Fusarium_ Anamorph
A	_F. moniliforme_
	F. verticillioides
B	_F. subglutinans_
	F. sacchari
	F. neoceras
C	_F. fujikuroi_
	F. proliferatum
D	_F. proliferatum_
E	_F. subglutinans_
	F. sacchari var. _subglutinans_
F	_F. moniliforme_
G	_F. nygamai_
Unknown	_F. anthophilum_
	F. annulatum
	F. beomiforme
	F. dlamini
	F. napiforme
	F. proliferatum var. _minus_
	F. sacchari var. _elongatum_
	F. succisae

mating populations both share the *F. sacchari/F. subglutinans* name and additional varieties and groups of strains that belong to this group morphologically remain without a known sexual stage, suggesting that additional mating populations remain to be defined within this group.

Investigators who used strains of "*F. moniliforme*" or "*F. subglutinans*" in their experiments could have very different results because the strains they were working with were from different mating populations and they were misled by the biologically inaccurate nomenclatural identity. For example, identifying the predominant mating population within a niche is probably crucial in breeding for resistance and in obtaining reliable results. Unfortunately much of the work that has been previously done with *Fusarium* strains that carry the *moniliforme* or *subglutinans* names probably needs to be reevaluated and/or repeated in light of the differences that are now known to exist.

Sexual Distinctions Between Biological Species of *Gibberella fujikuroi*

An understanding of the differences between the mating populations and the origins of these differences will greatly aid our understanding of these organisms and their capability for genetic change. In our studies we have used the criterion that a fertile cross is one in which a cirrus of ascospores can be seen emerging from a mature perithecium. Apparently well-developed perithecia that lack this cirrus of spores are scored, for our purposes, the same as those on which no perithecia are formed at all. Thus, positive crosses for us are very strongly fertile and if these crosses occurred under field conditions, measurable levels of recombination would probably be occurring. This high level of fertility is not a strict requirement for the definition of mating populations or of intersterility, (see Perkins [1995] for a more detailed discussion of this point). Desjardins [A. E. Desjardins, 1994, personal communication) has noted several instances in which strains from the C and D mating populations can cross with one another, and we have found some isolates which seem to mate well-enough to exude spores from perithecia in crosses with both the C and the D tester strains. These crosses should offer a means to move genes from one of these mating populations to another via classical genetic methods, to determine the degree of colinearity of the genomes, and to identify characters that are responsible for the differentiation of the strains into different mating populations. Crosses such as these, which are relatively infertile, may be important in evolutionary terms, since they represent a means by which rare, but potentially significant, genetic exchanges can occur. These crosses are much less important from a population genetic or epidemiological standpoint since too few progeny will be produced to have an immediate impact on the population.

At present the most reliable way to distinguish the different mating populations is by crosses on carrot agar as described in Leslie [1991] and Klittich and Leslie [1988]. These crosses take approximately five weeks to complete and are definitive only when they are positive. Sterile crosses with members of known mating populations combined with fertile crosses when members of the new mating population are intercrossed are the only definitive criteria by which a new mating population can be designated. However, any methods that reduce the number of intercrosses (which increase as a square of the number of strains being examined) greatly reduce the amount of time and effort required to reach a satisfactory conclusion. Once mating populations are defined, other tools may be appropriate as quicker methods for identifying members of existing mating populations.

The known mating populations within *G. fujikuroi* are all heterothallic, with a one locus/two allele mechanism that is common to many ascomycetes. Strains within the same mating population may be either of two mating types, "+" or "-", and are usually designated as X^+ or X^- where X is the mating population being examined. Fertile crosses are made by crossing members of the same mating population but of opposite mating type. In laboratory practice, the strain that will serve as the female parent is grown on media and then fertilized (spermatized) with a spore suspension from the strain serving as the male parent. Strain roles are reversed to assess male and female fertility. Standard testers for each mating type in the six well-described mating populations are available from the Fungal Genetics Stock Center, Department of Microbiology, University of Kansas Medical Center, Kansas City, KS.

Mating type is not directly tied to either male or female fertility. Under laboratory conditions, field strains are used as males in crosses with standard testers of both mating types from all six mating populations. A successful cross identifies the mating population to which the field strain belongs (same as the tester) and the mating type of the field strain (opposite that of the tester). A fertile cross indicates that the field strain can serve as the male parent in a cross. Female fertility, functionally hermaphrodism, is tested only in crosses in which the field strain is already known to be male fertile. Strains that are fertile in both crosses are described as either female-fertile or hermaphrodites that can function as either the male or the female parent in a cross. In cases where it is important to know which parent was the male and which was the female, strains carrying the *pal1* mutation [Chaisrisook and Leslie, 1990] can be used to distinguish perithecia formed by the two different parental strains. We have screened numerous strains that were male sterile for female fertility, but have yet to find strains that were male sterile and female fertile. We suggest that this class is missing because the formation of a complex perithecium requires a complicated developmental process that can be disrupted in numerous steps by a mutant allele (e.g. the hundreds of mutants identified by Leslie and Raju [1985, 1992] in natural populations of *N. crassa*). Loss of male function, however, might require the loss of ability to form any asexual spores - a loss that is likely to be heavily selected against under field conditions.

The relative frequencies of strains that are male-only and those that are hermaphrodites may provide a good indication of how often sexual reproduction is occurring under field conditions [Leslie and Klein, 1995]. Based on existing data [Leslie, 1995], it is likely that sexual reproduction occurs under field conditions at a significant rate, but that there are fertility barriers to gene flow in addition to the barriers posed by mating type.

Asexual Differences Between Mating Populations of *Gibberella fujikuroi*

The mating populations of *G. fujikuroi* can be distinguished from one another using criteria other than sexual crosses. Some of these criteria are sufficient to resolve all of the mating populations, but others divide the six mating populations into fewer than six groups. The two mating populations that are the most difficult to distinguish are the C and D mating populations. This difficulty is not surprising since there is limited gene flow between some members of these two populations under laboratory conditions. It would perhaps be more accurate to consider these two mating populations as distinct at the subspecies rather than the species level. All of the differences described here are differences that are correlative, and there is no implication that these differences are responsible for the speciation which has occurred. The development of diagnostic tools within the existing biological species framework should be rapid and the resulting diagnostics should enable fast and reliable assignment of the members of the *G. fujikuroi* species complex to a mating population.

Fumonisin Production. In the context of these proceedings, fumonisin production is the most important trait that is known to differ between mating populations [Leslie and Plattner, 1991; Leslie *et al.*, 1992b]. Members of the A mating population are almost universally capable of synthesizing fumonisins. This ability is known to carry across strain differences defined by differences in vegetative compatibility group [Leslie, 1993; Leslie *et al.*, 1992a]. It is highly significant that no members of the F mating population, which shares the *F. moniliforme* anamorph with the A mating population, are known to produce fumonisins. Thus even the accurate identification of a strain to *F. moniliforme* using currently accepted morphological criteria is insufficient to determine whether the strain is likely to be a fumonisin producer or if a contaminated feedstuff is likely to be contaminated with this toxin. Members of the D mating population are known to synthesize significant quantities of fumonisins as well. The variability within this population for fumonisin synthesis is much higher than that found in the A mating population and nonproducing strains are much more common. The few members of the C mating population that have been tested do not produce significant levels of fumonisins, but the close relationship of the C and D mating populations suggests that strains in the C mating population might be able to produce fumonisins. No members of either the B or the E mating populations are known to produce fumonisins.

Host Preference. Host preference is a trait that can be used to distinguish some mating populations [Jardine and Leslie, 1992; Leslie and Plattner, 1991; Leslie *et al.*, 1992b]. The A mating population is most common on maize where it can compose up to 90% of the populations from maize seed [Campbell *et al.*, 1992, 1993]. Members of the D and E mating populations also can be recovered at significant frequencies from maize, especially the stalks, and members of the B and F mating populations are usually encountered at less-frequent levels. In sorghum, the F mating population predominates with mating populations D and B recovered at significant frequencies at some locations. From figs [Subbarao and Michailides, 1993], the A mating population predominates; the D mating population is most common on asparagus [Elmer, 1995; Elmer and Ferrandino, 1992]; and the C mating population is traditionally associated with high levels of gibberellic acid and the bakane disease of rice [Hsieh *et al.*, 1977; Kuhlman, 1982]. Isolates from pine and from mango are generally infertile when crossed with standard testers, but are associated with the B mating population on the basis of isozyme profiles [Shaw et al., 1993b]. Isolates from bananas can belong to all of the known mating populations and we have been unable to attach any significance to this association [Shaw et al., 1993a]. Isolates from sugar cane have not been closely examined. A listing of available strains, their hosts and geographic origins, and their mating populations (and some mating types) has been published elsewhere [Leslie, 1995].

Differences in secondary metabolite production and host preference probably are also significant. These differences suggest that these fungi now occupy different ecological niches. From a toxicological standpoint, knowing the distribution of these fungi can be important in evaluating potential contamination risks of different mycotoxins. For example, sorghum products will be much less likely to be contaminated with the fumonisin mycotoxins than will maize products since the dominant mating population on sorghum, F, produces little if any fumonisins, while the dominant mating population on maize, A, can produce large quantities of these toxins [Leslie *et al.*, 1992b].

Isozymes. Isozyme differences have been described for these fungi [Huss and Leslie, 1993; Kathariou, 1981], and these differences can be used to distinguish all six mating populations. Within a mating population nine of ten isozymes examined were monomorphic, with the F mating population containing a polymorphism for isocitrate dehydrogenase; three isozymes had the same band in all six of the biological species. It is possible to devise several keys to distinguish the mating populations from one another. The Cs and the Ds are very difficult to resolve however. The best strategy usually depends on an initial guess for the type that is thought to predominate in the population. In the context of fumonisin production, members of the A mating population have a distinctive triose phosphate isomerase band that is diagnostic for this population and is most useful when evaluating cultures recovered from maize or from fumonisin-contaminated feedstuffs.

Antibiotic Resistance. Native resistance levels to hygromycin B and benomyl can be used to distinguish the mating populations from one another [Yan *et al.*, 1993]. Of 48 isolates examined from the A and F mating populations (*F. moniliforme* anamorph), all but two would have been properly identified based on their sensitivity to hygromycin. In the *F. subglutinans* group, only 1/29 isolates would have been misidentified using hygromycin sensitivity as the distinguishing marker and 2/29 with benomyl sensitivity as the critical trait. Asexual isolates with affiliations to these groups have not been examined, however, and could distort the relatively simple system presently thought to exist.

DNA-Sequence Based Differences. Differences between mating populations based on differences in DNA amounts and sequences are known. Differences in DNA-DNA thermal renaturation rates have been described [Ellis, 1988, 1989], but the strains used in these experiments were identified only as morphological species and not as mating populations making these differences difficult to exploit without a more comprehensive repetition. Differences in restriction fragment length polymorphism (RFLP) and random amplified polymorphic DNA (RAPD) patterns have been reported, but neither has as yet been well developed as a diagnostic [DuTeau and Leslie, 1991; Xu, 1994]. Work in this area is now in progress and suitable primers and/or probes should be available for diagnostic purposes in the near future.

The internally transcribed spacer (ITS) of the ribosomal DNA (rDNA) repeat has been proposed as a taxonomic tool for distinguishing fungal species [White *et al.*, 1990]. Restriction maps were made of PCR-amplified ITS sequences from 102 isolates belonging to different mating populations in the *G. fujikuroi* species complex [Anderson, 1994]. Mating populations A and E had identical restriction maps as did populations C and D. Six of the 11 B isolates had the same map as the A and E mating populations; the remaining five B isolates had a map that was different from all of the others. The F population isolates were identical to one another and distinct from all of the other groups. The differences within the B mating population could be indicative of evolutionary divergence that is in progress but that has yet to be completely resolved.

Differences in electrophoretic karyotypes are known [Xu, 1994; Xu *et al.*, 1995] and are sufficient to resolve all six of the mating populations. In all cases 12 chromosomes have been identified that range in size from 0.7 Megabase pairs (Mbp) to > 10 Mbp. All strains from the same mating population have a similar electrophoretic karyotype, regardless of geographic or host origin, but each mating population has a distinctive karyotype. Comparison of karyotype profiles following Southern analysis using homologous and heterologous nuclear gene probes and single copy RFLP probes revealed some differences in hybridization between, but not within biological species. Estimated genome sizes are 45-50 Mbp for mating populations A, B, D and F, and 50-55 Mbp for mating populations C and E. The smallest of the 12 chromosomes varies the most between mating populations. This chromosome was present in all of the field strains examined, but can be lost following meiosis. The technical complexity of the technique is too high for these differences to be usable for routine diagnostics at this time. Differences in the dispensable chromosome and the sequences that are carried there could be usable in such a context.

FIELD BIOLOGY OF *FUSARIUM MONILIFORME*

Fusarium moniliforme on Maize

In maize, *F. moniliforme* is associated with root, stalk and ear rots that cause annual losses in Kansas of 4-8% of the crop [Jardine, 1986]. Infected plant tissues may rot or remain asymptomatic [Koehler, 1942]. Kernel infection may result from internal growth in the plant stalks [Foley, 1962; Kedera *et al.*, 1992], through wounds to the kernel (such as bird, insect or hail damage), or from growth along the silks through the tip end of the ear [Wicklow *et al.*, 1988]. Using scanning electron microscopy, Bacon *et al.* [1992] showed that *F. moniliforme* is sequestered in the tip cap of the kernels. This localization of the fungus has led some researchers to the hypothesis that *F. moniliforme* arrives at the kernel by the way of growth occurring on or within the cob, and that the infection of an entire cob may begin from a single infection point. If infection is systemic [Kedera *et al.*, 1992], then strains carried on the kernel would have an advantage when the seed germinates to form a new plant. Developing strains of *F. moniliforme* that are unable to synthesize fumonisins, but that can successfully colonize the plant, may be an effective method of reducing fumonisin contamination of the resulting grain. We are presently in the process of developing such strains.

Recently, maize plants were shown to be routinely infected by multiple genetically discrete isolates of *F. moniliforme* under both field and greenhouse conditions [Kedera *et al.*, 1994]. These workers concluded that most plants would be infected by 2-3 different strains but reported individual plants that were infected by as many as five strains and individual ears that were infected with at least four different strains. With multiple infections as the rule, many plants are likely to be infected with strains of opposite mating type that can cross to produce perithecia and complete the life cycle under field conditions [Leslie, 1991]. Multiple infections also make it more difficult to do defined pathogenicity assays, e.g. those of Jardine and Leslie [1992] with *F. moniliforme* in sorghum, because the background strains may be responsible for disease symptoms that are inadvertently attributed to the inoculated strain.

F. moniliforme also has been shown to reduce or eliminate the infection of maize kernels by other fungi [Rheeder *et al.*, 1990; Van Wyck *et al.*, 1988; Wicklow, 1988; Wicklow *et al.*, 1988]. Zummo and Scott [1992] showed that *F. moniliforme* infection can inhibit kernel infection by *A. flavus* in inoculated ears and results in lower levels of aflatoxin contamination in the kernels. If antagonistic strains of *F. moniliforme* can be developed, then it might be possible to reduce both aflatoxin and fumonisin contamination by inoculating maize seed with appropriate strains of *F. moniliforme*. It might also be possible to develop synergistic strain sets that would be more competitive than would any single strain by itself. Manipulation of the *F. moniliforme* - *Zea mays* relationship through the use of modified fungal strains has the potential to significantly alter the agricultural practices presently in place; however, more detailed knowledge of this relationship is needed to determine the range of changes that are possible and the ease with which they could be implemented.

Population Genetics of *Fusarium moniliforme*

Most of the intraspecific population studies with these fungi that have not focused on mating type and sexuality have focused instead on the vegetative incompatibility (*vic*) loci [Leslie, 1993], that are responsible for the vegetative compatibility group (VCG) phenotype. These loci govern stability of heterokaryons that have been formed, but do not directly affect the ability of the two strains to physically fuse with one another. In *G. fujikuroi* the known *vic* loci all act in an allelic manner [Leslie, 1993], i.e. alleles at a given locus interact with other alleles at the same locus but not with alleles at other loci. Strains that are identical at all *vic* loci are able to form a stable heterokaryon, are said to be vegetatively compatible and to belong to the same VCG. An assumption that has gone into much VCG testing is that strains in the same VCG are closely related, if not clones, of one another. This assumption would be true if sexual reproduction and migration were relatively rare since most VCGs would then represent local clones. The validity of this assumption has not been critically tested in the *G. fujikuroi* system, however.

Several preliminary studies [Campbell *et al.*, 1992; Campbell and Leslie, 1993; Farrokhi-Nejad and Leslie, 1990] have been made of a set of approximately 400 isolates collected from two maize hybrids grown at 12 locations in eight states from the upper midwest portion of the United States. Over 90% of the isolates were in the A mating population with a few isolates belonging to the D and E mating populations recovered as well. All 24 seed lots contained both A$^+$ and A$^-$ mating types, but not necessarily in equal frequencies. Female fertility also varied by seed lot, ranging from 13-74%, with an overall average of 46% female fertile.

Extensive VCG analyses have been conducted with this large set of strains. Complementary *nit*rate non-utilizing (*nit*) mutants [Correll *et al.*, 1987] have been generated in every strain, and all strains have been paired in all possible pairwise combinations. Fifteen strains were identified as being heterokaryon self-incompatible (HSI) [Correll *et al.*, 1989], the remaining strains were divided into 154 VCGs. Nine of the 15 HSI strains could be assigned to one of these VCGs. Single-member VCGs, containing only a single isolate, accounted for 88 of the observed VCGs and accounted for 12-47% of the VCGs found per site. The 66 multi-member VCGs (mVCGs) accounted for the remaining 300 isolates. At least 4 and as many as 12 VCGs were found at each site in each seed lot. Genetic diversity (number of VCGs/number of isolates) ranged from 0.25-0.80 in isolates from hybrid 3377 and from 0.27-0.69 in hybrid 3475. Of the 66 mVCGs, 20 were recovered from at least one seed lot of each hybrid.

At six locations, isolates in a common mVCG were recovered from both hybrids. No seed lot was completely isolated from another seed lot since 3-10 mVCGs per seed lot were also represented in a second seed lot. At one location, 67% of the isolates belonged to a single VCG. This VCG contained more isolates than any other VCG detected, but no members of this group were found in any of the 23 other seed lots. At least 12 VCGs, comprising approximately 6% of the isolates, from hybrid 3377 were found at more than one site. In hybrid 3475, at least eight VCGs, comprising approximately 4% of the isolates, were found in as many as five sites.

Of the mVCGs that we examined, 34% contained only A$^+$ strains and 61% contained only A$^-$ strains. The remaining 5 mVCGs contained isolates with A$^+$ and A$^-$ mating types. In two of these five mVCGs, all of the strains but one had the same mating type. Within the other three mixed mVCGs, the strains in the VCG could be divided into two distinct groups. Within the subgroups, all strains were the same mating type and all VCG pairing reactions were strong and clear. Between subgroups pairing reactions were weaker, but usually detectable, and often appeared as a mere dark line of pigmentation beneath the agar with no visible aerial mycelium. The genetic basis of these weak interactions is unknown.

This set of 400 strains is a good target for future studies. In the sense that the two maize hybrid sources may each constitute a single population there are two populations available each with approximately 200 isolates. If geographic locations are considered, then 12 populations are available - each with 30-40 isolates per population. The total number of isolates to be examined may be significantly reduced if strains in the same VCG from the same seed lot are considered to be clones. The genetic diversity observed within the fields, as well as the observation of numerically small VCGs that were represented at multiple widely separated sites implies that some of the observed variability within seed lots may be attributable to migration of the fungus in asymptomatic seed, providing a convenient mechanism for gene flow between fields. Comparisons between this population and others originating from plants from different seed companies and in different geographic locations should greatly expand our understanding of the population structure of this pathogen. There also is a need for more detailed studies of populations from hosts such as rice, sorghum and sugarcane which have not been as intensively examined.

A major component to future population studies with *G. fujikuroi* should employ neutral markers, e.g. the anonymous RFLPs used to recently construct a genetic map of this organism [Xu, 1994; Xu and Leslie, 1993], in analyses such as those described by Milgroom [1995]. It is possible that some of the mating populations have co-evolved with the host they are most commonly associated with, e.g. mating population A with maize in the Americas and mating population F with sorghum in Africa. If this hypothesis is correct and if there has not been much migration, then A populations from maize in the Americas and F populations from sorghum in Africa should be much more polymorphic than A populations from maize in Africa and F populations from sorghum in the Americas. This hypothesis can be tested in many ways using molecular markers.

CONCLUSIONS

F. moniliforme is a complicated entity, in a taxonomic sense, whose identification remains difficult for many investigators. Recent studies of the genetics and field biology of the organism have permitted the discrimination of different groups within the population, and have opened new avenues of investigation into the biology of this organism. Data obtained from these studies will help to determine the risk(s) this organism poses for both human and animal food supplies, and will guide the development of new methods to reduce these impacts.

ACKNOWLEDGMENTS

Contribution no. 95-491-B from the Kansas Agricultural Experiment Station, Manhattan.

Work in my laboratory is supported in part by the Kansas Agricultural Experiment Station, the Sorghum/Millet Collaborative Research Support Program (INTSORMIL) AID/DAN-1254-G-00-0021-00 from the US Agency for International Development, the USDA North Central Biotechnology Initiative, USDA Cooperative Research Agreement 58-3620-109, Pioneer Hi-Bred International, and Myco Pharmaceuticals, Inc.

REFERENCES

Anderson, C. Restriction mapping of the internal transcribed spacer of the ribosomal DNA of *Fusarium* section *Liseola*. M.S. Thesis, Program in Genetics, Kansas State University, Manhattan, KS, 1994.

Bacon, C. W.; Bennett, R. M.; Hinton, D. M.; Voss, K. A. Scanning electron microscopy of *Fusarium moniliforme* within asymptomatic corn kernels and kernels associated with equine leukoencephalomalacia. *Plant Dis.* **1992** *76*, 144-148.

Caldis, P. D. Etiology and transmission of endosepsis (internal rot) of the fruit of the fig. *Hilgardia* **1927**, *2*, 287-328.

Campbell, C. L.; Leslie, J. F.; Farrokhi-Nejad, R. Genetic diversity of *Fusarium moniliforme* in seed from two maize cultivars. *Phytopathology* **1992**, *82*, 1082.

Campbell, C. L.; Leslie, J. F. Using VCGs to determine genetic diversity of *Fusarium moniliforme* in 24 maize seed lots. *Phytopathology* **1993**, *83*, 1413.

Chaisrisook, C.; Leslie, J. F. A maternally expressed nuclear gene controlling perithecial pigmentation in *Gibberella fujikuroi* (*Fusarium moniliforme*). *J. Hered.* **1990**, *81*, 189-192.

Correll, J. C.; Gordon, T. R.; McCain, A. H. Genetic diversity in California and Florida populations of the pitch canker fungus *Fusarium subglutinans* f. sp. *pini*. *Phytopathology* **1992**, *82*, 415-420.

Correll, J. C.; Klittich; C. J. R.; Leslie, J. F. Nitrate nonutilizing mutants of *Fusarium oxysporum* and their use in vegetative compatibility tests. *Phytopathology* **1987**, *77*, 1640-1646.

Correll, J. C.; Klittich, C. J. R.; Leslie, J. F. Heterokaryon self-incompatibility in *Gibberella fujikuroi* (*Fusarium moniliforme*). *Mycol. Res.* **1989**, *93*, 21-27.

DuTeau, N. M.; Leslie, J. F. RAPD markers for *Gibberella fujikuroi* (*Fusarium* section *Liseola*). *Fung. Genet. Newsl.* **1991**, *38*, 37.

Ellis, J. J. Section *Liseola* of *Fusarium. Mycologia* **1988**, *80*, 255-258.

Ellis, J. J. An alignment of toxigenic *Gibberella* strains having anamorphs in section *Liseola* of *Fusarium. Mycologia* **1989**, *81*, 307-311.

Elmer, W. H. A single mating population of *Gibberella fujikuroi* (*Fusarium proliferatum*) predominates in asparagus fields in Connecticut, Massachusetts and Michigan. *Mycologia* **1995**, *87*, 68-71.

Elmer, W. H.; Ferrandino, F. J. Pathogenicity of *Fusarium* species (section *Liseola*) to asparagus. *Mycologia* **1992**, *84*, 253-257.

Farrokhi-Nejad, R.; Leslie, J. F. Vegetative compatibility group diversity within populations of *Fusarium moniliforme* isolated from corn seed. *Phytopathology* **1990**, *80*, 1043.

Foley, D. C. Systemic infection of corn by *Fusarium moniliforme. Phytopathology* **1962**, *52*, 870-872.

Frederiksen, R. A., Ed. *Compendium of Sorghum Diseases.* APS Press: St. Paul, MN, 1986.

Hsieh, W. H.; Smith, S. N.; Snyder, W. C. Mating groups in *Fusarium moniliforme. Phytopathology* **1977**, *67*, 1041-1043.

Huss, M. J.; Leslie, J. F. Isozyme variation among six different biological species within the *Gibberella fujikuroi* species complex (*Fusarium* section *Liseola*). *Fung. Genet. Newsl.* **1993**, *40A*, 26.

Jardine, D. J. Stalk rots of corn and sorghum. Kansas State University Cooperative Extension Service Bulletin L-741, Manhattan, KS, 1986.

Jardine, D. J.; Leslie, J. F. Aggressiveness of *Gibberella fujikuroi* (*Fusarium moniliforme*) isolates to grain sorghum under greenhouse conditions. *Plant Dis.* **1992**, *76*, 897-900.

Kathariou, S. Gene pool organization in *Fusarium moniliforme.* Ph.D. dissertation, Department of Genetics, University of California-Berkeley, 1981.

Kedera, C. J.; Leslie, J. F.; Claflin, L. E. Systemic infection of corn by *Fusarium moniliforme. Phytopathology* **1992**, *82*, 1138.

Kedera, C. J.; Leslie, J. F.; Claflin, L. E. Genetic diversity of *Fusarium* section *Liseola* (*Gibberella fujikuroi*) in individual maize plants. *Phytopathology* **1994**, *84*, 603-607.

Klaasen, J. A.; Nelson, P. E. Identification of a mating population within the *Fusarium nygamai* anamorph. *Proc. VIIth Internat. Fusarium Workshop (University Park, Pennsylvania)* **1993**, 59.

Klittich, C. J. R.; Leslie, J. F. Nitrate reduction mutants of *Fusarium moniliforme* (*Gibberella fujikuroi*). *Genetics* **1988**, *118*, 417-423.

Klittich, C. J. R.; Leslie, J. F. Identification of a second mating population within the *Fusarium moniliforme* anamorph of *Gibberella fujikuroi. Mycologia* **1992**, *84*, 541-547.

Koehler, B. Natural mode of entrance of fungi into corn ears and some symptoms that indicate infection. *J. Agric. Res.* **1942**, *64*, 421-442.

Kuhlman, E. G. Varieties of *Gibberella fujikuroi* with anamorphs in *Fusarium* Section *Liseola. Mycologia* **1982**, *74*, 759-768.

Leslie, J. F. Mating populations in *Gibberella fujikuroi* (*Fusarium* Section *Liseola*). *Phytopathology* **1991**, *81*, 1058-1060.

Leslie, J. F. Fungal vegetative compatibility. *Annu. Rev. Phytopath.* **1993**, *31*, 127-151.

Leslie, J. F. *Gibberella fujikuroi*: Available populations and variable traits. *Can J. Bot.* **1995**, *73*, S282-S291.

Leslie, J. F.; Doe, F. J.; Plattner, R. D.; Shackelford, D. D.; Jonz, J. Fumonisin B$_1$ production and vegetative compatibility of strains from *Gibberella fujikuroi* mating population "A" (*Fusarium moniliforme*). *Mycopathologia* **1992**a, *117*, 37-45.

Leslie, J. F.; Klein, K. K. Motherhood and the price of sex. *Fung. Genet. Newsl.* **1995**, *43A*, 80.

Leslie, J. F.; Pearson, C. A. S.; Nelson, P. E.; Toussoun, T. A. *Fusarium* species from corn, sorghum, and soybean fields in the central and eastern United States. *Phytopathology* **1990**, *80*, 343-350.

Leslie, J. F.; Plattner, R. D. Fertility and fumonisin B$_1$ production by strains of *Fusarium moniliforme* (*Gibberella fujikuroi*). *Proc. 17th Sorg. Improve. Conf. No. Amer.* (Lubbock, Texas) **1991**, 80-84.

Leslie, J. F.; Plattner, R. D.; Desjardins, A. E.; Klittich, C. J. R. Fumonisin B$_1$ production by strains from different mating populations of *Gibberella fujikuroi* (*Fusarium* section *Liseola*). *Phytopathology* **1992**b, *82*, 341-345.

Leslie, J. F.; Raju, N. B. Recessive mutations from natural populations of *Neurospora crassa* that are expressed in the sexual diplophase. *Genetics* **1985**, *111*, 759-777.

Marasas, W. F. O.; Nelson, P. E.; Toussoun, T. A. *Toxigenic* Fusarium *Species: Identity and Mycotoxicology*; Pennsylvania State University Press: University Park, 1984.

Martin, J. P.; Handojo, H.; Wismer, C. A. Pokkah boeng. In *Diseases of Sugarcane: Major Diseases*; Ricaud, C.; Egan, B. T.; Gillaspie, A. G., Jr.; Hughes, C. G., Eds; Elsevier: New York, 1989; pp 157-168.

Milgroom, M. Analysis of population structure in fungal plant pathogens. In *Disease Analysis Through Genetics and Biotechnology: Interdisciplinary Bridges to Improved Sorghum and Millet Crops*; Leslie, J. F.; Frederiksen, R. A., Eds.; Iowa State University Press: Ames, IA, 1995; pp 213-229.

Nelson, P. E.; Toussoun, T. A.; Marasas, W. F. O. Fusarium *Species: An Illustrated Guide for Identification*; Pennsylvania State University Press, University Park, PA; 1983.

Perkins, D. D. How should the infertility of interspecies crosses be designated? *Mycologia* **1995**, *86*, 758-761.

Phinney, B. O.; West, C. A. Gibberellins as native plant growth regulators. *Annu. Rev. Plant Physiol.* **1960**, *11*, 411-436.

Raju, N. B.; Leslie, J. F. Cytology of recessive sexual-phase mutants from wild strains of *N. crassa*. Genome **1992**, *35*, 816-825.

Rheeder, J. P.; Marasas, W. F. O.; Van Wyck, P. S. Fungal associations in corn kernels and effects on germination. *Phytopathology* **1990**, *80*, 131-134.

Rohrbach, K. G.; Pfeiffer, J. B. Susceptibility of pineapple cultivars to fruit diseases incited by *Penicillium funiculosum* and *Fusarium moniliforme*. Phytopathology **1976**, *66*, 1386-1390.

Shaw, S. F.; Elliott, V.; Leslie, J. F. Genetic diversity in *Gibberella fujikuroi* (*Fusarium* Section *Liseola*) from bananas. *Fung. Genet. Newsl.* **1993**a, *40A*, 25.

Shaw, S. F.; Elliott, V.; Mansour, I. M.; Leslie, J. F. Genetic diversity in *Fusarium* Section *Liseola* from mangoes. *Fung. Genet. Newsl.* **1993**b, *40A*, 25.

Shurtleff, M. C., Ed. *Compendium of Corn Diseases*, 2nd ed.; APS Press: St. Paul, MN, 1980.

Snyder, W. C.; Hansen, H. N. The species concept in *Fusarium* with reference to Discolor and other sections. *Amer. J. Bot.* **1945**, *32*, 657-666.

Subbarao, K. V.; Michailides, T. J. Virulence of *Fusarium* species causing fig endosepsis in cultivated and wild caprifigs. *Phytopathology* **1993**, *83*, 527-533.

Sun, S.-K.; Snyder, W. C. The bakane disease of the rice plant. In Fusarium: *Diseases, Biology and Taxonomy*; Nelson, P. E.; Toussoun, T. A.; Cook, R. J., Eds.; Pennsylvania State University Press: University Park, PA, 1981, pp 104-113.

Thiel, P. G.; Marasas, W. F. O.; Sydenham, E. W.; Shephard, G. S.; Gelderblom, W. C. A.; Nieuwenhuis, J. J. Survey of fumonisin production by *Fusarium* species. *Appl. Environ. Microbiol.* **1991**, *57*, 1089-1093.

Van Wyck, P. S.; Scholtz, D. J.; Marasas, W. F. O. Protection of maize seedlings by *Fusarium moniliforme* against infection by *Fusarium graminearum* in the soil. *Plant Soil* **1988**, *107*, 251-257.

Varma, A.; Lele, V. C.; Raychaudhuri, S. P.; Ram, A.; Sang, A. Mango malformation: A fungal disease. *Phytopathol. Z.* **1974**, *79*, 254-257.

White, T. J.; Bruns, T.; Lee, S.; Taylor, J. Amplification and direct sequencing of fungal ribosomal RNA genes for phylogenetics. In *PCR Protocols: A Guide to Methods and Applications*; Innis, M. A.; Gelfand, D. H.; Sninsky, J. J.; White, J. W., Eds.; Academic Press: New York, 1990, pp 315-322.

Wicklow, D. T. Patterns of fungal association within maize kernels harvested in North Carolina. *Plant Dis.* **1988**, *72*, 113-115.

Wicklow, D. T.; Horn, B. W.; Shotwell, O. L.; Hesseltine, C. W.; Caldwell, R. W. Fungal interference with *Aspergillus flavus* infection and aflatoxin contamination of maize grown in a controlled environment. *Phytopathology* **1988**, *78*, 68-74.

Xu, J.-R. Electrophoretic karyotype and genetic map of *Fusarium moniliforme* (teleomorph *Gibberella fujikuroi* mating population A); Ph.D. dissertation, Genetics Program, Kansas State University, Manhattan, KS, 1994.

Xu, J.-R.; Leslie, J. F. RFLP map and electrophoretic karyotype of *Gibberella fujikuroi* (*Fusarium moniliforme*). *Fung. Genet. Newsl.* **1993**, *40A*, 25.

Xu, J.-R.; Yan, K.; Dickman, M. B.; Leslie, J. F. Electrophoretic karyotypes distinguish the biological species of *Gibberella fujikuroi* (*Fusarium* Section *Liseola*). *Mol. Plant-Microbe Interactions* **1995**, *8*, 74-84.

Yan, K.; Dickman, M. B.; Xu, J. R.; Leslie, J. F. Sensitivity of field strains of *Gibberella fujikuroi* (*Fusarium* section *Liseola*) to benomyl and hygromycin B. *Mycologia* **1993**, *85*, 206-213.

Zummo, N.; Scott, G. E. Interaction of *Fusarium moniliforme* and *Aspergillus flavus* on kernel infection and aflatoxin contamination in maize ears. *Plant Dis.* **1992**, *76*, 771-773.

15

GENETIC AND BIOCHEMICAL ASPECTS OF FUMONISIN PRODUCTION

Anne E. Desjardins, Ronald D. Plattner, and Robert H. Proctor

Mycotoxin Research and Bioactive Constituents Research
National Center for Agricultural Utilization Research
USDA/ARS, 1815 N. University Street
Peoria, IL 61604

ABSTRACT

Fumonisin mycotoxins are produced by *Gibberella fujikuroi* (*Fusarium moniliforme*) mating population A, a major pathogen of maize and sorghum worldwide. Fumonisin biosynthetic genes are being identified by genetic crosses utilizing naturally occurring fumonisin production variants. Meiotic analysis has identified three putative fumonisin biosynthetic loci. *Fum1*, which can control the ability to produce fumonisins, is being localized by marker-based mapping. *Fum2* and *fum3*, which control hydroxylation of carbon-10 and carbon-5, respectively, appear to be linked. Additional experimental crosses should elucidate the linkage relationships among *fum1*, *fum2* and *fum3*. When genetic analysis has localized the position of the fumonisin biosynthetic genes to a particular chromosomal region or regions, the genes will be identified by complementation of function via DNA-mediated transformation. Understanding fumonisin biosynthesis and its regulation should facilitate development of measures to control fumonisin contamination.

INTRODUCTION

Fumonisins are amino-polyalcohols that can contaminate maize and maize based products worldwide. Several fungal species of the genus *Fusarium* can produce fumonisins in agricultural crops and commodities. Interest in these toxins is due primarily to the discovery that fumonisins inhibit sphingolipid biosynthesis and may thereby impair human and animal health. The most effective control strategy for fumonisin contamination is prevention of *Fusarium* infection and fumonisin production in the field and in storage. In the long term, understanding the molecular biology of fumonisin production should help the development of practical and specific controls. In recent years, rapid advances in the molecular genetics of filamentous fungi have opened the way for detailed genetic analysis

Fumonisins in Food, Edited by L. Jackson *et al.*
Plenum Press, New York, 1996

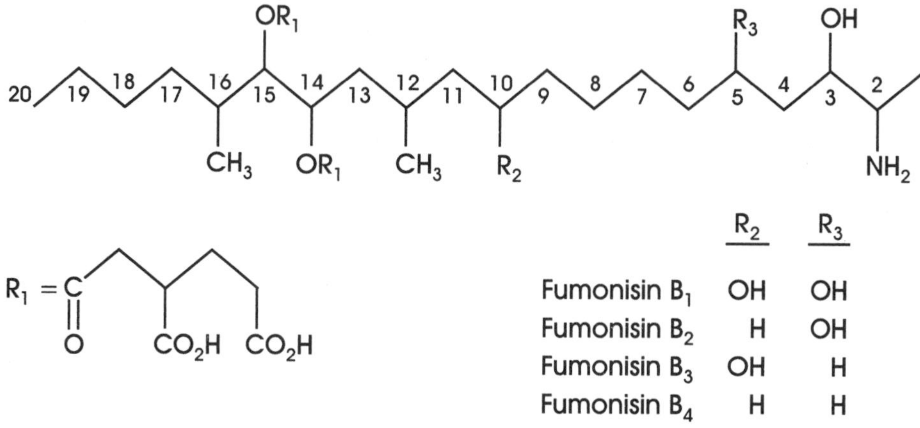

Figure 1. Structures of naturally occurring fumonisins.

of fumonisin biosynthesis in *Fusarium* species. This paper will describe recent progress in understanding the biochemistry and genetics of the fumonisin biosynthetic pathway.

Fumonisins were discovered in 1988 (Bezuidenhout et al., 1988) and their chemistry and toxicology are areas of active research. The most prevalent fumonisin homolog in naturally contaminated maize is fumonisin B_1 (FB_1), a propane-1,2,3-tricarboxylic diester of 2-amino-12,16-dimethyl-3,5,10,14,15-pentahydroxyicosane. Less oxygenated fumonisin homologs can also occur naturally. These include fumonisin B_2 (FB_2) which lacks the hydroxyl at carbon atom 10 (C-10), fumonisin B_3, which lacks the hydroxyl at C-5, and fumonisin B_4 which lacks hydroxyl groups at both C-5 and C-10 (Figure 1) (Bezuidenhout et al., 1988; Plattner et al., 1992).

Fumonisins are named after the fungus *Fusarium moniliforme* mating population A (teleomorph, *Gibberella fujikuroi*) from which FB_1 was first isolated. According to the taxonomic system of Nelson, Marasas and coworkers, at least some strains of six additional *Fusarium* species have been reported to produce fumonisins: *F. anthophilum, F. dlamini, F. napiforme, F. nygamai, F. proliferatum* and *F. subglutinans* (Nelson et al., 1993). Of the species tested, *G. fujikuroi* mating population A is probably the most agriculturally important producer of fumonisins because of its frequent association with food grains such as maize, millet and sorghum, and the ability of most strains of this species to produce high levels of fumonisins on maize.

FUMONISIN BIOSYNTHESIS IN *GIBBERELLA FUJIKUROI* MATING POPULATION A

Precursor Feeding Experiments

The structural similarity of fumonisins to the long chain sphingolipid bases suggested that fumonisin biosynthesis may be similar to sphingosine biosynthesis. The latter begins with the condensation of an amino acid with a fatty acyl-CoA which is catalyzed by the enzyme serine palmitoyltransferase. If fumonisins are synthesized in a similar manner, alanine would replace serine and an 18 carbon fatty acyl CoA would replace palmitoyl CoA. Stable isotope feeding studies of *F. moniliforme* determined that alanine is the biosynthetic precursor of C-1 and C-2 of FB_1 (Branham and Plattner, 1993; Blackwell et al., 1994). Further experiments with alanine labeled at multiple sites verified that alanine is incorporated directly and not via degradation to another precursor molecule (Plattner and Branham, 1994).

Sphingosine carbons other than C-1 and C-2 are derived from acetate via a fatty acid synthase. Addition of $[^{13}C]$ and $[^{14}C]$-acetate to cultures of *F. moniliforme* resulted in incorporation of label indicating that C-3 through C-20 of the FB_1 molecule are derived from acetate (Blackwell et al., 1994). Furthermore, even-numbered carbon atoms from C-4 through C-20 are derived from the methyl group of acetate, and odd-numbered carbon atoms from C-3 through C-19 are derived from the carboxyl group of acetate. Isotope feeding experiments have also shown that the methyl groups at C-21 and C-22 of FB_1 are derived from the S-methyl group of methionine (Plattner and Shackelford, 1992). These studies, however, were unable to determine whether methylation occurs before or after condensation with alanine.

Some major questions about fumonisin biosynthesis remain unanswered by the precursor feeding experiments to date. In particular, whether the fumonisin polyalcohol moiety is synthesized by a fatty acid synthase or by a polyketide synthase has yet to be determined. It can be difficult to determine whether a compound is biogenetically a fatty acid or a polyketide derivative. Apparent incorporation of labeled fatty acids can be the result of degradation to acetate prior to incorporation. Conversely, lack of incorporation of fatty acids can result from poor uptake or rapid degradation by fungal cells. If fumonisin biosynthesis is analogous to sphingosine biosynthesis, then alanine would be condensed with the CoA ester of an 18 carbon fatty acid, such as stearic acid. However, feeding experiments with isotope-labeled stearic acid failed to show incorporation of label into FB_1 (Plattner and Branham, 1994). These data may indicate that stearic acid is not a biosynthetic intermediate, or may simply reflect problems of fatty acid uptake or degradation.

In fatty acid biosynthesis, all carbonyls of the polyketone intermediate are reduced before detachment from the synthase enzyme complex. Functional groups can then be added by dehydrogenation and oxidation. The C-4 hydroxyl group of phytosphingosine, for example, originates from molecular-oxygen dependent hydroxylation (Kulmacz and Schroepfer, 1978). Conversion of a fatty acid intermediate to FB_4 would require oxygenations at C-14 and C-15; with further oxygenations at C-5 and C-10 to produce FB_3, FB_2 and FB_1. In this scheme, FB_4, FB_3, and FB_2 (or precursors of them) are likely intermediates in FB_1 biosynthesis.

Polyketide biosynthesis resembles fatty acid biosynthesis, but allows more variation in processing of the carbonyls before detachment from the enzyme complex to yield a greater diversity of products. Conversion of the polyketone intermediate to FB_4 would require deoxygenations at C-5, C-7, C-9,C-11, C-13, C-17 and C-19; and oxygenations at C-10 and C-14. In this scheme, FB_4, FB_3 and FB_2 (or precursors of them) are unlikely to be intermediates in FB_1 biosynthesis, because this would require deoxygenation and reoxygenation at C-5. At this time, precursor feeding experiments have not unambiguously distinguished between these alternative biosynthetic pathways. Incorporation studies with $[^{18}O_2]$ acetate, $[^{18}O]$ water and $^{18}O_2$ may help determine whether fumonisins are polyketides or fatty acid derivatives.

Classical Genetic Analysis

Experimental crosses offer an ideal approach to genetic dissection of fungal biochemistry and physiology. Hundreds of meiotic progeny can easily be obtained for analysis from each cross between a set of parent strains. Thus, complex genetic interactions can be dissected into monogenetic factors. *Gibberella fujikuroi* mating population A has many advantages as an experimental system for the genetic analysis of fumonisin biosynthesis.

This fungus is a heterothallic ascomycete that can complete its life cycle in a few weeks under laboratory conditions, and the products of meiosis are available in their original tetrads. *Gibberella fujikuroi* is haploid during almost all of its life cycle, which facilitates interpretation of segregation ratios. In addition, *G. fujikuroi* grows rapidly on defined media and can be transformed by using standard molecular biology techniques. Finally, Xu and Leslie (1993) have recently constructed a restriction fragment length polymorphism map of this species, and have correlated twelve linkage groups to twelve putative chromosomes separated using pulsed field gel electrophoresis (Xu et al., 1995).

Genetic analysis is based on the use of mutant strains that differ in the phenotype under study. Although mutations occur spontaneously in natural populations, most naturally occurring mutant phenotypes are not sufficiently distinct to be of use in genetic analysis. This may result because interacting genes are present, or because different alleles differ in their effect on the phenotype. However, extensive strain surveys, although laborious, should be able to identify mutants with discrete phenotypes, because these can occur with very low frequency among natural populations.

Most naturally occurring strains of *G. fujikuroi* mating population A produce predominately FB_1 (70-80%), and lower levels of FB_2 (10-20%), FB_3 (10-20%) and FB_4 (5-10%), when grown on maize substrate. Surveys of more than 350 genetically fertile members of mating population A isolated from various substrates in the Dominican Republic, Honduras, India, Mexico, Nepal, Taiwan, The Peoples' Republic of China, and the United States indicated that more than 95% of these strains produced predominantly FB_1 (Leslie et al., 1992a and 1992b; Desjardins et al., 1992 and 1995; Plattner, Desjardins, Leslie and Nelson, unpublished). Four rare types of FB_1 deficient mutants were identified in these strain surveys. Maize collected in Nepal yielded seven mutants that consistently produced no detectable FB_1, FB_2, FB_3 or FB_4 (Desjardins et al., 1992). Maize collected in South Carolina, USA, yielded two mutants that produced high levels of FB_2 and FB_4, with no detectable FB_1 or FB_3; and one mutant that produced high levels of FB_3 and FB_4, with no detectable FB_1 or FB_2 (Plattner, Desjardins and Leslie, unpublished). Maize and sorghum collected in Kansas, USA, yielded two mutants that were nonproducers of fumonisins in liquid culture and low producers of fumonisins (0-600 µg/g culture material) on maize substrate compared to standard strains (6000-16,000 µg/g) (Plattner, Desjardins and Leslie, unpublished). The first three types of FB_1 deficient mutants (fumonisin nonproducers; FB_2 and FB_4 producers; FB_3 and FB_4 producers) were easily and reliably scored, and thus were particularly useful for classical genetic analysis of fumonisin biosynthesis.

To determine the genetic basis of FB_1 deficient mutants, three types of mutants were crossed to wildtype strains that produce high levels of FB_1. Segregation of fumonisin production phenotypes was scored by high performance liquid chromatography of extracts of progeny cultures grown for four weeks on autoclaved maize. In all three cases, mutant and wildtype fumonisin production phenotypes segregated 1:1 among the progeny. First, strains M-5500 and M-5538, which produce no fumonisins, were crossed to strain M-3125, a producer of FB_1, FB_2, FB_3, and FB_4 (Crosses 57 and 65). Half of the tetrad progeny (Desjardins et al., 1992) and random ascospore progeny (Table 1) of Crosses 57 and 65 produced no fumonisins, and half produced high levels of fumonisins. Tests of allelism by crossing strain M-5538 (or one of its progeny) with a progeny of strain M-5500 (Crosses 97 and 98, Table 1) yielded no fumonisin producing recombinants among 278 progeny tested, indicating that these genes are tightly linked (Plattner, Desjardins and Leslie, unpublished). These data define a single locus, designated *fum1*, or a set of closely linked loci, that can determine whether fumonisins are produced.

Genetic analysis of mutant strains was continued by crossing strain A-0822, which produces only FB_2 and FB_4, to wildtype strain M-3120. Twenty-one of the random ascospore progeny of this Cross 109 produced only FB_2 and FB_4, and nineteen produced all four

Table 1. Segregation of fumonisin production in progeny of crosses of *Gibberella fujikuroi* mating population A

Cross number	Parents female male	fumonisin phenotypes	No. of random ascospore progeny in each fumonisin phenotype class				
			$B_1\,B_2\,B_3\,B_4$	$B_2\,B_4$	$B_3\,B_4$	No fumonisins	Total progeny
57	M-3125 M-5500	$B_1\,B_2\,B_3\,B_4$ no fumonisins	31	0	0	25	56[a]
65	M-3125 M-5538	$B_1\,B_2\,B_3\,B_4$ no fumonisins	13	0	0	7	20[a]
97	57-7-7 M-5538	no fumonisins no fumonisins	0	0	0	49	49[b]
98	57-7-7 65-4-5	no fumonisins no fumonisins	0	0	0	229	229[b]
109	M-3120 A-0822	$B_1\,B_2\,B_3\,B_4$ $B_2\,B_4$	19	21	0	0	40
397	M-3125 A-0819	$B_1\,B_2\,B_3\,B_4$ $B_3\,B_4$	23	0	16	0	39
459	109-R-14 A-0819	$B_2\,B_4$ $B_3\,B_4$	0	29	20	0	49
506	57-7-7 459-R-18	no fumonisins $B_3\,B_4$	0	0	5	16	21
510	57-7-7 109-R-20	no fumonisins $B_2\,B_4$	1	9	0	10	20

[a.] Data from (Desjardins et al., 1992 and 1995).
[b.] Liquid cultures of the strains from cross 97 and 98 were incubated for 2 weeks and analyzed for fumonisins by HPLC (Branham and Plattner, 1993). Maize cultures of strains from all other crosses in Table 1 were incubated for 4 weeks and analyzed for fumonisins by HPLC or GC-MS (Desjardins et al., 1992).

fumonisins (Table 1) (Desjardins and Plattner, unpublished). These data define a single locus, designated *fum2*, or a set of closely linked loci, that can determine whether fumonisins hydroxylated at C-10 are produced. Mutant strain A-0819, which produces only FB_3 and FB_4, was then crossed to wildtype strain M-3125. Sixteen of the random ascospore progeny of this Cross 397 produced only FB_3 and FB_4, and twenty-three produced all four fumonisins (Table 1) (Desjardins and Plattner, unpublished). These data define a single locus, designated *fum3*, or a set of closely linked loci, that can determine whether fumonisins hydroxylated at C-5 are produced.

To establish linkage relationships among fumonisin biosynthetic mutations, the three types of FB_1 deficient mutants (or their progeny) are being crossed to each other in various pairwise combinations. This genetic analysis is not complete, but preliminary results indicate that these mutations are all linked. Cross 459 between *fum2⁻* and *fum3⁻* mutations has yielded no wildtype recombinants among forty-nine progeny tested. Cross 506 between *fum1⁻* and *fum3⁻* mutations has also yielded no wildtype recombinants among 21 progeny tested. Cross 510 between *fum1⁻* and *fum2⁻* mutations has yielded one wildtype progeny among 20 progeny tested (Table 1) (Desjardins and Plattner, unpublished). These preliminary data are consistent with the hypothesis that *fum1*, *fum2* and *fum3* constitute a gene cluster encoding fumonisin biosynthetic enzymes in *G. fujikuroi* mating population A.

Because the fumonisin biosynthetic pathway is not known, we cannot yet assign gene enzyme relationships. The genetic data, however, are consistent with the possibility that *fum2* encodes a C-10 hydroxylase that can convert FB_4 to FB_3 and can also convert FB_2 to FB_1; and that *fum3* encodes a C-5 hydroxylase that can convert FB_4 to FB_2 and FB_3 to FB_1 (Figure 2). Preliminary efforts to test these hypotheses by feeding FB_2 and FB_3 to cultures of *G.*

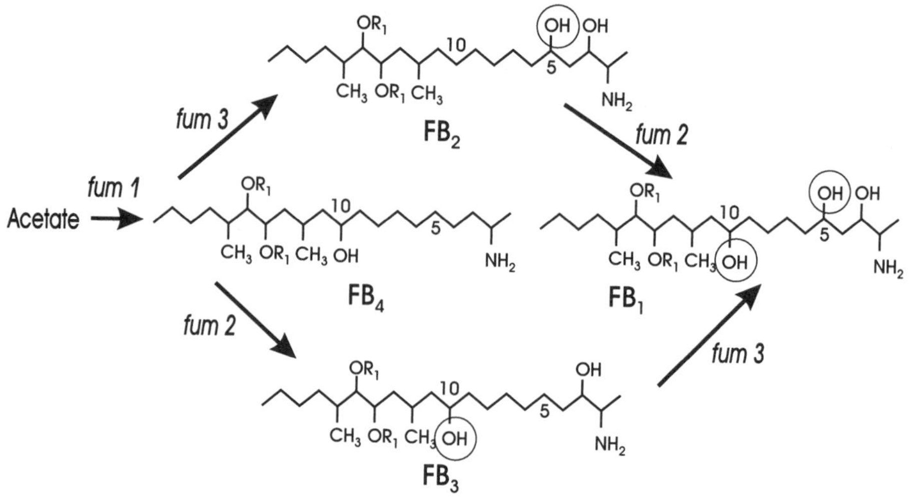

Figure 2. A proposed model for fumonisin biosynthesis based on the genetic data in Table 1

fujikuroi mating population A have been unsuccessful. The added compounds were neither degraded nor converted to FB₁ by the fungal cultures (Plattner, unpublished). These data may indicate that hydroxylation occurs prior to condensation with alanine, or may simply reflect poor uptake of fumonisins by fungal cells.

Marker-Based Mapping of *fum1*

An understanding of the precise function of *fum1*, *fum2*, and *fum3* in fumonisin biosynthesis requires that these loci be isolated and cloned. To this end, we are mapping the location of *fum1* in the genome of *G. fujikuroi* mating population A via rapid amplification of polymorphic DNA (RAPD) analysis (Williams et al., 1990). RAPD analysis is a polymerase chain reaction (PCR)-based technique used to identify genetic markers. The technique employs arbitrary 10-mer oligonucleotide primers to amplify DNA fragments from DNAs isolated from two or more individuals (e.g. fungal strains). The amplified fragments correspond to regions of DNA that are flanked at both ends by 10-nucleotide-long sites complementary to the oligonucleotide primers. Some amplified fragments from different individuals of the same species are the same, due to similarities in the genome of the individuals. However, because of genetic differences between the individuals, some of the amplified fragments should be different. These latter fragments constitute genetic (RAPD) markers. The identification of such markers that are closely linked to *fum1* should facilitate the isolation of this gene.

The experimental cross being used to map *fum1* utilized a fumonisin nonproducing (*fum1⁻*) parent (strain A04643) derived from field isolate M-5538, and a fumonisin producing (*fum1⁺*) parent (strain FKMA15) derived from field isolate M-3120. To speed the process of identifying RAPD markers linked to *fum1*, genomic DNAs from progeny from the cross between strain A04643 and FKMA15 (Xu and Leslie, 1992) were pooled, or bulked, according to a process known as bulked segregant analysis (Michelmore et al., 1991). The first bulk consisted of DNA from 20 *fum1⁻* progeny and the second bulk consisted of DNA from 20 *fum1⁺* progeny. Because the genome of each progeny results from recombination of the genomes of the two parental strains, the only region of DNA that should be the same for all individuals within each bulked DNA sample should be that in the vicinity of the respective *fum1* alleles. Thus the only DNA unique to the first bulk should be that in the region of *fum1⁻*. Likewise, the only DNA unique to the second bulk should be that in the vicinity of *fum1⁺*. All other regions of both parental genomes should be represented in each of the two bulks.

For RAPD-bulked segregant analysis, each 10-mer oligonucleotide primer was included in 4 PCR reactions, each with a different DNA template. In the first reaction, the template consisted of DNA isolated from the *fum1⁻* parent, in the second, the template consisted of DNA from the *fum1⁺* parent, in the third reaction, bulked DNA from *fum1⁻* progeny, and in the fourth reaction, bulked DNA from *fum1⁺* progeny. To date, 280 RAPD primers have been used to screen each of these DNA templates and we have identified two primers (OPA16 and OPH3) that amplify a fragment from the *fum1⁻* parent and bulked progeny DNA templates but not from the *fum1⁺* parent or bulked progeny templates (Figure 3). These data indicate that the *fum1⁻* allele and the two RAPD markers are linked. Additional RAPD analyses with DNA from individual progeny have facilitated the determination of recombination frequencies, and thus an estimation of the genetic distance between *fum1* and the two markers. These analyses indicate that the OPA16 and OPH3 markers are located on opposite sides of *fum1* at distances of 8.9 and 7.8 cM, respectively (Figure 4). In addition, the OPA16 and OPH3 markers are linked to other genetic markers on chromosome I that were previously shown to be linked to *fum1* (Xu and Leslie, 1993). However, a genetic

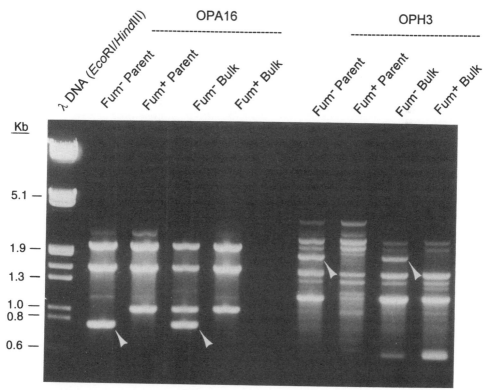

Figure 3. Agarose gel electrophoresis of PCR products from reactions with RAPD primer OPA16 or OPH3 and DNA templates from *fum⁻/⁺* parents and progeny bulks. Arrows indicate bands that are unique to the *fum⁻* parent and progeny bulks.

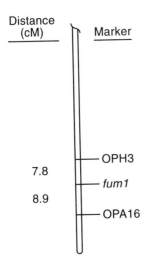

Distance
(cM)   Marker

7.8 ── OPH3

8.9 ── fum1

── OPA16

Figure 4. Linkage relationship between *fum1* and RAPD markers OPA16 and OPH3 (Operon Technologies, Inc.) on chromosome 1 of *Gibberella fujikuroi*. The position of *fum1* near one end of chromosome 1 was previously demonstrated by Xu and Leslie (1993).

distance of 7.8 cM could represent a physical distance of over 250 kilobase pairs, a distance too far to facilitate cloning *fum1*. Thus, we are currently attempting to identify additional markers more closely linked to *fum1*.

Once a putative *fum1* allele (*fum1$^+$*) that confers the ability to produce fumonisins has been isolated, its function will be confirmed by complementation analysis. In this analysis, the *fum1$^+$* allele will be transformed into a fumonisin nonproducing strain of *G. fujikuroi* that carries the alternative allele, *fum1$^-$*. If *fum1$^+$* does indeed confer the ability to produce fumonisins, the resulting transformants should be fumonisin producers. In addition, gene disruption of *fum1$^+$*, to render this gene nonfunctional, will be carried out to confirm that *fum1* is involved in fumonisin biosynthesis. Once this functional analysis of *fum1* has been completed, we can analyze the gene sequence to determine 1) the difference between the *fum1$^+$* and *fum1$^-$* alleles and 2) characteristics of the putative *fum1* gene product(s).

CONCLUSIONS

Biochemistry, classical genetics and molecular genetics are facilitating the dissection of the complex fumonisin biosynthetic pathway in *G. fujikuroi* mating population A. Although fumonisin biosynthesis appears to share some features with sphingolipid base biosynthesis, the extent of the similarities is not yet clear. It also appears likely that some of the fumonisin biosynthetic genes are physically linked and constitute a gene cluster on chromosome 1. Biosynthetic gene clusters are rather unusual but not unique in fungi. In particular, genes that control biosynthesis of aflatoxins by *Aspergillus* species (Skory et al., 1992), and trichothecenes by *Fusarium* species (Hohn et al., 1993) are also organized in complex clusters. The functional significance of gene clustering for these toxin biosynthetic pathways is not clear. Gene clustering may facilitate the formation of chromosome structures involved in the regulation of pathway gene expression, or may reflect unique evolutionary origins of these fungal toxin biosynthetic pathways[*].

_____

[*] Disclaimer: Names are necessary to report factually on available data; however, the USDA neither guarantees nor warrants the standard of the product, and the use of the name by USDA implies no approval of the product to the exclusion of others that may also be suitable.

REFERENCES

Bezuidenhout, S. C.; Gelderblom, W. C. A.; Gorst-Allman, C. P.; Horak, R. M.; Marasas, W. F. O.; Spiteller, G.; Vleggar, R. Structure elucidation of fumonisins, mycotoxins from *Fusarium moniliforme*. *J. Chem. Soc., Chem. Commun.* **1988**, 743-745.

Blackwell, B. A.; Miller, J. D.; Savard, M. E. Production of carbon 14-labeled fumonisin in liquid culture. *J. AOAC Int.* **1994**, *77*, 506-511.

Branham, B. E.; Plattner, R. D. Alanine is a precursor in the biosynthesis of fumonisin B_1 by *Fusarium moniliforme*. *Mycopathologia* **1993**, *124*, 99-104.

Desjardins, A. E., Plattner, R. D.; Shackelford, D. D.; Leslie, J. F.; Nelson, P. E. Heritability of fumonisin B_1 production in *Gibberella fujikuroi* mating population A. *Appl. Environ. Microbiol.* **1992**, *58*, 2799-2805.

Desjardins, A. E., Plattner, R. D.; Nelsen, T. C.; Leslie, J. F. Genetic analysis of fumonisin production and virulence of *Gibberella fujikuroi* mating population A (*Fusarium moniliforme*) on maize (*Zea mays*) seedlings. *Appl. Environ. Microbiol.* **1995**, *61*, 79-86.

Hohn, T. M.; McCormick, S. P.; Desjardins, A. E. Evidence for a gene cluster involving trichothecene-pathway biosynthetic genes in *Fusarium sporotrichioides*. *Curr. Genet.* **1993**, *24*, 291-295.

Kulmacz, R. J.; Schroepfer, Jr., G. J. Sphingolipid base metabolism. Concerning the origin of the oxygen atom at carbon atom 4 of phytosphingosine. *J. Am. Chem. Soc.* **1978**, *100*, 3963-3964.

Leslie, J. F.; Plattner, R. D.; Desjardins, A. E.; Klittich, C. J. R. Fumonisin B_1 production by strains from different mating populations of *Gibberella fujikuroi* (*Fusarium* section *Liseola*). *Phytopathology* **1992**, *82*, 341-345.

Leslie, J. F.; Doe, F. J.; Plattner, R. D.; Shackelford, D. D.; Jonz, J. Fumonisin B_1 production and vegetative compatibility of strains from *Gibberella fujikuroi* mating population "A" (*Fusarium moniliforme*). *Mycopathologia* **1992a**, *117*, 37-45.

Michelmore, R. W.; Paran, I.; Kesseli, R. V. Identification of markers linked to disease-resistance genes by bulked segregant analysis: a rapid method to detect markers in specific genomic regions by using segregating populations. *Proc. Natl. Acad. Sci. USA* **1991**, *88*, 9828-9832.

Nelson, P. E.; Desjardins, A. E.; Plattner, R. D. Fumonisins, mycotoxins produced by *Fusarium* species: biology, chemistry and significance. *Ann. Rev. Phytopathol.* **1993**, *31*, 233-252.

Plattner, R. D.; Weisleder, D.; Shackelford, D. D.; Peterson, R.; Powell, R. G. A new fumonisin from solid cultures of *Fusarium moniliforme*. *Mycopathologia* **1992**, *117*, 23-28.

Plattner, R. D.; Shackelford, D. D. Biosynthesis of labeled fumonisins in liquid cultures of *Fusarium moniliforme*. *Mycopathologia* **1992a**, *117*, 17-22.

Plattner, R. D.; Branham, B. E. Labeled fumonisins: production and use of fumonisin B_1 containing stable isotopes. *J. AOAC Int.* **1994**, *77*, 525-532.

Skory, C. D.; Chang, P.-K.; Cary, J.; Linz, J. E. Isolation and characterization of a gene from *Aspergillus parasiticus* associated with the conversion of versicolorin A to sterigmatocystin in aflatoxin biosynthesis. *Appl. Environ. Microbiol.* **1992**, *58*, 3527-3537.

Williams, J. G. K.; Kubelik, A. R.; Livak, K. J.; Rafalski, J. A.; Tingey, S. V. DNA polymorphisms amplified by arbitrary primers are useful as genetic markers. *Nucleic Acids Res.* **1990**, *18*, 6531-6535.

Xu, J.-R.; Leslie, J. F. RFLP map and electrophoretic karyotype of *Gibberella fujikuroi* (*Fusarium moniliforme*). *Fungal Genet. Newsl.* **1993**, *40A*, 25, (Abstract).

Xu, J.-R.; Yan, K.; Dickman, M. B.; Leslie, J. F. Electrophoretic karyotypes distinguish the biological species of *Gibberella fujikuroi* (*Fusarium* section *Liseola*). *Mol. Plant Microbe Int.* **1995**, *8*, 74-84.

16

FUSARIC ACID AND PATHOGENIC INTERACTIONS OF CORN AND NON-CORN ISOLATES OF *FUSARIUM MONILIFORME*, A NONOBLIGATE PATHOGEN OF CORN

C. W. Bacon and D. M. Hinton

Toxicology and Mycotoxin Research Unit, Russell Research Center
USDA, ARS, P.O. Box 5677
Athens, Georgia 30604-5677

ABSTRACT

Fusarium moniliform is a nonobligate parasite of corn, which exists as a complex of closely related fungi from different mating population or biological species. Strains of this fungus isolated from corn, have been determined to belong to mating populations A, although other populations have been isolated from corn. The ultrastructural association of the fungus with corn during growth, and the effects of the host on suppression of disease suppression are reviewed. This fungus enters a relationship with corn cultivars that is not always pathogenic. Pathogenesis is delayed, if it ever occurs. *F. moniliforme* can exist entirely as an endophyte, systemically colonizing kernels, remaining there until germination upon which the fungus infects the emerging seedlings. The symptomless association persists during the growth cycle of corn, and the resulting endophytic hyphae may be the source of mycotoxin production. The host's ability to suppress the fungus appears to be related to one class of compounds, the cyclic hydroxamic acids and their decomposition products, which can be catabolized by the fungi of mating population A but not C.

INTRODUCTION

Fusarium moniliforme Sheldon is a nonobligate pathogen that is commonly associated with corn where it spends most of its life cycle (Figure 1). This fungus also may exist as a facultative saprophyte during the remaining part of the life cycle (Nyvall and Kommedahl, 1970). As a saprophyte it can produce infective structures for primary and secondary disease establishment that may also serve as over wintering structures (Nyvall and Kommedahl, 1968; Ooka and Kommedahl, 1977; Liddell and Burgess, 1985). Further, it is only during the saprophytic stage that the teleomorphic state *Gibberella fujikuroi* (Sawada) Ito in

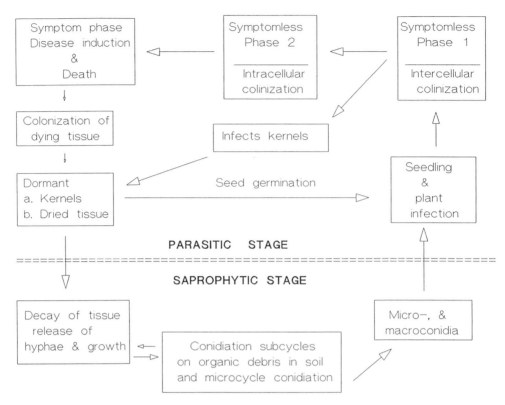

Figure 1. Life and disease cycles of *Fusarium moniliforme* on corn, expressed initially at the seedling stage of infection, where there is primarily a systemic route of infection resulting in endophytic colonization. The portal of entry for the infection may be either the kernel or seedling root. Although infection during the later stage of corn development is shown (plant infection), the events in this case is considered to entail the silk and silk channel resulting in infection only in the ear and kernels.

Ito and K. Kimura has been observed. Unlike most other nonobligate pathogens, i.e., those that kill their host cells in advance of infection and obtain nutrients from nonliving tissue, the mode of nutrition of *F. moniliforme* is complicated by the fact that a large percentage of the isolates get their nutrients from living host cells after establishing close contact with them. Disease symptoms and death of the infected corn plant may result from either the direct or indirect involvement of biotic and abiotic factors and or the depletion of nutrients. The disease cycle is not common on modern corn cultivars and is probably not a major portion of the life cycle of this fungus, but due to reduction in the grain yield, it is the disease cycle that has attracted the most attention. Detailed knowledge of both the disease and the normal life cycles is important since as long as the fungus is viable it is capable of producing toxins that can accumulate in the grain.

Various aspects of the fungus life cycle have been controversial in the past, which apparently reflects the diversity of fungi found within a population with no real attempt to distinguish between strains. *F. moniliforme* belongs to the section *Liseola* and can be separated into two reproductively isolated mating populations designated as populations A and F of *G. fujukuroi* (Klittich and Leslie, 1992). Four other mating populations (B, C, D, and E) have also been described (Leslie, 1991). Fungi of mating population A are primarily associated with corn, while fungi of mating population F are usually associated from sorghum. Species isolated from sorghum belong to group D and are designated as *F. proliferatum* (Matsushima) Nirenberg. Another group is referred to here as the 'rice group' since they are isolated from rice; others have placed fungi from this host into group C, and it is the intention to designate these isolates as *F. fujikuroi* (Paul Nelson, personal communication). Fungi within these mating populations vary relative to their specific mycotoxin content, especially fumonisin (Leslie, 1991; Leslie et al., 1992), and may consist of genetically diverse strains (Klittich and Leslie, 1992; Yan et al., 1993; Kedera et al., 1994; Desjardin et al., 1994).

As is true of most nonobligate pathogens, *F. moniliforme* is not host specific, and has been recovered from sorghum, wheat, rice, oats, beans, cotton, peanuts, pecans, bananas, sugar beets, green peppers, flax, soybean, figs, stone fruits, several forages and sugar cane (see review of Bacon and Nelson, 1994 and references contained therein). The listing of plant diseases caused by *F. moniliforme* indicates that it is either the primary or secondary incitant of seedling blights, foot rots, stem rots, pre- and postharvest fruit rots, stunting and hypertrophies in at least 32 plant families (Booth, 1971; Kommedahl and Windels, 1981). However, it is the parasitism of corn that is the primary focus since on corn this may fungus produce several different mycotoxins that belong to very different chemical classes. This is not surprising since nonobligate parasites are considered able to attack many different plants and plant parts because they produce a large number of nonspecific phytotoxins or enzymes that affect processes or substances that are common to all plants. We believe that many of these phytotoxins are also mycotoxins, of which the fumonisins, the major focus of current mycological, toxicological and chemical studies, constitutes only one class.

Fungi of the section *Liseola* produce at least one to all of five known biologically active compounds that may function as either a phytotoxin or mycotoxin (Table 1). The effects that these compounds produce on animals vary and their detailed animal toxicology have been reviewed (Marasas et al., 1984; Riley et al., 1993; Norred et al., 1992). Briefly, fusarin C is considered a mutagen (Wiebe and Bjeldanes, 1981; Cheng et al., 1985; Bjeldanes and Thomson, 1979). Fusaric acid was initially considered a hypotensive agent (Hidaka et al., 1969) but presently a moderately toxic metabolite (Porter and Bacon, 1993) and a potent synergist with other mycotoxins (Bacon et al., 1995; Dowd, 1988). Moniliformin is another weak mycotoxin and phytotoxin (Cole et al., 1973). Finally, beauvericin is an insect toxin (Vesonder and Hesseltine, 1981; Plattner and Nelson, 1994) whose toxicity to mammals is unknown. Other taxa outside the section *Liseola* produce these and related compounds, but many of these taxa are not considered pathogens of corn, which indicates again the nonspecific nature of the fusaria metabolites as phytotoxins. Indeed, no role for these toxins in the pathogenesis of specific corn diseases has been established, nor has the nature of the

Table 1. Biologically active compounds produced by *Fusarium moniliforme*

Toxins[a]	Phytotoxic	Mycotoxic
Fumonisin B$_1$	+[b]	+
Fusarin C	?	+
Beauvericin	?	+\-
Moniliformin	+	+
Fusaric acid	+	+
Dehydrofusaric acid	+	?
Hydroxyfusaric acid	+	?

[a] Fungi reported as *F. moniliforme* but mating population, i.e., biological species, not always reported; see text for references.

[b] +, active; -, inactive; ?, unknown.

host-parasite relationship been described for fungi within specific populations of *F. monili-forme*.

The nature of the association of *F. moniliforme* with corn varies and may reflect genetic variation within field populations. Some isolates are virulent, while others are apparently nonvirulent but highly infective and associated with corn without producing any signs of disease (Thomas and Buddenhajen, 1980; Foley, 1962; Kedera et al., 1994). The latter aspect of this association, referred to here as symptomless association, is a major form of this association and the only form that allows for the survival of the plant as a food or feed item with a potential for the accumulation of mycotoxin. However, very little is known about this aspect of the life cycle of the fungus. Present corn breeding practice is to select against the disease aspect and to ignore symptomless association. Major questions remain to be answered. What is the nature of the symptomless association found endophyticly in corn throughout the life of the plant? Do any of the phytotoxins facilitate disease expression by this infection? Are these endophytic hyphae latent? Are the endophytic hyphae metabo-lically active and do they contribute to the *in planta* or kernel mycotoxin content during plant growth? Of the known phytotoxins, fusaric acid is the only metabolite with known effects on pathogenesis. The ability of this toxin to cause any of several corn diseases is unknown and the relationship of this toxin to the symptomless association of *F. moniliforme* infection in corn is undefined.

This study has three major objectives that will be presented in three questions. First, what are the effects of fusaric acid in corn diseases, primarily diseases of seedlings? Second, what are the symptoms produced by corn and non-corn isolates of *F. moniliforme* will be discussed and what are the ultrastructural associations of between these isolates and the plant? Third, do stress metabolites of corn provide a potential means of control of the symptomless association by *F. moniliforme*? As a prelude to these effects a very brief description of the major corn diseases produced by *F. moniliforme* will be reviewed so that the symptoms observed during pathogenesis can be related to specific toxins and or infections.

SYMPTOMS OF *F. MONILIFORME*-INDUCED DISEASES OF CORN

Several major diseases of corn are attributed to *F. moniliforme*, including seed rot, seedling blight, root rot, stalk rot, and kernel or ear rot (Cook, 1981; Shurtleff, 1980; Kommedahl and Windels, 1981). In seed rot the kernels decay before and during germination and very few ever produce coleoptile and radicals. Seed rot is also caused by fungi other than *F. moniliforme*, and the extent of *F. moniliforme's* involvement in field cases of seed rot is unknown. Seedling blight, or damping-off, caused by *F. moniliforme* can be distinguished from diseases caused by other fungi by a white-to-pink mycelium with masses of spores in stem tissues, mainly the mesocotyl near the ground level with other tissues in the leaves appearing water-soaked. Root rot may be initiated as a part of the seedling blight disease but persists into the matured corn plant where late in the season symptoms similar to seedling blight appear on roots. These symptoms may be exaggerated during plant maturity and seed development. Root rot is not a major factor in reducing plant yield except under poor growing conditions that induce rapid senescence of root tissue. Stalk rot is very destructive and is a major factor responsible for yield reduction. It may be initiated from root rot or initiated directly by conditions favorable to root rot. The nature of losses due to stalk rot caused by *F. moniliforme* is difficult to define since under field conditions there is a large complex of contributing fungi (Shurtleff, 1980). Symptoms of stalk rot disease caused by *F. moniliforme* include a salmon-colored mycelium in the stem pith, which disintegrates resulting in stalk breakage or lodging. Under warm, moist conditions, a cottony-pink growth of mycelium with conidia appears on the leaf sheaths and at the node usually following lodging.

The above diseases are responsible for most of the economic losses on corn that are attributed to *F. moniliforme*. If the plant is killed during the seedling or germination stage, there is very little chance of certain isolates of the fungus being incorporated into the food chain. However, the symptomless association described below may be very significant to human and livestock health since the nature of this relationship appears to be compatible during most of the life cycle of the fungus in corn.

PHYTOTOXICITY OF FUSARIC ACID TO CORN

Fusaric acid (5-butylpicolinic acid) was first discovered during the laboratory culture of *F. heterosporum* Nees by Yabuta et al. (1937). This toxin was one of the first fungal metabolites implicated in the pathogenesis of wilt symptoms of tomatoes caused by *F. oxysporum* Schlecht. emend. Snyd & Hans. (Gaumann, 1957). Subsequently, this toxin has been isolated from *F. moniliforme*, *F. subglutinans* (Wollenw. & Rink.) Nelson, Toussoun & Marasas, and *F. solani* (Mart.) Appel & Wollenw. emend Snyd. & Hans. It is now known to universally occur in several populations of *F. moniliforme*, *F. proliferatum*, and in the newly erected species *F. napiforme* (Table 2, Bacon, C., Porter, J.K. and Leslie, J. F., 1995, unpublished data). There are at least two closely related analogues, dehydrofusaric acid (Stoll and Renz, 1957) and hydroxy-fusaric acid (Table 1) that co-occur with fusaric acid (Braun, 1960; Pitel and Vining, 1970). Both analogues are phytotoxic, although the latter is much weaker than the former. The variation and complex nature of the phytopathological response is due to the α-carboxyl group, and the nature and length of the alkyl side chain of fusaric acid (Backmann, 1956). The α-carboxyl group is responsible for inhibiting respiratory and polyphenol oxidase activities while the alkyl side chain affects membrane permeability (Fisher, 1965; Tamari and Kaji, 1954). Picolinic acid, a structurally related compound without the alkyl side chain, inhibits the growth of corn roots (Vesonder and Hesseltine, 1981), a process requiring high respiratory metabolism. However, picolinic acid has not been isolated from cultures of *F. moniliforme*. The lack of root development is part of the symptoms produced by fusaric acid when it is added to germinating kernels or when kernels are infected by specific isolates of *F. moniliforme*.

Table 2. Average production of fusaric acid by *Fusarium* species[a]

Species[b]	Mating population	Host	Total fusaric acid	
			Medium[c]	Corn
F. moniliforme	A	Corn	13.9	10.2
F. moniliforme	F	Corn	125.4	96.96
F. proliferatum	D	Sorghum	95.0	8.7
F. fujikuroi	C	Rice	615.9	987.9
F. subglutinans	B and E	Corn	367.2	98.5
F. crookwellense	none	Corn	467.2	234.9
F. napiforme	unknown	Corn	15.9	17.4

[a]Bacon, C.W., Porter, J.K. and Leslie, J. F. (unpublished).
[b]Data represent an average of at least 15 isolates, except 6 for *F. fujikuroi*, 7 for *F. subglutinans*, and 2 each for *F. crookwellense* and *F. napiforme*.
[c]Concentration in medium, mg/L after 30 days incubation; in corn, mg/kg after 4 weeks incubation.

The method used to determine fusaric acid production, cultivar resistance, and pathogenicity of each isolate was a modification of that used by Molot and Simone (1967). All inoculations and plantings were performed under aseptic conditions. Sterilized kernels (100 to 150) of each cultivar were placed in a 100 x 15 mm Petri dish and inoculated with 5 ml of each fungal inoculum. Inoculated kernels were incubated in the dark for two days at 25 °C and then for two to four days, at 2 to 6 °C. There was approximately 70-99% germination of all cultivars. Early coleorhiza protrusion was the germination criterion for most cultivars, although for some slow germinating cultivars enlargement of the embryo axis was used as the criterion.

The ability of strains from different mating populations of *F. moniliforme* to produce fusaric acid on laboratory media (Pitel and Vining, 1970; Paterson and Rutherford, 1991) and autoclaved cracked corn was measured. Samples were ground in methyl alcohol:1% KH_2PO_4, 1:1 (v/v, pH of 3.0). The fusaric acid was extracted into methylene chloride, which was then extracted with a 5% $NaHCO_3$ solution. The pH of the bicarbonate solution was lowered to 3.0 with HCl, and the solution was extracted with methylene chloride. The fusaric acid content was determined by HPLC. *F. fujikuroi*, the rice isolates, produced the highest amounts of fusaric acid on both substrates (Table 2). Of the two remaining mating populations, strains from mating population F produced more fusaric acid than different strains from population A.

The role of fusaric acid in the pathogenesis of corn seedling disease relative to the amount of toxin produced under subculture conditions was examined. Kernels were sterilized to remove surface and systemic seed borne pathogens (Bacon et al., 1994), and inoculated with conidia from one of several isolates of *F. moniliforme* (Table 3). The kernels were planted in sterile soil, and the seedlings and plants were watered daily, and maintained in a plant growth room under 16 hr of light (256 μEinstein m^{-2}) at 29 and 32 °C day and night, respectively. *Fusarium*-infected plants were harvested and analyzed for fusaric acid (Bacon et al., 1995 unpublished data). Pathogenesis, measured as the inability of seedlings to emerge (per cent germination) after 21 to 25 days, was actually a measure of both seed rot and seedling blight.

We found no correlation between pathogenesis and the *in vitro* and *in vivo* production of fusaric acid (Table 3). Further, *in planta* concentrations of fusaric acid were not consistent with either stalk rot or seedling blight. There was an inhibition of seedling growth and death when the virulent isolate, RRC 374, was added to seedlings. The topical application of fusaric acid to leaves of one- to four-week-old seedlings of five cultivars and inbred corn lines (Bacon et al., 1994) produced necrotic symptoms but there was no wilting and or water soaked symptoms similar to those described above. These results indicate that although there

Table 3. Production of fusaric acid *in planta*

Fungus[a]	% Seedling germination[b]	Fusaric acid[c]	
		Culture	*In planta*
RRC 374	10	21.2	11.2
RRC PAT	91	49.0	67.2
RRC 415	70	31.3	21.2
ATCC 14164	99	715.4	121.2

[a] ATCC 144164 was a rice isolate and presumed to belong to mating population C, i.e. *F. fujukuroi*; all other isolates belong to *F. moniliforme,* mating population A.
[b] Trucker's Favorite, a field corn.
[c] Concentration in medium, mg/L; concentration in plants mg/kg dry weight, plants harvested after 4-6 weeks growth.

is some phytotoxicity attributable to fusaric acid that extends beyond the seedling stage, this toxin is probably responsible for seed rot or seedling die-off.

The role of fusaric acid into the pathogenesis of several diseases in other plants caused by *F. oxysporum* (Davis, 1969; Heitefuss et al., 1960; Fisher, 1965) is also not clear. Studies utilizing non-fusaric acid producing mutants showed no correlation between pathogenicity and the ability to produce fusaric acid (Kuo and Scheffer, 1964). Further, highly virulent isolates of *F. oxysporum* could produce either high or very low levels of fusaric acid (Davis, 1969). Clearly, these data suggest that there are other factors than fusaric acid that are important in the ability of these fungi to induce plant diseases. Other than seed rot and the early aspects of seedling blight, the role of fusaric acid in the symptomless association in corn and corn diseases described above is not well supported by data.

HOST-PARASITE ASSOCIATION

The studies described in this section review the effects of *F. moniliforme* on the infection of kernels and seedling plants up to 6 weeks old, since beyond this age, it is very difficult to grow gnotobiotic corn plants. The isolates of fungi used in this study to determine the nature of the association between the fungus and the corn plant were obtained from corn (population A), sorghum (population D), and rice, (population C). Inoculum from each fungus was prepared by incubating the fungus on potato-dextrose agar (Difco, Detroit, MI) under a 12 h light/dark cycle at 25-27 °C for 15-21 days. Inocula consisted of both conidia and mycelium obtained by flooding a Petri plate with 10 ml of sterile distilled water. The conidial concentrations averaged 10^6-10^9 conidia per milliliter for most isolates, but this concentration was not critical, especially since mycelia also contributed to infection, and lower concentrations, i.e. 10^3, gave identical results. The kernels were double sterilized to completely kill both surface and internally borne microorganisms (Bacon et al., 1994). Thus, the data reflect only effects from the isolates placed on the seed. The six cultivars of corn used in this study, consisted of inbred and hybrid lines of field and sweet corn, some of which were established as resistant to *F. moniliforme*. All cultivars tried were susceptible to infection and the symptomless association (Bacon et al., 1994). Commercial cultivars included Truckers Favorite, a field corn, and Florida Staysweet, a sweet corn. The inbred and hybrid lines of field corn included the genetic cultivars PR and K61, and K61 x PR and PR x K61, respectively. Light microscopy and ultrastructural studies were performed on sterilized kernels (Bacon et al., 1994) or on gnotobioticly cultured corn genotypes described above. Plant tissue used for the light microscope study was stained first with 2,3,5-triphenyltetrazolium chloride and then with aniline blue. *F. moniliforme*-infected corn plants stained by this procedure had highly contrasting dark blue stained hyphae against a background of unstained plant tissue (Bacon, C.W. and D. M Hinton, 1995, unpublished data).

The ultrastructure of kernels that were naturally infected by *F. moniliforme* in the field showed that the fungus is systemically localized within the pedicel of each kernel (Bacon et al., 1992), and although not established, it is likely that from this location the systemic infection of seedlings occurs. This study also showed that the fungus was not associated with the vascular bundle tissue within the pedicel, and that it did not invade the vascular bundle during germination, which reinforces the conclusion of Pennypacker (1981) that *F. moniliforme* is not a vascular rot or wilt pathogen in corn. Micrographs of the root system did not reveal hyphae within any tissue of the stele, although occasionally a hypha was seen in the endodermis (Bacon and Hinton, 1995, unpublished data).

The fungi (Table 3) used to established the host-parasite relationship would either show symptoms of seedling blight or would remain symptomless on specific corn cultivars or inbred lines (Bacon et al., 1994). The exception to this rule was strain ATCC 14164 from rice. This isolate produced the bakanae disease in all corn genotypes and never produced seedling blight. All isolates infected the seedlings soon after germination since *F. monili-forme* could be recovered from 2 day-old gnotobiotic seedlings. The fungus was isolated from the leaf, stem, hypocotyl, and roots. The distribution of *F. moniliforme* was the same in both diseased and symptomless fungi-seedling combinations (Table 3). This distribution indicates that anatomical features within corn genotypes did not account for the lack of disease expression. The topical application of inoculum to gnotobiotic corn kernels and the recovery of that isolate from the plant, indicates the importance of the seedborne infection as both a disseminating agent and as a source of inoculum. Surface sterilized kernels, known to be systemically infected, also yielded *F. moniliforme*. This indicates that the systemically infected kernels are also important inoculum sources.

The hyphae within tissue of leaves and roots of two to six week-old plants not showing signs of a disease were intercellular (Figure 2, a-c). This endophytic habit was observed in cultivars and fungal strain combinations known to show both the symptomless association and the disease and is characteristic of phase 1 of the parasitic state of the life cycle (Figure 1).

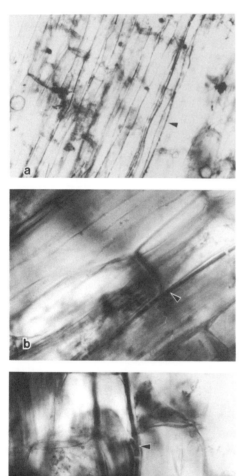

Figure 2. Light micrograph of the endophytic hyphae of *Fusarium moniliforme*. a) A view of the fungus (arrow) among root tissue showing several hyphae running parallel with the long axis of the cells. b) Higher magnification of corn tissue showing a hypha (arrow) between two cells walls. c) Higher magnification showing several hyphae (arrow) running parallel with cell.

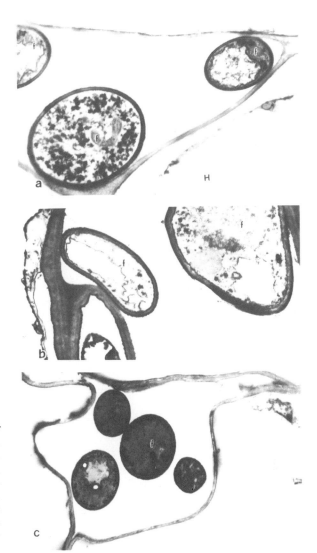

Figure 3. Transmission electron micrographs of the endophytic hyphae of *Fusarium moniliforme* in cross section. a) Three hyphae (f) in an intercellular space formed by three corn cells (H). b) and c) Hyphae in intercellular spaces of older corn seedlings; note in b) the hyphae have developed signs of degeneration; plants with such an infection never show symptoms of a disease.

Intercellular hyphae are in direct contact with the cell walls of the host (Figures 3a-c). At the ultrastructural level the walls of the fungus and host are not separated by an extracellular granular matrix reported for other intercellular pathogens and mutualist (Ehrlich and Ehrlich, 1971; Hadley et al., 1971; Hinton and Bacon, 1985). Both symptom and symptomless fungal-corn combinations separated the host cells along the middle lamella (Figure 2, b and c), which characterizes phase 2 of the parasitic stage and the initiation of the disease cycle (Figure 1). Although there was no extensive study of the fungus and plant interface, there was some evidence of cell wall disintegration, which if present, may be localized and would go unnoticed. Throughout the entire period of host colonization of the symptomless fungal-corn combination (phases 1 and 2) there was no evidence that the intercellular hyphae alter the integrity of host cytoplasmic membranes. This suggests that during the symptomless phases 1 and 2 there is no toxin being produced by the fungus disruptive to the host membrane, nor is there a host reaction indicative of a disease reaction, e.g. the hypersensitive response.

Host cell walls responding to the early symptoms of the disease often show wall appositions beneath the hyphal contact points (Ehrlich and Ehrlich, 1971; Bracker, 1967). There were wall appositions in the adjacent walls of infected corn plants early before disease symptoms are initiated (Figures 4, a and b). These wall appositions are several orders of

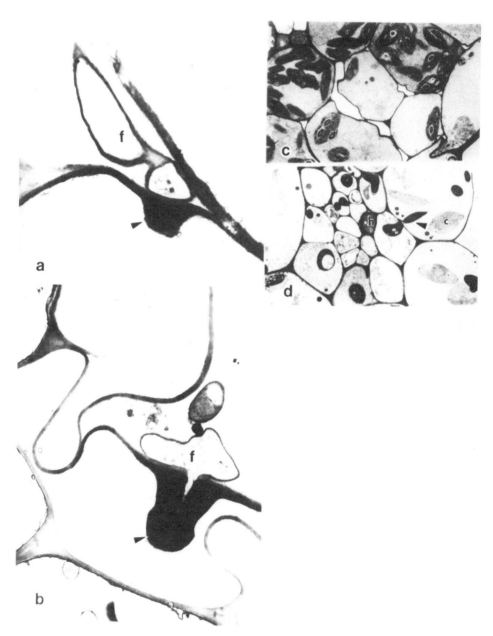

Figure 4. Transmission electron micrographs of three week old seedlings showing initial symptoms of seedling blight. a) and b) show host responses to the infection by the production of wall appositions (arrows) beneath hyphal contact points. In b) note that the fungus (f) has split the wall apposition just before its enters the cells. c) A normal leaf without the fungus. d) A leaf showing that the fungus (arrow) has invaded several host cells. Hyphae can be seen penetrating from one host cell to the next.

magnitude larger than the normal cell wall, and are highly electron dense. The appearance of the wall appositions is characteristic of the initiation of the disease state (Figure 1). Similar wall appositions occur in cells of host and nonhosts of several pathogens (Galum et al., 1970; Heath, 1972; Bracker and Littlefield, 1973). These large thickened areas are similar to those observed in certain lichen (Jacobs, 1969), and other mutualistic associations (Hinton and Bacon, 1985; Hadley et al., 1971). Early in the association there was no evidence of penetration of host cells by hyphae or any specialized infection structure such as appressoria or penetration pegs (phases 2 and 3). Three weeks following infection, hyphae became intracellular, invading moribund or dead corn tissue, often showing greatly enlarged penetration pegs (Figure 4d). This is the disease state of the symptom phase, which is further characterized by the destruction of the integrity of the host organelles (Figure 4d). Seedlings showing this state are visibly dying, with necrotic tissue representing over 50% of the plant. The initiation of the disease phase of the parasitic stage is not common, and probably reflects plant selection against this phase (King and Scott, 1981; Headrick and Pataky, 1989; De Leon and Pandey, 1989), and or an interaction with suitable environmental conditions, oftentimes late in the life of the plant (Ochor et al., 1987; Styer and Cantliffe, 1983).

THE SYMPTOMLESS ASSOCIATION AND FUMONISIN PRODUCTION

The precise timing of fumonisin formation and accumulation patterns in the vegetative parts of corn seedlings are unknown. Fumonisins have been detected in corn cobs, glumes of old corn florets (bee wings), kernels, and corn screenings. The highest concentrations have been detected in bee wings. However, corn screenings have higher concentrations than whole corn and have produced the most toxicological concern, particularly to swine and horses. This distribution suggests that the highest amount of fumonisins occurs in diseased tissue as evidenced by the high concentrations occurring in screenings, which often result from the shatter of decayed or poorly developed kernels. Such kernels might result from conditions favorable for fungal growth during kernel development, i.e., wet environmental conditions, and or improper storage conditions. Fumonisins also are found in corn and corn-based products for human consumption (Chu et al., 1995; Sydenham et al., 1991; Thiel et al., 1991; Richard et al., 1993), which indicates that sound corn, i.e. symptomless kernels, contains the fumonisins. The symptomless association might account for fumonisin contamination of these kernels.

The subject of symptom development is one of the least defined areas of plant pathology. Perhaps when this topic is very well defined, symptomless association can also be defined. There is no obvious explanation for what apparently is a basically compatible relationship lasting for an extended time period. Indeed, in most instances the host-parasite interaction never develops into a disease. To explain this interaction in the context of a pathological concept, we could consider that the symptomless association is the result of host suppression. This concept assumes that the host is responsible for the suppression. The suppressive substance(s) must therefore be produced by the host.

We have studied several natural substances in corn and found that the cyclic hydroxamic acids might be responsible for the suppression of this fungus (Niemeyer, 1988; Wahlroos and Virtanen, 1958; Whitney and Mortimore, 1959). These corn metabolites are low molecular weight compounds found in many grasses, and the highest concentration occurs during the early stages of seedling growth (Xie et al., 1991). The cyclic hydroxamic acids are defense chemicals because of their biological activity against fungi, insects, and bacteria. Thus, they are not necessarily phytoalexins, but are produced in response to a wide range of generalized stimuli, and may be more appropriately considered stress metabolites since they appear to have a role in the adaptation of the corn plant to changes and stresses in its environment. The major cyclic hydroxamic acids found in corn are 2,4-dihydroxy-6-methoxybenzoxazione (DIMBOA) and 2,4-dihydroxy-benzoxazinone (DIBOA) (Petersen and Böttger, 1991; Niemeyer, 1988). *In planta*, DIMBOA and DIBOA are sequestered and stabilized as the 2-β-0-D-glucosides, and upon injury or disease lesions, the highly toxic aglucones are rapidly degraded by the plant to the benzoxazolinones, 6-methoxy-benzoxazolinone (MBOA) and 2-benzoxazolinone (BOA). The decomposition products also have biological properties (Woodward et al., 1978; Xie et al., 1991), although much less than the parent metabolites, and they are found in corn stubble and debris. However, the role of BOA and MBOA in the control of the growth of *F. moniliforme* from stubble, debris and soil as illustrated during the saprophytic stage of the life cycle (Figure 1) is unknown.

The cyclic hydroxamic acids are found in varying concentrations in all corn lines, ranging from low to high, thus implying a role in corn resistance to fungi. Both DIMBOA and DIBOA are very unstable (half-life approximately 5.3 hr at 28 °C and pH of 6.8) (Woodward et al., 1978), which makes physiological studies of them difficult since the germination and growth phases of *F. moniliforme* are relatively long processes (requiring 16 to 36 hr). Since MBOA and BOA are very stable and found in nature, we designed a study of the benzoxazolinones to measure their direct involvement into, and to provide indirect evidence for the role played by the cyclic hydroxamic acids in controlling the growth of *F. moniliforme*.

Specifically, a series of studies was designed to establish the accumulation pattern during corn development relative to host suppression by a corn genotype and fungus interaction (Richardson and Bacon, 1993). We also examined the role played by the decomposition products of the cyclic hydroxamic acids on the growth of *F. moniliforme* (Richardson and Bacon, 1995). Analysis of DIMBOA, DIBOA and related metabolites was performed on plant or media extracts by high-performance liquid chromatography (Niemeyer, 1988). The results indicated that MBOA and BOA were fungistatic to the corn isolates

Table 4. Effects of *Fusarium moniliforme* culture age and exposure time on catabolism of various concentrations of MBOA[a]

Culture age/exposure time	% Original MBOA recovered (mM)[b]		
	2.5	5.0	7.5
3-day-old culture			
24 h exposure	13.9	35.0	42.8
	0.0	2.0	0.0
72 h exposure			
7 day-old culture	29.6	38.1	39.7
24 h exposure	0.0	0.3	31.8
72 h exposure			
Analysis of variance[c]			
Source of variation			
Culture age	ns	ns	*
Exposure time	**	**	**
Age x time interaction	ns	ns	**

[a]Richardson and Bacon (1995) accepted.
[b]Mean of 3 replicates.
[c]Analysis of variance of a completely randomized design, with significance levels of 0.05 (*), 0.01 (**), or not significant (ns).

Table 5. Percentage recovery of the benzoxazolinones, MBOA and BOA, from medium containing various isolates of *Fusarium moniliforme* following a 24 h incubation period

Population	Host	Treatment[a]	% Recovery[b]	
			MBOA	BOA
None	—	Control	89 ± 2.1	85 ± 1.7
A	Sweet corn	Live fungus	nd[c]	nd
		Heat-killed fungus	88 ± 1.1	85 ± 2.5
A	Field corn	Live fungus	nd	nd
		Heat-killed fungus	88 ± 2.0	84 ± 0.9
C	Rice	Live fungus	93 ± 3.6	83 ± 0.8
		Heat-killed fungus	90 ± 2.8	84 ± 2.9
D[d]	Field corn	Live fungus	nd	nd
		Heat-killed fungus	94 ± 0.5	96 ± 4.9

[a]Control treatments included either Czapek-Dox broth alone or heat-killed cultures; benzoxazolinone added after autoclaving; Richardson and Bacon (1995) accepted.
[b]Mean $\pm$ the standard deviation of 3 replicates.
[c]nd, not detected.
[d]Data on population D (Bacon et al., 1995).

but were fungitoxic to rice isolates. This study showed that corn isolates of this fungus completely catabolized both metabolites within 24 hr into nontoxic unknown compounds (Table 4). The benzoxazolinones were catabolized by enzymes induced in young hyphae and once the benzoxazolinones were detoxified, *F. moniliforme* resumed growth (Richardson and Bacon, 1995). However, the rice isolates (mating population C), could not degrade these metabolites, which may have special significance since rice species do not contain the benzoazolinones, and their fungal endophyte apparently did not evolve the necessary enzymes to detoxify these substances (Table 5). Further, conidia could not germinate nor catabolize MBOA or BOA (Table 6). Seedlings with bakanae disease did not grow much beyond the 4 to 8 week stage, at which time they were usually so spindly that they died. Although the rice isolates produce large amounts of fusaric acid (Table 2), the bakanae disease is considered to result from the excessive synthesis of gibberellins by the fungus (Marasas et al., 1984).

The presence of cyclic hydroxamic acids in young tissue might delay or prevent extensive colonization of the host by the fungus during the early stage of plant growth. At the latter stages of corn growth, maturity or early senescence, these compounds are no longer synthesized, and their decomposition products, the benzoazolinones, have accumulated and

Table 6. Recovery of MBOA or BOA from Czepek-Dox Medium inoculated with conidia of *Fusarium moniliforme*, mating population A[a]

Incubation (days)	% Recovery[b]	
	MBOA	BOA
0	87 ± 2.8	84 ± 1.6
3	87 ± 1.1	84 ± 2.4
7	89 ± 3.0	86 ± 2.3
14	88 ± 2.1	42 ± 39.1

[a]Richardson and Bacon (1995).
[b]Mean $\pm$ the standard deviation of 3 replicates.

remain in the corn stubble and debris. These results suggest that strains infecting corn have the potential to decompose the benzoazolinones and resume growth later when the plant reaches the mature stage of its life cycle. It is also not known if *F. moniliforme* can decompose residual benzoazolinones from stubble in the soil. These compounds, therefore resemble the general detoxification mechanisms of several phytoalexins by fungal pathogens (Van Etten et al., 1989). The final expression of a disease might result if one or all abiotic and biotic variables are present (Kommedahl and Windels, 1981). The cyclic hydroxamic acids and other stress metabolites might be the regulators for plant colonization by *F. moniliforme* and subsequent disease development.

SUMMARY

The corn pathogen *F. moniliforme* exists as a complex of closely related fungi, which we now know to consist of distinct mating populations. A second mating population, F, is also found in corn but it is mainly associated with sorghum. The fungi of the species that colonize corn belong predominantly to mating population A, and represent the valid species. Fungi from three mating populations can infect corn and enter a relationship that should not be described as pathogenic since pathogenesis is delayed, if it ever occurs. In addition to the parasitic mode of existence, these fungi can also exist as saprophytes and are best described nonobligate parasites. The nonobligate nature of *F. moniliforme* enables this pathogen to parasitize a large variety of hosts. On corn this fungus exists predominantly as an intercellular symptomless endophyte, colonizing leaves, stems, and roots. The fungus can exist entirely as an endophyte, systemically colonizing the kernel, remaining there until germination when the fungus then infects the emerging seedling and maturing plant, and finally infecting a new kernel. Concurrent with this endophytic mode of existence, the fungus also exists as a saprophyte, colonizing moribund or dead corn tissue that also serves as the dissemination and survival structure for this fungus. During the active part of the saprophytic stage, *F. moniliforme* colonizes plant debris during which time a considerable inoculum density of both microconidia and macroconidia results.

The diseases of corn are initiated by a variety of factors, most of which are unknown. The nonobligate nature of *F. moniliforme* and its large number of potential hosts indicates that it produces a large number of nonspecific phytotoxins or enzymes that enable it to colonize these hosts. Several of these phytotoxins have been established as being mycotoxic, which may be a secondary function. One phytotoxin, fusaric acid may be instrumental in the development of the seedling disease, but its significance in the pathogenesis of other corn diseases is inconclusive. This toxin is produced in corn seedlings and plants which showed no apparent sign of disease.

The production of low levels of mycotoxins, i.e. the fumonisins, by *F. moniliforme* is most likely to occur during the symptomless association of the fungus' life cycle. If the disease cycle is initiated very early, mycotoxins cannot enter the food chain. However, if the disease is initiated late in the life cycle of the corn plant (after kernel development), the concentrations of mycotoxin are expected to be higher than those produced during the symptomless stage. Nevertheless, the symptomless endophytic stage of infection is the most important phase to control since it may serve as inoculum for the disease state, and its endophytic habit makes it difficult to control with conventional fungicides.

The nature of the host's ability to suppress the fungus is complex, but one class of compounds, the cyclic hydroxamic acids, may be part of the complexity. The cyclic hydroxamic acids are produced early during the growth of corn, and they are very unstable, decomposing to less toxic metabolites. Although the cyclic hydroxamic acids can kill several insects, and most fungi and bacteria, they and their decomposition products are fungistatic to *F. moniliforme*. All corn isolates can detoxify the decomposition metabolites; the rice isolates cannot. The fungistatic nature of these substances suggests that they may be important metabolites to study relative to the symptomless association observed for most isolates of *F. moniliforme*. The nature of the data presented for the symptomless endophytic nature of *F. moniliforme* indicates its importance, and suggests that plant breeding efforts, which is primarily concerned with the disease cycle, should be also be directed at the symptomless aspect of plant colonization.

ACKNOWLEDGMENT

We thank J. F. Leslie for providing fungal isolates, R. M. Bennett for technical assistance, and J. K. Porter and E. Wray for providing unpublished supporting data useful for concepts expressed in this paper.

REFERENCES

Backmann, E. Der einfluss von fusarinsaure auf die wasserpermeabilitat von pflanzlichen protoplasten. *Phytopath. Z.* **1956**, 27, 255-288.

Bacon, C. W.; Bennett, R. M.; Hinton, D. M.; Voss, K. A. Scanning electron microscopy of *Fusarium moniliforme* within asymptomatic corn kernels and kernels associated with equine leukoencephalomalacia. *Plant Dis.* **1992**, 76, 144-148.

Bacon, C. W.; Nelson, P. E. Fumonisin production in corn by toxigenic strains of *Fusarium moniliforme* and *Fusarium proliferatum. J. Food Protec.* **1994**, 57, 514-521.

Bacon, C. W.; Hinton, D. M.; Richardson, M. D. A corn seedling test for resistance to *Fusarium moniliforme. Plant Dis.* **1994**, 78, 302-305.

Bacon, C. W.; Porter, J. K.; Norred, W. P. Toxic interaction of fumonisin B$_1$ and fusaric acid measured by injection into fertile chicken eggs. *Mycopathologia* **1995**, 129, 29-35.

Bjeldanes, L. F.; Thomson, S. V. Mutagenic activity of *Fusarium moniliforme* isolates in the *Salmonella typhimurium* assay. *Appl. Environ. Microbiol.* **1979**, 37, 1118-1121.

Booth, C. *The Genus Fusarium.* Commonwealth Mycological Institute: Kew, Surrey, England **1971**.

Bracker, C. E. Ultrastructure of fungi. *Ann. Rev. Phytopathol.* **1967**, 5, 343-374.

Bracker, C. E.; Littlefield, L. J. Structural concepts of host-pathogen interfaces. In *Fungal Pathogenicity and the Plant's Response*; Byrde, R.J.W. Cutting, C.V., Eds.; Academic Press: New York, 1973; pp 159-317.

Braun, R. Uber wirkungsweise und umwandlungen der fusarinsaure. *Phytopath. Z.* **1960**, 39, 197-241.

Cheng, S. J.; Jiang, Y. Z.; Li, M. H.; Lo, H. Z. A mutagenic metabolite produced by *Fusarium moniliforme* isolated from Linxian county, China. *Carcinogenesis* **1985**, 6, 903-905.

Chu, F. S.; Huang, X.; Maragos, C. M. Production and characterization of anti-idiotype and anti-anti-idiotype antibodies against fumonisin B$_1$. *J. Agric. Food Chem.* **1995**, 43, 261-267.

Cole, R. J.; Kieksey, J. W.; Cutler, H. G.; Doupnik, B. L.; Peckham, J. C. Toxin from *Fusarium moniliforme*: Effects on plants and animals. *Science* **1973**, 179, 1324-1326.

Cook, R. J. Water relations in the biology of Fusarium. In *Fusarium: Diseases, Biology, and Taxonomy*; Nelson, P. E.; Toussoun, T. A.; Cook, R. J., Eds.; The Pennsylvania State University Press: University Park, 1981; pp 236-244.

Davis, D. Fusaric acid in selective pathogenicity. *Phytopathology* **1969**, 59, 1391-1395.

De Leon, C.; Pandey, S. Improvement of resistance to ear and stalk rot and agronomic traits in tropical maize gene pools. *Crop Sci.* **1989**, 29, 12-17.

Desjardins, A. E.; Plattner, R. D.; Nelson, P. E. Fumonisin production and other traits of *Fusarium moniliforme* strains from maize in northeast Mexico. *Appl. Environ. Microbiol.* **1994**, 60, 1695-1697.

Dowd, P. F. Toxicological and biochemical interactions of the fungal metabolites fusaric acid and kojic acid with xenobiotics in *Heliothis zea* (F) and *Spodoptera frugiperda* (J.E. Smith). *Pestic. Biochem. Physiol.* **1988**, 32, 123-134.

Ehrlich, M. A.; Ehrlich, H. G. Fine structure of the host-parasite interfaces in mycoparasitism. *Ann. Rev. Phytopathol.* **1971**, 9, 155-184.

Fisher, K. D. Hydrolytic enzyme and toxin production by sweet potato fusaria. *Phytopathology* **1965**, 55, 396-398.

Foley, D. C. Systemic infection of corn by *Fusarium moniliforme*. *Phytopathology* **1962**, 52, 870-872.

Galum, M.; Paran, N.; Ben-Shaul, Y. Structural modifications of the phycobiont in the lichen thallus. *Protoplasma* **1970**, 69, 85-96.

Gaumann, E. Fusaric acid as a wilt toxin. *Phytopathology* **1957**, 47, 342-357.

Hadley, G.; Johnson, R. P.; John, D. A. Fine structure of the host-fungus interface in orchid mycorrhiza. *Planta* **1971**, 100, 191-199.

Heath, M. C. Ultrastructure of host and nonhost reactions to cowpea rust. *Phytopathology* **1972**, 62, 27-38.

Headrick, J. M. ; Pataky, J. K. Resistance to kernel infection by *Fusarium moniliforme* in inbred lines of sweet corn and the effect of infection on emergence. *Plant Dis.* **1989**. 73; 882-892.

Heitefuss, R.; Stahmann, M. A.; Walker, J. C. Production of pectolytic enzymes and fusaric acid by *Fusarium oxysporum F. sp. conglutinans* in relation to cabbage yellows. *Phytopathology* **1960**, 50, 367-370.

Hidaka, H.; Nagatsu, T.; Takeya, K. Fusaric acid, a hypotensive agent produced by fungi. *J. Antibiot.* **1969**, 22, 228-230.

Hinton, D. M.; Bacon, C. W. The distribution and ultrastructure of the endophyte of toxic tall fescue. *Can. J. Bot.* **1985**, 63, 36-42.

Jacobs, J. B.; Ahmadjian, V. The ultrastructure of lichens. I. A general survey. *J. Phycol.* **1969**; 5, 227-240.

Kedera, C. J.; Leslie, J. F.; Claflin, L. E. Genetic diversity of *Fusarium* section *Liseola (Gibberella fujikuroi)* in individual maize stalks. *Phytopathology* **1994**; 84; 603-607.

King, S. B.; Scott, G. E. Genotypic differences in maize to kernel infection by *Fusarium moniliforme*. Phytopathology **1981**, 71, 1245-1247.

Klittich, C. J.; Leslie, J. F. Identification of a second mating population within the *Fusarium moniliforme* anamorph of *Gibberella fujikuroi*. *Mycologia* **1992**, 84, 541-547.

Kommedahl, T.; Windels, C. E. Root-,stalk-, and ear-infecting *Fusarium* species on corn in the USA. In *Fusarium: Diseases, Biology and Taxonomy*; Nelson, P. E.; Toussoun, T. A.; Cook, R. J., Eds.; The Pennsylvania State University Press: University Park, 1981; pp 94-103.

Kuo, M. S.; Scheffer, R. P. Evaluation of fusaric acid as a factor in development of *Fusarium* wilt. *Phytopathology* **1964**, 54, 1041-1044.

Leslie, J. F. Mating populations in *Gibberella fujikuroi* (*Fusarium* section *Liseola*). *Phytopathology* **1991**, 81, 1058-1060.

Leslie, J. F.; Doe, F. J.; Plattner, R. D.; Shackelford, D. D.; Jonz, J. Fumonisin B$_1$ production and vegetative compatibility of strains from *Gibberella fujikuroi* mating population 'A' (*Fusarium moniliforme*). *Mycopathologia* **1992**, 117, 37-45.

Liddell, C. M.; Burgess, L. W. Survival of *Fusarium moniliforme* at controlled temperature and relative humidity. *Trans. Br. Mycol. Soc.* **1985**, 84, 121-130.

Marasas, W. F. O.; Nelson, P. E.; Toussoun, T. A. *Toxigenic Fusarium Species*; The Pennsylvania State University Press: University Park, 1984.

Molot, P. M.; Simone, J. Technique de contamination artificielle des semences de mais par les Fusarioses. *Rev. Zoo. Agricole Appl.* **1967**, 3, 29-32.

Niemeyer, H. M. Hydroxamic acids (4-hydro-1,4-benzozazin-3-ones), defense chemical in the Gramineae. *Phytochemistry* **1988**, 27, 3349-3358.

Norred, W. P.; Wang, E.; Yoo, H.; Riley, R. T.; Merrill, A. H., Jr. *In vitro* toxicology of fumonisins and the mechanistic implications. *Mycopathologia* **1992**, 117, 73-78.

Nyvall, R. F.; Kommedahl, T. Individual thickened hyphae as survival structures of *Fusarium moniliforme* in corn. *Phytopathology* **1968**, 58, 1704-1707.

Nyvall, R. F.; Kommedahl, T. Saprophytism and survival of *Fusarium moniliforme* in corn stalks. *Phytopathology* **1970**, 60, 1233-1235.

Ochor, T. E.; Trevatham, L. E.; King, S. B. Relationship of harvest date and host genotype to infection of maize kernels by *Fusarium moniliforme*. *Plant Dis.* **1987**, 71, 311-313.

Ooka, J. J; Kommedahl, T. Wind and rain dispersal of *Fusarium moniliforme* in corn fields. *Phytopathology* **1977**, 67, 1233-1235.

Paterson, R. R. M.; Rutherford, M. A. A simplified rapid technique for fusaric acid detection in *Fusarium* strains. *Mycopathologia* **1991**, 113, 171-173.

Pennypacker, B. W. Anatomical changes involved in the pathogenesis of plants by *Fusarium*. In *Fusarium: Diseases, biology, and taxonomy*; Nelson, P. E.; Toussoun, T. A.; Cook, R. J., Ed.; Pennsylvania State University Press: University Park, 1981; pp 400-408.

Petersen, W.; Böttger, M. Contribution of organic acids to the acidification of the rhizosphere of maize seedlings. *Plant Soil* **1991**, 132, 159-163.

Pitel, D. W.; Vining, L. C. Accumulation of dehydrofusaric acid and its conversion to fusaric and 10-hydroxy-fusaric acids in cultures of *Gibberella fujikuroi*. *Can. J. Biochem.* **1970**, 48, 623-630.

Plattner, R. D.; Nelson, P. E. Production of beauvericin by a strain of *Fusarium proliferatum* isolated from corn fodder for swine. *Appl. Environ. Microbiol.* **1994**, 60, 3894-3896.

Porter, J. K.; Bacon, C. W. Fusaric acid, a toxin produced by *Fusarium moniliforme*: Effects on brain and pineal neurotransmitters and metabolites in rats. *Proc.6th Colloq. Eur. Pineal Soc.* **1993**, June 23-27, E19. (Abstract)

Richard, J. L.; Bennett, G. A.; Ross, P. F.; Nelson, P. E. Analysis of naturally occurring mycotoxins in feedstuffs and food. *J. Anim. Sci.* **1993**, 71, 2563-2574.

Richardson, M. D.; Bacon, C. W. Cyclic hydroxamic acid accumulation in corn seedlings exposed to reduced water potentials before, during, and after germination. *J. Chem. Ecol.* **1993**, 8, 1613-1624.

Richardson, M. D.; Bacon, C. W. Catabolism of 6-methoxy-benzoxazolinone and 2-benzoxazolinone. *Mycologia* **1995**, 87, 510-517.

Riley, R. T.; Norred, W. P.; Bacon, C. W. Fungal toxins in foods: Recent concerns. *Annu. Rev. Nutr.* **1993**, 13, 167-189.

Shurtleff, M. C. *Compendium of Corn Diseases,* 2nd ed.; The American Phytopathological Society: St. Paul, MN, 1980.

Stoll, C.; Renz, J. Uber den Fusarinsaure- und dehydrofusarinsaure-stoffwechsel von *Gibberella fujikuroi*. *Phytopath. Z.* **1957**, 29, 380-387.

Styer, R. C.; Cantliffe, D. J. Relationship between environment during seed development and seed vigor of two endosperm mutants of corn. *J. Amer. Soc. Hort. Sci.* **1983**. 108, 717-720.

Sydenham, E. W.; Shephard, G. S.; Thiel, P. G.; Marasas, W. F. O. Stockenström, S. Fumonisin contamination of commercial corn-based human foodstuffs. *J. Agric. Food Chem.* **1991**, 39, 2014-2018.

Tamari, K.; Kaji, J. Studies on the mechanism of the growth inhibitory action of fusarinic acid on plants. *J. Bact.* **1954**, 41, 143-165.

Thiel, P. G.; Shephard, G. S.; Sydenham, E. W.; Marasas, W. F. O.; Nelson, P. E.; Wilson, T. M. Levels of fumonisins B$_1$ and B$_2$ in feeds associated with confirmed cases of equine leukoencephalomalacia. *J. Agric. Food Chem.* **1991**, 39, 109-111.

Thomas, M. D.; Buddenhagen, I. W. Incidence and persistence of *Fusarium moniliforme* in symptomless maize kernels and seedlings in Nigeria. *Mycologia* **1980**, 72, 882-887.

Van Etten, H. D.; Matthews, D. E.; Matthews, P. S. Phytoalexin detoxification: importance for pathogenicity and practical implications. *Ann. Rev. Phytopathol.* **1989**, 27, 143-164.

Vesonder, R. F.; Hesseltine, C. W. Metabolites of Fusarium. In *Fusarium: Diseases, Biology and Taxonomy*; Nelson, P.E.; Toussoun, T.A. Cook, R.J., Eds.; The Pennsylvania State University Press: University Park, 1981; pp 350-364.

Wahlroos, O.; Virtanen, A. I. On the antifungal effect of benzoxazolinone and 6-methoxybenzoxazolinone, respectively on *Fusarium nivale*. *Acta Chem. Scand.* **1958**, 12, 124-128.

Whitney, N. J.; Mortimore, C. G. An antifungal substance in the corn plant and its effect on growth of two stalk-rotting fungi. *Nature* **1959**, 183, 341.

Wiebe, L. A.; Bjeldanes, L. F. Fusarin C, a mutagen from *Fusarium moniliforme*. *J. Food Sci* **1981**, 46, 1424-1426.

Woodward, M. D.; Corcuera, L. J.; Helgeson, J. P.; Upper, C. D. Decomposition of 2,4-dihydroxy-7-methoxy-2H-1,4-benzoxazin-3(3H)-one in aqueous solution. *Plant Physiol.* **1978**, 61, 796-802.

Xie, Y. S.; Arnason, J. T.; Philogène, B. J. R.; Atkinson, J.; Morand, P. Distribution and variation of hydroxamic acids and related compounds in maize (*Zea mays*) root system. *Can. J. Bot.* **1991**, 69, 677-681.

Yabuta, T.; Kambe, K.; Hayashi, T. Biochemistry of the bakanae-fungus. I. Fusarinic acid, a new product of the bakanae fungus. *J. Agr. Chem. Soc. Japan* **1937**, 10, 1059-1068.

Yan, K.; Dickman, M. B.; Xu, J.-R.; Leslie, J. F. Sensitivity of field strains of *Gibberella fujikuroi* (*Fusarium* section *Liseola*) to benomyl and hygromycin B. *Mycologia,* **1993**, 85, 206-213.

17

FUMONISINS IN MAIZE GENOTYPES GROWN IN VARIOUS GEOGRAPHIC AREAS

Angelo Visconti

Institute of Toxins and Mycotoxins
National Research Council (CNR)
Viale Einaudi 51
70125 Bari, Italy

SUMMARY

Data on the occurrence of fumonisins B_1 and B_2 in maize genotypes (inbred lines and hybrids) cultivated in several countries from three continents will be presented. Samples originated at different times (from 1990 to 1994) from experimental stations in Argentina, Benin, Croatia, Poland, Portugal, Italy, Romania and Zambia. Fumonisin contamination was negligible in samples from eastern Europe (Croatia, Romania and Poland), whereas it was quite relevant and widespread in samples from Argentina, western Europe and Africa. A general trend of higher fumonisin levels was observed in Argentina, Zambia (limited to 1993 crop), Portugal, and Italy (mainly 1990 and 1991 crop years), with a significant percentage of samples showing more than 1000 ng/g, a level of concern. The data from the present study suggest that the environmental conditions in the specific area of cultivation play an important role in the accumulation of fumonisins in maize.

INTRODUCTION

Fumonisins are a group of mycotoxins produced mainly by *Fusarium moniliforme, F. proliferatum* and other species of the *Liseola* section. Maize is the crop most frequently infected by *F. moniliforme*, a cosmopolitan species that grows well saprophytically and survives effectively in live, asymptomatic kernels (Headrick and Pataky, 1991). Genotypic differences in maize to kernel infection by *F. moniliforme* have been shown to be under genetic control (King and Scott, 1981). Genetic resistance to kernel infection has been studied intensively (Boling and Grogan, 1965; Lunsford et al. 1976; Scott and King, 1984) and offers the most potential for control. Variable response of the host phenotype to *F. moniliforme* infection offers a real possibility of breeding for resistance to this fungus and the consequent formation of fumonisins into the kernels. An indirect approach, less effective from a phytopathological point of view, but of primary interest for mycotoxicologists,

Fumonisins in Food, Edited by L. Jackson *et al.*
Plenum Press, New York, 1996

consists in the selection of genotypes which are not prone to the formation of fumonisins under field conditions. In this direction a study has been carried out recently by Shelby et al. (1994) which showed differential production of fumonisins in 15 commercially available maize hybrids planted in 17 locations of the USA. The study suggests the possibility of using such a tool for selecting maize hybrids in areas where fumonisin is a problem. A similar study has been carried out in our laboratory during the past three years in collaboration with several scientists from various countries in order to establish a differential fumonisin accumulation in maize genotypes grown in various regions of the globe, other than the USA. Results of this study are presented in this paper.

EXPERIMENTAL

A total of 224 maize hybrids or inbred lines have been analyzed for fumonisins B_1 and B_2 (FB_1 and FB_2). These genotypes originated from eight countries in three continents, i.e. Europe, Africa and South America. Some hybrids have been tested for two or more consecutive years. Analyses have been performed according to a procedure described previously (Doko et al., 1994). Briefly, the extraction was performed with methanol-water (3:1), the clean-up with a strong anion exchange (SAX) minicolumn, and the final determination by reversed phase HPLC with fluorescence detector after derivatization with o-phthaldialdehyde (OPA). The name and country of origin of the maize genotypes considered in this presentation are listed in Table 1. Results of fumonisin analysis are reported in the text as combined amount of FB_1+FB_2, unless otherwise specified.

RESULTS

Argentina

Fifty one maize genopypes grown in Argentina in 1994 have been analyzed for fumonisins. With the exception of two dent-type hybrids (Prozea 10 and AX 746), all the tested genotypes had a flint-type endosperm, and the vegetative cycle, expressed in terms of days to reach relative maturity, varied from 104 to 132. No correlation was found between length of the vegetative cycle and the concentration of fumonisin, although a slight negative trend was observed. The minimum and maximum concentrations were recorded at 230 and 37000 ng/g, respectively; mean and median were calculated at 3040 and 1650, respectively. More than 60% of samples contained fumonisin levels higher than 1000 ng/g, and 5 samples were above 5000 ng/g, the maximum limit recommended by the Mycotoxin Committee of the American Association of Veterinary Diagnostician (Riley et al., 1993). Exceptional results were obtained with the two dent-type hybrids, the first (Prozea 10) showing the highest fumonisin concentration (37000 ng/g) and the second (AX 746) having more FB_2 (2990 ng/g) than FB_1 (1230 ng/g).

In addition, a project relevant to the contamination of maize genotypes from Argentina and the formation of fumonisins in relation to fungal infection has been performed in collaboration with the Universidad de Rio Cuarto, Cordoba (L. Ramirez and S. Chulze). The maize genotype Morgan 400 has been used to establish the concurrence of fungal growth and fumonisin formation at different ear maturity stages (Chulze et al. 1994). The predominant *Fusarium* species of the *Liseola* section were measured at 45, 60, 75, 90, and 105 days after flowering. The incidence of *Fusarium subglutinans* contamination was the highest during the first stage (up to 60 days after flowering) then decreased. *F. moniliforme* contamination increased consistently until the last harvesting

Table 1. Origin of maize genotypes considered in this study

Argentina			
Rodas	Atlas	A-257	A-258
Ax-845	Ax-905	A-967	Prozea 10
Prozea 20	Prozea 25	Dekalb 3F24	Dekalb 4F37
Dekalb 3S41	Dekalb 689	Dekalb 752	Dekalb 761
Dekalb 762	Dekalb 821	Tribrido 43	Cargill 4R 21
Cargill 34911	Boyero 3L-95	Boyero 6	Ici 8302
Ici 8330	Ici 8340	Ici 8398	Norkin 310
Norkintal 361	Norkin 366	Norkin 4-5284	Insu M-14
Insu M-20	Pioneer 3456	Pioneer 3478	Hib. 4-F-26-R2
Tribrido 92	Record 160	Ax 746	Ax 788
Ax 830	Dekalb 636	Dekalb 4F-91	Dekalb 648
Tresur Diodo	Boyero 5	Morgan 401 (T)	Pioneer 3362
Aurora	Norkin 331		

Benin			
Tz Esr	Dmr Esr W	Pool 16 Sr	Acr 87 Pool 16 Sr
Ev 8443 Sr	Ev 8422 Sr	Tzb Sr	Tzb Sr Se
Okomosa	unknown A	unknown B	

Croatia			
Bc 183 (=9201)	Bc 175 (=9202)	Eta 272 (=9203)	Bc 278 (=9204)
Nadezda (=9205)	Bc 388 9206)	Bc 3786 (=9209)	Bc 4692 (=9210)
Bc 492 (=9212)	Jumbo (=9213)	Bc 592 (=9214)	Dorado (=9215)
Bc 588B (=9216)	Bc 608R (=9217)	Bc 678 (=9218)	Bc 6661 (=9219)
Medimurec 38 (=9207)	Podravec 36 (=9208)	Bc 488 (=9211)	Bc 22811 (=9301)
Bc 222 (=9302)	Bc 318 (=9308)	Bc 408E (=9313)	Bc (=9321)
Bc (9322)	Bc (=9323)	Bc (=9324)	Bc (=9325)
Bc (=9326)	Bc (=9327)	Bc (=9328)	Bc (=9329)
Bc (=9330)	Bc (=9331)	Bc (=9332)	Bc (=9333)
Bc (=9334)	Bc (=9335)		

Italy			
W 153 R	W 182 R	A 660	H 99
OH 43	W 22	H 95	Mo 17
K 55	Ky 226	LO 876	R 805
33-16	C 103	B 14	B37
B 73	B 84	B 87	B 89
FR 27	Pa 884	Va 60	Va 85
K 816	TX 325		

Poland			
Zenit	Klg 2210	Betulisa	Rah Be 90102
Rah Be 86101	Smolimag	Ruten	

Portugal			
Aioa-Pioneer	Constanza-Pioneer	Atrix-Coop.do Pau	G-4507-Sapec
Corki-Coop.do Pau	Agencia-Coop.doPau	Dracma-Sapec	Prisma-Sapec
Xl-75-A-Dekalb			

Romania			
Fundulea 2671-92	Fundulea 2727-92	Fundulea 2659-92	Fundulea 2639-92
Fundulea 2643-92	Fundulea 2603-92		

Zambia			
MM 501	MM 502	MM 504	MM 601
MM 603	MM 604	MM 612	MM 752
MM 501-4	MM 509	MM 505	MM 605
MM 602-2	MM 601-4	MM 705	MM 609
MM 701-1	MM 2519	MM 608	MM 701

(105 day after flowering), whereas *F. proliferatum* reached a peak at 75 to 90 days after flowering. A good agreement was found between the occurrence of fumonisin producing species (*F. proliferatum* and *F. moniliforme*) and fumonisin contamination at different maturity stages. Fumonisin B_1 could be detected from 45 days after flowering, whereas fumonisins B_2 and B_3 could be detected only after the second sampling (60 days from flowering). The sampling at 75 days from flowering showed the highest fumonisin concentrations (ranging from 500 to 16700 ng/g), and included a subsample with an extremely high concentration of FB_2 (11300 ng/g), even higher than FB_1 (4970 ng/g). The FB_2 to FB_1 ratio higher than 1 is quite unusual in naturally contaminated maize as well as in *Fusarium* cultures (see also below). In the present study this phenomenon occurred only in 2 samples from Argentina (1 in the group of the above 51 genotypes, and 1 out of 35 samples analyzed for the experiments at different maturity stage). This behavior has also been found for some atypical strains of *Fusarium* (Plattner, 1995), which have been isolated with a significant frequency in Argentina, particularly within strains of *F. proliferatum* (Chulze et al., 1995, unpublished data).

Benin

Nine out of 11 samples obtained from Benin were fumonisin positive with only one exceeding 1000 ng/g. The mean levels in positive samples and the median were found at 700 and 190 ng/g (FB_1+FB_2), respectively. The FAO maturity class (400 to 700) was known for 9 of the 11 samples; for the remaining two samples, obtained from a local commercial source, no information was available.

Croatia

A total of 54 samples have been analyzed from Croatia, representing 40 hybrids, 14 of which have been replicated for two consecutive years (1992 and 1993). Samples were obtained from B. Palversic, Institute for Breeding and Production of Field Crops, Zagreb. Most of them had dent endosperm with FAO maturity class ranging from 200 to 600. Of the 19 hybrids tested in 1992, 11 were positive for fumonisin contamination, but the levels of contamination were negligible, lower than 70 ng/g. The 14 hybrids replicated for the second year in 1993 confirmed the insignificant level of contamination recorded in 1992 (35% positives with less than 100 ng/g) (Table 2). Of the remaining 21 hybrids tested in 1993, only 5 samples (24%) were fumonisin positive, with levels between 100 and 1000 ng/g. A linear correlation between the natural occurring levels of fumonisins and the ear rot caused by artificial inoculation of *Fusarium moniliforme* in the 14 hybrids planted again in 1993 could not be established because of the low variation of data for both parameters (Table 2). Nevertheless, the low grade of fungal infection recorded for all hybrids but one (Bc 4692, ear rot grade 5) is in agreement with the insignificant levels of fumonisin contamination. Although it would be desirable to produce more of these data for several additional years, these results clearly indicate that there is a good chance of finding maize hybrids with very low tendency to accumulate fumonisins in this particular area (at least under climatic conditions of the above mentioned seasons).

Italy

Maize samples from 26 inbred lines were obtained from Dr. E. Pè of the Department of Genetics, University of Milan. Six, 19 and 1 samples were taken from the 1989, 1990 and 1991 crop years, respectively. With the exception of the inbreds Mo 17 and C 103 with semident type endosperm, all the remaining samples represented dent type endosperm. All

Table 2. Natural occurrence of Fumonisins (FB$_1$+FB$_2$) in maize hybrids from Croatia (1992 and 1993 crop years) and ear infection after artificial inoculation with *Fusarium moniliforme* under field conditions

Hybrid	1992 Fumonisins (ng/g)	1993 Fumonisins (ng/g)	1993 ear rot (1-9) Fusarium moniliforme
Eta 272 (9203)	10	n.d.[a]	2.2 C[b]
Bc 278 (9204)	n.d.	n.d.	3.4 BC
Nadezda (9205)	10	100	2.0 C
Bc 388 (9206)	10	n.d.	4.2 AB
Bc 3786 (9209)	n.d.	n.d.	3.3 BC
Bc 4692 (9210)	20	70	5.0 A
Bc 492 (9212)	n.d.	n.d.	2.4 C
Jumbo (9213)	50	n.d.	2.4 C
Bc 592 (9214)	20	n.d.	3.0 BC
Dorado (9215)	n.d.	10	2.5 BC
Bc 588B (9216)	n.d.	100	2.0 C
Bc 608R (9217)	n.d.	n.d.	3.3 BC
Bc 678 (9218)	n.d.	n.d.	3.0 BC
Bc 6661 (9219)	70	10	3.4 BC

[a] n.d. = not detected, < 10 ng/g
[b] Values followed by the same letter are not significantly different at the 0.05 level based on t-test. (Data from B. Palaversic, personal communication).

the tested samples were contaminated with levels of fumonisins (FB$_1$ + FB$_2$) ranging from 10 to 2850 ng/g. In particular, the one sample from 1991 and three samples from 1992 crop years contained more than 1000 ng/g fumonisins. The mean and median values were 450 and 70 ng/g, respectively. FAO maturity class varied from 300 to 700, but no correlation was found with the length of the vegetative cycle and the fumonisin contamination within this group of inbreds.

Poland

Results of the analysis of 7 hybrids from Poland showed only 2 samples contaminated at levels lower than 30 ng/g. Samples derived from the 1992 crop, and flint and flint-dent endosperms were represented. Three high lysine hybrids (Betulisa, RAH BE 90102 and RAH BE86101), that were included in this group, were found to be fumonisin-free.

The same hybrids, after artificial inoculation with *Fusarium moniliforme* under field conditions, showed a different degree of ear rot with very high contents of fumonisins (from 28 to 261 µg/g) in the damaged kernels (Chelkowski et al., 1994). The artificial infection of four of these hybrids for two consecutive years gave rise to varying degrees of fumonisin accumulation and ear rot disease, as shown in Table 3. Although none of the four genotypes examined could be considered severely attacked by the fungus (ear rot grade lower than 5 on a 0-10 scale), the order of ear rot intensity and fumonisin contamination in relation to the specific hybrids was in some extent reproduced during the two-year experiments. The fumonisin content reported in Table 3 was estimated by separate analysis of *Fusarium* damaged kernels and symptomless kernels, and varied from 500 to 12600 ng/g. These results suggest that the possibility exists of selecting hybrids with different tendencies to accumulate fumonisin in the field (Chelkowski et al. 1994). Nevertheless this point should not be emphasized due to the limited number of hybrids (4) and seasons (2) considered; further confirmation is needed.

Table 3. Ear infection and Fumonisins (FB_1+FB_2) accumulation in maize hybrids from Poland inoculated with *Fusarium moniliforme* under field conditions

Hybrid	1992 crop		1993 crop		average 1992-1993 crop	
	Ear rot (0-10)	Fumonisins (μg/g)*	Ear rot (0-10)	Fumonisins (μg/g)*	Ear rot (0-10)	Fumonisins (μg/g)*
RAHBE 90102	4.8	5.9	4.6	9.2	4.7	7.6
ZENIT	2.8	3.3	4.6	12.6	3.7	8.0
KLG 2210	0.8	0.5	2.6	0.5	1.7	0.5
SMOLIMAG	1.2	1.1	1.2	0.4	1.2	0.8

* Derived from separated analysis of *Fusarium* damaged kernels and symptomless kernels. Fumonisin content in *Fusarium* damaged kernels varied from 24 to 214 μg/g (FB_1+FB_2).

Portugal

Samples from Portugal, provided by the Agricultural High School of Coimbra, originated from the 1992 crop. The FAO maturity class varied from 500 to 700, whereas no information was available about endosperm caracteristics. All 9 samples were contaminated with both FB_1 and FB_2 at levels ranging from 90 to 4450 ng/g, and the mean and median values were found at 1930 and 1670 ng/g. Most hybrids (67%) cultivated in this country contained more than 1000 ng/g.

Romania

The 6 hybrids tested in this study were obtained from Dr. M. Ittu, Research Institute for Cereals and Industrial Crops, Fundulea. The vegetative cycle varied from 230 to 630 (FAO) representing either dent, dent-flint or flint maize endosperm. The incidence of fumonisin positive samples was 50%, and the levels of contamination were lower than 30 ng/g.

Zambia

Samples from 20 maize hybrids, planted in Zambia for three consecutive seasons, were obtained from L.D. Ristanovic, Golden Valley Regional Research Station, Chisamba.

Table 4. Correlation of FB_1 and FB_2 contents in maize genotypes grown in different countries

Country	n	r	slope (FB_2/FB_1)
Europe			
Italy	26	0.9836	0.21
Portugal	9	0.9760	0.33
Africa			
Benin	11	0.9980	0.26
Zambia (1992)	20	0.9918	0.21
Zambia (1993)	20	0.9752	0.41
South America			
Argentina	51	0.9454	0.38

n = no. of tests
r = correlation coefficient

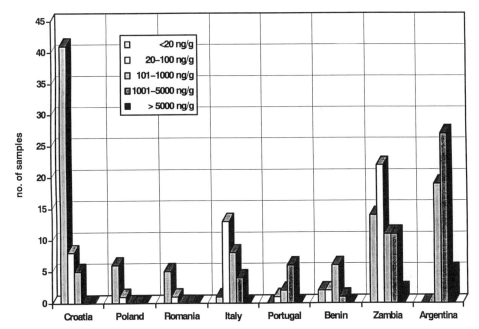

Figure 1. Fumonisin content (FB$_1$ + FB$_2$) in maize genotypes grown in different countries. Mean and median values are represented.

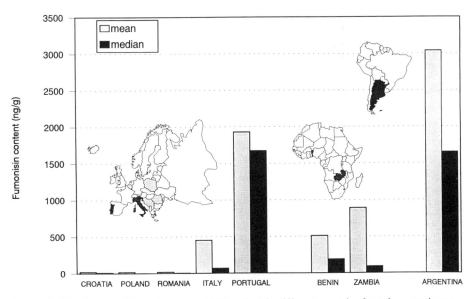

Figure 2. Distribution of fumonisin content (FB$_1$ + FB$_2$) in different countries from three continents.

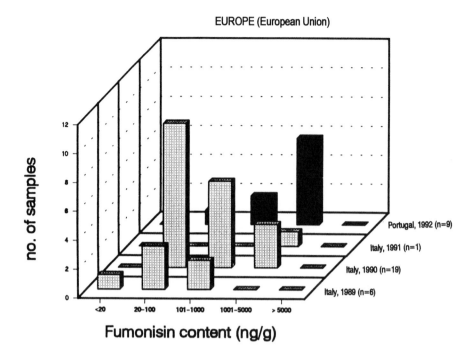

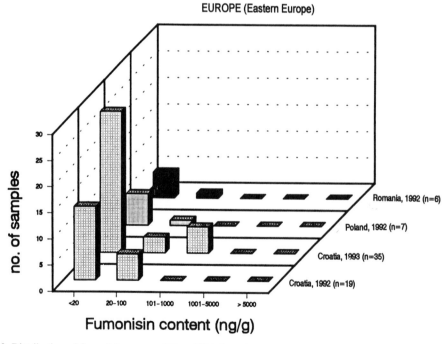

Figure 3. Distribution of fumonisin content (FB$_1$ + FB$_2$) in maize genotypes grown in different geographic areas. Individual countries and crop year are represented within four separate groups: European Union, Eastern Europe, South America and Africa.

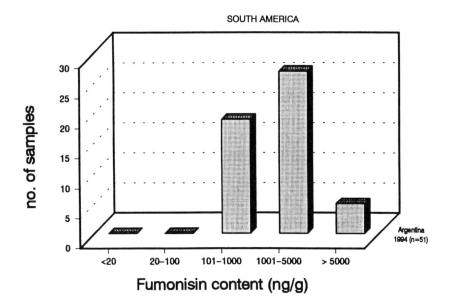

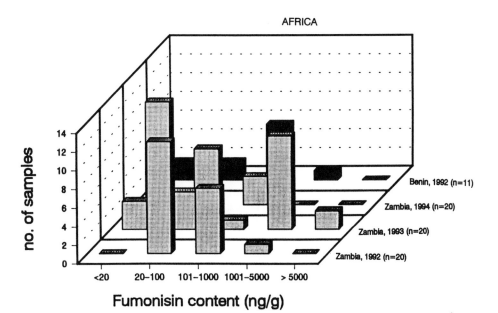

Figure 3. Continued.

In 1992, all samples were contaminated with fumonisins at levels ranging from 20 to 1710 ng/g, with mean and median values at 610 and 130 ng/g, respectively. In 1993, the level of contamination for the same hybrids was considerably higher, showing an even higher dispersion of data. Three samples were fumonisin-free, 12 samples had more than 1000 ng/g, and one sample had 13050 ng/g fumonisins. The mean and median values were at 2400 and 1630 ng/g, respectively. The third year of experiments (1994) showed very low levels of fumonisins (< 200 ng/g) in the 9 samples with positive results. No correlation was found between hybrid and season, nor between hybrids and fumonisin content. Considering all measurements performed during the three-year experiments, the mean and median values were 890 and 90 ng/g, respectively. These results are quite surprising and no conclusions or explanations could be found for them. The experiments are being repeated for an additional year and results will be part of a PhD thesis (Johanne E. Schjoth, NPPI, Aas, Norway), together with other experiments intended to correlate fumonisin formation, fungal contamination, and different planting dates and locations in Zambia.

GENERAL CONSIDERATIONS

Although all data represent the amount of combined FB_1 and FB_2, the individual concentration of FB_1 and FB_2 can be roughly estimated in each sample on the basis of the results of the linear regression analysis, which are reported in Table 4.

An extremely good correlation (coefficient of correlation very close to 1) was observed for all countries and seasons considered (countries with very low positive levels and incidence were not considered for a good statitical evaluation). The FB_2 to FB_1 ratio (slope of the regression line) varied fron 0.21 for Italy and Zambia (1992 crop) to 0.41 in Zambia (1993 crop), with a general trend to higher values with higher fumonisin average concentrations.

Of the countries considered in this study, Argentina showed the highest incidence and level of fumonisin contamination, followed by Portugal, Zambia (limited to 1993 crop) and Italy (mainly 1990 and 1991 crop years). The distribution and the values of mean and median of fumonisin content in maize genotypes from individual countries are shown in Figures 1 and 2. Based on the geographic areas and/or the levels of fumonisin contamination, four groups of countries could be identified, i.e. Eastern Europe (Poland, Romania and Croatia), the European Union (Italy and Portugal), Africa (Benin and Zambia) and South America (Argentina). The distribution of fumonisin content in these four groups are represented individually and summarized in Figure 3 and 4, respectively. Only maize genotypes cultivated in the first group of countries (Eastern Europe) showed no or negligible fumonisin contamination. An intermediate degree of contamination was found in the African countries, with about 20% of samples containing more than 1000 ng/g fumonisins. In Benin, 82% of samples were positive and in Zambia, 100, 70 and 45% were positive for 1992, 1993 and 1994 crop years, respectively. In the remaining two groups of countries (European Union and Argentina) the incidence of fumonisin positive samples was 100%, and the level of contamination was higher. In particular the contamination levels in Argentina were significantly higher (P <0.001) than any other country or group of countries.

In a previous publication (Doko et al., 1994) presenting part of these data, a trend for higher contamination levels with longer vegetative cycles was observed when several countries were considered together, but no trend was observed within each country. Reconsidering the statistical analysis with the additional data produced herein, in particular those relevant to the numerous hybrids from Argentina, no interaction between hybrid and length of the vegetative cycle, nor between hybrid and endosperm characteristics could be detected.

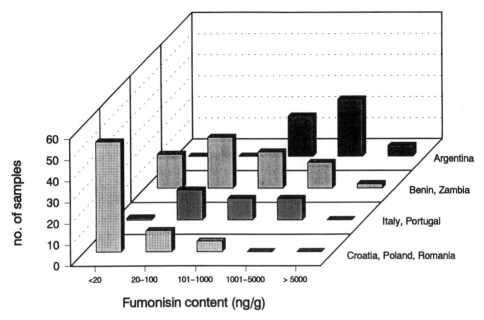

Figure 4. Distribution of fumonisin content (FB_1 + FB_2) in maize genotypes grown in different geographic areas.

These data add to those relevant to the lack of correlation with other genotype characteristics, such as percent of protein, oil, starch and fiber, previously reported by Shelby et al. (1994).

Repeated experiments were carried out with the same genotypes for consecutive seasons with a different degree of success. In particular, while in Zambia no conclusion could be drawn about a possible correlation between hybrid, fumonisin contamination and season, the experiments carried out in Croatia confirmed for two consecutive years the negligible level of contamination of the 14 hybrids tested. The latter result is of great relevance because it provides a good evidence for genotypes "resistant" to the formation of fumonisins in the field, at least under the environmental conditions of that specific area. This assumes even more significance if compared with data relevant to several genotypes in the USA (Shelby et al., 1994), which showed higher fumonisin concentrations (0.5 to 50 µg/g) than those found in the positive samples of this region (<0.1µg/g).

In conclusion, the environmental conditions of the specific area of cultivation seem to play an important role in the formation of fumonisin in maize. Despite the unclear results obtained in some regions (data from Zambia), there is significant difference in fumonisin levels among hybrids grown in different geographic areas, suggesting the possibility of selecting low-fumonisin hybrids which are adapted to a particular area. Further investigations in this direction are necessary and highly recommended to prove the consistent resistance to fumonisin formation of the most promising hybrids also in relation to the particular area of cultivation.

AKNOWLEDGEMENTS

Results of the present study have been partially published in scientific journals (Doko et al. 1994) or presented at conferences (Chelkowski et al., 1994; Chulze et al., 1994). The work has been made possible by the collaboration with the European Commission, the Norwegian Plant Protection Institute, Aas, Norway (L. Sundheim), the Universidad de Rio Cuarto, Cordoba, Argentina (S. Chulze), the Institute for Breeding and Production of Field Crops, Zagreb, Croatia (B. Palaversic), the Polish Academy of Sciences, Poznan, Poland (J. Chelkowski) and the enthusiasm of several visiting scientists and fellows who contributed at different extents to the accomplishment of the work, namely M.B. Doko, S. Rapior, M. Pascale, L. Ramirez and J. E. Schjoth. The author acknowledge the institutions and scientists mentioned in the text for providing the maize genotypes for the analyses.

REFERENCES

Boling, M.B.; Grogan, C. O. Gene action affecting host resistance in to *Fusarium* ear rot of maize. *Crop Sci.* **1965**, *5*, 305-307.

Chelkowski, J.; Pronczuk, M.; Visconti, A.; Doko, M. B. Fumonisins B_1 and B_2 accumulation in maize kernels inoculated under field conditions with *Fusarium moniliforme* Sheldon and in naturally infected cobs. Poland, *Genet. Pol.* 1994, 35B, 333-338.

Chulze, S.; Ramirez, L.; Farnochi, C.; Pascale, M.; Visconti, A. *Fusarium* and fumonisins in Argentina. *V International Mycological Congress*, Vancouver, British Columbia, Canada, 14-21 August 1994. Abs. 37.

Doko, M.B.; Rapior, S.; Visconti, A.; Schjoth J. E. Incidence and levels of fumonisin contamination in maize genotypes grown in Europe and Africa. *J. Agric. Food Chem.*, **1995**, *43*, 429-434.

Headrick, J.M., Pataky, J.K. Maternal influence on the resistance of sweet corn lines to kernel infection by *Fusarium moniliforme*. *Phytopathology*, **1991**, *81*, 268-274.

King S.B.; Scott, G.E. Genotypic differences in maize infection by *Fusarium moniliforme*. *Phytopathology*, **1981**, *71*, 1245-1247.

Lunsford, J. N.; Futrell, M. C.; Scott, G. E. Maternal effects and type of gene action conditioning resistance to *Fusarium moniliforme* seedling blight in maize. *Crop Sci.* **1976**, *16*, 105-107.

Plattner, R. D. Detection of fumonisins in *Fusarium moniliforme* cultures by HPLC with electrospray MS and evaporative light scattering detectors. *Natural Toxins*, **1995**, *Special Issue on "Mycotoxins and Toxic Plant Components"*, 3, 394-398.

Riley, R.T; Norred, W.P; Bacon, C.V.. Fungal toxins in foods: Recent concerns. *Annu. Rev. Nutr.*, **1993**, *13*, 167-189.

Scott, G.E.; King, S.B. Site of action of factors for resistance to *Fusarium moniliforme* in maize. *Plant Dis.* **1984**, *68*, 804-806.

Shelby, R.A.; White, D.G.; Bauske, E.M. Differential fumonisin production in maize hybrids. *Plant Dis.*, **1994**, *78*, 582-584.

18

LIQUID CULTURE METHODS FOR THE PRODUCTION OF FUMONISIN

Susanne E. Keller and Theodore M. Sullivan

Biotechnology Studies Branch
US Food and Drug Administration
National Center for Food Safety and Technology
6502 S. Archer Rd.
Summit-Argo, IL 60501

ABSTRACT

Currently, fumonisin B_1 is obtained primarily by using solid culture methods. Although fumonisin B_1 concentrations obtained in solid culture are typically quite high, subsequent extraction and purification present problems. In addition, current methods utilize complex media which makes analysis of biosynthetic pathways and control mechanisms difficult. Liquid culture methods of production could eliminate many problems associated with production in solid culture. However, in the past, concentrations obtained in liquid culture have been relatively low. In this work, factors affecting the production of fumonisin B_1 from a shake flask scale of 100 ml to a fermenter scale of 100 liters were examined. Best results were obtained by using a fed batch method that is nitrogen limited, with pH control. With this method, concentrations in excess of 1000 ppm can be obtained.

INTRODUCTION

Currently, fumonisins are obtained primarily by using solid culture methods. The method generally employed involves the use of autoclaved corn kernels (Leslie et al., 1992a,b; Nelson et al., 1991, 1992, 1993; Thiel et al., 1991; Ross et al., 1990). Nelson et al. (1993) reported the use of 500 g of yellow corn kernels and 500 ml of distilled water in an autoclavable polyethylene bag for the large-scale production of fumonisin. Concentrations of fumonisins obtained in this manner were 6-10 mg per gram of culture material (Nelson et al., 1994). Important factors related to high production were reported to be temperature control, with an optimal temperature range of 20-25°C (Alberts et al., 1990; LeBars et al., 1994), moisture, and aeration. Although substantial quantities of fumonisins can be obtained by this method, there are some disadvantages. Recovery and analysis of the fumonisins require considerable quantities of hazardous solvents and other potentially dangerous chemicals. Nelson et al. (1993) reported the use of chloroform/acetone (50/50, V/V) to kill the fungus prior to extraction of the fumonisins. Extraction of the fumonisins is generally accomplished with either methanol/water or acetonitrile/water. In addition, the examination of the metabolism and biochemistry involved in the production of fumonisins is difficult with complex solid media.

By contrast, liquid culture methods greatly reduce the need for solvents during isolation and purification since the fumonisin is found in the broth (Blackwell et al., 1994; Branham and Plattner, 1993; Jackson and Bennett, 1990; Plattner and Shackelford, 1992). Culture material can be removed by filtration; extraction with solvents is not required. Culture broth can be assayed directly for fumonisin concentration using any of the current analytical methods. In addition, liquid culture techniques, particularly with defined media, have the added advantage of providing a convenient means to study the metabolism and biochemistry involved with the production of fumonisins. Although production of fumonisins by liquid culture methods would appear to be more beneficial when viewed in terms of isolation and purification, yields reported in the literature appear to be significantly less than is obtained from solid culture with highest values reported to be approximately 0.5 to 0.6 mg fumonisin/ml of culture (Nelson et al., 1993). These yields were obtained using a semi-defined liquid medium reported by Gilchrist and Grogan (1976) for the production of a host-specific toxin from *Alternaria alternata* f. sp. *lycopersici* (AAL-toxin). A defined medium has been described by Jackson and Bennett (1990) but yields were not as high. Blackwell et al. (1994) described a semi-defined medium for growth of *Fusarium monili-forme* with a defined medium for subsequent toxin production. In their study, toxin production required high inoculum levels. The highest toxin concentrations obtained were from 0.4 to 0.5 mg/ml. Blackwell et al. (1994) also reported a requirement for relatively low O_2 tension for optimal biosynthesis in liquid culture.

BATCH CULTURE GROWTH OF MICROORGANISMS

In a typical batch culture for the production of a secondary metabolite, the development of the culture can be divided into two phases: a growth phase, sometimes referred to as the trophophase (Bu'Lock, 1967), and a production phase referred to as the idiophase. Primary metabolism, which occurs during the trophophase, is described by Martin and Demain (1980) as essentially identical for all living things. During primary metabolism the metabolic reactions are finely balanced and metabolic intermediates other than those required for cell survival rarely accumulate. Most cellular growth occurs during the trophophase, whereas the production of secondary metabolites, also referred to as idiolites, occurs in the idiophase. In complex media these two phases can be quite distinct; however, in defined media, where growth can be much slower, the two phases often overlap (Demain, 1992; Martin and Demain, 1980). Demain (1992) defines a secondary metabolite as secondary because it is not necessary for exponential growth of the producing culture. This distinction is important since the two phases can overlap with production of the secondary metabolite while growth still occurs.

The beginning of the production phase, or idiophase, of a batch culture usually occurs when the culture becomes nutrient limited and primary growth ceases. In commercial operations this is usually accomplished by limiting either carbon, nitrogen or phosphate levels (Demain, 1992, DeWitt et al., 1989). The carbon source, particularly glucose, is often limited when it is known that catabolite repression affects the production of the secondary metabolite. Catabolite repression can also be avoided by the use of an alternative carbon source such as lactose or by feeding glucose so that its level is never high enough to interfere with the production of the metabolite. Often a secondary metabolite will be controlled by nitrogen and/or phosphate levels instead of, or in addition to, catabolite repression. In the case of nitrogen repression the use of alternative nitrogen sources such as amino acids in the place of ammonium ions will increase production. Phosphate suppression of secondary metabolite production can be controlled by limiting quantities used or by the use of complexing agents (DeWitt et al., 1989). Phytic acid could also be substituted as an alternative source of phosphate since inorganic phosphate may be a more potent suppressor. For a more complete discussion of effectors of idiolite biosynthesis, Demain (1992) provides a good overview.

CONTROL OF FUMONISIN PRODUCTION

The literature on the growth of Fusaria and the production of fumonisins, indicates that fumonisins are secondary metabolites. They appear generally after primary growth, which can occur without the production of the fumonisin. Therefore, fumonisin production might be controlled by methods similar to those used in the production of antibiotics.

To determine possible metabolic control mechanisms for the production of fumonisin, the defined medium of Jackson and Bennett (1990) was used with some modifications as follows: glucose, 90 g/l; $(NH_4)_2SO_4$, 3.5 g/l; KH_2PO_4, 0.1 g/l; K_2HPO_4, 0.1 g/l; $MgSO_4 \bullet 7H_2O$, 0.3 g/l; $CaCl_2 \bullet 2H_2O$, 0.4 g/l; $MnSO_4 \bullet H_2O$, 0.016 g/l; thiamine, riboflavin, pantothenate, niacin, pyridoxamine, and thioctic acid, all at 1000 µg/ml; folic acid, biotin, and B12 at 100 µg/ml. Antifoam (Sigma Chemical Co., St. Louis, MO) was also added as needed to prevent excessive foaming. The culture used in this work was *Fusarium proliferatum* (M5991), kindly provided by P. Nelson (Pennsylvania State University). Inoculum was prepared from frozen stock grown 5 days in the above media. Cultures were grown using 100 ml media in 500 ml baffled Erlenmeyer flasks in a shaking incubator at 25°C. Under these conditions, growth occurred as dispersed mycelial fragments. Microscopic examination of the culture showed separate and small clumps of mycelial fragments and numerous microconidia. Culture growth was monitored by following the decrease in glucose, nitrogen, and pH. In actively growing shake flask cultures, the pH dropped rapidly below 2.0 within the first week. In most studies, only the concentration of fumonisin B_1 (FB1) was determined. Analysis of FB1 was by the HPLC method of Shephard et al. (1990), after first filtering the culture (0.25 micron filter) and diluting appropriately with acetonitrile:H_2O (50:50, v/v).

Several carbon sources were initially examined for their effect on FB1 production. Of those examined, only growth on glucose produced substantial quantities of FB1. Results using glucose as the sole carbon source showed that substantial quantities remained in the media at the onset and at the end of fumonisin production. The average concentration of glucose remaining in 22 separate shake flasks at the onset of FB1 production was 45.0 ± 17.6 g/l (initial glucose was 90 g/l). The average concentration of glucose remaining in the same flasks at the time of harvest was 20.2 ± 15.5 g/l. This would indicate that fumonisin production is not catabolite repressed.

In addition to the lack of catabolite repression by glucose, phosphate levels did not repress production of fumonisin. In an experiment with six flasks each at high (15 mM KH_2PO_4) and low (1.5 mM KH_2PO_4) phosphate levels, the highest FB1 levels measured were 180 ± 100 and 62.7 ± 26.4 ppm respectively. Lower phosphate levels appeared to result in slightly lower FB1 levels, although, this may be more a result of less buffering capacity. When phytic acid was used to replace KH_2PO_4 in the media, growth was observed but no FB1 was detected.

Nitrogen levels, however, appeared to be related to FB1 production. Figure 1 shows the disappearance of nitrogen (measured as free ammonium) and the corresponding increase in FB1 content in two separate trials. This same relationship was observed regardless of the

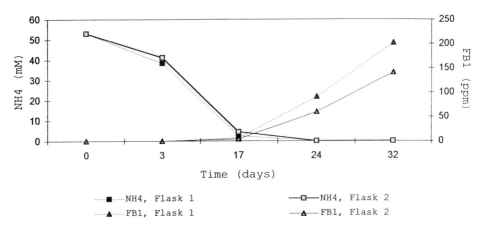

Figure 1. Nitrogen disappearance and corresponding appearance of fumonisin B₁.

rate of ammonium disappearance. The ammonium levels at the onset of FB1 production were measured in 67 separate shake flask trials and are shown in Figure 2. The frequency at which FB1 was first detected (within 5 mM ammonium range levels) was plotted against the ammonium levels. From this graph it appears that FB1 production began sometime after ammonium was reduced to at least 35 mM. In most cases, FB1 was not detected until ammonium concentrations were reduced to less than 20 mM. Although additional evidence is required to prove a repressive effect of nitrogen levels on FB1 production, the data are sufficient to develop a control strategy for scale-up in stirred bioreactors.

In addition to the examination of carbon, nitrogen, and phosphate effects on FB1 production, other effectors were also examined. The effect of detergents on the growth and production of FB1 was tested using three concentrations of Triton X100 and Tween 80 (Table 1). Both Triton X100 and Tween 80 exhibited negative effects on the production of FB1.

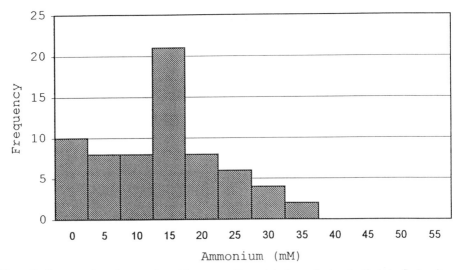

Figure 2. Concentration of ammonium at the onset of fumonisin B₁ production in 67 shake flask cultures.

Table 1. Effect of Tween 80 and Triton X100 on fumonisin B$_1$ production

	Flask 1 ppm FB1(final)	Flask 2 ppm FB1(final)
Control	357	371
0.005% Tween 80	366	337
0.01% Tween 80	255	233
0.1% Tween 80	209	166
0.005% Triton X100	42	58

The effect of Triton X100 was more dramatic than that of Tween 80. Only the lowest concentration of Triton X100 is shown. Higher levels of Triton X100 appeared to inhibit the growth of the culture. At 0.01% and 0.1% Triton X100, the culture grew as pellets of various sizes up to approximately 4 mm.

Lastly, the effect of trace minerals on the growth of *F. proliferatum* and the production of FB1 was also tested. In *F. moniliforme*, zinc, iron and cobalt have been examined with respect to their effects on growth and the production of fusarin C (Jackson et al., 1989; Madan, 1978; Thind and Madan, 1977). Notably, zinc has been found to increase ammonium assimilation and dry weight accumulation. The effect of these minerals on FB1 production is not known. To examine their effect on growth and FB1 production, an experiment was designed with zinc, iron, and cobalt added singly and in combination at two different concentrations. The results showed that iron and cobalt did not affect growth or fumonisin production (Table 2). In contrast, zinc greatly increased dry weight and ammonium assimilation. These findings support the results of Jackson et al. (1989). FB1 production started once nitrogen levels were reduced to approximately 30 mM. The faster the ammonium levels dropped to 30 mM, the sooner FB1 production started. Final concentrations of FB1 did not vary much between high or low zinc levels; therefore, the effect appears to be limited to the

Table 2. Effect of metals on growth and fumonisin B$_1$ production by *Fusarium proliferatum*.

Condition	average mM NH$_4$ remaining after 17 days	average ppm FB1 after 17 days	Dry weight (g/l) at harvest
control	50.1	N.D.	3.29
9.0 mg/l Co	49.8	N.D.	3.27
90 mg/l Co	48.0	N.D.	2.28
100 mg/l Fe	51.2	N.D.	1.29
200 mg/l Fe	47.2	N.D.	1.50
Co and Fe(low)	49.5	N.D.	2.61
Co and Fe(high)	48.3	N.D.	5.43
3.2 mg/l Zn	16.7	11.7	12.78
32 mg/l Zn	12.2	55.7	15.22
Co and Zn(low)	15.8	10.4	14.64
Co and Zn(high)	15.9	41.2	13.92
Fe and Zn(low)	14.2	52.0	14.24
Fe and Zn(high)	11.2	46.6	15.71
Fe, Co, Zn(low)	28.9	3.9	13.77
Fe, Co, Zn(high)	11.6	43.2	15.42

N.D.= none detected
(low)=low concentration
(high)=high concentration

growth and uptake of nitrogen rather than a direct effect of zinc on FB1 synthesis. In flasks without zinc added, where ammonium was not reduced below 30 mM, no FB1 production was detected.

PRODUCTION OF FB1 IN BIOREACTORS

Basic information on the growth of *F. proliferatum* and effectors of FB1 production allows development of appropriate fermentation strategies to be applied to bioreactor systems. The use of bioreactor systems has distinct advantages over shake flask systems. In general, bioreactors allow greater control of growth parameters than is possible in shake flask systems and production can often be improved.

For these bioreactor studies, inoculum was prepared from frozen stock grown approximately 5 days, as before, then transferred to 2-liter glass bioreactors (Biostat MD, B. Braun Biotech, Allentown, PA). These cultures were then grown approximately 2-5 days or until nitrogen was exhausted, then used to inoculate 10-liter bioreactors (Biostat UD10, B. Braun Biotech). Most studies were completed at the 10-liter scale. For the 100-liter scale (Biostat UD150, B. Braun Biotech), inoculum was prepared from 10-liter bioreactors. Media in each case were the standard formulation without added zinc. Initially, the 10-liter stainless-steel bioreactors were run in a manner to duplicate shake flask conditions. Dissolved oxygen was maintained at approximately 100% using a stir rate of 500 RPM and an air flow rate of 2 standard liter per minute (SLPM). The temperature was held at 25°C and the pH was allowed to change as it does in shake flasks. Under these conditions, results were similar to what was observed in shake flasks, with a typical growth curve and maximum cell mass of approximately 10-15 g/l dry weight. As is the case in shake flasks, pH reduction was dramatic and occurred in the first 24-48 hours with subsequent reductions in the metabolic rate as indicated by a reduction in the rate of decrease of glucose and nitrogen concentrations. Fumonisin production was slow and did not exceed 100 ppm.

Since the reduction of pH appeared to slow the metabolism significantly; further studies in bioreactors were done with pH controlled at 3.5 using NaOH. Controlling pH at 3.5 resulted in greater glucose consumption. In these studies, glucose became limiting before nitrogen, and no FB1 was produced. Therefore, the fermentation was changed to a fed-batch method with glucose added to maintain levels at 30-40 g/l. The results of one such fermentation are shown in Figure 3. Dry weight was higher than achieved previously, leveling at approximately 20 g/l. FB1 production started after approximately 10 days and continued to the end of the fermentation, reaching approximately 1200 ppm after 40 days.

The same fermentation was repeated at the 100 liter scale using the standard medium formulation with added zinc (32 mg/l). The effects of zinc were similar to those found in shake flask. The fermentation was dramatically faster with greater cell growth. Ammonium in the medium was exhausted in the first 24 hours; FB1 production started within 48 hours after inoculation. FB1 concentrations of 600-800 ppm were reached after 7 days. Cell concentrations reached 40-50 g/l dry weight.

FUTURE RESEARCH

Current data indicate fumonisins can be successfully produced at a rate and concentration acceptable for production and metabolic studies in submerged culture using stirred bioreactors. The primary controlling factor appears to be nitrogen levels, although additional roles for zinc and high glucose cannot be ruled out. However, additional research is required to understand the mechanisms by which nitrogen may repress fumonisin production. Further

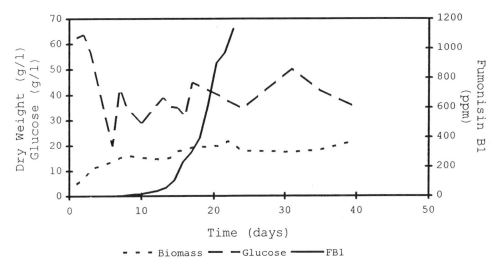

Figure 3. Biomass, glucose, and fumonisin B₁ in a fermentation with pH control.

work is also needed to determine if the sole effect of zinc in fumonisin production is to accelerate growth and nitrogen assimilation or whether it plays an additional, more direct, role in fumonisin production. Elucidating the mechanisms by which these effectors control fumonisin production could aid in a better understanding of field conditions that lead to fumonisin-contaminated corn. This, in turn, could help in the development of strategies for control of fumonisin production in the field.

ACKNOWLEDGMENTS

The authors thank Steven Gendel, US Food and Drug Administration (FDA) for critical evaluation and discussion. The authors also thank Lauren Jackson, Robert Eppley, and Steven Musser, all of the FDA, for their analytical assistance. This publication was partially supported by a Cooperative Agreement, No. FD-000431, from the FDA and the National Center for Food Safety and Technology.

REFERENCES

Alberts, J. F.; Gelderblom, W. C. A.; Thiel, P. G.; Marasas, W. F. O.; Van Schalkwyk, D. J.; Behrend, Y. Effects of temperature and incubation period on production of fumonisin B1 by *Fusarium moniliforme. Appl. Environ. Microbiol.* **1990**, 56, 1729-1733.

Blackwell, B. A.; Miller, J. D.; Savard, M. E Production of carbon 14-labeled fumonisin in liquid culture. *J. AOAC Int.* **1994**, 77, 506-511.

Branham, B. E.; Plattner, R. D. Alanine is a precursor in the biosynthesis of fumonisin B1 by *Fusarium moniliforme. Mycopathologia.* **1993**, 124, 99-104.

Bu'Lock, J. D. Essays in biosynthesis and microbial development. John Wiley & Sons, Inc., New York, 1967.

Demain, A. L. Regulation of secondary metabolism. In *Biotechnology of Filamentous Fungi, Technology and Products.* Eds. D. B. Finkelstein and C. Ball. Butterworth-Heinemann, Stoneham, MA., 1992; 89-112.

DeWitt, J. P.; Jackson, J. V.; Paulus, T. J. Actinomycetes. In *Fermentation Process Development of Industrial Organisms.* Ed. J. O. Neway. Marcel Dekker, Inc. New York, 1989; 1-52.

Gilchrist, D. G.; Grogan, R. G. Production and nature of a host-specific toxin from *Alternaria alternata* f. sp. *lycopersici. Physiol. Biochem.* **1976**, 66, 165-171.

Jackson, M. A.; Slininger, P. J.; Bothast, R. J. Effects of zinc, iron, cobalt, and manganese on *Fusarium moniliforme* NRRL 13616 growth and fusarin C biosynthesis in submerged cultures. *Appl. Environ. Microbiol.* **1989**, 55, 649-655.

Jackson, M. A.; Bennett, G. A. Production of fumonisin B1 by *Fusarium moniliforme* NRRL 12616 in submerged culture. *Appl. Environ. Microbiol.* **1990**, 56, 2296-2298.

LeBars, J.; LeBars, P.; Dupuy, J.; Boudra, H.; Cassini, R. Biotic and abiotic factors in fumonisin B1 production and stability. *J. AOAC. Int.* **1994**, 77, 517-521.

Leslie, J. F.; Doe, F. J.; Plattner, R. D.; Jonz, J. Fumonisin B_1 production and vegetative compatibility of strains from *Gibberella fujikuroi* mating population "A" (*Fusarium moniliforme*). *Mycopathologia* **1992**a, 117, 37-45.

Leslie, J. F.; Plattner, R. D.; Desjardins, A. E.; Klittich, C. J. R. Fumonisin B_1 production by strains from different mating populations of *Gibberella fujikuroi* (*Fusarium* section Liseola). *Mycotoxicology* **1992**b, 82, 341-345.

Maden, M. Trace element studies on four species of *Fusarium. Microbios* **1978**, 23, 161-172.

Martin, J. F.; Demain A. L. Control of antibiotic biosynthesis. *Microbiol. Rev.* **1980**, 44, 230-251.

Nelson, P. E.; Plattner, R. D.; Schackelford, D. D.; Desjardins, A. E. Production of fumonisins by *Fusarium moniliforme* strains from various substrates and geographic areas. *Appl. Environ. Microbiol.* **1991**, 57, 2410-2412.

Nelson, P. E.; Plattner, R. D.; Schackelford, D. D.; Desjardins, A. E. Fumonisin B_1 production by *Fusarium* species other than *F. moniliforme* in section Liseola and by some related species. *Appl. Environ. Microbiol.* **1992**, 58, 984-989.

Nelson, P. E.; Desjardins, A. E.; Plattner, R. D. Fumonisins, mycotoxins produced by *Fusarium* species: biology, chemistry, and significance. *Annu. Rev. Phytopathol.* **1993**, 31, 233-252.

Nelson, P. E.; Huba, J. J.; Ross, P. F.; Rice, L. G. Fumonisin production by *Fusarium* species on solid substrates. *J. AOAC Int.* **1994**, 77, 522-525.

Plattner, R. D.; Shackelford, D. D. Biosynthesis of labeled fumonisins in liquid cultures of *Fusarium moniliforme. Mycopathology* **1992**, 117, 17-22.

Ross, P. F.; Nelson, P. E.; Richard, J. L.; Osweiler, G. D.; Rice, L. G.; Plattner, R. D.; Wilson, T. M. Production of fumonisins by *Fusarium moniliforme* and *Fusarium proliferatum* isolates associated with equine leukoencephalomalacia and a pulmonary edema syndrome in swine. *Appl. Environ. Microbiol.* **1990**, 56, 3225-3226.

Shephard, G. S.; Sydenham, E. W.; Thiel, P. G.; Gelderblom, W. C. A. Quantitative determination of fumonisins B_1 and B_2 by high-performance liquid chromatography with fluorescence detection. *J. Liq. Chromatogr.* **1990**, 13, 2077-2087.

Thiel, P. G.; Marasas, W. F. O.; Sydenham, E. W.; Shephard, G. S.; Gelderblom, W. C. A.; Nieuwenhuis, J. J. Survey of fumonisin production by *Fusarium* species. *Appl. Environ. Microbiol.* **1991**, 57, 1089-1093.

Thind, K. S.; Madan, M. Effect of various trace elements on the growth and sporulation of four fungi. *Proc. Indian Natl. Sci. Acad.* **1977**, 43, 115-124.

19

BIOSYNTHESIS OF FUMONISIN AND AAL DERIVATIVES BY *ALTERNARIA* AND *FUSARIUM* IN LABORATORY CULTURE

C. J. Mirocha[1], Junping Chen[1], Weiping Xie[1], Yichun Xu[1], H. K. Abbas[2], and L. R. Hogge[3]

[1] Department of Plant Pathology
University of Minnesota
St. Paul, MN 55108
[2] USDA, ARS Southern Weed Laboratory
Stoneville, MS 38776; and
[3] National Research Council Canada
Saskatoon, Saskatchewan, Canada

ABSTRACT

Cultures of *Fusarium moniliforme* and *Alternaria alternata* f. sp. *lycopersici* were grown in the laboratory and analyzed for various fumonisin derivatives. Analyses were made by continuous flow fast atom bombardment and ionspray mass spectrometry interfaced to microcapillary HPLC. Besides FB1, FB2 and FB3 derivatives, two isomers of the one-armed FB1 (protonated molecular ion at m/z 564) and two isomers of the one-armed FB2 (m/z 548) were found. Two different isolates of *A. alternata* when grown in culture yielded FB1, FB2 and FB3. One of them also yielded the one-armed FB1 which was identical to that metabolite found in *Fusarium* and naturally infected corn. FB1, FB2 and FB3 have been found in 2 different isolates of *Alternaria alternata*, both obtained from tomato.

INTRODUCTION

Fumonisin is a natural toxic product produced by certain isolates of *Fusarium moniliforme* (Marasas et al., 1988) and also by an isolate of *Alternaria alternata* f. sp. *lycopersici* (Chen et al., 1992). It is of economic importance because of its implication in animal health, i.e., it causes leukoencephalomalacia in equine (Marasas et al., 1988), pulmonary edema in swine (Ross et al., 1990) and ill thrift resembling viral enteritis in broiler chicks (Brown et al., 1992). It is commonly found in corn and most abundantly

Fumonisins in Food, Edited by L. Jackson *et al.*
Plenum Press, New York, 1996

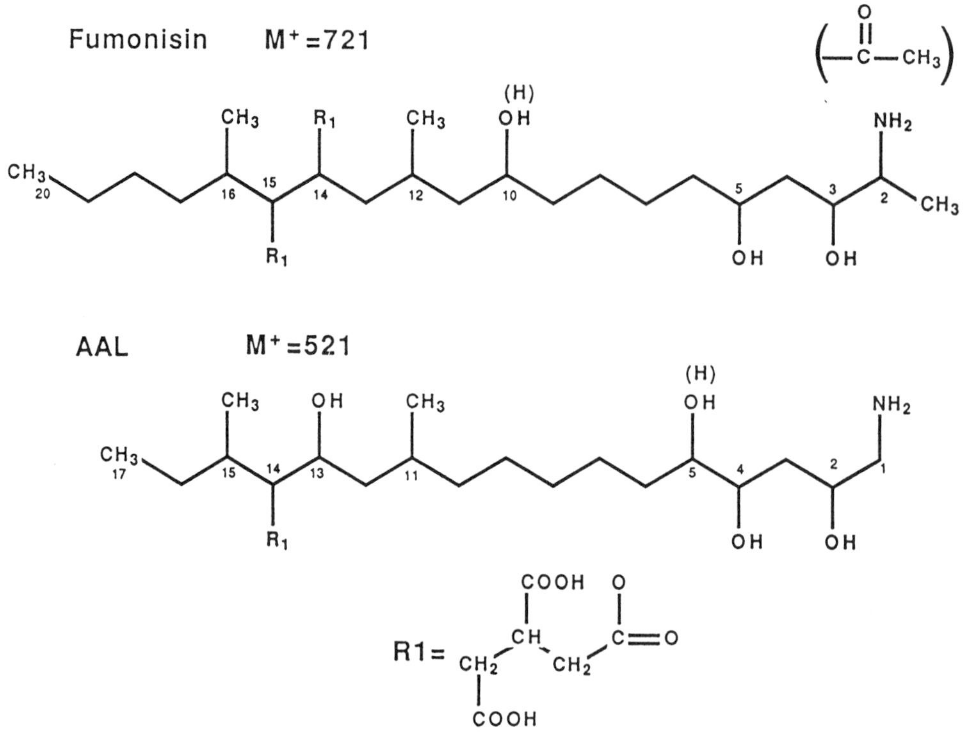

Figure 1. Structures of fumonisin B1 (8 chiral centers) and AAL toxin (6 chiral centers).

in corn screenings. It has also been found in pasture grass in New Zealand (Mirocha et al., 1992).

FB1 is chemically related to AAL toxin (Figure 1) which is produced by *Alternaria alternata* f. sp. *lycopersici* (herein referred to as *A. alternata*) first described by Gilchrist et al. (1976) as the host specific pathogen on tomato. Its chemistry was first reported by Bottini et al. (1981).

The hydrolyzed products of both FB1 and AAL toxin are structurally related to sphingosine and both interfere with sphingosine metabolism, i.e., ceramide synthesis in the rat (Wang et al., 1991) and sphingolipid metabolism in plants (Abbas et al., 1994). Both toxins reduce growth in duckweed (Tanaka et al., 1993). The hydrolysis products of AAL and FB1 are also related to myriocyn (thermozymocidin), an antibiotic produced by the ascomycete *Myriococcum albomyces*. The latter was reported to be too toxic to animals for use as a therapeutant (Kluepfel et al., 1972).

The objective of this research is to describe the various naturally occurring derivatives of fumonisin and AAL toxins as they were found in a laboratory culture of *Alternaria alternata* f. sp. *lycopersici*. Only microgram quantities of the metabolites in a biological mixture were available for characterization and hence they were resolved on microcapillary

HPLC columns interfaced to a continuous flow fast atom bombardment mass spectrometer (CF/FAB/MS).

MATERIALS AND METHODS

Analysis by Mass Spectrometry

Most of the analytical procedures were done by continuous flow fast atom bombardment (CF/FAB/MS) or electrospray mass spectrometry (ES/MS) and/or ionspray mass spectrometry (IS/MS). The IS/MS and ES/MS were interfaced to HPLC via a 4.6 mm reverse phase C18 column. The CF/FAB/MS was interfaced to HPLC via a microcapillary column. The latter was packed with 3 micron particle size C-18 and has an inside diameter of 320 microns and a length of 10 to 20 cm. The solvent system for microcapillary HPLC in CF/FAB/MS consisted of acetonitrile:water: 2% glycerol and 0.1% trifluoroacetic acid after running for one minute isocratic with 0% acetonitrile to 64% in 14 min with a flow rate of 3 to 4 microliters per min. Two Shimadzu LC-600 solvent delivery system pumps were used.

Analyses were done on VG70EQ by CF/FAB/MS using a xenon gun and on a VG70SEQ (secondary ion mass spectrometry) with a cesium gun. Additional analyses were

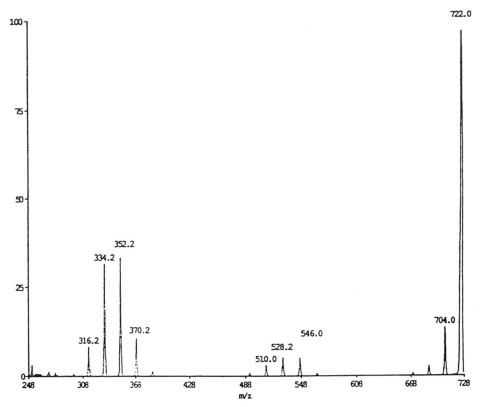

Figure 2. Parent/product ion ionspray mass spectrum of fumonisin B1 run on a Sciex Taga 6000 quadrupole MS/MS. The protonated molecular ion at m/z 722 fragments into the following product ions: 704 (loss of water); 546 (loss of propane tricarboxylic group); 3 losses of water (528, 510 and 492); 370 (loss of second propane tricarboxylic group); 3 consecutive losses of water (352, 334 and 316).

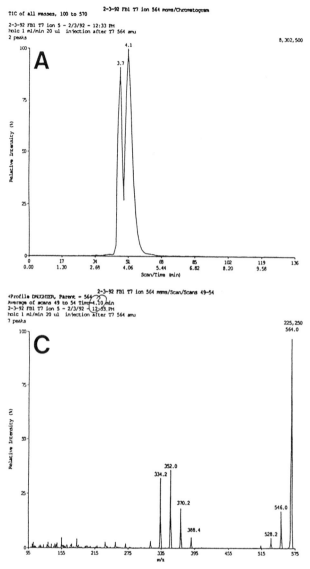

Figure 3. Parent/product ion fragments obtained from MS/MS experiments run on a Sciex Taga 6000. **A**. Total ion chromatogram of two isomers of the one-armed FB1 derivative (M + H)+ = 564 obtained from a culture of *F. moniliforme*. **B**. Product ion fragments of isomer 1 (3.7 min). Note loss of only one propane tricarboxylic acid group (176 amu). **C**. Product ion fragment of isomer 2 found at 4.1 min. **D**. Parent/Product ion fragment of the one-armed FB2 [(M + H)+ = 548] found at retention time of 8 min. The latter had 3 isomers; only the major one is shown.

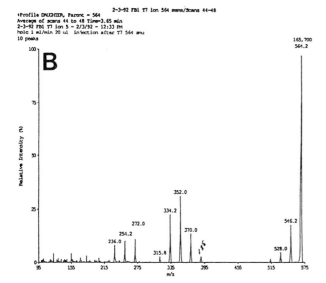

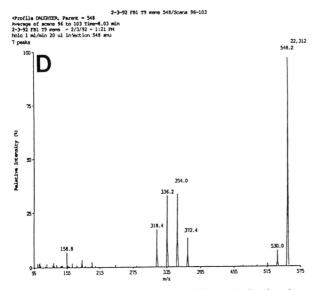

Figure 3. Continued.

done with microcapillary columns on a Trio 2000 equipped with electrospray (Fisons Instruments, Ltd.) and a Sciex Taga 6000 MS/MS.

A pure culture of *Alternaria alternata* f. sp. *lycopersici* was seeded into flasks containing a liquid nutrient medium used for the production of AAL toxins and fumonisin FB1 as described by Chen et al. (1992). The culture was extracted and purified as described by Chen et al. (1992). The culture filtrate concentrate was dissolved in 250 µl of a solvent system consisting of acetonitrile:water:trifluoroacetic acid (1:1:0.1%). An appropriate amount of the sample was injected into the microcapillary HPLC unit.

Identity of the various mycotoxins and their derivatives is based on interpretation of the mass spectra and retention times on a microcapillary column based on authentic standards when available or solely on interpretation of mass spectra.

RESULTS AND DISCUSSION

Fumonisin and its derivatives do not readily avail themselves for analysis by electron impact mass spectrometry; most of the information available comes from CF/FAB/MS as well as ionspray and electrospray mass spectrometry, i.e., after interfacing with HPLC. In IS/MS/MS, the molecule readily shows a pseudo molecular ion at m/z 722 (721 + H) with concomitant losses of 176 amu for the tricarballylic group (TCA) to yield m/z 546 (Figure 2). The second TCA group (minus 176) yields a prominent fragment at m/z 370. Losses of either or both TCA groups is followed by three successive losses of water (depending on the number of hydroxyl groups) i.e. m/z 546 yields 528, 510 492 and m/z 370 yields 352 and 334 and 316. The protonated molecular ion (722) also loses a water to yield m/z 704.

The loss of a TCA group is significant in the interpretation of the FB1 derivative which contains only one tricarballylic group (designated as 1TCAFB1) (M = 563 + H) found in Figure 3. Two well resolved isomers are shown in the mass chromatogram of m/z 564, i.e., retention time of 3.7 and 4.1 min (Figure 3A) when analyzed by IS/MS. The latter are authentic standards (fraction T7) taken from a culture of *Fusarium moniliforme*. IS/MS/MS experiments show a loss of 176 amu to yield m/z 388, a loss of 4 waters to account for 4 hydroxyl groups and a loss of 116 amu to account for the fragment made from the terminal carbons 1 through 5. Both isomers have identical parent/product ion fragments (Figures 3A, B and C).

Fraction T9 of the *F. moniliforme* extract contained 3 isomers of 1TCAFB2 with a pseudomolecular ion of m/z+ 548 (Figure 3D). Analysis by MS/MS of the major isomer (only 1 isomer is shown) found at retention time 8.0 min shows a loss of 176 amu (TCA), from the protonated molecular ion at m/z 548, to yield a product ion fragment at m/z 372 with subsequent loss of 3 moles of water indicating only 3 hydroxyl groups in the molecule. The latter is analogous to the m/z 564 derivative but with one less hydroxyl and apparently

---

Figure 4. CF/FAB/MS mass spectra of fumonisin and AAL toxin derivatives from *A. alternata* f. sp. *lycopersici* and *F. moniliforme*. **A.** Chromatographic resolution of authentic N-acetyl-AAL toxin and the one-armed FB1 derivative; the FAB spectrum of the N-acetyl AAL is shown. **B.** Total ion chromatogram of the FAB spectrum of the one-armed FB1 (2 isomers) found in *A. alternata* culture; its FAB spectrum is shown in bottom box. **C.** Total ion mass chromatogram of authentic FB1 (m/z 722) and its one armed derivative (m/z 564). The retention time found in the microcapillary C18 column is compared with the identical products found in the A. alternata culture. The FAB mass spectrum of the one-armed FB1 found in the *Alternaria* culture is shown in the bottom box. Note a trace of fumonisin FB1 found in the fraction. **D.** FAB mass spectrum of authentic one-armed FB1 found in the *F. moniliforme* culture used as a comparison.

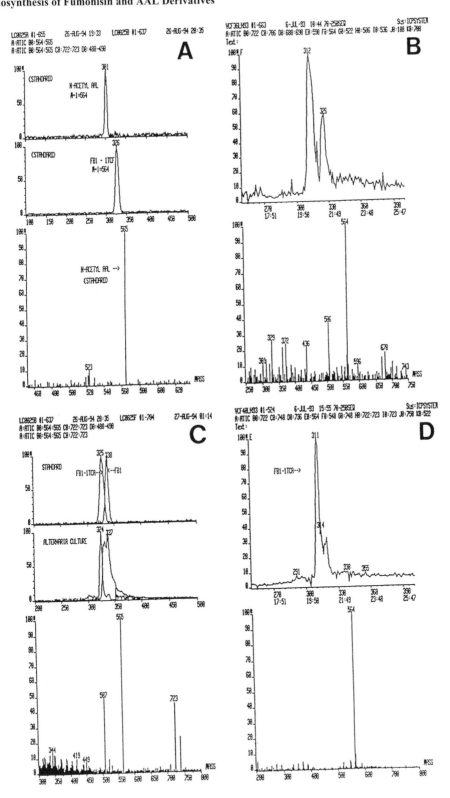

arises by fungal biosynthesis. This derivative is similar but not identical to the N-acetyl derivative of AAL that Caldas et al. (1994) described as a product of *A. alternata* in culture.

It should be noted that the FB1 derivative of *Fusarium* with a pseudomolecular ion of 564 has the same protonated molecular ion as the N-acetyl of AAL toxin; however, the retention times when resolved on microcapillary columns are different (Figure 4A).

Two isomers of the one-armed FB1, i.e., (M + H)+ = m/z 564 exist both in culture and naturally in *F. moniliforme* infected corn. Both isomers were found in a corn sample (FS#870) which contained about 30 ppm FB1 when analyzed by HPLC.

A comparison was made of the mass chromatogram of 1TCAFB1 (m/z 564) obtained by partial hydrolysis of FB1 with a mass chromatogram of the same metabolite obtained from the *A. alternata* culture (Figure 4C). The retention time (scan 326) of the standard from *F. moniliforme* is found just before FB1 (scan 338) and is comparable to that of scan 324 found in the *A. alternata* culture (Figure 4C). The full scan FAB mass spectrum of the *Alternaria* derivative with a protonated molecular ion at m/z 564 is shown in Figure 4C. The FAB spectrum shows a molecular ion at m/z 565 (mass marker is off and actual mass was m/z 564). This same spectrum shows a small amount of FB1 with a protonated molecular ion at m/z 723 (actual is 722).

Figure 4C also compares the retention time of authentic FB1 from *F. moniliforme* and that of FB1 from *A. alternata*. The retention time and FAB mass spectra from the two cultures were identical. The FB1 in the *Alternaria* culture (scan 337) was found in great

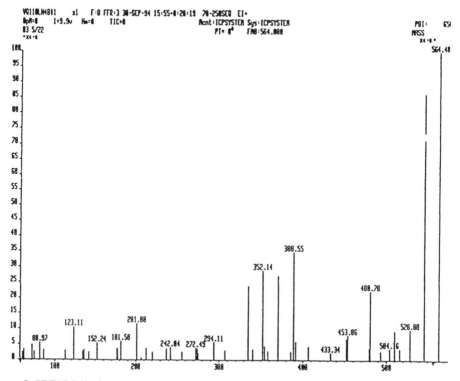

Figure 5. CF/FAB/MS of a parent/product ion mass spectrum of the one-armed FB1 found in the *A. alternata* culture run on a VG70SEQ. Note the loss of one propane tricarboxylic group at m/z 388 and the fragment ion at m/z 272.

abundance and hence the column was overloaded accounting for the large chromatographic peak in Figure 4C.

The N-acetyl-AAL derivative which also has a protonated molecular ion at m/z 564, has a different retention time than the m/z 564 from *Fusarium* (Figure 4A) when resolved by microcapillary HPLC. The full FAB mass spectrum of the N-acetyl of AAL is shown in Figure 4A. The protonated molecular ion is shown at m/z 565 (actual is 564).

1TCAFB1 with a protonated molecular ion at m/z 564 (scan 311) and found in a *F. moniliforme* culture is shown in Figure 4D. It was analyzed on a VG70SEQ instrument with CF/FAB/MS with a microcapillary column. The mass chromatogram of m/z 564 and full mass spectrum is shown in Figure 4D. The same product (scan 312) was found in a culture of *A. alternata* and is shown in Figure 4B. Two isomers of 1TCAFB1 were found in the *A. alternata* culture; scan 312 of *A. alternata* corresponds to scan 311 of *F. moniliforme* (Figure 4B and 4D).

The m/z 564 metabolite, i.e., protonated molecular ion at m/z 564, from *A. alternata* was analyzed by MS/MS with a VG70SEQ mass spectrometer using CF/FAB/MS/MS; the parent/product ion mass spectrum is shown in Figure 5. The fragment ion at m/z 388 represents a loss of 176 amu representing the loss of the TCA group. The fragment ions at m/z 370, 352 and 334 represent a loss of water from the hydroxyl groups. The m/z 272 fragment ion is a loss of 116 amu units (388 - 116 = 272) and represents the terminal 5 carbon fragment (carbons 1 through 5) of the molecule. This fragment is also found in the MS/MS analysis of 1TCAFB1 found in a *Fusarium* culture and shown in Figure 3B.

The mass chromatograms of authentic FB1 and FB2 and FB3 derived from a *Fusarium* culture is shown in Figure 6A for comparison of their respective retention times. Besides 1TCAFB1, the *A. alternata* culture also yielded both FB3 and FB2 shown in Figure 6. Figure 6C shows the mass chromatograms of AAL toxin (m/z 522, scan 299), fumonisin B1 (m/z 722, scan 314) and FB3 and FB2 (m/z 706, scans 350 and 366) respectively, obtained from a culture of *A. alternata*. The latter mass chromatograms have identical retention times as those found in the *Fusarium* culture in Figure 6A. The full FAB mass spectra of FB1, FB3 and FB2 (scans 314, 366 and 350 respectively) are shown in Figure 6B.

Finally, a culture extract of *A. alternata* (SWSL#1) isolated from tomato grown in Mississippi, and cultured on corn meal agar, was analyzed by CF/FAB/MS in the selected ion recording mode. AAL toxin, fumonisin B1 and fumonisins B3 and B2 were found (Figure 6D). Their retention times on the microcapillary column were identical to the respective authentic standards.

The information shown here as well as that reported by Chen et al., 1992 support the fact that fumonisin derivatives are produced by certain isolates of *A. alternata*. No exhaustive attempt was made to test numerous isolates; however, of the three isolates tested, two produced fumonisin. The quantity of fumonisin FB1 as well as the derivatives described are not produced in large amounts as found in the *F. moniliforme* cultures. Moreover, they are produced only in older cultures, i.e., the earliest found were in cultures from 10 to 20 days old and the largest amounts found in cultures from 36 to 56 days old. The important information is that apparently *Alternaria* spp. have a biosynthetic system similar to that of *F. moniliforme* that differs only in the regulatory mechanisms that control quantity and time of biosynthesis of the metabolites. Although *Alternaria* can produce the fumonisins, we at present have only limited data to suggest that AAL toxin is also biosynthesized by *Fusarium* spp.

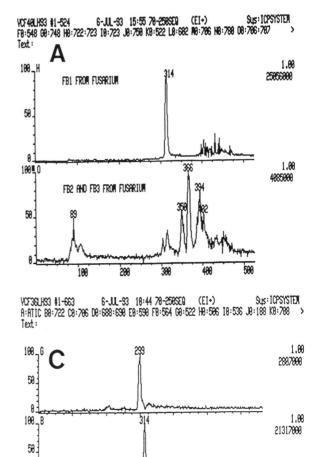

Figure 6. CF/FAB/MS total ion mass chromatograms of fumonisin derivatives (obtained by CF/FAB/MS) found in cultures of *F. moniliforme* and *A. alternata*. **A**. Mass chromatograms of m/z 722 (FB1) and 706 (FB2 and FB3) found in a culture of *F. moniliforme* used as a comparison of the retention time for products found in the *A. alternata* culture. **B**. FAB mass spectrum of scan 314 found in *A. alternata* and shown in box C (mass chromatogram B). **C**. Mass chromatogram of AAL toxin (G), FB1 (B) and FB2 and FB3 shown in box C obtained from a culture of *A. alternata*. **D**. Mass chromatogram of AAL (m/z 522), FB1 (722), FB2 & FB3 (706) found in an *Alternaria* culture isolated from tomato.

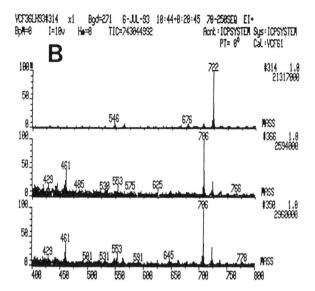

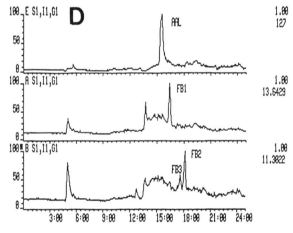

Figure 6. Continued.

ACKNOWLEDGEMENT

Published as paper No. 21,970 of the contribution series of the Minnesota Agricultural Experiment Station based on research conducted under Project 22-34H.

REFERENCES

Abbas, H. K.; Tanaka, T.: Duke, S. O.; Porter, J. K.; Wray, E. M.; Hodges, L.; Sessions, A. E.; Wang, E.; Merrill, A. H.; Riley, R. T. Fumonisin and AAL toxin induced disruption of sphingolipid metabolism with accumulation of free sphingoid bases. *Plant Physiol.* **1994**, 106, 1085-1093.

Bezuidenhout, S. C.; Gelderblom, W. C. A.; Gorst-Allman, C. P.; Horak, R. M.; Marasas, W. F. O.; Spiteller, G.; Vleggaar, R. J. Structure elucidation of the fumonisins, mycotoxins from *Fusarium moniliforme*. *Chem. Soc. Chem. Comm.* 1988, 743-745.

Bottini, A. T.; Bowen, J. R.; Gilchrist, D. G. Phytotoxins II: Characterization of a phytotoxic fraction from *Alternaria alternata* f. sp. *lycopersici. Tetrahedron Letters* **1981**, 22, 2723-2726.

Brown, T. P.; Rottinghaus, G.; Williams, M. E. Fumonisin mycotoxicosis in broilers: performance and pathology. *Avian Diseases* **1992**, 36, 450-454.

Caldas, E. D.; Jones, A. D.; Ward, B.; Winter, C. K.; Gilchrist, D. G. Structural characterization of three new AAL toxins produced by *Alternaria alternata* f. sp. *lycopersici. J. Agric. Food Chem.* **1994**, 42,327-333.

Chen, Junping; Mirocha, C. J.; Xie, W.; Hogge, L.; Olson, D. Production of the mycotoxin fumonisin B1 by *Alternaria alternata* f. sp. *lycopersici. Appl. & Environ. Microbiol.* **1992**, 58, 3928-3931.

Gilchrist, D. G.; Grogan, R. G. Production and nature of a host-specific toxin from *Alternaria alternata* f. sp. *lycopersici. Phytopathology* **1976**, 66, 165-171.

Kluepfel, D.; Bagli, J.; Baker, H.; Charest, M.; Kudelski, A.; Sehgal, S. N.; Vezina, C. Myriocyn, a new antibiotic from *Myriococcum albomyces. J. Antibiotics* **1972**, 34, 109-115.

Marasas, W. F. O.; Kellerman, T. S.; Gelderblom, W. C. A.; Coetzer, J. A. W.; Thiel, P. G.; van der Lugt, J. J. Leukoencephalomalacia in a horse induced by fumonisin B1 isolated from *Fusarium moniliforme*. *Onderstepoort J. Vet. Res.* **1988**, 55, 197-203.

Mirocha, C. J.; Mackintosh, C. G.; Mirza, U. A.; Xie, W.; Xu, Y.; Chen, J. Occurrence of fumonisin in forage grass in New Zealand. *Appl. & Environ. Microbiol.* **1992**, 58, 3196-3198.

Ross, P. F.; Nelson, P. E.; Richard, J. L.; Osweiler, G. D.; Rice, L. G.; Plattner, R. D.; Wilson, T. M. Production of fumonisins by *Fusarium moniliforme* and *Fusarium proliferatum* isolates associated with equine leukoencephalomalacia and pulmonary edema syndrome in swine. *Appl. Environ. Microbiol.* **1990**, 56, 3225-3226.

Tanaka, T.; Abbas, H. K.; Duke, S. O. Structure-dependent phytotoxicity of fumonisins and related compounds in a duckweed bioassay. *Phytochem.* **1993**, 33, 779-785.

Wang, E.; Norred, W. P.; Bacon, C. W.; Riley, R. T.; Merrill, A. H., Jr. Inhibition of sphingolipid biosynthesis by fumonisins: implications for diseases associated with *Fusarium moniliforme*. J. Biol Chem. **1992**, 266, 14486-14490.

20

FUMONISIN TOXICITY AND METABOLISM STUDIES AT THE USDA

Fumonisin Toxicity and Metabolism

William P. Norred, Kenneth A. Voss, Ronald T. Riley and
Ronald D. Plattner*

Toxicology and Mycotoxins Research Unit
Richard B. Russell Agricultural Research Center
ARS/USDA
P. O. Box 5677
Athens, Georgia 30604-5677 and
*National Center for Agricultural Utilization Research
ARS/USDA
1815 N. University St.
Peoria, IL 61604

ABSTRACT

Fumonisins are responsible for many of the toxic effects of the common corn fungus, *Fusarium moniliforme*. They are acute renal and liver toxins in rats, and have tumor promoting activity. Fumonisin B_1 is poorly absorbed, rapidly excreted, and persists in small amounts in the liver and kidney. Fumonisins are specific inhibitors of ceramide synthase, and the toxic effects they produce may be related to their ability to disrupt sphingolipid metabolism, resulting in a myriad of problems in cell regulation and communication. In this paper, research that has been conducted on *F. moniliforme* and the fumonisins at the USDA's Russell Research Center is reviewed.

INTRODUCTION

With the discovery of the fumonisins in 1988 (Bezuidenhout *et al.*, 1988; Gelderblom *et al.*, 1988), a new era in mycotoxicology began. In the years since, there has been an escalating degree of interest in these toxins by scientists, corn producers, commodity groups, food processors and regulatory agencies. See Riley *et al.* (1993a), Norred (1993), and Voss *et al.* (1995a) for recent reviews on the subject. The reason for the concern, which exceeds even that which occurred with the discovery

Fumonisins in Food, Edited by L. Jackson *et al.*
Plenum Press, New York, 1996

of the aflatoxins in the 1960's, is twofold. First, the fungi that are the primary producers of fumonisins, *Fusarium moniliforme* and *F. proliferatum* (Marasas et al., 1984; Nelson *et al.*, 1993), are extremely common, and are associated with corn grown throughout the world. Second, the toxicological properties of the fumonisins in both domestic and laboratory animals are well documented, and include fatal illnesses in horses and pigs, and both acute toxicity and tumor promotion (and possibly initiation) in rats. The involvement of fumonisins in human disease is at this point speculative, but there is epidemiological evidence that the toxins may be a contributing factor in the high incidence of esophageal cancer in South Africa (Sydenham *et al.*, 1990; Rheeder *et al.*, 1992), China (Chu and Li, 1994; Yoshizawa *et al.*, 1994), and northern Italy (Doko *et al.*, 1995; Franceschi *et al.*, 1990). Because of these factors, there is little doubt that these mycotoxins will continue to have a significant, adverse impact on the corn industry, including export markets.

The explosion of research activity on fumonisins has resulted, given the short time since their discovery, in a surprising number of significant findings. Surveys of corn and corn products, including products for human consumption, have consistently revealed the presence of fumonisins in foods and feeds (Thiel *et al.*, 1991; Ross *et al.*, 1991; Sydenham *et al.*, 1991; Sydenham *et al.*, 1992; Ueno *et al.*, 1993; Murphy *et al.*, 1993; Chamberlain *et al.*, 1993). Fumonisin is now known to be the causative agent of several fatal diseases in animals, including equine leucoencephalomalacia (ELEM) (Marasas *et al.*, 1988) and porcine pulmonary edema (Harrison *et al.*, 1990; Colvin *et al.*, 1993). In early 1995, an outbreak of ELEM killed at least 38 horses in Kentucky and Virginia, and fumonisin was identified in the corn-based horse feed the horses had consumed (House, 1995). Fumonisin is also known to cause liver and kidney toxicity in rodents (Voss *et al.*, 1993; Voss *et al.*, 1994), and to produce liver tumors in rats (Gelderblom *et al.*, 1991), apparently able to both initiate and promote tumor formation (Gelderblom *et al.*, 1992). Fumonisin has been found to have specific, potent activity as an inhibitor of sphingolipid biosynthesis by blocking the conversion of sphinganine to ceramide (Wang *et al.*, 1991; Yoo *et al.*, 1992; Norred *et al.*, 1992), a finding that has potentially far-reaching implications for understanding the mechanism of fumonisin toxicity as well as the role of sphingolipids in cell function and regulation (Riley *et al.*, 1994a). Greatly needed is more toxicological research to fully define the levels of fumonisins required to affect animal or human health, chemical research to find more rapid and accurate analytical methods, techniques for decontaminating corn, and biochemical research to more fully elucidate the apparent mechanism of toxicity of fumonisin, i. e. the disruption of sphingolipid biosynthesis.

Since 1985, USDA scientists at the Russell Research Center have studied *Fusarium moniliforme*, the diseases associated with it in both plants and animals, and the toxins that are produced by this organism. Shortly after publication of the structure of the fumonisins, which were isolated by South African scientists from culture material (autoclaved corn inoculated with *F. moniliforme* and incubated for several weeks) (Bezuidenhout *et al.*, 1988), we reported the first finding of the toxins in a naturally contaminated sample of corn which had been associated with an outbreak of ELEM in Illinois (Norred *et al.*, 1989.; Voss *et al.*, 1989; Plattner *et al.*, 1990). Since then, an intense effort to characterize the toxicological properties of fumonisins and related compounds has been underway in several laboratories, including our own. This paper will be limited to a review of the *in vivo* and *in vitro* investigations that have been conducted in the Toxicology and Mycotoxins Research Unit of the Russell Research Center, including a summary of some on-going, as yet unpublished, research results.

IN VIVO INVESTIGATIONS

Studies with Naturally Contaminated Corn and with *F. moniliforme* Culture Material

In 1987 we were provided with a quantity (about 25 kg) of corn from Illinois that had been fed to horses, of which 10 had died with symptoms typical of ELEM. The corn was examined for mycoflora, and *F. moniliforme* was the predominant fungal species present. Rats were fed the contaminated corn for a 4-week trial, and were compared to rats fed high quality seed corn that was free of fungal contamination (Voss *et al.,* 1989). Toxicity of the ELEM-associated corn to the rats was clearly evident, and was manifested by lowered body weights, increased serum enzyme activities and elevated serum bilirubin levels, decreased liver and kidney weights, and microscopic hepatic and renal lesions. The histopathologic lesions in livers included biliary hyperplasia, single cell degenerations and necroses, mitotic figures, and mild inflammatory infiltration and periportal fibrosis. In kidney, changes observed included tubular basophilia, focal cytoplasmic vacuolation, sloughing of tubular epithelial cells into luminae, and epithelial regeneration. The toxic signs produced were essentially identical to those reported previously when culture material prepared from a particularly toxic strain of *F. moniliforme*, MRC 826, was fed to rats (Kriek *et al.,* 1981; Voss, 1990) or monkeys (Jaskiewicz *et al.,* 1987).

In an effort to identify the toxic principals produced by *F. moniliforme* which were responsible for the lesions observed, feeding studies were undertaken in which extracts of culture material were dried, incorporated into feed, and fed to rats (Voss *et al.,* 1990). The culture material itself, water extracts, chloroform/methanol extracts, and the residues of culture material after extraction with either water or chloroform/methanol were fed for 4 weeks. The results clearly indicated that the agent produced by *F. moniliforme* that was responsible for the hepatotoxicity in rats was a water soluble compound. Water extract of the culture material (but not sound corn) was toxic, as was the residue remaining after organic solvent extraction. However, neither the chloroform/methanol extract nor the residue after water extraction were toxic. While these studies were concluding, Gelderblom *et al.* (1988) published their paper establishing the identity of the water soluble component which had the ability to induce γ-glutamyl transpeptidase positive foci in rat liver, and which they named fumonisins. Shortly thereafter, French scientists isolated and characterized the same compound from *F. moniliforme*, and gave it the name macrofusine (Laurent *et al.,* 1988). We then analyzed our culture material and extracts by gas chromatography/mass spectrometry and thin layer chromatography and identified fumonisins in the culture material, the water extract and the residue of culture material after organic extraction - the same components that produced toxic signs in rats after feeding for 4 weeks. We also reexamined the samples of naturally contaminated corn from Illinois, analyzed them for fumonisins, and found 120 ppm in one batch, and 20 ppm in another (Plattner *et al.,* 1990). Attention was then turned to investigations of purified fumonisins, primarily fumonisin B_1, and a concerted effort was made in our laboratory to produce and purify sufficient amounts of the toxin for acute and subchronic toxicity studies.

Recently, we have investigated the reproductive toxicity of diets augmented with *F. moniliforme* culture material to male and female rats. A preliminary report of the findings was recently made (Voss *et al.,* 1995), and a manuscript describing the findings in detail is in preparation. Briefly, diets that contained 0, 1, 10 or 55 ppm of fumonisin B_1 were fed to male and female rats beginning 9 and 2 weeks, respectively, prior to mating. No significant reproductive effects were found in the male rats, dams and fetuses examined on gestation day 15 (G15), or dams or litters examined on day 21 *post partum*, although there was nephrosis in males fed ≥ 10 ppm fumonisin B_1 and in females fed 55 ppm. On G15, free sphingoid bases were not elevated in tissues from the fetuses (whereas the bases were elevated in the dams), thus suggesting *in utero* exposure to fumonisin B_1 did not occur. In a second experiment, radiolabelled fumonisin B_1 was not found in the fetuses after administration of an intravenous dose of $[^{14}C]$fumonisin B_1 to dams on G15. This is also considered strong evidence that *in utero* exposure to fumonisin B_1 did not occur. A similar conclusion was reported in a recent developmental study with purified fumonisin B_1 in CD1 mice (Reddy *et al.*, 1995).

Studies with Purified Fumonisin B_1

The first study we conducted used fumonisin B_1 that was judged to be 99% pure based on comparison with a standard obtained from the South African Medical Research Council (Voss *et al.*, 1993). Male and female Sprague-Dawley rats were fed for 4 weeks with diets amended with fumonisin B_1 to provide final concentrations of 0, 15, 50 or 150 ppm. During this period, overt signs of toxicity, such as weight loss or decreased feed consumptions, were not observed. Both males and females fed 150 ppm, however, experienced hepatotoxicity. Additionally, males fed ≥ 15 ppm and females fed ≥ 50 ppm had cortical nephrosis, which indicated that in rats the kidney may be a more sensitive target organ to the effects of fumonisin B_1 than the liver. In addition to the histopathological findings, elevated levels of sphinganine and sphinganine/sphingosine ratio were found in both liver and kidney samples from these animals (Riley *et al.*, 1994). Importantly in this study, the low dose level used, 15 ppm, is well within the range of concentrations that have been reported in naturally contaminated samples of corn or corn-based feeds, and toxic effects were observed when this level was fed. Additionally, this study indicated that the types and severity of toxic lesions that have been observed by numerous investigators when *F. moniliforme* culture material was fed to rats can be unequivocally replicated by feeding purified fumonisin B_1.

Ultrastructural features of the livers and kidneys of rats fed 15 ppm fumonisin B_1 or more revealed mitochondrial swelling and increased numbers and sizes of both clear and electron dense cytoplasmic vacuoles (Riley *et al.*, 1994). Lamella membranous whorls in bile canaliculi, and disruptions of the normal architecture of the deep infoldings of the basal lateral membrane in kidney cells were also observed. Although the cause for the ultrastructural changes produced by fumonisin are as yet unknown, the degree of disruption of sphingolipid biosynthesis was closely correlated with the changes. Thus disruption of sphingolipid bisynthesis was hypothesized as the molecular mechanism of nephrotoxicity of fumonisin.

Results obtained from feeding rats fumonisin B_1 for 4 weeks (Voss *et al.*, 1993) provided the basis for dose selection for a subchronic feeding study conducted by Voss *et al.* (1994). Groups of male and female B6C3F1 mice and Fischer 344 rats (15 animals per group) were fed 0, 1, 3, 9, 27, or 81 ppm of purified fumonisin B_1 for 13 weeks. Mice were more resistant to the toxic effects of fumonisin B_1 than rats. In female mice, hepatotoxicity (elevated serum chemical profiles and hepatopathy) was seen in the group fed 81 ppm fumonisin B_1, but not in mice fed ≤ 27 ppm. No toxic signs were seen in any of the groups of male mice fed the toxin. In rats, on the other hand, kidney toxicity occurred in both sexes in groups fed ≥ 9 ppm fumonisin B_1. Hepatotoxicity was not observed in the rats, supporting earlier findings (Voss *et al.*, 1989; Voss *et al.*, 1993) that in rats fumonisin B_1 is nephrotoxic at doses that do not cause liver toxicity. Gelderblom *et al.* (1991) found hepatocellular carcinoma in rats fed diets containing 50 ppm purified fumonisin B_1 for 26 weeks. Greatly needed are long term feeding studies in several species with lower levels of the toxin, so that a reasonable estimation of the carcinogenic potential of fumonisins can be made. Such investigations are currently underway by scientists at the FDA's National Center for Toxicological Research.

Neurotoxicity of *Fusarium moniliforme*-Contaminated Corn and Fumonisin

In an effort to develop a laboratory animal model for *F. moniliforme*-induced leucoencephalomalacia (LEM) in horses, Porter *et al.* (1990) fed rats corn naturally infected with the fungus, or culture material prepared with *F. moniliforme* MRC 826, and examined brains and brain neurotransmitters. Levels of brain 5-hydroxyindoleacetic acid (5-HIAA), a major metabolite of serotonin (5-HT), were increased in the treated rats compared to rats fed fungal-free corn, and the ratio of 5-HIAA to 5-HT was also elevated. It was suggested that these elevations in brain neurotransmitters might be a useful bioassay for the neurotoxic effects of *F. moniliforme* toxins. However, a second study using purified fumonisin B_1 instead of culture material revealed that the changes in levels of brain 5-HIAA and 5-HIAA:5-HT could not be reproduced by fumonisin B_1 alone (Porter *et al.*, 1993). This study suggested that some other toxin produced by *F. moniliforme* must be responsible for the brain effects in rats. Furthermore, since LEM has been produced experimentally in horses by administration of purified fumonisin B_1 (Marasas *et al.*, 1988; Kellerman *et al.*, 1990), then the effects on brain neurotransmitters produced in rats by *F. moniliforme* is most likely unrelated to the LEM effects in horses.

Toxicity of *Fusarium* Metabolites to Fertile Chicken Eggs

The toxicity of *F. proliferatum* extracts and of purified fumonisin B_1 to chicken embryos was shown by Javed *et al.* (1993). Besides mortality, which depended both on dose given and whether the dose was administered on day 1 or day 10, embryonic changes including hydrocephalus, enlarged beaks, and elongated necks. These studies were extended by Bacon *et al.* (1995), who also demonstrated the lethality of fumonisin B_1 to chicken embryos, but who also co-administered another secondary metabolite of *F. moniliforme*, fusaric acid. Whereas fusaric acid alone was relatively non-toxic to the eggs, a synergistic response was obtained when fusaric acid and fumonisin B_1 were co-administered. Although these studies demonstrated the lethality of fumonisin if chicken embryos are exposed, there is some question as to the relevance, since Vudathala *et al.* (1994) dosed laying hens with radiolabelled fumonisin B_1, and found that there was no distribution of the toxin to eggs.

Distribution and Excretion of Radiolabelled Fumonisin B_1 in Rats

[^{14}C]fumonisin B_1 was prepared by the addition of [^{14}C]methionine to a liquid culture of *F. moniliforme*, and administered either orally or intravenously to rats (Norred *et al.*, 1993). Up to 80% of an oral dose was eliminated in feces, and up to 35 % was found in feces after intravenous (iv) dosing, indicating that once absorbed, the toxin is readily eliminated in bile. Renal excretion (3% of the oral dose; 10% of the iv dose) also occurred. Radiolabelled fumonisin was rapidly cleared from the blood after iv administration, with 90% clearance within 10 min, and 95% clearance within 30 min. Most tissues and organs did not appear to be significant distribution sites for fumonisin, however the target organs, liver and kidney, had measurable amounts of radiolabel shortly after dosing, and the levels were persistent. Ninety-six hours after oral dosing, liver retained 0.4% of the dose, and kidney 0.2 %. When the labelled toxin was given iv, liver had about 30% and kidney 8% of the dose, and these levels remained high from 0.5 to 96 hours after administration. Other investigators have conducted similar type distribution studies, including studies in swine and laying hens, and similar results have been obtained (Shephard *et al.,* 1992; Prelusky *et al.,* 1994; Vudathala *et al.,* 1994). All of these studies indicate that fumonisin B$_1$ is a poorly absorbed compound that is rapidly cleared from blood, undergoes both biliary and urinary excretion, and tends to persist in relatively small quantities in liver and kidney for extended periods (>96 hrs).

Effects of Fumonisins on Plants

Increased free sphinganine appears to play an important role in the phytotoxicity of fumonisin B$_1$ and AAL-toxin, a mycotoxin with structural similarities to fumonisins, and which is produced by a fungal tomato pathogen (Abbas *et al.,* 1994). For example, susceptible varieties of tomato plants (*asc/asc*) exposed to pure fumonisin B$_1$ or AAL-toxin show a time-dependent increase in free sphinganine concentration which parallels the onset of disease symptoms. In resistant varieties (*Asc/Asc)* of tomato, the free sphinganine concentration decreases after 24 hr of exposure, and at 72 hr no disease symptoms are observed and the free sphinganine concentration is only a fraction of that in the susceptible varieties. Recently, corn seedlings have been shown to respond to fumonisin exposure with elevation in free sphingoid base concentrations (Riley *et al.,* 1995).

IN VITRO INVESTIGATIONS

As a part of our research mission to determine mechanisms of toxicity of mycotoxins, to develop bioassay methods for detection of known toxins, to aid isolation and identification of new toxins, and to develop potential therapeutic strategies and commercial applications, we have utilized a number of *in vitro* techniques to study *F. moniliforme*. The following is a summary of these studies. Relatively few of our studies of the effects of fumonisin on sphingolipid biosynthesis will be presented, since this topic is being covered by our collaborator, Dr. A. E. Merrill, in another chapter.

Studies with Extracts of *F. moniliforme* Culture Material and Naturally Contaminated Corn

Primary rat hepatocytes have a number of advantages as a bioassay tool, because they are easy to obtain and maintain, they retain xenobiotic metabolizing enzymes with higher activities than found in cultured cell lines, and are useful for studying cytotoxicity, geno-toxicity, induction of biochemical lesions, and biotransformations of toxic substances (Rauckman and Padilla, 1987). We used primary hepatocytes to investigate toxicity of extracts of corn naturally contaminated with *F. moniliforme* and which had killed a number of horses in Illinois (Norred *et al.,* 1990). (This is the same sample of corn referred to earlier that was used for our initial *in vivo* investigations.) Organic (chloroform/methanol) extracts of the corn did not cause release of lactate dehydrogenase from hepatocytes, nor did they stimulate unscheduled DNA synthesis (UDS), indicating low cell lethality and lack of genotoxicity. However, the extracts were very potent as inhibitors of valine incorporation into hepatocytes, indicative of activity as protein synthesis inhibitors. Similar activity was found in extracts of *F. moniliforme* MRC 826 and RRC 408 (an isolate obtained from the Illinois corn). Separation of the extracts into acidic and neutral fractions revealed that the protein inhibition activity was associated with the neutral fraction. Several known mycotox-ins, including moniliformin, fusarin C and zearalenone were tested for inhibition of valine incorporation and were devoid of activity. A methanol/water extract of *F. moniliforme* RRC

408 culture material, which was found to contain fumonisins, was also inactive as an inhibitor of protein synthesis.

Trichothecenes are mycotoxins produced by a number of *Fusaria* species, and are potent inhibitors of protein synthesis. *F. moniliforme* is generally not believed to produce trichothecenes (Marasas, 1986). Indeed, an examination of the extracts we used by tandem mass spectrometry failed to reveal the presence of trichothecenes or their precursors. Application of the extracts to the shaved skin of rabbits did not cause the characteristic dermal irritation that is a well known toxic property of trichothecenes. Although the identity of the agent produced by *F. moniliforme* that inhibits protein synthesis is as yet unknown, the toxin(s) do not appear to be fumonisins, fusarin C or trichothecenes. Further studies are needed to identify the agent, and to determine if it is involved in any of the field intoxications attributed to *F. moniliforme* contamination of corn.

Primary hepatocytes were used in another series of studies to examine various isolates of *F. moniliforme* (Norred *et al.,* 1991). Aqueous and organic extracts of corn cultures of 10 different isolates of the fungus were tested for their ability to kill cells (measured by release of lactate dehydrogenase) or inhibit protein synthesis (measured by inhibition of valine incorporation). Each isolate was also evaluated for its ability to produce fumonisin B_1 and fusarin C. The fungal isolates differed greatly in their ability to produce the mycotoxins, and in their ability to cause cytotoxic effects by the two methods measured. For the 10 isolates used in this investigation, there appeared to be an inverse relationship between the ability to produce fumonisin and the ability to produce fusarin C. Cell death did not occur to a significant degree when hepatocytes were exposed to aqueous extracts, but organic extracts of some of the isolates were cytotoxic. There appeared to be some correlation between ability of the isolate to produce fusarin C and lethality. Protein synthesis was inhibited by aqueous extracts of all the isolates, although some isolates were more potent than others. Protein synthesis inhibition produced by organic extracts was apparently coincidental to cell death caused by the extracts. There was no correlation between inhibition of protein synthesis and fumonisin production by the isolate. The conclusions of this study were that there are effects produced in hepatocytes by extracts of *F. moniliforme* isolates that cannot be accounted for on the basis of known mycotoxins, and that other, unknown agents must be responsible.

Studies with Purified Toxins

Because of reports of the ability of *F. moniliforme*-contaminated corn, and later, fumonisins, to induce hepatocellular carcinoma in rats (Jaskiewicz *et al.,* 1987a; Gelderblom *et al.,* 1991), we investigated whether selected purified secondary metabolites of *F. moniliforme* and *F. proliferatum* are genotoxic to primary hepatocytes (Norred *et al.,* 1992). Hepatocytes were exposed to the test compounds overnight at different concentrations as follows: moniliformin (5, 10, 50, 100 or 500 µM), fumonisin B_1 (0.5, 2.5, 5, 25, 50, 250 µM), fusarin C (1, 5, 10, 50, 100 µM) or bikaverin (1, 5, 10, 50, 100, 500 µM). Cytotoxicity was observed in cells exposed to the highest dose of fusarin C and bikaverin, but none occurred at any doses of moniliformin or fumonisin B_1. Fumonisin B_1, bikaverin and moniliformin were not genotoxic to primary hepatocytes, as measured by the UDS assay. Results of the UDS procedure with fusarin C were inconclusive since a marginal effect on UDS occurred. Fusarin C is a potent mutagen in the Ames' test (Gelderblom *et al.,* 1983), but undergoes rapid conjugation with glutathione (Gelderblom *et al.,* 1988a) and may not be available *in vivo* or in primary hepatocytes for interaction with macromolecules. The failure of fumonisin B_1 to induce UDS in primary hepatocyte cultures is also consistent with its inability to produce a positive result in the Ames' mutagenicity assay (Gelderblom and Snyman, 1991), thus it appears that if fumonisin B_1 is a complete carcinogen, it must act by a non-genotoxic mechanism.

Effects of Fumonisins on Sphingolipid Biochemistry

As indicated earlier, a thorough treatment of this topic and its ramifications will be covered in another chapter of these proceedings. We examined the structure of fumonisins when they first appeared in the literature, and noted the structural similarity of the fumonisin "backbone" (with the tricarballylic groups removed) to sphinganine. The consultation of a noted sphingolipid biochemist, Dr. A. E. Merrill, Jr. at Emory University, was sought out, he was given a small quantity of purified fumonisin B_1, and within a short period of time the discovery was made that the toxin is a potent inhibitor of ceramide N-acyltransferase, a key enzyme in the metabolic pathway that results in the conversion of sphinganine to ceramide, and in the turnover of complex sphingolipids to sphingosine (Wang et al., 1991; Norred et al., 1992a). The result of this inhibition is an elevation of free sphingoid bases, especially sphinganine, in tissues, serum and in cells sloughed off into urine. The end result of the inhibition is an elevation in the sphinganine (Sa)-to-sphingosine (So) ratio, and measurement of this parameter has been proposed as a biomarker for fumonisin exposure (Riley et al., 1994b). Unlike fumonisin, which is rapidly eliminated from serum, free sphinganine levels remained elevated after ponies were removed from fumonisin-containing diets (Wang et al., 1992). Similar results have been seen in vitro where free sphinganine remains elevated for >96 hr after 95% of the fumonisin B_1 has left the cells (Riley and Yoo, 1995). We have recently used precision-cut rat liver and kidney slices, and found that elevated sphinganine levels occur with exposure of the slices to as little as 0.1 μM fumonisin B_1 (Norred et al., 1995).

The inhibition of sphingolipid biosynthesis is correlated with cytotoxicity of the toxin in a proliferating cell line, LLC-PK$_1$ cells (Yoo et al., 1992). Concentrations of fumonisin B_1 that inhibited sphingolipid biosynthesis after 7 hours exposure were cytotoxic after 3- to 5-days exposure. The increase in free sphinganine concentration was maximal at 48 hr. After 24 hr, many cells began to develop a fibroblast-like appearance, with loss of cell-cell contact and an elongated, spindle shape. If fumonisin was removed, the cells that survived resumed growth and had a normal epithelial morphology. In recent studies (Yoo et al., 1995), it was found that there is a close association in the the the amount of sphinganine that accumulates in the cells, the depletion of complex sphingolipids, and the decrease in cell growth and increased cell death (as measured by reduction in cell protein and release of lactate dehydrogenase into the medium). β-chloroalanine, an inhibitor of the first enzyme in sphingolipid biosynthesis, in the presence of fumonisin B_1 reduced the intracellular concentration of free sphinganine by about 90%, and also reduced the inhibition of cell growth and reduced the extent of cell death. The two most likely explanations for the increased cell death after inhibition of sphingolipid biosynthesis by fumonisins are: 1) that the free sphinganine is growth inhibitory and cytotoxic for the cells, as has been seen in many other systems (Stevens et al., 1990; Hannun et al., 1991), and 2) that complex sphingolipids are required for cell survival, as has been proven with mutants lacking serine palmitoyltransferase (Hanada et al., 1990; 1992). There is no doubt that the loss of complex sphingolipids plays a role in the abnormal behavior in fumonisin treated cells, however, the accumulation of free sphinganine (or a metabolite) appears to be at least an equally important contributing factor with regards to its acute effects. Supporting these in vitro results, ponies intoxicated by feeding fumonisin-containing feed were found to have elevated Sa/So ratios (Wang et al., 1992), and in a more controlled feeding study with pigs it was found that the ratio was elevated early during the study by feed that contained as little as 23 ppm total fumonisins (Riley et al., 1993). Other species that have been studied include rats, mice, rabbits, catfish, mink, chickens, and turkeys. Thus, elevations in sphingoid bases in response to fumonisin exposure is an event that occurs both in vitro and in vivo, precedes overt toxicity, and is a useful biomarker of fumonisin or fumonisin-like exposure.

FUTURE RESEARCH

Fumonisin research has lead to many significant findings, but is in relative infancy. Many questions remain as to the role these toxins play in diseases. Perhaps more importantly are the questions that have arisen based on the apparent mechanism of toxicity of fumonisins. How do elevated sphingoid bases and altered complex sphingolipids lead to toxic lesions? Are there other fungal metabolites (or other naturally occurring substances) that have similar biological effects? Are there therapeutic (or other) uses of fumonisins or fumonisin-like compounds that can be made based on these biochemical phenomena? What parameters of cellular physiology are affected by altered sphingolipid biochemistry, and what are the long term effects of these changes? Greatly needed is a more thorough understanding of the ultimate consequences of altered sphingolipid biochemistry, and whether these alterations are the ultimate, underlying cause for the various toxic manifestations of fumonisins that are observed in different animal species. It is clear that much more research is needed on these important mycotoxins and the fungi that produce them.

REFERENCES

Abbas, H.K.; Tanaka, T.; Duke, S.O.; Porter, J.K.; Wray, E.M.; Hodges, L.; Sessions, A.E.; Wang, E.; Merrill, A.H.,Jr.;Riley, R.T. Fumonisin- and AAL-toxin-induced disruption of sphingolipid metabolism with accumulation of free sphingoid bases. *Plant Physiol.* **1994,** *106,* 1085-1093.

Bacon, C. W.; Porter, J.K; Norred, W. P. Toxic interaction of fumonisin B$_1$ and fusaric acid measured by injection in fertile chicken eggs. *Mycopathologia* **1995,** 129, 29-35.

Bezuidenhout, S.C.; Gelderblom, W.C.A.; Gorst-Allman, C.P.; Horak, R.M..; Marasas, W.F.O.; Spiteller, G.; Vleggaar, R. Structure elucidation of the fumonisins, mycotoxins from *Fusarium moniliforme*. *J. Chem. Soc. ,Chem. Commun.* **1988,** 743-745.

Chamberlain, W.J.; Voss, K.A.; Norred, W.P. Analysis of commercial laboratory rat rations for fumonisin B$_1$, a mycotoxin produced on corn by *Fusarium moniliforme*. *Cont. Topics Lab. Anim. Sci.* **1993,** 32, 26-28.

Chu, F.S.; Li, G.Y. Simultaneous occurrence of fumonisin B$_1$ and other mycotoxins in moldy corn collected from the People's Republic of China in regions with high incidences of esophageal cancer. *Appl. Environ. Microbiol.* **1994,** 60, 847-852.

Colvin, B.M.; Cooley, A.J.; Beaver, R.W. Fumonisin toxicosis in swine: clinical and pathologic findings. *J. Vet. Diagn. Invest.* **1993,** 5, 232-241.

Doko, M. B.; Rapior, S.; Visconti, A.;Schjoth, J. E. Incidence and levels of fumonisin contamination in maize genotypes grown in Europe and Africa. *J. Agric. Food Chem.* **1995,** 43, 429-434.

Franceschi, S.; Bidoli, E.; Baron, A.E.; La Vecchia, C. Maize and risk of cancers of the oral cavity, pharynx, and esophagus in northeastern Italy. *J. Nat.Cancer Inst.* **1990,** 82, 1407 1411.

Gelderblom, W.C.A.; Thiel, P.G.; van der Merwe, K.J.; Marasas, W.F.O.;Spies, H.S.C. A mutagen produced by *Fusarium moniliforme*. *Toxicon* **1983,** 4, 467-473.

Gelderblom, W.C.A.; Jaskiewicz, K.; Marasas, W.F.O.; Thiel, P.G.; Horak, R.M.; Vleggaar, R.; Kriek, N.P.J. Fumonisins - novel mycotoxins with cancer-promoting activity produced by *Fusarium moniliforme*. *Appl. Environ. Microbiol.* **1988,** 54, 1806-1811.

Gelderblom, W.C.A.; Thiel, P.G.; van der Merwe, K.J. The chemical and enzymatic interaction of glutathione with the fungal metabolite, fusarin C. *Mutation Res.* **1988a,** 199, 207-214.

Gelderblom, W.C.A.; Snyman, S.D. Mutagenicity of potentially carcinogenic mycotoxins of *Fusarium moniliforme*. *Mycotox. Res.* **1991,** 7, 46-52.

Gelderblom, W.C.A.; Kriek, N.P.J.; Marasas, W.F.O.; Thiel, P.G. Toxicity and carcinogenicity of the *Fusarium moniliforme* metabolite, fumonisin B$_1$, in rats. *Carcinogenesis* **1991,** 12, 1247-1251.

Gelderblom, W.C.A.; Semple, E.; Marasas, W.F.O.;Farber, E. The cancer-initiating potential of the fumonisin-B mycotoxins. *Carcinogenesis* **1992,** 13, 433-437.

Hanada, K.; Nishijima, M.; Akamatsu, Y. A temperature-sensitive mammalian cell mutant with thermolabile serine palmitoyltransferase for the sphingolipid biosynthesis. *J. Biol. Chem.* **1990,** 265, 22137-22142.

Hanada, K.; Nishijima, M.; Kiso, H. Jr.; Hasegawa, A.; Fujita, S.; Ogawa, T.; Akamatsu, Y. Sphingolipids are essential for the growth of Chinese hamster ovary cell. Restoration of the growth of a mutant defective in sphingoid base biosynthesis with exogenous sphingolipids. *J. Biol. Chem.* **1992,** 267, 23527-23533.

Hannun, Y. A.; Merrill, A. H., Jr.; Bell, R. M. Use of sphingosine as an inhibitor of protein kinase C. *Meth. Enzymol.* **1991,** 210, 316-328.

Harrison, L.R.; Colvin, B.M.; Greene, J.T.; Newman, L.E.; Cole, J.R. Pulmonary edema and hydrothorax in swine produced by fumonisin B_1, a toxic metabolite of *Fusarium moniliforme*. *J. Vet. Diagn. Invest.* **1990,** 2, 217-221.

House, C. Moldy corn kills several horses in Kentucky, Virginia. *Feedstuffs* **1995,** 67, 1-3.

Jaskiewicz, K.; Marasas, W.F.O.; Taljaard, J.J.F. Hepatitis in vervet monkeys caused by *Fusarium moniliforme*. *J. Comp. Path.* **1987,** 97, 281-291.

Jaskiewicz, K.; van Rensburg, S.J.; Marasas, W.F.O.; Gelderblom, W.C.A. Carcinogenicity of *Fusarium moniliforme* culture material in rats. *J. Nat. Cancer Inst.* **1987a,** 78, 321-325.

Javed, T.; Richard, J.L.; Bennett, G.A.; Dombrink-Kurtzman, M.A.; Bunte, R.M;., Koelkebeck, K.W.; Côté, L.M.; Leeper, R.W.; Buck, W.B. Embryopathic and embryocidal effects of purified fumonisin B_1 or *Fusarium proliferatum* culture material extract on chicken embryos. *Mycopathologia* **1993,** 123, 185-193.

Kellerman, T.S.; Marasas, W.F.O.; Thiel, P.G.; Gelderblom, W.C.A.; Cawood, M.; Coetzer, A.W. Leucoencephalomalacia in two horses induced by oral dosing of fumonisin B1. *Onderstepoort J. Vet. Res.* **1990,** 57, 269-275.

Kriek, N.P.J.; Marasas, W.F.O.; Thiel, P.G. Hepato- and cardiotoxicity of *Fusarium verticilloides* (*F. moniliforme*) isolates from southern African maize. *Fd. Cosmet. Toxicol.* **1981,** 19, 447-456.

Laurent, D.; Pellegrin, F.; Kohler, F.; Lambert, C.; Fouquet, L.; Domenech, J.;Boccas, B. *Fusarium moniliforme* du maïs en Nouvelle-Caledonie: Toxicologie animale. *Microbiol.Aliments Nutr.Microbiol.Foods Feeds Nutr.* **1988,** 6, 159-164.

Marasas W.F.O.; Nelson P.E.; Toussoun T.A. *Toxigenic Fusarium Species: Identity and Mycotoxicology;* The Pennsylvania State University Press: University Park and London,1984.

Marasas, W.F.O., *Fusarium moniliforme*: a mycotoxicological miasma. In *Mycotoxins and Phycotoxins;* Steyn, P. S; Vleggaar, R., Eds.; Elsevier Science Publishers: Amsterdam,1986.

Marasas, W.F.O.; Kellerman, T.S.; Gelderblom, W.C.A.; Coetzer, J.A.W.; Thiel, P.G.; van der Lugt, J.J. Leukoencephalomalacia in a horse induced by fumonisin B_1 isolated from *Fusarium moniliforme*. *Onderstepoort J. Vet. Res.* **1988,** 55,197-203.

Murphy, P.A.; Rice, L.G.; Ross, P.F. Fumonisin B_1, fumonisin B_2, and fumonisin B_3 content of Iowa, Wisconsin, and Illinois corn and corn screenings. *J. Agr. Food Chem.* **1993,** 41, 263-266.

Nelson, P.E.; Desjardins, A.E.; Plattner, R.D. Fumonisins, mycotoxins produced by *Fusarium* species: Biology, chemistry, and significance. *Annu. Rev. Phytopathol.* **1993,** 31, 233-252.

Norred W.P.; Plattner R.D.; Voss K.A.; Bacon C.W.; Porter J.K. Natural occurrence of fumonisins in corn associated with equine leukoencephalomalacia (ELEM). *Toxicologist* **1989,** 9, 258.

Norred, W.P.; Bacon, C.W.; Porter, J.K.; Voss, K.A. Inhibition of protein synthesis in rat primary hepatocytes by extracts of *Fusarium moniliforme*-contaminated corn. *Fd. Chem. Toxicol.* **1990,** 28, 89-94.

Norred, W.P.; Bacon, C.W.; Plattner, R.D.; Vesonder, R.F. Differential cytotoxicity and mycotoxin content among isolates of *Fusarium moniliforme*. *Mycopathologia* **1991,** 115, 37-43.

Norred, W.P.; Plattner, R.D.; Vesonder, R.F.; Bacon, C.W.; Voss, K.A. Effects of selected secondary metabolites of *Fusarium moniliforme* on unscheduled synthesis of DNA by rat primary hepatocytes. *Food Chem.Toxicol.* **1992,** 30, 233-237.

Norred, W.P.; Wang, E.; Yoo, H.; Riley, R.T.; Merrill, A.H. In vitro toxicology of fumonisins and the mechanistic implications. *Mycopathologia* **1992a,** 117, 73-78.

Norred, W.P. Fumonisins - mycotoxins produced by *Fusarium moniliforme*. *J. Toxicol. Environ. Health* **1993,** 38, 309-328.

Norred, W.P.; Plattner, R.D.; Chamberlain, W.J. Distribution and excretion of [^{14}C]fumonisin B_1 in male Sprague-Dawley rats. *Nat. Toxins* **1993,** 1, 341-346.

Norred W.P.; Riley R.T.; Malcom P.J.; Meredith F.I. Dose-response characteristics of sphingoid base elevations induced in liver and kidney slices by fumonisin B_1. *Toxicologist* **1995,** 15, 215.

Plattner, R.D.; Norred, W.P.; Bacon, C.W.; Voss, K.A.; Peterson, R.; Shackelford, D.D.; Weisleder, D. A method of detection of fumonisins in corn samples associated with field cases of equine leukoencephalomalacia. *Mycologia* **1990,** 82, 698-702.

Porter, J.K.; Voss, K.A.; Bacon, C.W.; Norred, W.P. Effects of *Fusarium moniliforme* and corn associated with equine leukoencephalomalacia on rat neurotransmitters and metabolites. *Proc. Soc. Exp. Biol. Med.* **1990,** 194, 265-269.

Porter, J.K.; Voss, K.A.; Chamberlain, W.J.; Bacon, C.W.; Norred, W.P. Neurotransmitters in rats fed fumonisin B_1. *Proc. Soc. Exp. Biol. Med.* **1993**, 202, 360-364.

Prelusky, D.B.; Trenholm, H.L.; Savard, M.E. Pharmacokinetic fate of ^{14}C-labelled fumonisin B_1 in swine. *Nat. Toxins* **1994**, 2, 73-80.

Rauckman, E. J.; G.M. Padilla (eds). *The isolated hepatocyte: Use in toxicology and xenobiotic biotransformations;* Academic Press: Orlando, 1987.

Reddy, R. V.; Reddy, C. S.; Johnson, G. C.; Rottinghaus, G. E.; Casteel, S. W. Developmental effects of pure fumonisin B_1 in CD1 mice. *The Toxicologist* **1995**, 15, 157.

Rheeder, J.P.; Marasas, W.F.O.; Thiel, P.G.; Sydenham, E.W.; Shephard, G.S.; Vanschalkwyk, D.J. *Fusarium moniliforme* and fumonisins in corn in relation to human esophageal cancer in Transkei. *Phytopath.* **1992**, 82, 353-357.

Riley, R.T.; An, N.-H.; Showker, J.L.; Yoo, H.-S.; Norred, W.P.; Chamberlain, W.J.; Wang, E.; Merrill, A.H., Jr.; Motelin, G.; Beasley, V.R.; Hashek, W. M.. Alteration of tissue and serum sphinganine to sphingosine ratio: an early biomarker in pigs of exposure to fumonisin-containing feeds. *Toxicol. Appl. Pharmacol.* **1993**, 118, 105-112.

Riley, R.T.; Norred, W.P.; Bacon, C.W. Fungal toxins in foods: recent concerns. *Annu. Rev. Nutr.* **1993a**, 13, 167-189.

Riley, R.T.; Hinton, D.M.; Chamberlain, W.J.; Bacon, C.W.; Wang, E.; Merrill, A.H., Jr.; Voss, K.A. Dietary fumonisin B_1 induces disruption of sphingolipid metabolism in Sprague-Dawley rats: A new mechanism of nephrotoxicity. *J. Nutr.* **1994**, 124, 594-603.

Riley, R.T.; Voss, K.A.; Yoo, H.S.; Gelderblom, W.C.A.; Merrill, A.H.,Jr. Mechanism of fumonisin toxicity and carcinogenesis. *J.Food Protect.* **1994a**, 57, 638-645.

Riley, R.T.; Wang, E.; Merrill, A.H., Jr. Liquid chromatographic determination of sphinganine and sphingosine: Use of the free sphinganine-to sphingosine ratio as a biomarker for consumption of fumonisins. *J. Assoc. Off. Analyt. Chem. Int.* **1994b**, 77, 533-540.

Riley, R. T.; Wang, E.; Schroeder, J. J.; Smith, E. R.; Plattner, R. D.; Abbas, H.; Yoo, H.-S.; Merrill, A. H., Jr. Evidence for disruption of sphingolipid metabolism as the biochemical lesion responsible for the toxicity and carcinogenicity of fumonisins. *Nat. Toxins*, **1995**, submitted.

Riley, R. T.; Yoo, H.-S. Time and dose relationship between the cellular effects of fumonisin B_1 (FB_1) and the uptake and accumulation of [^{14}C]FB_1 in LLC-PK_1 cells. *The Toxicologist* **1995**, 15, 290.

Ross, P.F.; Rice, L.G.; Plattner, R.D.; Osweiler, G.D.; Wilson, T.M.; Owens, D.L.; Nelson, H.A.; Richard, J.L. Concentrations of fumonisin B_1 in feeds associated with animal health problems. *Mycopathologia* **1991**, 114, 129-135.

Shephard, G.S.; Thiel, P.G.; Sydenham, E.W.; Alberts, J.F.; Gelderblom, W.C.A. Fate of a single dose of the ^{14}C-labelled mycotoxin, fumonisin B_1 in rats. *Toxicon* **1992**, 30, 768-770.

Stevens, V. L.; Nimkar, S.; Jamison, W. C.; Liotta, D. C.; Merrill, A. H., Jr. Characteristics of the growth inhibition and cytotoxicity of long-chain (sphingoid) bases for Chinese hamster ovary cells: Evidence for an involvement of protein kinase C. *Biochem. Biophys. Acta* **1990**, 1051, 37-45.

Sydenham, E.W.; Thiel, P.G.; Marasas, W.F.O.; Shephard, G.S.; Van Schalkwyk, D.J.; Koch, K.R. Natural occurrence of some *Fusarium* mycotoxins in corn from low and high esophageal cancer prevalence areas of the Transkei, South Africa. *J. Agric. Food Chem.* **1990**, 38, 1900-1903.

Sydenham, E.W.; Shephard, G.S.; Thiel, P.G.; Marasas, W.F.O.; Stockenström, S. Fumonisin contamination of commercial corn-based human foodstuffs. *J. Agric. Food Chem.* **1991**, 39, 2014-2018.

Sydenham, E.W.; Marasas, W.F.O.; Shephard, G.S.; Thiel, P.G.; Hirooka, E.Y. Fumonisin concentrations in Brazilian feeds associated with field outbreaks of confirmed and suspected animal mycotoxicoses. *J. Agr. Food Chem.* **1992**, 40, 994-997.

Thiel, P.G.; Shephard, G.S.; Sydenham, E.W.; Marasas, W.F.O.; Nelson, P.E.; Wilson, T.M. Levels of fumonisins B_1 and B_2 in feeds associated with confirmed cases of equine leukoencephalomalacia. *J. Agric. Food Chem.* **1991**, 39, 109-111.

Ueno, Y.; Aoyama, S.; Sugiura, Y.; Wang, D.-S.; Lee, U.-S.; Hirooka, E.U.; Hara, S.; Karki, T.; Chen, G.; Yu, S.-Z. A limited survey of fumonisins in corn and corn-based products in Asian countries. *Mycotox. Res.* **1993**, 9, 27-34.

Voss, K.A.; Norred, W.P.; Plattner, R.D.; Bacon, C.W. Hepatotoxicity and renal toxicity in rats of corn samples associated with field cases of equine leukoencephalomalacia. *Fd. Chem. Toxicol.* **1989**, 27, 89-96.

Voss, K.A. Toxins from *Fusarium moniliforme*, a common fungus in corn. *Vet. Hum. Toxicol.* **1990**, 32, 57-63.

Voss, K.A.; Plattner, R.D.; Bacon, C.W.; Norred, W.P. Comparative studies of hepatotoxicity and fumonisin B_1 and B_2 content of water and chloroform/methanol extracts of *Fusarium moniliforme* strain MRC 826 culture material. *Mycopathologia* **1990**, 112, 81-92.

Voss, K.A.; Chamberlain, W.J.; Bacon, C.W.; Norred, W.P. A preliminary investigation on renal and hepatic toxicity in rats fed purified fumonisin B_1. *Nat. Toxins* **1993**, 1, 222-228.

Voss, K.A.; Chamberlain, W.J.; Bacon, C.W.; Herbert, R.A.; Walters, D.B.; Norred, W.P. Subchronic feeding study of the mycotoxin fumonisin B_1 in B6C3F1 mice and Fischer 344 rats. *Fund. Appl. Toxicol.* **1994**, 24, 102-110.

Voss K.A.; Bacon C.W.; Norred W.P.; Chapin R.E.; Chamberlain W.J. Reproductive toxicity study of *Fusarium moniliforme* culture material in rats. *Toxicologist* **1995**, 15, 215.

Voss, K. A.; Riley, R. T.; Bacon, C. W.; Chamberlain, W. J.; Norred, W. P. Subchronic toxic effects of *Fusarium moniliforme* and fumonisin B_1 in rats and mice: a review. *Nat. Toxins* **1995a**, in press.

Vudathala, D.K.; Prelusky, D.B.; Ayroud, M.; Trenholm, H.L.; Miller, J.D. Pharmacokinetic fate and pathological effects of ^{14}C-fumonisin B_1 in laying hens. *Nat. Toxins* **1994**, 2, 81-88.

Wang, E.; Norred, W.P.; Bacon, C.W.; Riley, R.T.; Merrill, A.H., Jr. Inhibition of sphingolipid biosynthesis by fumonisins: implications for diseases associated with *Fusarium moniliforme*. *J. Biol. Chem.* **1991**, 266, 14486-14490.

Wang, E.; Ross, P.F.; Wilson, T.M.; Riley, R.T.; Merrill, A.H. Increases in serum sphingosine and sphinganine and decreases in complex sphingolipids in ponies given feed containing fumonisins, mycotoxins produced by *Fusarium moniliforme*. *J. Nutr.* **1992**, 122, 1706-1716.

Yoo, H.-S.; Norred, W.P.; Wang, E.; Merrill, A., Jr.; Riley, R.T. Fumonisin inhibition of de novo sphingolipid biosynthesis and cytotoxicity are correlated in LLC-PK1 cells. *Toxicol. Appl. Pharmacol.* **1992**, 113, 9-15.

Yoo, H.-S.; Showker, J. L.; Riley, R. T. Partitioning of increased sphinganine and decreased complex sphingolipids as causative factors in fumonisin B_1-induced cytostatic and cytotoxic effects in LLC-PK_1 cells. *The Toxicologist* **1995**, 15, 290.

Yoshizawa, T.; Yamashita, A.; Luo, Y. Fumonisin occurrence in corn from high- and low-risk areas for human esophageal cancer in China. *Appl. Environ. Microbiol.* **1994, 60, 1626-1629.**

21

THE MYCOTOXIN FUMONISIN INDUCES APOPTOSIS IN CULTURED HUMAN CELLS AND IN LIVERS AND KIDNEYS OF RATS

William H. Tolleson[1], Kenneth L. Dooley[1], Winslow G. Sheldon[2],
J. Dale Thurman[2], Thomas J. Bucci[2], and Paul C. Howard[1,3]

[1] National Center for Toxicological Research, Food and Drug Administration
3900 NCTR Road, Jefferson, Arkansas 72079, and
[2] Pathology Associates Incorporated
3900 NCTR Road, Jefferson, Arkansas 72079
[3] To whom correspondence should be addressed:
Division of Biochemical Toxicology
HFT-110
National Center for Toxicological Research
Jefferson, Arkansas 72079-9502
telephone (501)543-7672
FAX (501)543-7136
E-Mail: PHOWARD@fdant.nctr.fda.gov

ABSTRACT

Fumonisin B_1 is a mycotoxin produced by *Fusarium moniliforme*, a fungus that infects corn and other grains in the U.S. Fumonisin ingestion causes a variety of effects including equine leukoencephalomalacia and porcine pulmonary edema, and has been associated epidemiologically with human esophageal cancer. Fumonisin B_1 produces growth inhibition and increased apoptosis in primary human keratinocyte cultures and in HET-1A cells. In order to set the doses for a 2-year tumor bioassay, male and female F344 rats were fed fumonisin B_1 (99, 163, 234, and 484 ppm) for 28 days and the organs examined histologically. There was a dose dependent decrease in liver and kidney weights in the rats. The liver weight loss was accompanied by the induction of apoptosis and hepatocellular and bile duct hyperplasia in both sexes, with the female rats being more responsive at lower doses. The induction of tubular epithelial cell apoptosis was the primary response of the kidneys to dietary fumonisin B_1. Apoptosis was present at all doses in the kidneys of the male rats, and occurred in the females only at 163, 234, and 484 ppm fumonisin B_1. These results demonstrate that fumonisin B_1 treatment causes a similar increase in apoptosis both *in vivo* and *in vitro*.

Fumonisins in Food, Edited by L. Jackson *et al.*
Plenum Press, New York, 1996

INTRODUCTION

Fumonisin B_1 and related compounds occur in grain and grain products as the result of infestation and growth of fungi of the *Fusarium* species (Nelson *et al.*, 1991; Thiel *et al.*, 1991). Fumonisin B_1 was identified in 1988 as the major carcinogenic substance produced by cultures of *Fusarium moniliforme* (Bezuidenhout *et al.*, 1988; Gelderblom *et al.*, 1992a). Since 1988, several isomers of fumonisins have been characterized (*e.g.* fumonisin B_2, fumonisin B_3), and several species other than *F. moniliforme* have been shown to produce fumonisins (Cawood *et al.*, 1991; Thiel *et al.*, 1991; Ross *et al.*, 1992).

To determine the overall human cancer risk resulting from exposure to fumonisins, one needs to understand: (1) the exposure and bioavailability of the fumonisins; (2) the mechanism(s) of action of the fumonisins; (3) the effects of other compounds to exacerbate or antagonize the effects; (4) the comparative toxicity of the fumonisin isomers; and (5) the carcinogenic potential of the fumonisin(s) in a representative animal bioassay. The studies in our laboratories have focused on three aspects of the possible role of fumonisin B_1 in cancer including (a) the mechanism of action of fumonisin B_1 in cultured human cells, (b) the mechanism of action of fumonisin B_1 *in vivo* in rodents, and (c) the tumorigenicity of fumonisin B_1 in a chronic feeding study in rats and mice. We describe in this manuscript the effects of fumonisin B_1 in cultures of human cells, and the toxicity of fumonisin B_1 in a 28-day feeding study with F344 rats. The latter study was used to set the dose-range for a two-year fumonisin B_1 feeding study which is in progress.

METHODS

Fumonisin B_1

Fumonisin B_1 was provided by the Division of Natural Products, Center for Food Safety and Applied Nutrition, Food and Drug Administration (FDA), Washington, DC, as part of an interagency agreement between the FDA and the National Toxicology Program. The fumonisin B_1 was isolated as the free acid and was determined by HPLC, mass spectrometry, and NMR techniques to be 92.5% pure. The impurities were predominantly of hydrolyzed fumonisin B_1. The concentrations of fumonisin B_1 in the dosed feed were determined using HPLC techniques (Holcomb *et al.*, 1993a,b).

Keratinocytes

Primary human keratinocytes (NHKc) were isolated from neonatal human foreskins obtained at a local hospital. The specimens were collected in Keratinocyte SFM (Gibco, Grand Island) supplemented with 5% calf serum, 200 units/mL penicillin, and 200 µg/mL streptomycin sulfate. Keratinocytes were isolated by a modified trypsin-float technique and cultured as described previously (Pirisi *et al.*, 1987). Briefly, after overnight digestion in 0.2% trypsin buffer (4 mL 0.25% trypsin + 1 mL Keratinocyte SFM) at 4°C, NHKc were dissociated from the epidermis by maceration with scalpel blades, and resuspended in 5 mL Keratinocyte SFM. NHKc were pelleted at ~500 g for 4 min, and 10^4-10^5 cells plated in 10 mL Keratinocyte SFM in 100 mm tissue culture dishes. NHKc were maintained in Kerati-nocyte SFM supplemented with 2 µg/mL epidermal growth factor (EGF), 15 µg/mL bovine pituitary extract (BPE), and 50 µg/mL gentamicin (KSFM^{++}). For certain treatments, EGF and BPE were omitted (KSFM=). Cells were passaged 1:10 prior to confluency (approximately once weekly) and were maintained at 37°C in a humidified 5% CO_2, 95% air atmosphere.

Transformed Human Esophageal Epithelial Cells

HET-1A are SV40 large T-antigen immortalized human esophageal epithelial cells (Stoner *et al.*, 1991), and were obtained from Gary D. Stoner (Ohio State University, Columbus, Ohio). The HET-1A were maintained in EPM2 medium (BREFF, Ijamsville, MD) supplemented with 50 µg/mL gentamicin (Stoner *et al.*, 1991). Cultures were passaged 1:5 prior to confluency as previously described by Iype *et al.* (1993) and were maintained at 37°C in a humidified 5% CO_2, 95% air atmosphere.

Keratinocyte Clonal Growth Assays

Early passage NHKc were plated at 700 cells per 60 mm plate in KSFM^{++}; HET-1A were plated in EPM2 medium. After allowing one day for cell attachment, the media were changed to either KSFM^{++} or KSFM= containing the indicated levels of fumonisin B_1. After the cells were maintained on this medium for 5 days, the medium was removed, and the cells were washed with Dulbecco's phosphate-buffered saline (D-PBS, Gibco), fixed with methanol, and stained with Giemsa. Clonal growth was determined by the percent area of the plate covered by the cells using a video camera and commercial software package (Optimas Corporation, Bothell, WA).

Twenty-Eight Day Feeding Study with F344 Rats

Male and female F344 rats were obtained from the NCTR breeding colony (NCTR strain A) at 3 weeks of age, and kept two of the same sex per cage in polycarbonate cages with hardwood chip bedding. The rats were given access to autoclaved NIH-31 powdered feed (Purina, St. Louis, MO; <0.06 ppm fumonisin B_1) and water *ad libitum*. The rats were divided into five dose groups per sex (10 rats per group) in a random manner to control for weight and group assignment bias. They were maintained on the control diet until six weeks of age when the diets were changed to autoclaved NIH-31 powdered feed containing 0, 99, 163, 234, or 484 ppm (µg/g) fumonisin B_1. On a weekly basis, the feed, water, cages, and bedding were changed, and feed consumption and animal weights were determined.

After 28 days on the dosed feed, the rats were transferred to cages containing bedding and water but no food. The next morning the rats were sacrificed by carbon dioxide-asphyxiation, and whole body and liver, kidney, brain, testes, and heart weights were determined. Gross necropsy was performed, and tissues (National Toxicology Program, 1991; U.S. Food and Drug Administration, 1993) were processed to the level of wet tissue in 10% neutral buffered formalin. The tissues from the control and highest dose group, and livers and kidneys from all rats, were trimmed, mounted in Paraplast blocks, sectioned at 5 µm, and stained with hematoxylin and eosin for microscopic analysis.

Statistics

Statistical significance was determined using Students' two-tailed t-test (Huntsberger and Billingsley, 1977).

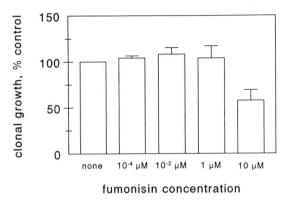

Figure 1. Effect of fumonisin B$_1$ on the clonal growth of normal human keratinocytes (NHKc). NHKc were plated at a density of 700 cells per 60 mm dish, and allowed to grow clonally for 5 days following administration of 0 - 10 μM fumonisin B$_1$ in KSFM^{++} medium. Afterwards, the medium was removed, cells washed, fixed in methanol, and stained with Giemsa. Clonal growth was determined by the area of plates occupied by NHKc. The data are the mean and standard deviation of three replicate experiments.

RESULTS

To determine the effect on cell growth *in vitro*, NHKc cells were exposed to fumonisin B$_1$ at concentrations up to 10 μM (Figure 1). Fumonisin B$_1$ had no effect on keratinocyte cell growth at concentrations up to 1 μM; however, 10 μM fumonisin B$_1$ suppressed clonal growth by 42%. This growth inhibition was accompanied by an increase in cellular apoptosis as determined by morphological examination, loss of clonogenicity, electrophoretic detection of DNA ladders, detection of intracellular DNA fragmentation using ApopTag®, and electron microscopic examination of sorted cells (Tolleson *et al.*, in press).

Figure 2 shows the results of exposure of HET-1A cells to fumonisin B$_1$. As with the NHKc, 1 μM fumonisin B$_1$ did not affect clonal growth of the HET-1A. However, clonal growth was inhibited 75% with 100 μM fumonisin B$_1$. The induction of apoptosis in the HET-1A was indicated morphologically.

Fumonisin B$_1$ was fed to rats for twenty-eight consecutive days. The mean body weights of the male and female rats during the course of this feeding study are presented in Figure 3. In the week prior to administration of the fumonisin B$_1$-containing feed, there was no statistical difference in the weight of the male or female rats in each of the groups. Following administration of the diet to male rats, the rats in the high dose group (484 ppm) gained less weight (14.9% at four weeks) than those in the other groups, which were statistically indistinguishable from one another. All male rats consumed essentially the same amount of feed over the course of the study, except the high dose group, which consumed less feed (data not shown). This may explain the decreased weight gain in the high dose male group. There was an apparent dose-dependent decrease in the animal weights in the female

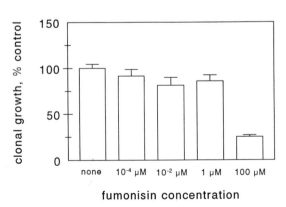

Figure 2. Effect of fumonisin B$_1$ on the clonal growth of SV-40 large T-antigen immortalized human esophageal epithelial cells (HET-1A). HET-1A were plated at a density of 800 cells per 60 mm dish, and allowed to grow clonally for 5 days following administration of 0 - 100 μM fumonisin B$_1$ in EPM2 media. Afterwards, the medium was removed, cells washed, fixed in methanol, and stained with Giemsa. Clonal growth was determined by the area of plates occupied by NHKc. The data are the mean and standard deviation of three replicate experiments.

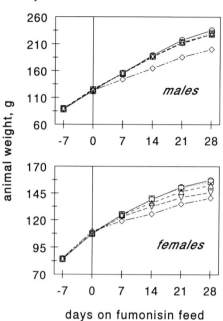

Figure 3. The body weights of the male (upper panel) and female (lower panel) F344 rats are shown for animals given autoclaved NIH-31 feed containing 0 ppm (O, solid line), 99 ppm (□, dotted line), 163 ppm (Δ, short dashed line), 234 ppm (∇, long dashed line), and 484 ppm (◊, mixed dashed line) fumonisin B₁ for 28-days. There were 10 rats per group, and the data are presented as the mean ± standard error of observation. The final body weights were taken the last day of feed exposure.

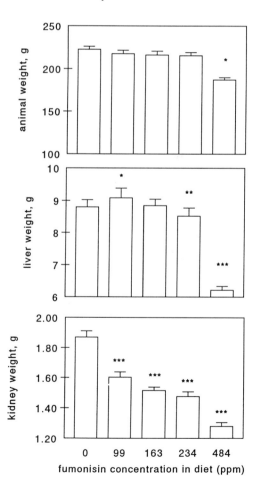

Figure 4. Liver, kidney, and total body weights of male F344 rats fed fumonisin B₁ in feed for twenty-eight consecutive days and fasted overnight prior to necropsy. Ten rats were included in each group, and statistical significance was determined by two-tailed t-test (*, p<0.10; **, p<0.05; ***, p<0.01).

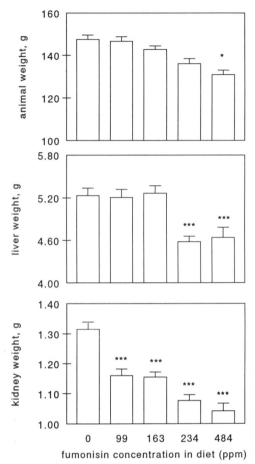

Figure 5. Liver, kidney, and total body weights of female F344 rats fed fumonisin B$_1$ in feed for twenty-eight consecutive days and fasted overnight prior to necropsy. Ten rats were included in each group, and statistical significance was determined by two-tailed t-test (*, p<0.10; **, p<0.05; ***, p<0.01).

rats exposed to fumonisin B$_1$. This did not correlate with any changes in the feed consumption by the female rats (data not shown).

After twenty-eight consecutive days of consuming fumonisin B$_1$-containing feed, the rats were fasted overnight and necropsied. In Figure 4 is shown the results of the examination of the liver, kidney, and total body weight for the male rats. A decrease (-17.4%, p<0.10) in total body weight was noted only in the high dose group. Compared to the liver weights in the male rats consuming control diets, the liver weights increased in the 99 ppm group (+3.2%, p<0.10), were equivalent in the 163 ppm group, and decreased in the 234 ppm (-3.2%, p<0.05) and 484 ppm (-29.4%, p<0.01) groups. The effect of fumonisin B$_1$ exposure was a decrease (p<0.01) in the total weight of the kidneys in the male rats, with decreases of 14.2%, 18.9%, 21.0%, and 31.6% for the 99, 163, 234, and 484 ppm fumonisin B$_1$ groups, respectively. Essentially the same results were seen when the data were normalized to total body weight, brain weight, or heart weight (not shown).

The results of continuous feeding of fumonisin B$_1$ to the female rats is shown in Figure 5. The total body weight of the female rats significantly decreased (-11.3%, p<0.10) only in the highest dose group when compared to rats exposed to control diet (Figure 5). The female rat liver weights decreased by 12.4% (p<0.01) and 11.3% (p<0.01) in the rats consuming 234 ppm and 484 ppm fumonisin B$_1$. As with the male

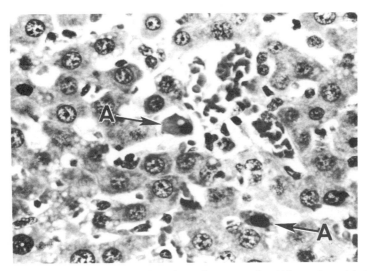

Figure 6. Hematoxylin and eosin stained section of a rat liver exposed to 484 ppm fumonisin B_1-containing diet. Apoptotic cells (A) are indicated.

rats, there were decreases (p<0.01) in the total kidney weight in all of the dose groups, with decreases of 11.8%, 12.2%, 18.2%, and 20.8% in the kidney weights in the 99, 163, 234, and 484 ppm groups, respectively. Essentially the same results were seen when the data was normalized to total body weight, brain weight, or heart weight (not shown).

Microscopic examination of the hematoxylin and eosin sections revealed extensive changes in the livers of the rats exposed to fumonisin B_1 (Figure 6). Most pronounced was the induction of apoptosis in the centrolobular region of the liver, and increasing severity to all parts of the liver with increasing dose. The apoptotic cells were characterized by cytoplasmic shrinkage and basophilia, condensation of the chromatin and deposition to one side of the nucleus, eventual formation of apoptotic bodies, and in some cases phagocytosis by neighboring hepatocytes. The presence of apoptotic hepatocytes was confirmed using ApopTag® (data not presented). Bile ductal epithelial cell hyperplasia was noted in the portal areas of the liver, and was consistently seen only in the high dose groups. There was a general disruption of the lobular structure of the liver (degeneration), with decreased definition of the sinusoids and increased cellularity in the centrolobular region. Hepatocytes had less cytoplasm, and vacuolization was increased in the highest fumonisin B_1 dose groups. A quantitative presentation of the effects of fumonisin B_1 on livers is presented in Figure 7 for the male and female rats. Hepatocellular apoptosis in males was noted in 90% and 100% of the rats at 234 and 484 ppm, respectively, with an associated increase in severity from minimal to mild. The liver degeneration followed the same severity, affecting all rats at 234 and 484 ppm. Bile duct hyperplasia was minimal and occurred in all rats at the highest dose. Mitotic figures were noted only in the high dose group. It seems apparent that a no observed effect level (NOEL) for fumonisin B_1 on male rat livers after 28-days of feeding is between 163 and 234 ppm.

Induction of hepatocyte apoptosis in the female rats was apparent at 99 ppm, and increased in prevalence and severity with increasing dose (Figure 7). The cells were confirmed as apoptotic using ApopTag® (data not presented). Essentially similar effects

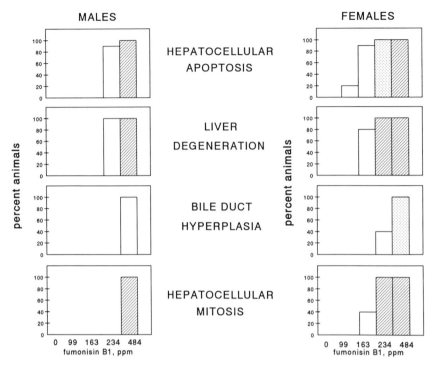

MALES   FEMALES

percent animals

percent animals

HEPATOCELLULAR
APOPTOSIS

LIVER
DEGENERATION

BILE DUCT
HYPERPLASIA

HEPATOCELLULAR
MITOSIS

0   99 163 234 484
fumonisin B1, ppm

0   99 163 234 484
fumonisin B1, ppm

Figure 7. Quantitative presentation of the results from microscopic examination of the livers of male and female F344 rats fed 0-484 ppm fumonisin B_1 for 28-days. The height of the bar represents the percent of rats with the indicated condition, and the shading of the bar indicates the severity of the condition. The conditions were graded and given the following numerical value: minimal, 1; mild, 2; moderate, 3; severe, 4. The severity of the conditions were averaged for the affected animals, and are represented as follows: 1 - 1.49, open bar; 1.5 - 1.99, dotted bar; 2.0 - 2.49, hatched bar; 2.5 - 2.99, cross-hatched bar; 3.0 - 4, solid bar.

were seen with liver degeneration, bile duct hyperplasia, and hepatocellular mitosis. Taking into consideration all the hepatic effects, the data indicated a threshold of effect (NOEL) after 28-days of feeding is between 99 and 163 ppm fumonisin B_1 in the female rat liver.

Exposure of rats to fumonisin B_1 resulted in the induction of tubular epithelial apoptosis in the inner cortex of the kidneys (Figure 8). The apoptosis was characterized by nuclear condensation and cytoplasmic eosinophilia, and was accompanied by degenerative changes in the tubular epithelium, including tubular epithelial cell hypertrophy. Mitotic cells were apparent but were not quantified. All of the male rats treated with fumonisin B_1 had essentially the same level of apoptosis, and except for the 99 ppm dose group, an equivalent severity of kidney degeneration (Figure 9). The results indicated the threshold of fumonisin B_1 effects (NOEL) on the male kidney after 28-days of feeding was below 99 ppm.

The kidneys of the female rats were less sensitive to the fumonisin B_1 than the male rats (Figure 9). Inner cortex tubular epithelial cell apoptosis was minimal in all the rats at 162 and 234 ppm, and increased in severity at 484 ppm fumonisin B_1. Similarly, the degeneration in the structure of the kidney was noted at 163 ppm, and increased in severity with increasing fumonisin B_1. Therefore, differing from the males, the kidneys of the female rats demonstrated no effect at 99 ppm fumonisin B_1.

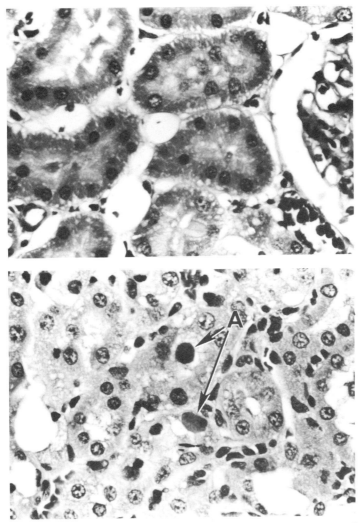

Figure 8. Hematoxylin and eosin stained section of rat kidneys exposed to control (upper panel) and fumonisin B$_1$-containing (lower panel) diets. Apoptotic (A) inner cortex tubular epithelial cells are indicated.

DISCUSSION

The mycotoxin fumonisin B$_1$ is undergoing rigorous examination with respect to its mechanism of action *in vitro* and *in vivo*, and its potency in animal tumor bioassays. Much of the interest is due to its potent toxicity in horses and swine, worldwide contamination, possible role in interruption of sphingolipid synthesis and signal transduction pathways, and epidemiological association in areas of the world with high rates of human esophageal cancer.

In a 28-day feeding study, male and female Sprague-Dawley rats were fed 0, 15, 50, or 150 ppm fumonisin B$_1$ (Voss *et al.*, 1993). Serum from the males contained increased alanine aminotransferase, aspartate aminotransferase, alkaline phosphatase, cholesterol, and triglycerides at 150 ppm, but not 50 ppm fumonisin B$_1$, indicating a no observed effect level

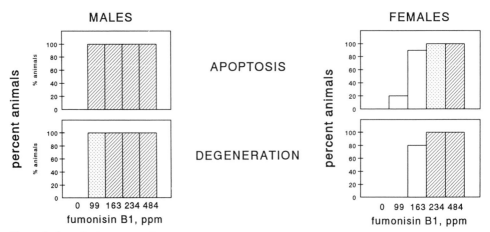

Figure 9. Quantitative presentation of the results from microscopic examination of the kidneys of male and female F344 rats fed 0-484 ppm fumonisin B$_1$ for 28-days. The height of the bar represents the percent of rats with the indicated condition, and the shading of the bar indicates the severity of the condition. The conditions were graded and given the following numerical value: minimal, 1; mild, 2; moderate, 3; severe, 4. The severity of the conditions were averaged for the affected animals, and are represented as follows: 1 - 1.49, open bar; 1.5 - 1.99, dotted bar; 2.0 - 2.49, hatched bar; 2.5 - 2.99, cross-hatched bar; 3.0 - 4, solid bar.

(NOEL) between 50 and 150 ppm. Serum from the female rats had elevations only in alanine aminotransferase, alkaline phosphatase, cholesterol, and triglycerides at 150 ppm fumonisin B$_1$, indicating a NOEL between 50 and 150 ppm. Examination of the whole body and organ weights in the males and females indicated a decrease only in absolute, but not relative kidney weights, at 150 ppm as compared to the controls. Interestingly, there was an increase in the absolute and relative liver weights in the males at 15 and 50 ppm fumonisin B$_1$ compared to the control rats. Microscopic analysis of the livers revealed scattered single cell hepatocellular necrosis, characterized by cytoplasmic vacuolization and variability in nuclear size and staining. The effect was most prominent in the females at the highest dose. In the kidneys, fumonisin B$_1$ induced single cell necrosis or pyknosis and proximal tubule epithelia basophilia and hyperplasia.

In a 90-day feeding study, Voss *et al.* (1995) used lower doses of fumonisin B$_1$ (0, 1, 3, 9, 27, 81 ppm) to more accurately determine the NOEL in F344 rats. Absolute and relative kidney weights were decreased at both 4 and 13 weeks in males at 27 and 81 ppm, and only after 13 weeks in the females at 27 and 81 ppm. The nephrosis was characterized by individual tubular cell degeneration and necrosis, with pyknotic nuclei, and sporadic mitotic figures. Cells with eosinophilic cytoplasm and pyknotic nuclei were sloughed into the tubular lumina.

In our present studies we have confirmed and extended these observations concerning the effect of fumonisin B$_1$ on animals. The previous studies of Voss and colleagues referred to single cell necrosis in the liver and kidneys, which we have confirmed as apoptosis. The apoptosis is not restricted to the animals *in vivo*, but also occurred in cultures of human primary keratinocytes and esophageal epithelial cells. This is consistent with *in vitro* observations by others demonstrating fumonisin B$_1$ inhibited cell proliferation (cytostasis) or induced either single cell necrosis or apoptosis in cultured turkey lymphocytes (Dombrink-Kurtzman *et al.*, 1994), cultured African green monkey kidney cells (CV-1; Jones *et al.*, 1995), and LLC-PK$_1$ cells (Yoo *et al.*, 1994; 1995). In our hands, the primary and transformed human cell lines responded to fumonisin B$_1$ by the induction of apoptosis. This is consistent with the *in vivo* observations and substantiates the use of normal human cell cultures to investigate the probable role of fumonisin B_1 in human cancer.

Apoptosis is a programmed form of cell death that was described as early as 1885 by Flemming (1885). This death occurs as the result of various stimuli including DNA damage, hyperthermia, or hormone withdrawal (Wyllie, 1992; Binder and Hiddemann, 1994; Kane 1995; Polakowska and Haake, 1994; Korsmeyer, 1995). Apoptosis is also required for the developmentally related loss of important tissues or organs such as tadpole tails or interdigit webs in mammalian embryos (Kerr *et al.*, 1974; 1987). It also plays a role in thymus atrophy following glucocorticoid administration (Wyllie and Morris, 1982) and in normal maintenance of hepatocyte populations (Kerr 1971; Benedetti *et al.*, 1988). Apoptosis is also involved in the surveillance of cells with DNA damage. Cells with accumulated DNA damage signal apoptosis through the p53 protein, thereby eliminating a cell whose replication could be hazardous to the organism. Apoptosis is characterized by shrinkage of cytoplasm, the condensation of the chromatin to the periphery of the nucleus, and endonuclease activation. Nuclease activation causes hydrolysis of the chromosomal DNA into 180-200 bp nucleosomal fragments. With the loss of chromosomal DNA, there follows a change in the RNA and protein levels in the cell, which eventually result in the fragmentation of the cell into smaller pieces generally referred to as apoptotic bodies. At this stage the apoptotic cell, or the smaller apoptotic bodies, can be phagocytized.

The net result of apoptosis is the controlled loss of functional cells. In the case of hepatocellular apoptosis, there could be subsequent hepatocellular proliferation to replace the apoptotic cells. This may explain the increase in mitotic cells in our studies (Figures 6 and 7) and those of Voss *et al.* (1993, 1995) following the loss of hepatocytes through apoptosis. Therefore, the organ weight will reflect the balance of the rate of apoptosis loss versus the rate of hepatocellular regeneration. Evidently, the higher levels of fumonisin B_1 rates of apoptosis that can not be overcome by mitotic replacement. This is the case with the livers and kidneys of F334 rats (Figures 4 & 5; Voss *et al.*, 1995), livers and kidneys of male Sprague-Dawley rats (Voss *et al.*, 1993), and kidneys of female Sprague-Dawley rats (Voss *et al.*, 1993) where the organs displayed apoptosis and decreased in size relative to time and dose of fumonisin B_1 exposure.

The induction of kidney tubular epithelial cell regeneration (mitosis) was not uniformly apparent in our studies with F344 rats; however, mitosis was quite obvious in the livers of the rats and paralleled increases in apoptosis. We have estimated with antibodies to proliferating cell nuclear antigen (PCNA) that the mitotic index in male rats exposed to 484 ppm fumonisin B_1 increased from 0.02% (controls) to 0.5% (data not presented). This 25-fold increase in mitosis in the liver could indicate the role of fumonisin B_1 in liver neoplasms. Fumonisin B_1 interrupts the ability of cells to synthesize complex sphingolipids, resulting in the accumulation of intracellular sphinganine and sphingosine (Wang *et al.*, 1991), each of which have been shown to down-regulate protein kinase C (Hannun *et al.*, 1986; Stevens *et al.*, 1990; Merrill and Jones, 1990). There should likewise be a concomitant decrease in cellular ceramides and complex sphingolipids, each of which have likewise been shown to participate in signal transduction pathways involved in cellular mitosis or apoptosis. Therefore, regardless of whether the apoptotic mechanism involves (a) the elevations of sphinganine or sphingosine, (b) changes in ceramides or sphingolipids, or (c) other yet identified cellular changes, the net effect is the induction of apoptosis and a resulting compensatory increase in cell proliferation.

The role of cellular proliferation in liver neoplasms is far from a settled issue in carcinogenesis. While there is compelling evidence that many tumor promoting stimuli result in increases in cell proliferation, other studies do not support a direct role. Fumonisin B_1 has not been shown to damage DNA (Gelderblom *et al.*, 1992b; P.C. Howard, unpublished data), therefore, its role as a tumor promoter through the induction of cell proliferation is provocative, and raises issues concerning the appropriate models for assessing the human risk following exposure to low levels of fumonisin B_1.

Our results have described the NOEL for fumonisin B_1 in male F344 rats fed fumonisin B_1 for 28-days. The NOEL for total body weight effects was between 234 and 484 ppm, and the liver weight effects indicated a NOEL between 163 and 234 ppm. Microscopic examination of the liver indicated the NOEL for hepatocellular apoptosis and liver regeneration were between 163 and 234 ppm. In the same livers, the induction of bile duct hyperplasia and hepatocellular mitosis occurred only in the highest dose of fumonisin B_1. In the feeding studies with F344 rats, 4 weeks of fumonisin B_1 feeding demonstrated that the NOEL in male rat livers is greater than 81 ppm (Voss et al., 1995), while the feeding studies with male Sprague-Dawley rats indicated a NOEL between 50 and 150 ppm fumonisin B_1 (Voss et al., 1993). When combined, these studies indicate that the dose required for a demonstrable fumonisin B_1 effect on the liver of male rats with 28-day exposure is, for the most part, between 100 and 250 ppm.

The kidneys of male rats were comparatively more sensitive than the livers to dietary fumonisin B_1, having effects on the kidney weights at all dose levels. Microscopic examination of the kidneys also revealed a NOEL less than 99 ppm, with induction of apoptosis and tubular structural degeneration at all dose levels. This is in agreement with the studies of Voss et al. where the NOEL for Sprague-Dawley male rat kidneys was less than 15 ppm (Voss et al., 1993) and the NOEL for F344 male rat kidneys was between 9 and 27 ppm fumonisin B_1 at 4 and 13 weeks (Voss et al., 1995). When combined, these results indicate that the NOEL for fumonisin B_1 effects on male rat kidneys is at a very low fumonisin B_1 concentration, between 9 and 15 ppm.

Decreases in the total body weight of female F344 rats fed fumonisin B_1 were noted at the highest dose, while decreases in liver weights were found at 234 and 484 ppm. Microscopic examination of the female livers revealed hepatocellular apoptosis at all doses, hepatocellular mitosis at 163, 234, and 484 ppm, and bile duct hyperplasia at the two highest doses. In studies with female Sprague-Dawley rats fed fumonisin B_1 for 28-days (Voss et al., 1993), the NOEL for serum chemistry effects indicative of liver damage were between 50 and 150 ppm; however, absolute and relative liver weights and microscopic examination of the tissues indicated the NOEL was above 150 ppm. When combined, these two studies indicate that the NOEL for livers in female rats fed fumonisin B_1 for 28-days is between 99 and 163 ppm.

The effect of fumonisin B_1 on the kidneys of the female rats was the loss of kidney weight at all doses (Figure 5), and indication of apoptosis and kidney tubular degeneration beginning at 163 ppm. The studies of Voss et al. demonstrated kidney NOELs for Sprague-Dawley rats (Voss et al., 1993) or F344 rats (Voss et al., 1995) fed fumonisin B_1 for 28 days to be between 15 and 50 ppm and above 81 ppm, respectively. Therefore, it seems that the kidneys of the female rats are sensitive to the effects of fumonisin B_1, and taking into consideration the three studies, probably have a NOEL below 100 ppm.

Studies are in progress to determine the apoptogenic mechanism of fumonisin B_1 in our cultured human cells. Additionally, a two-year chronic feeding study is in progress to determine the carcinogenicity of fumonisin B_1, and to study the role of sphingolipids and sphingolipid precursors in the in vivo response to fumonisin B_1 exposure.

ACKNOWLEDGEMENTS

These studies were supported by an interagency agreement between the National Center for Toxicological Research and National Toxicology Program (IAG 224-93-0001).

The authors thank G.D. Stoner for the HET-1A. W.H. Tolleson was supported by an Interagency Agreement with the Veteran's Administration Hospital, Little Rock, AR.

REFERENCES

Bendetti, A.; Jézéquel, A.M.; Orlandi, F. Preferential distribution of apoptotic bodies in acinar zone 3 of the normal human and rat liver. *J. Hepatol.* **1988**, *7*, 319-324.

Bezuidenhout, S.C.; Gelderblom, W.C.A.; Gorst-Allman, C.P.; Horak, R.M.; Marasas, W.F.O.; Spiteller, G.; Vleggaar, R. Structure elucidation of the fumonisins, mycotoxins from *Fusarium moniliforme*. *J. Chem. Soc. Chem. Commun.,* **1988**, 743-745.

Binder, C.; Hiddemann, W. Programmed cell death - many questions still to be answered. *Ann. Hematol.* **1994**, *69*, 45-55.

Cawood, M.E.; Gelderblom, W.C.A.; Vleggaar, R.; Behrend, Y.; Thiel, P.G.; Marasas, W.F.O. Isolation of the fumonisin mycotoxins: A quantitative approach. *J. Agric. Food Chem.*, **1991**, *39*, 1958-1962.

Dombrink-Kurtzman, M.A.; Bennett, G.A.; Richard, J.L.. An optimized MTT bioassay for determination of cytotoxicity of fumonisins in turkey lymphocytes. *J. A.O.A.C. International*, **1994**, 77, 512-516.

Flemming, W. Uber die bildung von richtungsfiguren in säugethiereiern beim untergang graafscher follikel. *Arch. Anat. Entwgesch.*, **1885**, 221-224.

Gelderblom, W.C.A.; Marasas, W.F.O.; Vleggaar, R.; Thiel, P.G.; Cawood, M.E. Fumonisins: Isolation, chemical characterization and biological effects. *Mycopatholia*, **1992a**, *117*, 11-16.

Gelderblom, W.C.A.; Semple, E.; Marasas, W.F.O.; Farber, E. The cancer-initiating potential of the fumonisin B mycotoxins. *Carcinogenesis*, **1992b**, *13*, 433-437.

Hannun, Y.A.; Loomis, C.R.; Merrill, Jr., A.H.; Bell, R.M. Sphingosine inhibition of protein kinase C activity and of phorbol dibutyrate binding in vitro and in human platelets. *J. Biol. Chem.*, **1986**, *261*, 12604-12609.

Holcomb, M.; Sutherland, J.B.; Chiarelli, M.P.; Korfmacher, W.A.; Thompson, Jr., H.C.; Lay, Jr., J.O.; Hankins, L.J.; Cerniglia, C.E. HPLC and FAB mass spectrometry analysis of fumonisins B_1 and B_2 produced by *Fusarium moniliforme* on food substances. *J. Agric. Food Chem.*, **1993a**, *41*, 357-360.

Holcomb, M.; Thompson, Jr., H.C.; Hankins, L.J. Analysis of fumonisin B_1 in rodent feed by gradient elution HPLC using precolumn derivatization with FMOC and fluorescence detection. *J. Agric. Food Chem.*, **1993b**, *41*, 764-767.

Huntsberger, D.V.; Billingsley, P. Elements of statistical inference, fourth edition; Allyn and Bacon, Inc.: Boston, MA, 1977.

Iype, P.T.; Gabriel, B.W.; Stoner, G.D.; Kaighn, M.E. A serum-free medium for human epidermal-like cells. *In Vitro Dev. Biol.*, **1993**, *29A*, 94-96.

Jones, C.; Huang, H.; Dickman, M.; Henderson, G.; Wang, H.; Gilchrist, D. Analysis of a carcinogen, fumonisin, which is a fungal toxin. *Proc. Amer. Assoc. Cancer Res.*, **1995**, *36*, 668.

Kane, A.B. Redefining cell death. *Amer. J. Pathol.*, **1995**, *146*, 1-18.

Kerr, J.F.R. Shrinkage necrosis: a distinct mode of cellular death. *J. Pathol.*, **1971**, *105*, 13-20.

Kerr, J.F.R.; Harmon, B.; Searle, J. An electron-microscope study of cell deletion in the anuran tadpole tail during spontaneous metamorphosis with special reference to apoptosis of striated muscle fibres. *J. Cell Science*, **1974**, *14*, 571-585.

Kerr, J.F.R.; Searle, J.; Harmon, B.V.; Bishop, C.J. Apoptosis, IN *Perspectives on Mammalian Cell Death*; Potten, C.S., Ed; Oxford University Press: Oxford, 1987.

Korsmeyer, S.J. Regulation of cell death. *Trends in Genetics*, **1995**, *11*, 101-105.

Merrill, Jr., A.H.; Jones, D.D. An update of the enzymology and regulation of sphingomyelin metabolism. *Biochim. Biophys. Acta*, **1990**, *1044*, 1-12.

National Toxicology Program. *Specifications for the Conduct of Studies to Evaluate the Toxic and Carcinogenic Potential of Chemical, Biological and Physical Agents in Laboratory Animals for the National Toxicology Program*: National Toxicology Program, 1991.

Nelson, P.E.; Plattner, R.D.; Shackelford, D.D.; Desjardins, A.E. Production of fumonisins by *Fusarium moniliforme* strains from various substrates and geographic areas. Appl. Environ. Microbiol. 1991, 57, 2410-2412.

Pirisi, L.; Yasumoto, S.; Feller, M.; Doniger, J.; DiPaolo, J.A. Transformation of human fibroblasts and keratinocytes with human papillomavirus type 16 DNA. *J. Virology*, **1987**, *61*, 1061-1066.

Polakowska, R.R.; Haake, A.R. Apoptosis: the skin from a new perspective. *Cell Death and Differentiation*, **1994**, *1*, 19-31.

Ross, P.F.; Rice, L.G.; Osweiler, G.D.; Nelson, P.E.; Richard, J.L.; Wilson, T.M. A review and update of animal toxicoses associated with fumonisin-contaminated feeds and production of fumonisins by *Fusarium* isolates. *Mycopathologia*, **1992**, *117*, 109-114.

Stevens, V.L.; Nimkar, S.; Jamison, W.C.L.; Liotta, D.C.; Merrill, Jr., A.H. Characteristics of the growth inhibition and cytotoxicity of long-chain (sphingoid) bases for Chinese hamster ovary cells: Evidence for an involvement of protein kinase C. *Biochim. Biophys. Acta*, **1990**, *1051*, 37-45.

Stoner, G.D.; Kaighn, M.E.; Reddel, R.R.; Resau, J.H.; Bowman, D.; Naito, Z.; Matsukura, N.; You, M.; Galati, A.J.; Harris, C.C. Establishment and characterization of SV40 T-antigen immortalized human esophageal epithelial cells. *Cancer Res.*, **1991**, *51*, 365-371.

Thiel, P.G.; Marasas, W.F.O.; Sydenham, E.W.; Shephard, G.S.; Gelderblom, W.C.A.; Nieuwenhuis, J.J. Survey of fumonisin production by *Fusarium* species. *Applied and Environmental Microbiology*, **1991**, *57*, 1089-1093.

Tolleson, W.H.; Melchior, W.B., Jr.; Morris, S.M.; McGarrity, L.J.; Domon, O.E.; Muskhelishvili, L.; James, S.J.; Howard, P.C. Apoptosis and antiproliferative effects of fumonisin B$_1$ in human keratinocytes, fibroblasts, esophageal epithelial cells, and hepatoma cells. *Carcinogenesis*, in press.

Toxicological Principals for the Safety Assessment of Direct Food Additives and Color Additives Used in Foods, "Redbook II". United States Food and Drug Administration, Center for Food Safety and Applied Nutrition, 1993.

Voss, K.A.; Chamberlain, W.J.; Bacon, C.W.; Norred, W.P. A preliminary investigation on renal and hepatic toxicity in rats fed purified fumonisin B1. *Natural Toxins*, **1993**, *1*, 222-228.

Voss, K.A.; Chamberlain, W.J.; Bacon, C.W.; Herbert, R.A.; Walters, D.B.; Norred, W.P. Subchronic feeding study of the mycotoxin fumonisin B1 in B6C3F1 mice and Fischer 344 rats. *Fund. Appl. Toxicol.*, **1995**, *24*, 102-110.

Wang, E.; Norred, W.P.; Bacon, C.W.; Riley, R.T.; Merrill, Jr., A.H. Inhibition of sphingolipid biosynthesis by fumonisins. *J. Biological Chemistry*, **1991**, *266*, 14486-14490.

Wyllie, A.H. Apoptosis and the regulation of cell numbers in normal and neoplastic tissues: an overview. *Cancer and Metastasis Reviews*, **1992**, *11*, 95-103.

Wyllie, A.H.; Morris, R.G. Hormone-induced cell death: purification and properties of thymocytes undergoing apoptosis after glucocorticoid treatment. *Am. J. Pathol.*, **1982**, *109*, 78-87.

Yoo, H.-S.; Showker, J.L.; Riley, R.T. Relationship between fumonisin (FB)-induced cytotoxicity and the elevation of free sphinganine (S$_a$) in LLC-PK$_1$ cells. *The Toxicologist*, **1994**, *14*, 772.

Yoo, H.-S.; Showker, J.L.; Riley, R.T. Partitioning of increased sphinganine (S$_a$) and decreased complex sphingolipids (CSL) as causative factors in fumonisin B$_1$ (FB$_1$)-induced cytostatic and cytotoxic effects in LLC-PK$_1$ cells. *The Toxicologist*, **1995**, *15*, 1555.

22

FUMONISIN B$_1$ TOXICITY IN MALE SPRAGUE-DAWLEY RATS

G. Bondy, M. Barker, R. Mueller, S. Fernie, J. D. Miller[*], C. Armstrong,
S. L. Hierlihy, P. Rowsell and C. Suzuki

Toxicology Research Division
Food Directorate, Health Canada
Ottawa, Canada
[*] Plant Research Centre
Agriculture and Agri-Food Canada
Ottawa, Canada

ABSTRACT

Male rats were gavaged with fumonisin B$_1$ (FB$_1$) once daily for 11 consecutive days at doses of 0, 1, 5, 15, 35, and 75 mg FB$_1$/kg body weight. Urine osmolality (at 5-75 mg FB$_1$/kg) and organic ion transport in kidney slices (at 5-75 mg FB$_1$/kg) were reduced. Urinary excretion of protein (at 15-75 mg FB$_1$/kg) and of the enzymes LDH (at 5-75 mg FB$_1$/kg), NAG (at 5-75 mg FB$_1$/kg) and GGT (at 15-75 mg FB$_1$/kg) were increased. These findings were indicative of glomerular and tubular toxicity. Histopathologic changes in the kidney consisted of necrosis of tubular epithelia of variable extent accentuated in the inner cortex. These changes were present at 1 and 5 mg FB$_1$/kg and were more pronounced at 15-75 mg FB$_1$/kg. Serum enzymes indicative of hepatotoxicity (ALT, GGT) were elevated compared to controls at 75 mg FB$_1$/kg only. There were noticeable increases in mitotic figures in hepatocytes at 35-75 mg FB$_1$/kg, while single cell necroses were increasingly numerous from 15-75 mg FB$_1$/kg. The kidneys were considered to be the primary target organs in this study.

INTRODUCTION

In previous studies the toxicity of FB$_1$ was evaluated in male Sprague-Dawley rats using intraperitoneal (ip) injection as the route of exposure (Bondy et al., 1995; Suzuki et al., 1995). Within a four day period, with daily doses of 7.5 or 10 mg FB$_1$/kg body weight (bw), all treated rats displayed clinical and histopathological signs of nephrotoxicity and hepatotoxicity. General signs of FB$_1$ toxicity in the ip study included reduced food consumption and body weights, as well as dehydration due to increased urine output. Since the primary means of exposure to FB$_1$ is ingestion, the present study was initiated to evaluate the effects of FB_1 on the same parameters using gavage as the route of toxin administration. Male Sprague-Dawley rats were administered FB_1 by gavage for 11 consecutive days, during which time general metabolic parameters such as body weight, food consumption, and urine and feces production were monitored. Markers of renal function, including urine volume, osmolality, proteinuria and enzymuria, were also evaluated. On the final day, blood was collected for hematology and serum biochemistry analyses, and organic ion transport in renal cortical slices was measured. Tissue samples were prepared for histopathological evaluation. The renal, and to a lesser extent, hepatic lesions seen in this study were consistent with those seen in Sprague-Dawley rats receiving feed experimentally contaminated with purified FB_1 (Voss et al., 1993).

MATERIALS AND METHODS

Chemicals

Fumonisin B_1 was produced at the Plant Research Centre, Agriculture Canada, as described by Miller *et al.* (1994).

Animals

Male Sprague-Dawley rats (133.1 ± 15.0 g) were obtained from Charles River Canada Inc. (Montreal, Canada). Rats were housed individually in plastic cages (Health Guard System, Research Equipment Company, Inc., Bryan, TX) under conditions meeting the requirements of the Canadian Council for Animal Care and were acclimatised for one week prior to commencement of the study. Rat chow and water were provided *ad libitum* before and throughout the study.

Experimental Design

Rats were divided randomly into five dose groups and one control group (n=6 per group). Body weights were monitored daily. For 11 consecutive days each rat received a single daily gavage dose of FB_1 (in sterile saline) at one of the following dose levels: 0 (controls), 1, 5, 15, 35 or 75 mg FB_1/kg body weight. Food consumption, water consumption, urine output and feces output were monitored at the beginning, middle and end of the study (days 1,6-8,11-12) by transferring four rats from each group to Nalgene metabolic cages for 24 hour time periods. Urine samples were collected into vessels containing 1 mL of sodium azide (0.5%) over ice. On the final day (day 12) all rats were anesthetized with isofluorene and exsanguinated via the abdominal aorta. Organ weights were recorded for the liver, kidneys, spleen, thymus and brain of each rat.

Hematology and Serum Biochemistry

At the time of necropsy blood was drawn from the abdominal aorta into 5-mL syringes using a 25 gauge needle, and transferred to Becton Dickinson Vacutainer tubes containing EDTA (7.5%; tripotassium salt) for hematology, or glass test tubes for serum. For hematology, blood was processed with a Coulter Counter Model S-PLUS IV system (Coulter Electronics, Inc., Hialeah, Florida). Smears for leukocyte differentials were stained with Wright's Giemsa. For reticulocyte counts, blood was mixed with new methylene blue, smears were made and then counterstained with Wright's Giemsa.

Clotted blood was centrifuged at 700 xg for 15 min to prepare serum. A Beckman Synchron CX5 Clinical System (Beckman Instruments Canada Inc., Mississauga, Ontario) and Beckman reagent kits were used for serum clinical chemistry. Serum total IgG and IgM were quantified by sandwich ELISA (Bondy and Pestka, 1991) using unconjugated and horseradish peroxidase-conjugated goat anti-rat IgG and IgM polyclonal antibodies from Pierce (Rockford, IL).

Bone marrow was removed from the femur. Smears were air-dried, fixed in methanol and stained with modified Wright's Giemsa. For differentials, 500 cells total were counted.

Urinalysis

The methodology for urinalysis has been described previously (Suzuki *et al.*, 1995). Briefly, urine volume, osmolality and total protein were measured in whole urine. A 1 mL aliquot of each urine sample was applied to Sephadex G-25 PD-10 columns (Pharmacia LKB, Baie D'Urfe, Canada) and eluted through with 8 mL of saline to remove low molecular weight enzyme inhibitors. The first 4 mL fraction collected from the column was analyzed for creatinine using the method of Heinegard and Tiderstrom (1973). The enzymes, N-acetyl-β-D-glucosaminidase (NAG), γ-glutamyltranspeptidase (GGT) and lactate dehydrogenase (LDH) were measured in the next 4 mL fraction collected from the column by the methods of Leaback and Walker (1961), Dierickx (1980) and an LDH Sigma Diagnostic kit (Sigma Chemical Co., St. Louis, MO), respectively.

Organic Ion Transport in Kidney Slices

Immediately after exsanguination a kidney was removed from each rat, weighed, and prepared for organic ion transport studies. Transport of the anion, p-aminohippuric acid (PAH) and the cation, tetraethylammonium (TEA) into renal cortical slices was measured as previously described (Suzuki *et al.*, 1995). Transport was expressed as a ratio between the amount of ion transported into the slice and the amount remaining in the medium (S/M). A decrease in this ratio represents reduced transport function.

Pathology

For light microscopy, tissues were fixed in 10% neutral buffered formalin (pH 7.2). Paraffin sections (4 μm) of tissues were stained with haematoxylin and eosin or Giemsa. Photomicrographs were taken with a Zeiss Axiophot using Kodak Plus-X pan (PX 135-24) film.

Statistics

Parametric and nonparametric statistical analyses were performed with Sigmastat (Jandel Scientific, San Rafael, California). Data comparisons were considered significant if $p < 0.05$.

RESULTS

General and Metabolic Parameters

All rats in the control group and in the 1, 5 and 15 mg/kg dose groups remained healthy in appearance for the duration of the study. Rats in the 35 and 75 mg/kg dose groups

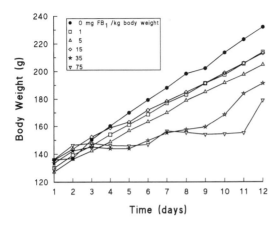

Figure 1. Body weight changes in rats treated with sterile saline or FB$_1$ at doses of 1, 5, 15, 35 or 75 mg/kg bw. Rats received a single gavage dose of saline or toxin daily for 11 consecutive days. Each point represents a mean of n=6 rats, except at the 75 mg FB$_1$/kg dose where n=5. Body weights of rats in the 5 mg FB$_1$/kg dose group are significantly different from corresponding controls (p < 0.05) on day 12 only. For the 35 and 75 mg FB$_1$/kg dose groups data are significantly different from corresponding controls (p < 0.05) from day 6 to day 12.

were slightly more lethargic towards the end of the dosing period. One rat in the highest dose group began to lose weight rapidly on day 7 following gavage and died on day 11. Data from this animal were excluded from analyses.

Total weight gain in control rats was consistently higher compared to weight gain in treated rats (Figure 1). Body weights were significantly lower compared to controls in treatment groups receiving 35 or 75 mg FB$_1$/kg starting on day 6 and continuing through day 12. Rats in the 5 mg/kg dose group had significantly reduced total body weights compared to controls on day 12 only. Total body weights in the 1 and 15 mg FB$_1$/kg dose groups were not significantly different from control body weights. In conjunction with body weights significantly lower than controls, rats in the 35 and 75 mg/kg dose groups demonstrated reduced food consumption which was significant on days 6 and 7 but which returned to control levels by days 11 and 12 (Figure 2). Although there were no statistically significant differences in food consumption between the control and the 5 mg/kg dose groups, food consumption in rats receiving 5 mg FB$_1$/kg was, on average, less than control food consumption by 1 g/day over the 12 day period, which resulted in a significant difference in body weight by day 12. Feces output was unaffected by oral administration of FB$_1$ (data not shown). With the exception of a significant increase in urine volume on day 8 in the 75 mg/kg dose group (Figure 3A), water consumption (data not shown) and urine output were unaffected in treatment rats.

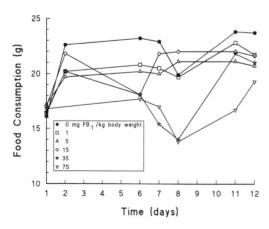

Figure 2. Food consumption in rats gavaged with sterile saline or FB$_1$ at doses of 1, 5, 15, 35 or 75 mg/kg. Each point represents a mean of n=4 rats, except at the 75 mg FB$_1$/kg dose where n=3. On days 6 and 7 data are significantly different from corresponding controls for the 35 and 75 mg FB$_1$/kg dose groups (p < 0.05).

Table 1. Absolute and relative organ weights for male Sprague-Dawley rats gavaged with fumonisin B_1 (FB_1) or sterile saline (controls; 0 mg FB_1/kg). Data are expressed as mean (standard error) for n=6 rats, except for the 75 mg FB_1/kg dose group, where n=5 rats.

FB_1 mg/kg	Organ weight (g)					Relative organ weight (% body weight)				
	Liver	Kidney[A]	Spleen	Thymus	Brain	Liver	Kidney	Spleen	Thymus	Brain
0	11.1	2.18	0.63	0.63	1.86	4.77	0.94	0.28	0.36	0.76
	(0.57)	(0.13)	(0.04)	(0.10)	(0.03)	(0.07)	(0.03)	(0.02)	(0.03)	(0.01)
1	9.81	2.01	0.71	0.70	1.85	4.59	0.94	0.33	0.33	0.84
	(0.47)	(0.07)	(0.07)	(0.02)	(0.02)	(0.13)	(0.03)	(0.03)	(0.02)	(0.02)
5	9.57	1.89[B]	0.70	0.66	1.74	4.67	0.93	0.34	0.33	0.83
	(0.35)	(0.04)	(0.02)	(0.02)	(0.04)	(0.11)	(0.03)	(0.01)	(0.02)	(0.02)
15	9.74	1.96	0.69	0.68	1.86	4.56	0.92	0.32	0.32	0.84
	(0.61)	(0.06)	(0.05)	(0.04)	(0.03)	(0.18)	(0.03)	(0.02)	(0.01)	(0.03)
35	8.21[B]	1.77[B]	0.62	0.53[B]	1.72	4.27	0.93	0.32	0.28[C]	0.86[C]
	(0.59)	(0.07)	(0.04)	(0.03)	(0.07)	(0.15)	(0.05)	(0.01)	(0.02)	(0.02)
75	7.64[B]	1.99	0.50	0.50[B]	1.80	4.25	1.11	0.28	0.28[C]	0.97[C]
	(0.57)	(0.14)	(0.04)	(0.04)	(0.00)	(0.19)	(0.11)	(0.02)	(0.01)	(0.03)

[A] weight of both kidneys; [B] significantly different from control organ weight (p<0.05); [C] significantly different from control relative organ weight (p<0.05)

Absolute weights of liver, kidney and thymus were reduced in some of the FB_1 dose groups (Table 1). Absolute liver and thymus weights were significantly reduced in the 35 and 75 mg/kg dose groups. In addition, decreases in absolute kidney weight were significant in the 5 and 35 mg/kg dose groups. Absolute spleen and brain weights were unaffected by FB_1 treatment. When organ weights were expressed relative to total body weight, only relative thymus and brain weights were affected (Table 1). Relative thymus weight was significantly reduced in the 35 and 75 mg/kg dose groups, while relative brain weights were significantly increased at the same doses.

Clinical Chemistry and Hematology

Most changes in clinical parameters occurred only in the two highest dose groups. Serum glucose was significantly reduced and alanine aminotransferase (ALT), aspartate aminotransferase (AST), γ-glutamyltransferase (GGT), cholesterol, and creatinine were significantly elevated at 75 mg FB_1/kg (Table 2). ALP was elevated in the 35 and 75 mg FB_1/kg dose groups. Serum magnesium was elevated in the 15, 35 and 75 mg/kg dose groups; calcium was increased in the 15 and 75 mg/kg dose groups only. Of the serum parameters measured, triglyceride (TG) levels were the most sensitive to FB_1, with significant reductions at 5 through 75 mg FB_1/kg. The following serum parameters remained unchanged in all dose groups (data not shown): total protein, albumin, globulin, amylase, creatine kinase, blood urea nitrogen, total bilirubin, uric acid, chloride, potassium, sodium, and total IgG and IgM.

There were few changes in circulating blood cells as a result of FB_1 treatment (Table 3). Reticulocyte numbers were significantly reduced at 35 mg FB_1/kg, but not at the higher dose of 75 mg FB_1/kg. White blood cell (WBC) numbers were significantly elevated at the highest dose only. Trends toward increased circulating lymphocytes, neutrophils and monocytes in the highest dose group were not significant. The following hematological parameters were not affected by oral administration of FB_1 (data not shown): red blood cells, hemoglobin, hematocrit, mean corpuscular volume, mean corpuscular hemoglobin concentration, mean corpuscular hemoglobin, red cell distribution width, platelets, mean platelet volume, basophils and eosinophils.

Table 2. Serum parameters for male Sprague-Dawley rats gavaged with fumonisin B_1 (FB_1) or sterile saline (controls; 0 mg FB_1/kg). Data are expressed as mean (standard error) for n=6 rats, except for the 75 mg FB_1/kg dose group, where n=5 rats.

FB_1 mg/kg	Glucose mmol/L	ALT[A] IU/L[B]	AST IU/L	ALP IU/L	GGT IU/L	Chol mmol/L	TGs IU/L	Creat μmol/L	Ca mmol/L	Mg mmol/L
0	11.02	48.67	112.3	539.2	2.33	1.25	1.44	36.5	2.59	1.00
	(0.32)	(2.12)	(6.1)	(63.5)	(0.49)	(0.07)	(0.18)	(0.9)	(0.02)	(0.03)
1	10.48	50.33	117.2	489.0	2.67	1.16	1.14	37.0	2.51	0.98
	(0.32)	(2.77)	(8.2)	(32.8)	(0.67)	(0.10)	(0.17)	(2.5)	(0.03)	(0.04)
5	9.97	45.17	109.3	411.8	1.83	1.35	0.70^C	36.8	2.56	1.04
	(0.61)	(2.30)	(6.9)	(59.3)	(0.40)	(0.09)	(0.07)	(3.0)	(0.03)	(0.04)
15	9.55	58.33	155.0	515.0	2.17	2.08	0.95^C	45.3	2.71^C	1.20^C
	(0.61)	(4.90)	(29.2)	(54.0)	(0.54)	(0.32)	(0.13)	(2.7)	(0.03)	(0.04)
35	8.80^C	74.67	167.3	836.7^C	3.17	2.04	0.52^C	45.7	2.66	1.19^C
	(0.52)	(6.09)	(30.7)	(86.3)	(0.31)	(0.25)	(0.05)	(1.5)	(0.03)	(0.03)
75	7.68^C	131.60^C	255.6^C	1700.6^C	6.80^C	2.73^C	0.54^C	50.0^C	2.72^C	1.23^C
	(0.23)	(11.77)	(18.2)	(115.7)	(1.53)	(0.30)	(0.09)	(3.0)	(0.05)	(0.02)

[A] Abbreviations: alanine aminotransferase (ALT), aspartate aminotransferase (AST), alkaline phosphatase (ALP), gamma glutamyltransferase (GGT), cholesterol (Chol), triglycerides (TGs), creatinine (Creat); [B] 1 International Unit (IU) of enzyme catalyzes 1 μmole of product/min under defined conditions; [C] significantly different from controls ($p < 0.05$).

Although there were no changes in the numbers or ratios of bone marrow cells (data not shown), there were dose-related increases in the number of vacuolated myeloid cells in bone marrow which were significant at 35 and 75 mg FB_1/kg. The number of vacuolated lymphoid cells was significantly increased at 75 mg FB_1/kg (Figure 4). Total numbers of vacuolated bone marrow cells were significantly elevated at both 35 and 75 mg FB_1/kg bw.

Urinalysis and Organic Ion Transport Studies

Urine osmolality was significantly decreased in all but the lowest dose group between days 6 and 8, but was not significantly different from controls by the end of the treatment period (Figure 3B). Urinary creatinine excretion was unchanged in all groups (data not shown), but urinary protein was increased on days 6 to 8 in the 15-75 mg FB_1/kg dose groups (Figure 3C). With the exception of the 5 mg/kg group, urine protein was not significantly different from controls by day 12. Urinary excretion of the enzyme LDH was elevated in all groups receiving FB_1, reaching a maximum between days 6 and 8 (Table 4). Urinary GGT and NAG excretion were also increased, peaking between days 6 and 8 in all but the lowest dose group (Tables 5,6). By the final two study days urine enzyme levels in all dose groups were not significantly different from control levels, except in the 15 mg/kg group where all three enzymes remained slightly elevated compared to controls. The dose-dependent effects

Figure 3. Urine volume (A), osmolality (B) and protein (C) in rats gavaged with sterile saline or FB_1 at doses of 1, 5, 15, 35 or 75 mg/kg bw. Each point represents a mean of n=4 rats, except at the 75 mg FB_1/kg dose where n=3. For urine volume, data are significantly different from the corresponding control on day 8 for the 75 mg FB_1/kg bw dose group ($p < 0.05$). For urine osmolality data are significantly different from corresponding controls on the following days: day 7 only for 5 and 15 mg FB_1/kg; days 7 and 8 for 35 mg FB_1/kg; days 6 through 8 for 75 mg FB_1/kg ($p < 0.05$). For urine protein data are significantly different from the corresponding controls on the following days: days 6, 7 and 12 for 15 mg FB_1/kg; days 7 and 8 for 35 mg FB_1/kg; days 6 through 8 for 75 mg FB_1/kg ($p < 0.05$).

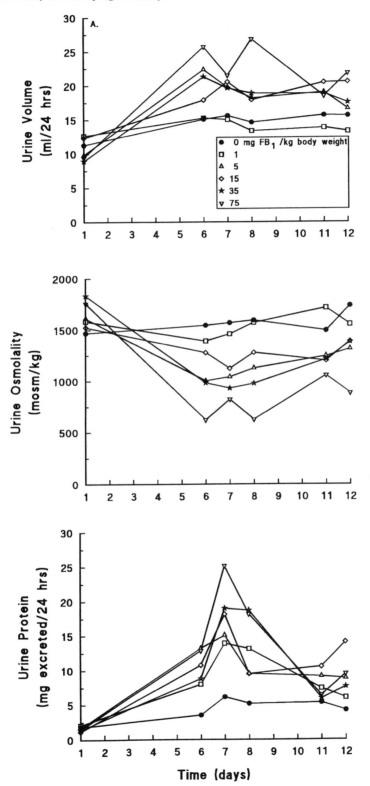

Table 3. Summary of hematology data for male Sprague-Dawley rats gavaged with fumonisin B_1 (FB_1) or sterile saline (controls; 0 mg FB_1/kg). Data are expressed as mean (standard error) for n=6 rats for the 0, 1, 5 and 35 mg FB_1/kg dose groups, n=5 rats for the 15 mg FB_1/kg dose group and n=4 rats for the 75 mg FB_1/kg dose group.

FB_1 (mg/kg)	Reticulocytes $(x10^9/L)$	WBC $(x10^9/L)$	Lymphocytes $(x10^9/L)$	Neutrophils $(x10^9/L)$	Monocytes $(x10^9/L)$
0	671.5	7.5	6.5	0.73	0.20
	(34.9)	(1.0)	(0.9)	(0.11)	(0.04)
1	632.2	7.2	6.5	0.50	0.15
	(19.4)	(0.7)	(0.7)	(0.09)	(0.04)
5	605.3	7.9	6.7	0.94	0.19
	(41.3)	(1.1)	(0.9)	(0.21)	(0.07)
15	546.2	7.1	6.4	0.53	0.10
	(7.9)	(0.6)	(0.5)	(0.06)	(0.02)
35	391.3[A]	7.8	6.5	0.90	0.27
	(41.7)	(1.1)	(0.9)	(0.21)	(0.06)
75	602.5	13.1[A]	9.3	3.02	0.47
	(23.8)	(1.8)	(0.8)	(1.0)	(0.18)

[A] significantly different from controls (p < 0.05).

of FB_1 on the transport of organic ions into renal cortical slices are shown in Figure 5. Uptake of the anion PAH was decreased significantly at 35 and 75 mg FB_1/kg, while uptake of the cation TEA was decreased significantly at 5 mg FB_1/kg and higher.

Histopathology

Light microscopical lesions in the liver and kidneys of treated animals were dose-related. In livers of rats from the 15-75 mg FB_1/kg dose groups there was a moderate increase in the incidence of mitosis and marked increase in the incidence of single cell necrosis of hepatocytes (Figure 6). There was also a slight atrophy of hepatocytes and dilation of sinusoids in zone 3 in treated animals. Changes in the kidneys of FB_1-treated rats were evident starting at 1 mg FB_1/kg and were more marked at 15 to 75 mg FB_1/kg. Necrosis and sloughing of tubular epithelial cells in the inner cortex, as well as anisokaryosis, cytoplasmic basophilia and atrophy of tubular epithelial cells were evident (Figure 7). In addition to renal and hepatic changes there was a slight increase in cytoplasmic vacuolation in adrenal cortical cells, changing from mild to moderate in some animals in the 15 to 75 mg FB_1/kg dose

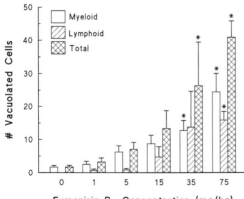

Figure 4. Numbers of vacuolated myeloid and lymphoid cells in bone marrow of rats gavaged with sterile saline or FB_1 at doses of 1, 5, 15, 35 or 75 mg/kg bw. * indicates that data are significantly different from corresponding controls (p < 0.05).

Table 4. Urine lactate dehydrogenase (LDH) in male Sprague-Dawley rats gavaged with fumonisin B_1 (FB_1) or sterile saline (controls; 0 mg FB_1/kg). Data are expressed as nmol/min excreted over 24 hours, mean (standard error) for n=4 rats, except for the 75 mg FB_1/kg dose group where n=3 rats.

FB_1 mg/kg	Day of study					
	1	6	7	8	11	12
0	58.8	79.9	101.9	120.1	113.2	74.2
	(11.6)	(11.0)	(24.8)	(19.4)	(20.1)	(5.9)
1	64.6	303.5	781.7	1108.9	275.5	383.7
	(7.6)	(97.4)	(312.7)	(210.8)	(122.2)	(117.1)
5	103.3	921.2	2131.0[A]	1546.3	247.0	605.7
	(31.1)	(325.3)	(291.5)	(247.2)	(110.7)	(170.9)
15	118.1	1459.3[A]	2497.0[A]	1965.6	795.8	2171.1[A]
	(61.4)	(288.0)	(327.0)	(88.6)	(435.6)	(469.5)
35	95.6	1427.4[A]	1666.1	3202.4[A]	477.2	851.9
	(16.3)	(207.8)	(436.7)	(635.0)	(248.8)	(213.6)
75	39.5	1772.3[A]	2204.9	4839.9[A]	1001.4	615.3
	(0.2)	(513.3)	(707.3)	(264.4)	(176.9)	(123.6)

[A] significantly different from same day control (p < 0.05).

Table 5. Urine N-acetyl-β-D-glucosaminidase (NAG) in male Sprague-Dawley rats gavaged with fumonisin B_1 (FB_1) or sterile saline (controls; 0 mg FB_1/kg). Data are expressed as nmol/min excreted over 24 hours, mean (standard error) for n=4 rats, except for the 75 mg FB_1/kg dose group where n=3 rats.

FB_1 mg/kg	Day of study					
	1	6	7	8	11	12
0	172.2	205.4	278.2	205.4	211.5	260.7
	(11.3)	(25.0)	(49.5)	(14.1)	(33.2)	(23.1)
1	169.4	291.5	463.0	331.2	267.5	234.3
	(9.5)	(33.1)	(39.1)	(30.9)	(22.3)	(30.0)
5	141.0	511.9	672.5[A]	403.4[A]	267.4	232.6
	(2.5)	(97.2)	(30.8)	(37.1)	(41.0)	(23.9)
15	215.3	574.7[A]	612.9[A]	476.5[A]	423.7[A]	463.4
	(67.2)	(29.8)	(72.7)	(34.7)	(63.8)	(133.4)
35	152.1	491.9	944.4[A]	870.0[A]	371.7[A]	333.2
	(9.7)	(72.1)	(70.8)	(38.6)	(30.3)	(59.9)
75	208.8	972.0[A]	1427.1[A]	1098.6[A]	444.5[A]	381.6
	(24.4)	(138.7)	(145.4)	(55.4)	(5.5)	(66.6)

[A] significantly different from same day control (p < 0.05).

Table 6. Urine γ-glutamyltransferase (GGT) in male Sprague-Dawley rats gavaged with fumonisin B$_1$ (FB$_1$) or sterile saline (controls; 0 mg FB$_1$/kg). Data are expressed as μmol/min excreted over 24 hours, mean (standard error) for n=4 rats, except for the 75 mg FB$_1$/kg dose group where n=3 rats.

FB$_1$ mg/kg	Day of study					
	1	6	7	8	11	12
0	1.86 (0.36)	3.52 (0.96)	6.15 (1.57)	5.19 (1.40)	5.33 (1.12)	4.23 (1.01)
1	2.53 (0.31)	8.02 (1.71)	13.93 (3.57)	13.14 (2.21)	7.40 (1.04)	6.04 (0.91)
5	2.81 (0.17)	13.21 (2.41)	15.09 (2.13)	9.55 (2.95)	9.15 (2.25)	8.86 (1.13)
15	2.33 (0.48)	15.77[A] (1.24)	18.18 (1.91)	9.52 (0.63)	10.55 (1.20)	14.13[A] (2.48)
35	3.12 (0.25)	12.49[A] (1.81)	19.02[A] (4.23)	18.71[A] (4.31)	5.94[A] (0.67)	7.65 (0.49)
75	4.19 (0.24)	16.58[A] (2.19)	25.17[A] (4.85)	18.18[A] (3.46)	6.43[A] (0.71)	9.22 (1.11)

[A] significantly different from same day control (p < 0.05).

groups. Mild lymphocytolysis in the thymic cortex of FB$_1$-treated rats, compared to minimal changes in control rats, was first evident at 5 mg FB$_1$/kg and persisted through the highest dose groups.

DISCUSSION

The kidneys were the most sensitive targets of FB$_1$ toxicity in male Sprague-Dawley rats, with changes in kidney histopathology (Figure 7), urinalysis parameters (Figures 3B,3C, Tables 4-6) and organic ion transport in isolated kidney slices (Figure 5) occurring at the lowest dose. Standard serum chemistry indices of kidney toxicity were comparatively insensitive to FB$_1$, with elevated creatinine only in the 75 mg FB$_1$/kg dose group (Table 2) and no changes in BUN in any dose group. Elevated serum calcium and magnesium, both starting at 15 mg FB$_1$/kg (Table 2), have been associated with renal dysfunction and

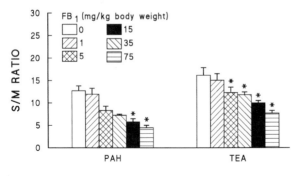

Figure 5. Organic anion (PAH) and cation (TEA) accumulation in renal cortical slices from rats gavaged for 11 days with sterile saline or FB$_1$ at doses of 1, 5, 15, 35 or 75 mg/kg bw. Values represent a mean of n=6 for all groups except the 75 mg FB$_1$/kg dose group, where n=5. * indicates that data are significantly different from corresponding controls.

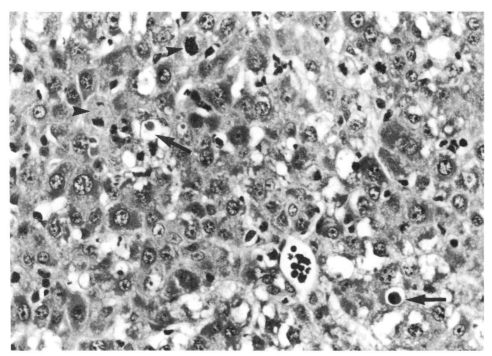

Figure 6. Photomicrograph of a liver section from a rat gavaged with 75 mg FB₁/kg bw. Arrows indicate single cell necrosis; arrowheads indicate mitotic figures (hematoxylin and eosin, x 40).

decreased glomerular filtration (Riley and Cornelius, 1989). Urine osmolality was transiently decreased in treatment rats without concurrent reductions in fluid intake and therefore without dehydration, implying a temporary disruption of renal fluid homeostasis. Additionally, both glomerular and tubular functions appear to be affected in FB₁-treated rats. Intraperitoneal administration of FB₁ has been shown to induce increased excretion of high molecular weight protein into urine (Suzuki et al., 1995), suggesting increased glomerular permeability. Significant increases in urinary protein were also observed in the present study (Figure 3C). Tubular damage was evident from transient elevations in urine GGT, LDH and NAG in all FB₁ dose groups (Tables 4-6). Despite the transience of urinary changes, the significant reductions in PAH and TEA transport at the time of necropsy (Figure 5) indicated that nephrotoxicity was still evident at the end of the dosing period (day 12). Histopathological changes in the kidney also confirmed the continued presence of nephrotoxicity (Figure 7). These responses are consistent with transient enzymuria accompanied by persistent renal damage in chronic studies of other nephrotoxins (Smith and Hook, 1984). The possibility that compensatory responses or resistance to FB₁-induced nephrotoxicity contributed to the transience of changes in urine volume, osmolality, protein and enzyme levels requires further investigation.

With respect to the parameters measured in this study, the liver appeared to be less sensitive to FB₁ toxicity than the kidneys. Serum enzymes (ALT, GGT) were significantly higher than control levels only in the highest dose group (Table 2), while dose-dependent light microscopic lesions were present starting at 15 mg FB₁/kg body weight. Increased serum lipids have been associated with hepatotoxicity in rats treated with FB₁ (Voss et al., 1993), as well as with toxins such as carbon tetrachloride (Romero et al., 1994). In the present

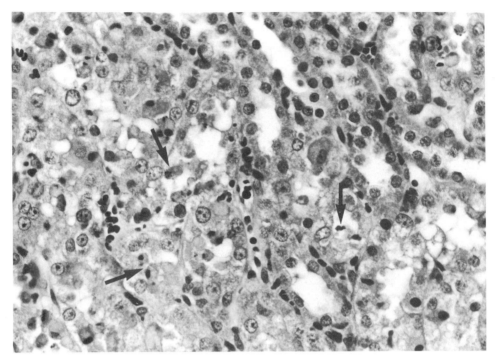

Figure 7. Photomicrograph of a kidney section from a rat gavaged with 75 mg FB₁/kg bw. Arrows indicate single cell necrosis (hematoxylin and eosin, x 40).

study serum cholesterol was elevated in the 75 mg FB₁/kg dose group but TGs were significantly depressed compared to controls starting at 5 mg FB₁/kg. Serum TGs are influenced by dietary fat intake and by endogenous synthesis in the liver and intestine (Sacher and McPherson, 1991). Reduced intake does not completely account for the decreases in serum TGs because food consumption was reduced only transiently and only in the two highest dose groups. Since the direct or indirect effects of FB₁ on endogenous TG biosynthesis have not been determined, the events contributing to depressed serum TGs in this study are unknown.

The dose-dependent increase in bone marrow lymphoid and myeloid cell vacuolation (Figure 4) is a reflection of FB₁-induced toxicity to bone marrow precursors. It is likely that vacuolation is a manifestation of cellular degeneration, which may either be reversible or precede cell death (Rebar, 1993).

Total IgM and IgG levels in serum were unaffected; however, thymus weights were lower in the two highest dose groups (Table 1) and accompanied by lymphocytolysis. The trend towards increased numbers of circulating lymphocytes, neutrophils and monocytes in the highest dose group indicates that the increase in total WBCs was not specific to a single cell subset (Table 3). Although increased numbers of circulating phagocytic cells such as neutrophils are generally a response to tissue damage, the functional effects of changes in the thymus of treated rats have yet to be determined. While the kidney and liver are targets of FB₁ in this model, there appears to be a potential for FB₁-induced alterations in immune function which deserves further consideration.

Many of the changes seen in this study were also seen in rats receiving FB₁ intraperitoneally for 4 consecutive days (Bondy et al., 1995; Suzuki et al., 1995), including effects on the kidneys, liver, thymus, and bone marrow as well as reduced food consumption and body weight. However, FB_1 was more toxic when administered ip than when administered by gavage, with more changes occurring within a shorter time span in the ip study. In pharmacokinetic studies with gavaged rats, FB_1 bioavailability was very low because the toxin was poorly absorbed from the gut and the majority entered the feces in predominantly the unmetabolized form (Shephard et al., 1992a, 1994). In intraperitoneally dosed rats, 32% of ^{14}C-labelled FB_1 appeared in urine within 24 hours, of which 78% was in the unmetabolized form (Shephard et al., 1992a, 1992b). An additional 66% of the dose was recovered from feces, having been eliminated from plasma by biliary excretion (Shephard et al., 1992b, 1994). The lower bioavailability of FB_1 in rats dosed by the gavage route accounts for reduced toxicity in the present study in comparison to FB_1 toxicity in ip studies (Bondy et al., 1995; Suzuki et al., 1995). The targets of FB_1 in this study, primarily the kidneys and liver, are consistent with data indicating that $[^{14}C]$-FB_1 persists in the liver and kidneys of orally dosed rats (Norred et al., 1993).

In summary, changes in kidney function and morphology were evident at doses as low as 1 mg FB_1/kg, whereas overt toxicity (reduced body weight and food consumption) was observable only at 35 mg FB_1/kg or higher. Urinalysis and transport studies indicated that both glomerular and tubular functions were compromised. Under the conditions of this study histopathological alterations to the liver were observed at doses as low as 15 mg FB_1/kg bw, even though clinical chemistry indices were less sensitive. The functional implications of morphological changes in bone marrow, adrenals and thymus are currently unknown.

The authors gratefully acknowledge the skilled technical support of Ms. Donna Beauchamp, Mr. Pierre Huard, Mr. Charles Séguin and Mr. Peter Smyth.

REFERENCES

Bondy, G.S.; Pestka, J.P. Dietary exposure to the trichothecene vomitoxin (deoxynivalenol) stimulates terminal differentiation of Peyer's patch B cells to IgA secreting plasma cells. *Toxicol. Appl. Pharmacol.* **1991**, *108*, 520-530.

Bondy, G.; Suzuki, C.; Barker, M.; Armstrong, C.; Fernie, S.; Hierlihy, L.; Rowsell, P.; Mueller, R. Toxicity of fumonisin B₁ administered intraperitoneally to male Sprague-Dawley rats. *Fd. Chem. Toxicol.* **1995**, 33, 653-665.

Dierickx, P.J. Urinary gamma-glutamyl transferase as a specific marker for mercury after heavy metal treatment of rats. *Toxicol. Lett.* **1980**, *6*, 235-238.

Heinegard, D.; Tiderstrom, G. Determination of serum creatinine by a direct colorimetric measurement. *Clin. Chem. Acta* **1973**, *43*, 305-310.

Leaback, D.H.; Walker, P.G. Studies on glucosaminidase 4. The fluorometric assay of N-acetyl-β-D-glucosaminidase. *Biochem. J.* **1961**, *78*, 151-156.

Miller, J.D.; Savard, M.E.; Rapior, S. Production and purification of fumonisins from a stirred jar fermenter. *Nat. Toxins* **1994**, *2*, 354-359.

Norred, W.P.; Plattner, R.D.; Chamberlain, W.J. Distribution and excretion of $[^{14}C]$ fumonisin B₁ in male Sprague-Dawley rats. *Nat. Toxins* **1993**, *1*, 341-346.

Rebar, A.H. General responses of the bone marrow to injury. *Toxic. Pathol.* **1993**, *21*, 118-129.

Riley, J.H.; Cornelius, L.M. Electrolytes, blood gases, and acid base balance. In *The Clinical Chemistry of Laboratory Animals*; Loeb, W.F., Quimby, F.W., Eds.; Pergamon Press: Toronto, 1989.

Romero, G.; Lasheras, B.; Sainz Suberviola, L.; Cenarruzabeitia, E. Protective effects of calcium channel blockers in carbon tetrachloride-induced liver toxicity. *Life Sci.* **1994**, *55*, 981-990.

Sacher, R.A.; McPherson, R.A. In *Widmann's Clinical Interpretation of Laboratory Tests*, Edition 10.; F.A. Davis Company: Philadelphia, 1991.

Shephard, G.S.; Thiel, P.G.; Sydenham, E.W. Initial studies on the toxicokinetics of fumonisin B₁ in rats. *Fd. Chem. Toxicol.* **1992a**, *30*, 277-279.

Shephard, G.S.; Thiel, P.G.; Sydenham, E.W.; Alberts, J.F.; Gelderblom, W.C.A. Fate of a single dose of the ^{14}C-labelled mycotoxin, fumonisin B₁, in rats. *Toxicon* **1992b**, *30*, 768-770.

Shephard, G.S.; Thiel, P.G.; Sydenham, E.W.; Alberts, J.F. Biliary excretion of the mycotoxin fumonisin B$_1$ in rats. *Fd. Chem. Toxicol.* **1994**, *32*, 489-491.

Smith, J.H.; Hook, J.B. Experimental nephrotoxicity *in vivo*. In *Nephrotoxicity Assessment and Pathogenesis*; Bach, P.H., Bonner, F.W., Bridges, J.W., Lock, E.A., Eds.; John Wiley and Sons: Chichester, England, 1984.

Suzuki, C.A.M.; Hierlihy, S.L.; Barker, M.; Curran, I.; Mueller, R.; Bondy, G.S. The effects of fumonisin B$_1$ on several markers of nephrotoxicity in rats. *Toxicol. Appl. Pharmacol.* **1995**, 133, 207-214.

Voss, K.A.; Chamberlain, W.J.; Bacon, C.W.; Norred, W.P. A preliminary investigation on renal and hepatic toxicity in rats fed purified fumonisin B$_1$. *Nat. Toxins* **1993**, *1*: 222-228.

23

BIOLOGICAL FATE OF FUMONISIN B₁ IN FOOD-PRODUCING ANIMALS

D.B. Prelusky[1], H.L. Trenholm[1], B.A. Rotter[1], J.D. Miller[2], M.E. Savard[2],
J.M. Yeung[3], and P.M. Scott[3]

[1] Centre for Food and Animal Research
   Agriculture and Agri-Food Canada
   Ottawa, Ontario, Canada K1A 0C6
[2] Plant Research Centre
   Agriculture and Agri-Food Canada
   Ottawa, Ontario, Canada K1A 0C6
[3] Food Research Division
   Health Canada
   Ottawa, Ontario, Canada K1A 0L2

ABSTRACT

The presence of mycotoxins in grains and feedstuffs causes not only animal health problems, but also a valid concern about the transmission of potentially toxic residues into animal-derived products intended for human consumption. In a series of studies at Agriculture and Agri-Food Canada, we investigated the biological fate of fumonisin B₁ (FB₁) in several food-producing animals (grower pigs, laying hens, dairy cattle), as well as monitored various parameters for evidence of toxicity in these species. In several experiments involving either single-dose protocols (iv, po) or longer-term feeding trials, the pharmacokinetic profiles of FB₁ (purity >95%) in these species were determined, including tissue accumulation and transmission of residues. Toxicological (and economical) implications such as performance (feed consumption, growth), productivity, and carcass quality were also measured when appropriate.

INTRODUCTION

During the past fifteen years, researchers in Canada have made considerable progress in trying to reduce the detrimental effects of mycotoxins on the agri-food industry, and in assuring a safe food supply. Most of the work has been carried out investigating deoxynivalenol (DON) and zearalenone (ZEN) which are recognized as a problem in animal feeds produced in eastern Canada. However, the more recent identification of fumonisin and its

Fumonisins in Food, Edited by L. Jackson *et al.*
Plenum Press, New York, 1996

subsequent detection in many parts of the world, including the USA (Sydenham *et al.*, 1990; Thiel *et al.*, 1992; Ross *et al.*, 1992), has raised concerns by agriculture and health agencies in Canada concerning potential domestic contamination and the implications in terms of both animal and human food safety. Levels of fumonisin in corn grown in Ontario are very low under normal weather conditions, however, high levels can exist in imported grains and feedstuffs (Miller *et al.*, 1995b).

At present, little is really known about the activity of FB_1 in livestock and poultry at sub-acute levels. Very limited quantities of the pure toxin have been available to study the broad range of issues that must be studied to determine the safety of FB_1. Much of the work reported to date involves acute studies using naturally contaminated grain or culture extracts as the source of toxin. This paper summarizes the results of research undertaken at the Centre for Food and Animal Research (CFAR) over the past several years using the purified toxin to determine certain aspects of the toxicology and biological fate of FB_1 when ingested by food producing animals of economic importance (swine, poultry, dairy cows). This work could only have been carried out as a collaborative, multi-disciplinary effort with other research institutes. Both the purified FB_1 (> 98% pure) and radiolabelled ^{14}C-FB_1 (> 95% pure) were produced at the Plant Research Centre by methods described by Blackwell *et al.* (1994) and Miller *et al.* (1995a), respectively.

SWINE

Traditionally, swine have tended to be one of the more sensitive animal species to the *Fusarium* toxins (Prelusky *et al.*, 1994a). Studies have been carried out to determine the pharmacokinetic fate of FB_1 in pigs, including the disposition of residues in tissues. The effects of dietary FB_1 on feed consumption and weight gains have also been investigated.

Experimental

Exp. 1: Pharmacokinetics of FB_1. Yorkshire barrow pigs, 10-14 weeks old (15-20 kg), raised under barrier maintained minimal disease conditions were used. Individual animals were housed in metabolism cages and acclimatized to their surroundings prior to testing. A corn-soybean meal-based pig diet, (Ottawa Grower Ration) and water, were provided *ad libitum*. In order to facilitate intravenous (iv) or intragastric (ig) dosing, as well as the sampling of blood, urine, and bile, cannulas were surgically placed according to the following protocol:

i) Group IV (intravenous dosing): jugular vein, urinary bladder.
ii) Group IV/B (iv dosing with bile collection): as above, plus gall bladder.
iii) Group IG (intragastric dosing): jugular vein, stomach, urinary bladder.
iv) Group IG/B (ig dosing with bile collection): as above, plus gall bladder.

Following surgery, animals were allowed to recover for 4-5 days prior to dosing.

Five barrows were allocated to each treatment: groups IG and IG/B were dosed at 0.35 µCi (0.50 mg) ^{14}C-FB_1/kg b.wt., and groups IV and IV/B at 0.25 µCi (0.40 mg) 14C-FB_1/kg b.wt. [1.0 µCi = 2.22 x 10^6 dpm (disintegrations per minute)]. Blood, urine, faeces, (and bile), were collected at selected time intervals over 72 hr post-dosing, and assayed for specific activity. For collection of tissue samples for radioactive counting, pigs were sacrificed at 72 hr post-dosing by an anaesthetic dose of pentobarbital followed by exsanguination. Necropsies were performed on all animals to confirm correct placement of cannulas and ligation of bile ducts.

Exp. 2: Short Term Feeding Trial and Residue Localization. Twelve Yorkshire barrows, 10-12 weeks old (14-18 kg) were obtained and housed in individual metabolism cages as described previously. Over a 33 day feeding period, animals received a diet containing 3.0 mg (3.6 µCi) ^{14}C-FB$_1$/kg feed during days 1-11, and 2.0 mg (2.4 µCi) ^{14}C-FB$_1$/kg feed during days 12-24. This was followed by a 9 day "withdrawal" period where the pigs received clean (FB$_1$-free) feed only. Feed consumption and weight gains for individual animals were monitored throughout the study. Blood was taken periodically during the trial, and total urine and faeces collected. On days 3, 6, 12, 24, 27 and 33, two animals were sacrificed, and tissues sampled and analyzed for ^{14}C-residual levels.

Exp. 3: Longer Term Feeding Trial for General Performance. Thirty-two Yorkshire pigs (16 females, 16 barrows), 8-10 weeks old (11-12 kg), were obtained and placed in group pens and fed a standard pig grower diet (Ottawa Grower Ration). After 3 days, pigs were moved to individual floor pens, where they acclimatized for an additional 4 days. Starting at 15 kg body weight, pigs (4 females, 4 males) were randomly assigned to one of four treatments containing 0.0, 0.1, 1.0 or 10.0 mg FB$_1$/kg diet. Animals were fed the experimental diet for an eight-week period, during which time feed consumption and weight gains were monitored.

Results and Discussion

Plasma Kinetics. Following iv administration of ^{14}C-FB$_1$ to bile-intact pigs (group IV), plasma radioactivity (RA) declined in a triexponential manner. Calculated kinetic parameters are shown in Table 1. The average half-life (t½) for the rapid initial α-phase was less than 3 min (range 0.9-3.6 min), and that of the slower distribution β-phase was 10.5 min (range 6.0-19.0 min). The average terminal elimination γ-phase t½ was 182.6 min (range 142.6-224.3 min). The transition between the β and γ phases occurred about 30 min post-dosing, at which time the plasma concentration of FB$_1$ was less than 2% of the peak blood concentration (Cp°). FB$_1$-derived activity could not be detected 300 min after dosing (analytical detection limit equivalent to 18 ng FB$_1$, and/or metabolites per ml plasma). The estimated area under curve (AUC) values were similar between animals ranging from 53,990 to 70,657 dpm min/ml. The weight-normalized volume of distribution by area (Vdγ) and total plasma clearance (Clp) were 2.4 ± 0.55 l/kg and 9.1 ± 1.07 ml/min/kg, respectively.

In iv dosed animals where bile flow was interrupted (group IV/B), radioactivity was cleared much more rapidly from the blood than non-cannulated pigs (group IV). A 2-compartment pharmacokinetic model satisfactorily described the plasma kinetics. Compared to intact animals, bile-cannulated pigs had a much faster terminal elimination half-life value (17.0 ± 8.0 min) and increased clearance rate (16.9 ± 2.7 ml/min/kg), and significantly reduced AUC (33584 ± 5798 dpm/min/ml) and volume of distribution (Vdβ = 0.393 ± .126 l/kg) values.

After intragastric (oral) administration of ^{14}C-FB$_1$ to intact pigs (group IG), RA in plasma could be detected in 4 of 5 animals dosed (Table 2). Radioactivity was first detected 30-45 min post-dosing and reached peak concentrations in individual animals between 60 and 90 min (T*max*); the mean peak plasma conc (C*max*) for the 4 pigs was 46 ± 12 dpm/ml at approximately 70 min. After 180 min, no RA could be measured in plasma (detection limit equivalent to 16 ng FB$_1$ and/or metabolites per ml plasma). The estimated mean bioavailability for FB$_1$ in swine was a marginal 4.07 ± 1.2% (N=4).

Table 1. Pharmacokinetic parameters of ^{14}C-labelled Fumonisin B$_1$ (FB$_1$) following a single intravenous dose to pigs (0.25 µCi, 0.4 mg FB$_1$/kg b.wt.)

Parameter	IV Dosed[1] (Bile Intact)	IV Dosed[1] (Bile Interrupted)
Weight (kg)	20.0 ± 2.0	19.6 ± 1.3
A (dpm ml^{-1})[2]	5227 ± 2097	3261 ± 1464
B (dpm ml^{-1})	1584 ± 89	854 ± 333
C (dpm ml^{-1})	109 ± 35	—
t½ α (min)	2.2 ± 1.1	3.6 ± 1.4
t½ β (min)	10.5 ± 5.7	17.0 ± 8.0
t½ γ (min)	182 ± 38	—
AUC (dpm min ml^{-1})	61613 ± 7510	33604 ± 5886
Vdγ (1 kg^{-1})	2.4 ± 0.55	0.393 ± 0.126
Clp (ml min^{-1} kg^{-1})	9.1 ± 1.1	16.9 ± 2.7
% Recovery (72 hr)		
Bile	—	70.8 ± 3.1
Urine	22.2 ± 4.3	16.2 ± 2.9
Faeces	58.3 ± 3.0	1.5 ± 1.6
Total	79.5 ± 2.8	88.5 ± 3.5

[1] Average ± S.D. N = 5.
[2] 100 dpm equivalent to 72 ng FB$_1$ and/or metabolites.

Table 2. Pharmacokinetic parameters of ^{14}C-labelled Fumonisin B$_1$ (FB$_1$) following a single intragastric dose to pigs (0.35 µCi, 0.5 mg FB$_1$/kg

Parameter	IG Dosed[1] (Bile Intact)[2]	IG Dosed[1] (Bile Interrupted)[3]
Weight (kg)	16.6 ± 0.9	14.4 ± 1.1
Cmax (dpm ml^{-1})[4]	37.0 ± 23.0	25.2 ± 23.7
Tmax (min)	70 ± 14	70 ± 17
t½ β (min)	95.9 ± 19.3	56.0 ± 11.1
AUC (dpm min ml^{-1})	3007 ± 1853	1555 ± 1459
F (%)	3.3 ± 2.1	1.9 ± 1.8
% Recovery (72 hr)[5]		
Bile	—	
Urine	0.60 ± 0.58	0.80 ± 0.30
Faeces	90.8 ± 3.2	89.8 ± 3.1
Total	91.4 ± 3.6	91.7 ± 3.0

[1] Average ± S.D.
[2] N = 4. No radioactivity detected in 1/5 pigs.
[3] N = 3. No radioactivity detected in 2/5 pigs.
[4] 100 dpm equivalent to 64 ng FB$_1$ and/or metabolites.

The plasma profile following ig dosing to bile interrupted pigs (group IG/B) was similar to intact pigs, except values related to absorption, C_{max}, AUC, and F, were on average, marginally less; 42 ± 8 dpm/ml, 2592 ± 478 dpm min/ml, and $3.22 \pm 0.77\%$ (N=3), respectively. Plasma RA could be detected in only 3 of the 5 IG/B pigs.

Elimination Kinetics. After iv administration total recovery of radioactivity ranged from 76.3 to 83.4% (79.5 ± 2.8%); 22.2% in urine and 58.3% in faeces (Table 1). These proportions were similar to that reported with rats, where ^{14}C-FB₁ administered ip resulted in 32% of the RA in the urine, and 66% in the faeces (Shephard *et al.*, 1992). Interestingly, with swine the RA which appeared in the urine occurred within the initial 3 hr post-dosing, whereas recovery in faeces was delayed 24 hr, requiring an additional period of several days to clear. On average 20% (range 17-29%) of the administration dose was still not recovered even after 72 hr.

Following ig administration, a greater proportion of RA was eliminated in the faeces; only trace levels (0.60 ± 0.58%) were found in urine (Table 2). Overall, total recovery was somewhat higher (91.4 ± 3.6%) than that observed following iv dosing, although the faecal excretion pattern was very similar; at least 24 hr was necessary before detectable levels were initially found. Shephard *et al.* (1992) found no RA in the urine of rats dosed orally with ^{14}C-FB₁.

With gall bladder-cannulated iv-dosed pigs (IV/B), recovery of RA was predominantly in the bile (70.8 ± 3.11%), with a lesser amount in the urine (16.2 ± 2.87%) (Table 1). Total recovery over a period of 72 hours was higher (88.4 ± 3.54%) than that with bile-intact pigs (group IV), and marginally less than following oral dosing. Elimination through both bile and urine was relatively rapid with most of the RA appearing in the initial 4 hr post-dosing, but while urinary excretion was virtually complete by 8 hr, excretion in the bile continued slowly for 24-36 hr.

With bile exteriorized ig-dosed pigs (IG/B) results were very similar to the IG group. Approximately 1.2% (0.7-1.7%) of the dose was eliminated in the bile, and 0.80% (0.4-1.2%) in the urine, demonstrating that a small fraction of the dose was absorbed from the gastrointestinal tract. Most of the RA was found in the faeces (89.8 ± 3.1%) with total recovery (91.7 ± 3.0%), the same as with ig-dosed intact animals.

Tissue Residues. Following a single iv dose of ^{14}C-FB₁ (Exp 1), sufficient specific activity remained in all tissues at sacrifice (72 hr post-dosing) to be counted, indicating residues were distributed extensively and were persistent in most tissues (Table 3). Liver and kidney contained the highest levels, following by large intestine, brain, lung and adrenal tissue.

The total body burden, based on specific activity of tissues and estimated organ weights, indicated that 72 hr post - iv dosing, approximately 20% of the RA still remained in the carcass of intact pigs, which would account for almost all of the unrecovered dose (Table 1). Significantly, the liver still accounted for up to 10% of the iv dose at this time which supports reports that liver is a primary target for fumonisins (Casteel *et al.*, 1993; Riley *et al.*, 1993). Muscle, because of its large proportion in the body, also contained a comparably large fraction of the dose (6.5 ± 2.1%). Kidney, although it accounted for less than 1% of the dose due to its comparably small size, still maintained a relatively high specific activity by study's end. This adds evidence to recent studies which indicate kidney is also a sensitive target to these toxins (Riley *et al.*, 1994). Following ig dosing, tissue RA concentrations were still measurable 72 hr post-dosing, but levels were typically 10-20 fold less than that detected following iv dosing. Consequently only a very small fraction of the dose (<1.5%) remained within the tissues after 72 hours.

Table 3. Specific activity of tissues[1] and total body burden 72 hr following administration of a single dose of ^{14}C-labelled Fumonisin B$_1$ to pigs

Tissue	(%)[2]	Group iv	Group ig
heart	(0.50)	131 ± 36	13 ± 5
liver	(3.0)	1682 ± 337 (10.1%)[3]	139 ± 36 (0.49%)
kidney	(0.47)	760 ± 153 (0.72%)	49 ± 13 (0.03%)
spleen	(0.12)	113 ± 27	10 ± 3
l. intestine	(1.4)	365 ± 147	63 ± 20 (0.08%)
s. intestine	(2.4)	40 ± 18	ND[4]
muscle	(45.8)	78 ± 26 (6.5%)	ND
brain	(0.54)	265 ± 35	29 ± 4
adrenal	(0.09)	191 ± 66	19 ± 7
stomach	(0.75)	56 ± 25	21 ± 9
lung	(1.15)	187 ± 48	ND
fat	(20.8)	42 ± 20	9 ± 9 (0.18%)
skin	(4.9)	18 ± 8	ND
bone	(11.8)	ND	ND
Total remaining body burden (% dose)[5]		19.8 ± 4.6	1.30 ± 0.62

[1]dpm per g tissue volume, average ± SD (N = 5). 100 dpm equivalent to 72 or 64 ng FB$_1$ and/or metabolites following iv or ig dosing, respectively.
[2]Percent body weight.
[3]Percent dose remaining in tissue, in parenthesis.
[4]ND = not detected.
[5]Based on estimated weights of organs measured

Following a more prolonged period (24 d) of ingestion of the toxin (Exp 2), results indicate ^{14}C-FB$_1$ derived residues do accumulate in both liver and kidney (Table 4). Data show that at 3 ppm FB$_1$, radioactivity was measurable in both tissues after 3 days exposure, and continued to increase as long as the pigs consumed the contaminated diet. At 24 days, residues were equivalent to about 160 and 65 ng/g tissue for liver and kidney, respectively. Removal of the contaminated diet resulted in a rapid drop in tissue residue concentration.

Table 4. Radioactivity (^{14}C) measured in organs and tissues following the feeding of ^{14}C-labelled Fumonisin B$_1$ (FB$_1$)

	Average Specific Activity (dpm/g tissue wet weight)[2]					
	Day					
Tissue[3]	3	6	12	24	27	33
Liver	141	252	268	348	119	36
Kidney	66	80	111	147	54	30
Bile	151	186	104	167	37	29

Radioactivity not detected in plasma, spleen, muscle, brain, adrenals, fat, or skin. Trace (25-40 dpm/g), sporadic radioactivity detected in heart, lung and bone.

As expected varying levels of radioactivity were also found in stomach, and small and large intestine.

[1]FB$_1$-spiked diet: 3 mg (3.0 μCi)[14]C-FB$_1$/kg feed during days 1-12, 2 mg (2.0 μCi)[14]C-FB$_1$/kg feed during days 13-24, and clean, non-spiked feed during days 25-33.
[2]Detection limit 25 dpm/g tissue. 100 dpm equivalent to ≈ 46 ng FB$_1$ and/or metabolites.
[3]Tissues from 2 pigs/day analyzed.

After 3 days on clean diet, levels dropped to approximately 35% peak levels (55 ng/g and 25 ng/g, liver and kidney, respectively). After 9 days on clean feed, levels were only marginally above detection limits ($\approx$ 12 ng/g) in both tissues. This confirms other recent studies which have reported the disposition for FB_1 and/or metabolites to target these particular two organs (Shephard *et al.*, 1992; Vudathala *et al.*, 1994). To date the chemical nature of the residue has not been determined.

Feed Consumption and Weight Gains. Interestingly, the daily consumption of a diet containing 2-3 ppm FB_1 (Exp 2) was found to increase feed consumption (FC), and consequently weight gains (WG), in barrows housed in the individual metabolism cages (Table 5). Differences in feed intake were most dramatic during the initial stages of the trial, averaging 160% and 140% control values during the periods 1-3 days and 4-6 days, respectively. Differences were reduced thereafter, although over 24 days animals on the FB_1-diet had still consumed 25% more feed than control pigs (1.17 vs 0.93 kg/day, respectively). Once returned to the clean, control diet (days 25-33) feed intake by treated (F) animals was no longer significantly different from control (C). There was an accompanying increase in weight gains in the FB_1-treated animals as well, particularly during the first 3 days of the study (Table 5). This trend in higher WG continued for most of the duration of the study.

There was no real change in feed efficiency over the 24 day period the pigs received the FB_1-diet (data not shown). Although there was an apparent relative improvement in weight gain/feed ratio with the FB_1-treated group following being switched to the clean feed at day 24, this appears to result from a decrease in control pig weight gains during the 25-33 day period. The reason for this decline in control animals gains was not evident.

Pigs in Exp 3 which were housed in standard floor pens also displayed deviations in their feeding behaviour when fed a FB_1-contaminated diet. With females, overall there were no significant effects on total FC, WG, or feed efficiencies over the eight-week period (Table 6). However, there appeared to be a positive effect of the toxin on FC and WG during the initial 5 weeks pigs consumed the mycotoxin. During this period, WG were up a similar level (12-15%), regardless of the concentration of FB_1. This was similar to the positive effect observed with the pigs confined to the individual metabolism cages in the previous study. It was only after the sixth week that a dramatic drop (8-16%) in performance relative to controls occurred; sufficient enough to essentially cancel out the gains made during the previous five weeks. The reason for the rapid manifestation of toxicity is not readily apparent at this time.

Table 5. Effect of Fumonisin FB_1 - contaminated diet on performance of grower pigs

Period (days)	Daily Feed Consumption (kg/day)		Daily Weight Gain (kg/day)	
	C[1]	F1	C[1]	F1
1 - 3	0.48 ± .052[2]	0.77 ± .059**[3]	0.30 ± .041	0.57 ± .065*
4 - 6	0.70 ± .035	0.99 ± .081*	0.42 ± .028	0.51 ± .046
7 - 12	0.98 ± .058	1.06 ± .093	0.56 ± .037	0.60 ± .070
13 - 24	1.08 ± .064	1.39 ± .049*	0.60 ± .022	0.68 ± .022*
25 - 33	1.45 ± .075	1.63 ± .044	0.53 ± .034	0.70 ± .053*
1 - 24	0.93 ± .051	1.17 ± .071*	0.52 ± .042	0.64 ± .015*
1 - 33	1.08 ± .056	1.27 ± .031	0.51 ± .026	0.67 ± .035*

[1]Diet : C = control (0 ppm FB_1); F = spiked diet (3 ppm FB_1 days 1-12, 2 ppm FB_1 days 13-24, 0 ppm FB_1 days 25 - 33).
[2]Average ± S.E.M.
[3]Significantly different from control animals (**P < 0.01, * P < 0.05).

Table 6. Effect of dietary Fumonisin B$_1$ (FB$_1$) on feed consumption (and weight gains) in swine, relative to controls, over an eight-week feeding trial

Females	Dietary FB$_1$ (ppm)			
Week	0	0.1	1.0	10.0
1-2	—	+10 (+9)[1]	+7 (+7)	+5 (+11)
3-5	—	+13 (+13)	+9 (+15)	+4 (+17)
Total 1-5	—	+12 (+12)	+9 (+12)	+4 (+15)
6-8	—	+3 (−9)	+2 (-8)	-2 (-16)
Total 1-8	—	+7 (+4)	+5 (+4)	+1 (+3)

Males (Barrows)	Dietary FB$_1$ (ppm)			
Week	0	0.1	1.0	10.0
1-2	—	+6 (0)[1]	-5 (-10)	-8 (-14)
3-5	—	0 (0)	-6 (0)	-9 (-8)
Total 1-5	—	+2 (0)	-6 (-4)	-9 (-10)
6-8	—	-6 (-9)	-10 (-15)	-13 (-11)
Total 1-8	—	-2 (-4)	-8 (-8)	-10 (-11)*

[1]Percent change relative to controls.
*Significantly different (P < 0.05).

At 190 ppm dietary FB$_1$ fed to gilts, Casteel *et al.* (1993) noted after a clinically normal initial week, that feed intake levels then dropped 40-60%, and at 200 ppm FB$_1$ fed to pigs (sex not indicated) Colvin *et al.* (1993) reported it required 4 days before a decrease in FC was observed.

With barrows, performance problems were evident early in the study (Table 6). At 1.0 and 10.0 ppm dietary FB$_1$, reduced rates in FC and WG were manifested during the first week on trial, and continued for the duration of the study. Overall, FC and WG were 8-10% and 8-11% lower, respectively, than controls. At 0.1 ppm essentially no overall effect was apparent during the initial 5 weeks, but during wks 6-8, there was a marginal drop in both FC and WG, which was similar to that observed with females during the same period of the feeding trial. Notably, eating and weight gain patterns for individual males during the initial 5 week period tended to be much more erratic than either female pigs or control animals (individual data not shown). Measurements which were down one week, were dramatically up the next, and then dropped again the following week, which continued for the 5 weeks. Such an irregular growth rate could significantly affect the carcass quality (Wood, 1989).

POULTRY (LAYING HENS)

Experimental

Exp. 4: Pharmacokinetics. White Leghorn laying hens 30 weeks old (1.30-1.68 kg) were housed in individual cages. During this acclimatization period birds were provided with water and corn-based laying ration (Ottawa Hatching Ration) *ad libitum* for a week prior to administration of fumonisin. Twelve birds were selected and weighed and the wing vein was cannulated. Six birds were given a single oral dose of 2.0 mg (1.28 µCi)/kg b.wt of ^{14}C-fumonisin dissolved in 0.9% saline (1 mg FB$_1$/ml) by intubation into the crop. Six different birds received an iv dose injected into the wing vein of 2.0 mg (0.64 µCi)/kg b.wt ^{14}C-FB$_1$ dissolved in normal (0.9%) heparinized saline (1 mg FB$_1$/0.1 ml). Following dosing, blood and excreta were collected at specific times intervals over 24 hours. At the end of 24 hours, birds were sacrificed by cervical dislocation and tissues sampled for analysis. If available, eggs were collected and separated into yolk, albumin and shell.

Exp 5: Feeding Study for Pathology/General Performance. Twenty-four mature laying hens 30 weeks old (1.45 - 1.75 kg) were housed as above. Feed intake was calculated during a 1 week acclimatization period in order to determine the amount of ration which would assure complete consumption during the trial. This level was established at 110 g feed/day. Birds were randomly allocated to two treatments (12 per group), receiving ration containing either 4.0 mg FB$_1$/kg feed (test) or a clean, non-contaminated diet (control, 0.0 mg FB$_1$/kg), for a 28 day period . Body weights, feed consumption, and egg production were monitored weekly. On days 7, 14, 21 and 28, 3 test and 3 control hens were euthanized by cervical dislocation and examined. Portions of brain, heart, lung, liver, kidney, bursa of Fabricius, esophagus, ovary, caecum, proventriculus, small intestine and breast muscle were collected and prepared for pathological examination.

Results and Discussion

Pharmacokinetics. After iv dosing to laying hens, plasma radioactivity (RA) underwent a very rapid bi-exponential decline. Calculated pharmacokinetic parameters are given in Table 7. The model indicates a very rapid distribution phase (α) with a half-live value (t$^1\!/_2$ α) of 2.6 ± 0.36 min followed by a slower terminal phase (t$^1\!/_2$ β) 45.6 ± 8.4 min. Only trace levels of radioactivity could be detected in plasma 180 min post-dosing. Of interest was the very small distribution volume (Vdβ = 0.063 - 0.105 l/kg) calculated for ^{14}C-FB$_1$ in poultry. Assuming a total blood volume in these hens of approximately 6-9% of body weight, distribution roughly equalled plasma volume; consequently it appeared RA was confined almost exclusively to this space. While this could be expected to result from binding of ^{14}C-FB$_1$ to certain plasma components, subsequent *in vitro* binding studies have indicated neither uptake by red blood cells nor strong binding to plasma proteins of over 10,000 molecular weight (data not shown). It is possible that any FB$_1$ interaction with components found in blood possesses only a very weak attraction and any such complex formed is rapidly dissociated, although this has not been reproduced *in vitro*.

Table 7. Pharmacokinetic parameters in laying hens after a single intravenous dose of ^{14}C-labelled Fumonisin B$_1$ (FB$_1$) (2.0 mg, 0.64 µCi ^{14}C-FB$_1$/kg b.wt.)[1]

Body wt. (kg)	1.53 ± 0.10
A (dpm ml^{-1})[2]	260038 ± 32414
B (dpm ml^{-1})	3358 ± 1006
t$^1\!/_2$ α (min)	2.6 ± 0.36
t$^1\!/_2$ β (min)	45.6 ± 8.40
AUC (dpm min ml^{-1})	1184999 ± 147714
Vd p (ml kg^{-1})	5.4 ± 0.78
Vd β (ml kg^{-1})	81 ± 14
Cl (ml min^{-1} kg^{-1})	1.20 ± 0.15
% Recovery (24 hr)	98.7 ± 4.1

[1] Based on average ± S.D. of 6 birds.
[2] 100 dpm ≈ 142 ng FB$_1$ and/or metabolites.

Table 8. Pharmacokinetic parameters in laying hens after a single oral dose of ^{14}C-labelled Fumonisin B$_1$ (FB$_1$) (2.0 mg, 1.28 µCi ^{14}C-FB$_1$/kg b.wt.)[1]

Body wt (kg)	1.53 ± 0.13
C *max* (dpm ml^{-1})[2]	85.6 ± 42.7
T *max* (min)	130 ± 16.4
lag-time (min)	39.8 ± 16.0
t$\frac{1}{2}$ β (min)	101.9 ± 43.5
AUC (dpm min ml^{-1})	17277 ± 12156
F (%)	0.70 ± 0.47
% Recovery (24 hr)	96.6 ± 3.3

[1] Based on average ± S.D. of 5 birds.
[2] 100 dpm ≈ 71 ng FB$_1$/metabolites.

Systemic absorption of FB$_1$ in poultry appeared to be very limited (Table 8). After a lag time of 14-58 min, estimated levels of absorption following oral dosing accounted for less than 2% (0.70 ± 0.47%) of the administered dose. Peak plasma RA levels ranged from 28 to 103 ng FB$_1$-equivalents (FB$_1$ and/or metabolites); the last detectable levels were measured at either the 240 or 360 min sampling time, post-dosing. The average terminal slope following oral administration, as compared to iv dosing, was dramatically increased (t$\frac{1}{2}$ β = 102 min vs 46 min, respectively) indicating that the terminal t$\frac{1}{2}$ was probably not entirely related to the excretion of the toxin, but also to its delayed absorption.

Elimination of radioactivity was nearly quantitative after 24 hour post-dosing (92-104%) following both oral and iv dosing (Tables 7 & 8), and excretory patterns varied only very slightly between the two routes of administration. Following both iv and oral dosing, the highest fraction of RA excreted (45-85%) occurred during the 2-6 hour period post-dosing, which is consistent with the rapid passage rate of ingesta through the digestive tract of the laying hen.

Tissue Residues. Analysis of tissue samples collected 24 hr post iv-dosing revealed an absence of residues in most tissue samples (Table 9; values based on average ± SD of 6 birds). Liver and kidney had trace amounts of radioactivity. No radioactivity was detected in the other tissues sampled. The caecum had the maximum amount of radioactivity present, which was less than 1% of the administered dose. In orally dosed birds, 24 hr post-dosing, trace levels of radioactivity were detectable in crop, liver, small intestine, caecum, and kidney (Table 9). No detectable level of radioactivity was present in the remaining tissues. Essentially no ^{14}C-FB$_1$-derived radioactivity was found in the eggs of dosed birds. Very slight residue levels detected in several of the shells were probably due to surface contamination with excreta.

General Performance/Pathology. No mortality or evidence of clinical illness was apparent during the 28 day feeding period in birds consuming the FB$_1$-spiked diet. Although the study design made statistical comparison with control birds difficult, there was no strong indication of reduced feed intake below the 90% restricted diet, of weight loss, or of reduced egg production (data not shown). At necropsy, all birds examined were in good body condition with adequate fat stores. No gross lesions were observed in either the control or test birds. Histologically, there were no lesions that could be related to the FB$_1$ in any of the birds sacrificed at 7-day intervals, to the end of the 28-day feeding period.

Table 9. Level of radioactivity in various tissues collected 24 hr post-dosing from laying hens given a single IV or single oral po dose of ^{14}C-Fumonisin B₁ (FB₁)[1]

	iv	po
Dose (mg/kg)	2.0	2.0
(μCi/kg)	0.64	1.28
(dpm x 10^6)	2.15±0.14	4.14±0.42
Crop	ND	956±1233 (≈680 ng/g)
Liver	374±277 (≈530 ng/g)[2]	299±146 (≈210 ng/g)
Small Intestine	ND	362±422 (≈255 ng/g)
Cecum	20145±6875 (≈28 μg/g)	85420±119095 (≈60 μg/g)
Kidney	457±119 (≈650 ng/g)	117±119 (≈83 ng/g)

(Not detected in the esophagus, proventriculus, gizzard, lung, spleen, heart, brain, fat, muscle, ovary or oviduct tissues, following either iv or po dosing.)

Yolk	ND[3]	ND
Albumin	ND	ND
Shell	ND[4]	ND[4]

[1] Detection limit ≈ 10-15 ng FB₁/metabolites per g tissue.
[2] FB₁/metabolite equivalents in parenthesis.
[3] Trace Level (< 10 dpm over blank) in one sample.
[4] 25-35 dpm found on one shell sample.

DAIRY COWS

Experimental

Exp. 6: Pharmacokinetics and Transmission of Residues. Four Holstein cross-bred cows (first-lactation heifers), weighing 445-630 kg (< 3 years old) were used. Animals were penned in individual stalls, provided with appropriate amounts of standard dairy ration, and water *ad libitum*. Two cows were dosed orally (1.0 or 5.0 mg FB₁/kg b.wt.) and two dosed iv (0.05 or 0.20 mg FB₁/kg b.wt.).

Following dosing, serial blood samples were taken during the initial 8 hours, and daily thereafter. Milk for FB₁/metabolite analysis was sampled for several days post-dosing. Analysis for FB₁-derived residues in milk was carried out initially by an HPLC detection method, and subsequently by a recently developed ELISA technique.

Results and Discussion

No overt adverse effects were noted in cows receiving FB₁ as either an oral or iv dose.

Pharmacokinetics. After iv dosing, plasma FB₁ underwent a very rapid bi-exponential decrease. Kinetic parameters are given in Table 10. The similarity in values between the two iv-dosed animals indicate the kinetics are not dose-dependent at this level. The profile shows an extremely rapid distribution phase ($t\frac{1}{2}\alpha$ = 1.7 min) followed by a relatively rapid elimination phase ($t\frac{1}{2}\beta$ = 15.1 and 18.7 min, low and high dose, respectively). No FB₁ could be detected in plasma beyond 120 min post-dosing. The volume of distribution (Vdβ) indicates that the toxin is confined to the extracellular space, and is not taken up extensively into tissues.

Table 10. Pharmacokinetic parameters in dairy cows after the single
intravenous administration of Fumonisin B_1 (FB_1)

	1 (0.05 mg/kg)	2 (0.20 mg/kg)
Body wt (kg)	630	452
Dose (mg)	31.5	89
A (ng ml^{-1})	800	4015
B (ng ml^{-1})	120	330
t$^{1}\!/_{2}$ α (min)	1.7	1.7
t$^{1}\!/_{2}$ β (min)	15.1	18.7
AUC (ng min ml^{-1})	4512	18711
Vd p (l kg^{-1})	.054	.046
Vd β (l kg^{-1})	.241	.288
Cl (ml min^{-1} kg^{-1})	11.1	10.7

[1] One cow per dose level.

Following oral administration of the toxin, no FB_1 or metabolites (AP_1, conjugates) could be found in the plasma (detection limit 4 and 8 ng/ml respectively). While this may indicate a negligible bioavailability of FB_1 in ruminants, it is very possible that its extremely rapid plasma clearance (demonstrated following iv dosing), would make it difficult to detect any oral absorption under 2% of the doses used. An alternative hypothesis is that a high liver first-pass effect, as indicated for FB_1 in other species (Vudathala et al., 1994; Prelusky et al., 1994b), would reduce considerably the amount of toxin which reached systemic circulation and was available for analysis.

Ruminants have traditionally been relatively tolerant to the *Fusarium* toxins, due in a large part to their large capacity to detoxify the toxins via the rumen microflora. However, a recent study carried out in our lab, in which FB_1 was incubated for 24 hr in the presence of protozoa-free or protozoa-containing rumen fluid, demonstrated that no metabolism occurred (data not shown). In comparison for example, the same *in vitro* system (protozoa-intact only) metabolized deoxynivalenol 100% within 6 hr (Prelusky, unpublished data). Undeveloped rumen microflora in calves has been implicated for their lack of tolerance to many mycotoxins (Prelusky et al., 1994a). Interestingly, though, Osweiler et al. (1993) reported that it required 148 ppm dietary FB_1 fed to feeder calves for 31 days to produce some toxic effects. This further suggests that the tolerance of ruminants to fumonisins is not dependent on its detoxification in the rumen.

Residues. HPLC analysis of milk collected for up to 3 days post-dosing detected no FB_1 (detection limit 3-7 ng/ml) or metabolites (AP_1, detection limit 20-25 ng/ml) during this period. A 5.0 mg FB_1/kg b.wt. dose would be approximately equivalent to a single days consumption of 125 ppm contaminated feed, whereas, based on a 1% oral bioavailability, an iv dose of 0.2 mg FB_1/kg b.wt. would be roughly equivalent to about 500 ppm dietary FB_1 for one day. Milk samples obtained for the 7 days post-dosing have subsequently been re-analyzed using an ELISA procedure using rabbit polyclonal antibodies raised against a fumonisin B_1-conjugate (J.M. Yeung, personal communications). The detection limit for FB_1 in milk was 0.5 ng/ml, with the antibodies 11% cross-reactive to AP_1 (4.5 ng/ml). Even with the approximately 10- and 5-fold improvements in detection capability for FB_1 and AP_1, respectively, no milk sample tested positive for the carry-over of FB_1-derived residues.

CONCLUSION

In **swine** it would seem that although low levels of dietary fumonisin are not acutely toxic, there exists both an economic issue, as well as a food safety concern. Low levels of dietary FB_1 appear to adversely affect feed intake patterns in males (barrows) in a manner not clearly understood, but suggesting that prolonged dietary exposure could alter growth rates and possibly carcass quality. Overall effects on females were less, but problems appear to occur after several weeks on a diet containing 1.0 - 10.0 ppm toxin. Furthermore, although FB_1 has a poor bioavailability in swine, the little which is absorbed undergoes extensive distribution and can remain in the body for an extended period of time. There appears to be a particular affinity for liver and kidney, which show rapid accumulation of residues even at low dietary levels (2-3 ppm); lower residual levels are also likely to be found in the other tissues as well. Accumulation appears to increase with exposure time, suggesting market weight pigs could contain significant residual levels even if exposed to only low dietary levels. Results further indicate that a withdrawal period of several weeks would be required to assure only minimal residual levels remain. It is suggested that the long biological half-life of FB_1 in swine is due primarily to enterohepatic circulation (EHC), and to a lesser extent the uptake and slow release of the toxin by certain tissues, particularly liver and kidney. This is supported by the disappearance of the long terminal elimination (γ) phase in bile-exteriorized animals, in which bile flow was removed and EHC interrupted.

In **laying hens**, at the level of 2 mg/kg bw, FB_1 is poorly absorbed (following oral administration) and is quickly eliminated from both orally and intravenously dosed birds. Twenty-four hours post-dosing, the level of ^{14}C-FB_1-derived residues in different organs and tissues was near negligible and most of the radioactivity was recovered in the excreta. Even feeding 4 mg/kg/feed daily over a period of 28 days did not cause any apparent adverse effects in the birds. This suggests that laying hens are not sensitive to low levels of fumonisin.

In **dairy cows**, FB_1 has an apparent negligible oral absorption, although this has not been confirmed since circumstances limited detection to 1-2% of the dose administered. Using an ELISA method, detection of FB_1-derived residues in milk was negligible for all dosing protocols down to 0.5 ng/ml for FB_1 and 4.5 ng/ml for AP_1.

REFERENCES

Casteel, S.W.; Turk, J.R.; Cowart, R.P.; Rottinghaus, G.E. Chronic toxicity of fumonisin in weanling pigs. *J. Vet. Diagn. Invest.* **1993**, 5,413-417.

Colvin, B.M.; Cooley, A.J.; Beaver, R.W. Fumonisin toxicosis in swine: clinical and pathological findings. *J. Vet. Diagn. Invest.* **1993**, 5,232-241.

Blackwell, B.A.; Miller, J.D.; Savard, M.E. Production of carbon-14 labelled fumonisin in liquid culture. *Journal AOAC International* **1994**, 77,506-511.

Miller, J.D.; Savard, M.E.; Rapior, S. Production and purification of fumonisins from a stirred jar fermenter. *Nat. Toxins* **1995a**, 2,354-359.

Miller, J.D.; Savard, M.E.; Schaafsma, A.W.; Seifert, K.A.; Reid, L.M. Mycotoxin production by *Fusarium moniliforme* and *Fusarium Proliferatum* from Ontario and occurrence of fumonisin in the 1993 corn crop. *Can. J. Plant Pathol.* **1995b** (in press).

Osweiler, G.D.; Kehrli, M.E.; Stabel, J.R.; Thurston, J.R.; Ross, P.F.; Wilson, T.M. Effects of fumonisin-contaminated corn screenings on growth and health of feeder calves. *J. Anim. Sci.* **1993**, 71,459-466.

Prelusky, D.B.; Rotter, B.A.; Rotter, R.G. Toxicology of Mycotoxins. In *Mycotoxins in grain: compounds other than aflatoxin.* Miller, J.D.; Trenholm, H.L. Eds; Eagan Press: St-Paul, MN, **1994a**; pp 359-403.

Prelusky, D.B.; Trenholm, H.L.; Savard, M.E. Pharmacokinetic fate of ^{14}C-labelled fumonisin B[1] in swine. *Nat. Toxins* **1994b**, 2,73-80.

Riley, R.T.; An, N-H.; Showker, J.L.; Yoo, H.-S.; Norred, W.P.; Chamberlain, W.J.; Wang, E.; Merrill, A.H.; Motelin, G.; Beasley, V.R.; Haschek, W.M. Alteration of tissue and serum sphinganine to sphingosine ratio: an early biomarker of exposure to fumonisin-containing feeds in pigs. *Toxicol. Appl. Pharmacol.* **1993**, 118,105-112.

Riley, R.T.; Hinton, D.M.; Chamberlain, W.J.; Bacon, C.W.; Wang, E.; Merrill, A.H.; Voss, K.A. Dietary fumonisin B[1] induces disruption of sphingolipid metabolism in Sprague-Dawley rats: a new mechanism of nephrotoxicity. *J. Nutr.* **1994**, 124,594-603.

Ross, P.F.; Rice, L.G.; Osweiler, G.D.; Nelson, P.E.; Richard, J.L.; Wilson, T.M. A review and update of animal toxicoses associated with fumonisin-contaminated feeds and production of fumonisins by *Fusarium* isolates. *Mycopathologia* **1992**, 117,109-114.

Shephard, G.S.; Thiel, P.G.; Sydenham, E.W.; Alberts, J.F.; Gelderblom, W.C.A. Fate of a single dose of the ^{14}C-labelled mycotoxin, fumonisin B[1], in rats. *Toxicon* **1992**, 30,768-770.

Sydenham, E.W.; Gelderblom, W.C.A.; Thiel, P.G.; Marasas, W.F.O. Evidence for the natural occurrence of fumonisin B[1], a mycotoxin produced by *Fusarium moniliforme* in corn. *J. Agric. Food Chem.* **1990**, 38,285-290.

Thiel, P.G.; Marasas, W.F.O.; Sydenham, E.W.; Shephard, G.S.; Gelderblom, W.C.A. The implications of naturally occurring levels of fumonisins in corn for human and animal health. *Mycopathologia* **1992**, 117,3-9.

Vudathala, D.K.; Prelusky, D.B.; Ayroud, M.; Trenholm, H.L.; Miller, J.D. Pharmacokinetic fate and pathological effects of ^{14}C-fumonisin B[1] in laying hens. *Nat. Toxins* **1994**, 2,81-88.

Wood, J.D. Meat quality, carcass composition and intake. In *The voluntary food intake of pigs*. Forbes, J.M.; Varley, M.A.; Lawrence, T.L.J., Eds; BSAP Occasional Publ. No. 13: Edinburgh, U.K., **1989**; pp. 79-86.

24

HEPATOTOXICITY AND -CARCINOGENICITY OF THE FUMONISINS IN RATS

A Review Regarding Mechanistic Implications for Establishing Risk In Humans

W.C.A. Gelderblom[1], S.D. Snyman[1], S. Abel[1], S. Lebepe-Mazur[1],
C.M. Smuts[1*], L. Van der Westhuizen[1], W.F.O. Marasas[1], T.C. Victor[2],
S. Knasmüller[3], and W. Huber[3]

[1] Programme on Mycotoxins and Experimental Carcinogenesis and
* National Research Programme for Nutritional Intervention
P.O. Box 19070, Tygerberg, 7505, South Africa
[2] Department of Medical Physiology and Biochemistry
MRC Centre for Molecular and Cellular Biology
University of Stellenbosch Medical School
P.O. Box 19063, Tygerberg, 7505, South Africa
[3] Institut für Tumorbiologie-Krebsforschung
Universität Wien, Borschkegasse 8a
A-1090 Vienna, Austria

ABSTRACT

Cancer induction by the non-genotoxic mycotoxin, fumonisin B_1, has been investigated by studying the mechanisms involved during cancer initiation and promotion in rat liver. Cancer initiation is effected through a toxic-proliferative response while the inhibitory effect on hepatocyte cell proliferation appears to be a key aspect determining cancer promotion. Dose-response effects of the fumonisins on the induction of early neoplastic lesions in both long- and short-term animal experiments have been established. The biphasic response of FB_1 on hepatocyte proliferation will be discussed in relation to the known mechanisms of cancer induction by the genotoxic hepatocarcinogens. Recent investigations regarding the effect of the fumonisins on lipid biosynthesis and its inhibitory effect on hepatocyte growth stimulatory responses *in vitro* will be highlighted. Integration of our current knowledge regarding the carcinogenic potential of the fumonisins in setting a realistic and applicable risk assessment model for this non-genotoxic carcinogen will finally be addressed.

Fumonisins in Food, Edited by L. Jackson *et al.*
Plenum Press, New York, 1996

INTRODUCTION

The mechanisms involved in cancer induction by chemicals have been investigated extensively during the past 20 years. The dominant view that emerges, is that the major driving force in the development of neoplasia is the interaction of carcinogenic genotoxic chemicals with the cellular genome of the target cells (Bishop, 1987). The importance of these DNA lesions has been recognized and is currently widely utilized in molecular epidemiological studies to assess the extent of human exposure to specific natural and industrial genotoxic carcinogens (Perera, 1987). Although the results accurately reflect carcinogen exposure at the individual level, the impact and actual risk of such an exposure is not yet known, as exposure to genotoxic carcinogens at a specific time point does not reflect the outcome of cancer in a specific population and *vice versa* (Butterworth and Goldsworthy, 1991). However, these data are often used as part of the exposure assessment in establishing risk assessment models, although the correlation of associated changes, some of which might occur in cellular oncogenes and/or tumor suppressor genes, and the ultimate cancer has not been established unequivocally (Farber, 1989; Pitot, 1993).

When considering the steps and mechanisms involved, it becomes clear that genetic changes, cellular adaptive reactions and responses and/or dynamic interaction between cells and tissues are important factors that underline the long and multistep nature of cancer development (Farber and Rubin, 1991; Farber, 1993). The latter considerations form part of the epigenetic way of cancer development, an aspect that is largely ignored when the mechanism of cancer induction by chemicals is considered (Weinstein, 1991). Furthermore, recent investigations indicated that an increasing number of non-genotoxic chemicals that apparently lack the ability to interact with cellular DNA are carcinogenic (Rao and Reddy, 1991; Gelderblom *et al.*, 1992). Therefore, the position of non-genotoxic chemicals in the spectrum of chemical carcinogenesis remains an open one. In this context it has to be realized that the chemical and/or biological basis of cancer induction, whether via genotoxic or non-genotoxic systems still remains unknown.

When considering the above arguments concerning the mechanisms of cancer development, the present paper will focus on fumonisin-related biological effects that could provide information on the possible mechanism involved in the toxic and carcinogenic effects of these apparently non-genotoxic compounds in rat liver.

FUMONISINS AS CANCER INITIATORS

(i) Non-Genotoxicity

The fumonisins are non-mutagenic when tested in the *Salmonella* mutagenicity test (Gelderblom *et al.*, 1991) and lack genotoxicity in *in vitro* DNA repair assays in primary rat hepatocytes (Gelderblom *et al.*, 1992; Norred *et al.*, 1992). Genotoxicity of fumonisins has also been the subject of another very recent series of studies involving different *in vitro* systems (Knasmuller *et al.*, unpublished data). The lack of mutagenesis by the fumonisins was confirmed when different concentrations, ranging from 0.7 to 500 μg per plate, were tested against strains TA 100 and TA 98 in the presence and absence of Aroclor 1254 induced S9 ensyme fraction. The fumonisins also failed to induce any genotoxic effects in the DNA-repair assay with *Eschericia coli* (strains 3431753. uvrB/vecA and 3431765.uvr+/vec+) and the umu-microtest with *Salmonella* TA 1535/pKS 1002) in the absence and presence of rat liver S9 mix. Fumonisin B_1 (FB_1) also failed to induce micronuclei and did not alter the mitotic activity of primary rat hepatocytes. The effect of

FB_1 on the induction of chromosomal aberrations in primary hepatocytes is presently under investigation.

(ii) Short-Term Assays in Rat Liver

Despite the fact that the fumonisins were negative in various genotoxicity and mutagenicity assays, short-term *in vivo* studies have shown that FB_1 mimics genotoxic carcinogens with respect to the induction of resistant hepatocytes in rat liver (Gelderblom *et al.*, 1992; 1994). This was substantiated by the observation that FB_1 induces two important enzymes, gamma glutamyltranspeptidase (GGT) and the placental form of glutathione-S-transferase (GSTP), which are accepted histological markers for putative preneoplastic lesions initiated by genotoxic carcinogens. At present it is not known whether the characteristic enzyme phenotype that is associated with the resistant phenotype (Farber, 1990) is also induced by the fumonisins. Feeding experiments with the fumonisins in rats indicated that an increase in cell proliferation is also likely to play a critical role in the induction of the "resistant" phenotype as hepatoxicity, and the resultant regenerative cell proliferation is a prerequisite for inititation (Gelderblom *et al.*, 1994). In this regard regenerative cell proliferation, known to be a prerequisite for cancer initiation by the genotoxic carcinogens (Columbano *et al.*, 1991), is also a critical event during FB_1-induced carcinogenesis. The only difference noticed thus far in the induction of the "resistant" phenotype between the fumonisins and other genotoxic carcinogens lies in the kinetics of the cancer initiation step. It is known that, with genotoxic carcinogens, cancer initiation is normally completed within a matter of hours or a few days (Farber *et al.*, 1989). However, single and/or multiple dosages of the fumonisins in the presence of a stimulus for regenerative cell proliferation fail to effect initiation (Gelderblom *et al.*, 1992). Prolonged exposure to a dietary level (250 mg FB_1/kg of the diet) that induced hepatotoxicity over a period of 14 to 21 days does effect initiation. A recent investigation indicated that another non-genotoxic carcinogen, the peroxisome proliferator clofibrate, induced resistant hepatocytes after prolonged exposure of up to six months (Nagai *et al.*, 1993). As in the case of the fumonisins, the induced hepatocyte nodules stained positively for GSTP and GGT.

(iii) Long-Term Experiments

Male BD IX rats treated with a diet containing 25 mg FB_1/kg over a period of 2 years showed a significant increase (P<0.01) in the amount of GSTP positive foci in the liver (unpublished data). The number of foci in the liver of the rats treated with a 10 mg FB_1/kg diet was also noticeably (however not significantly) increased as compared to the control and 1 mg FB_1/kg diet treated groups. When the number of cells per focus were considered, there was a significant (P<0.01) increase in the size of the foci between the rats treated with the 10 mg FB_1/kg diet and the control and 1 mg FB_1/kg groups, implying that cancer promotion was effected. However, the presence of these foci indicated that, as for the 25 mg FB_1/kg treated group, promotion (see next section) and most probably cancer initiation did occur after prolonged feeding at this low dietary level. At the lowest dietary level where "initiation" and promotion (10 mg/FB_1/kg diet) were effected, only a few necrotic cells were observed as a result of the fumonisin treatment. Whether this modest induction of cytotoxicity could result in an increased regenerative cell proliferation to an extent to support cancer initiation (see previous section) is not known at present.

In this regard, the presence of "spontaneous" initiated cells and their subsequent selection (promotion) by the fumonisins need to be considered. Especially since cancer promotion by the fumonisins occurs at dietary levels well below the level that is required for cancer initiation (see next section). There were no significant differences in the number of GSTP positive (GSTP$^+$) single cells in the liver of rats fed the control diet and at three different dietary levels of FB$_1$ (1, 10 and 25 mg FB$_1$/kg diet) over a period of two years (unpublished data). The presence of the GSTP$^+$ cells has been regarded as early evidence that cancer initiation was effected following carcinogen treatment (Satoh et al., 1989). A recent study suggested that a certain number of these cells could give rise to hepatocyte nodules when animals are treated with the cancer promoting regimen consisting of 2-acety-laminofluorene (AAF) and partial hepatectomy (Kato et al., 1993). Studies concerning the peroxisome proliferator, nafenopin, showed that another subclass of hepatocyte foci, which stained negatively for GGT and specific sub-units of GST, are induced after chronic exposure (Grasl-Kraupp, et al., 1993). As these altered hepatocyte foci normally occur in older rats, it was suggested that cancer induction by these compounds originated from initiated cells that occur "spontaneously" in the liver (Kraupp-Grasl et al., 1991). This was deduced from the fact that the ultimate cancer also exhibited a similar characteristic as these foci.

With this as a background, the following arguments concerning the role of the fumonisins in promoting initiated cells that occur "spontaneously" within the liver need to be considered:

- (i)Cell proliferation is required for cancer initiation by the fumonisins (Gelderblom et al., 1994), which is characteristic of cancer initiation by genotoxic agents (Columbano et al., 1991).
- (ii)In a short-term carcinogenesis assay in male Fischer rats there were no hepato-cyte nodules in the control animals fed only the initiating dosage of FB$_1$ (up to 500 mg FB$_1$/kg diet over a period of 21 days) and/or selected with the strong selecting 2-acetylamino-fluorene/partial hepatectomy (AAF/PH) promoting regimen (Gelderblom et al., 1994).
- (iii)Long-term studies in rats treated with a relatively high dietary level (25 mg FB$_1$/kg) of FB$_1$ revealed that the number of GSTP$^+$ single cells was not increased above that of the controls while the number of foci were enhanced significantly. This is in contrast with certain genotoxic carcinogens (Satoh et al., 1989) which significantly increased the number of single GSTP$^+$ cells.
- (iv)None of the control male BD IX rats developed hepatocellular carcinomas during the long-term studies (Gelderblom et al. 1991; unpublished data) where the carcinogenicity of FB$_1$ was evaluated.
- (v)Male Fischer rats subjected to the different cancer promoting regimens such as AAF/PH and/or chronically treated with phenobarbital (PB) did not develop hepatocyte nodules and/or cancer (Ghoshal et al., 1987).
- (vi)It is well accepted that initiated hepatocytes exist within the liver of older rats but there is no evidence that they are the precursors for hepatocyte nodules and/or hepatocellular carcinoma (Ghoshal, et al., 1987).
- (vii)Studies concerning clofibrate and the fumonisins indicated that non-genotoxic carcinogens have the ability to induce "resistant" hepatocytes similar to genotoxic carcinogens. An investigation on the carcinogenicity of ciprofibrate in 24 and 52 week old rats indicated that there was no difference in the tumor incidence and numbers of tumors per liver as would be expected since the number of "spontane-ously" initiated cells increased with age (Ward et al., 1987). It was suggested that the latter clearly established the role of ciprofibrate as an initiator (Rao et al., 1991).

Based on these arguments it would appear that the presence of hepatocyte foci in the liver of the rats fed the 10 mg FB$_1$/kg diet can be ascribed to cancer initiation and not due to the promotion of "spontaneous" initiated cells.

FUMONISINS AS CANCER PROMOTERS

The fumonisins promote the induction of hyperplastic foci in rat liver treated with diethylnitrosamine (Gelderblom *et al.*, 1988), a known genotoxic carcinogen. Subsequent studies were directed to the investigation of the mechanisms possibly involved that could effect the selective outgrowth of initiated cells. Prolonged exposure of diethylnitrosamine-initiated rats over a period of 21 days showed that dietary levels of 50 mg FB_1/kg and above induced the formation of hyperplastic foci and/or nodules. No cancer promoting activity was observed at a dietary level of 10 mg FB_1/kg indicating that, similar to cancer initiation, cancer promotion is also dose-dependent, which in turn is determined by the length of exposure.

Recent investigations showed that the fumonisins, like many genotoxic carcinogens, inhibit cell proliferation (Gelderblom *et al.*, 1994). Dietary levels up to 50 mg FB_1/kg that promoted cancer in DEN-initiated rats, also significantly inhibited and/or delayed hepatocyte cell proliferative stimulus induced by partial hepatectomy (unpublished data). Even a single dose of FB_1 (by gavage) following partial hepatectomy effectively inhibited regenerative cell proliferation (Gelderblom., 1992). The exact mechanism for this inhibitory effect is not known but *in vitro* studies in primary hepatocytes indicated that the interruption of growth stimulatory responses is likely to play a role (Gelderblom *et al.*, 1995). The mitoinhibitory effect on the epidermal growth factor (EGF)-induced DNA synthesis occurs at a level well below (10-15X) concentrations that induced a cytotoxic effect. Binding studies using ^{125}I EGF indicated that the interaction of the growth factor with its receptor was not effected and that inhibition occurs beyond the growth factor mediated events. The effect of FB_1-induced inhibition was shown to be reversible, because hepatocytes retained very little "memory" of FB_1 exposure. Subsequent studies concerning the accumulation of sphinganine also failed to indicate any effect that could be associated with this inhibition of the EGF mitogenic response (Gelderblom *et al.*, 1995). Inhibition of cell proliferation by the fumonisins has also been reported in renal epithelial ($LLC-PK_1$) cells (Yoo, *et al.*, 1992) and provides a reasonable hypothesis for the mechanism of cancer promotion by these compounds. This is further supported by the fact that many other cancer promoters, such as orotic acid (Pichiri-Coni *et al.*, 1990) and PB (Manjeshwar *et al.*, 1992) also exhibit a mitoinhibitory effect in primary hepatocytes.

In general it became apparent that, based on the short-term studies, the fumonisins are strong cancer promoting agents. This needs to be seen against the background that cancer initiation is effected in rat liver at a dietary level of 250 mg/kg over a period of 21 days (Gelderblom *et al.*, 1994). Initiation by FB_1 was also associated with distinct hepatotoxic effects. Cancer promotion, on the other hand, is effected in the absence of excessive hepatotoxicity at a dietary level 5-fold less (50 mg FB_1/kg) as compared to cancer initiation (unpublished data).

However, when considering the long-term experiments on FB_1-induced carcinogenesis it would appear, as discussed above, that at low dietary levels (10 mg FB_1/kg) cancer initiation is likely to occur in the absence of excessive hepatotoxic effects and seems to occur at the same dietary level where promotion is effected. It is clear that short-term effects induced by relatively high dosage levels cannot be directly related to those noticed during long-term experiments using low dietary levels.

HEPATOTOXICITY AND METABOLISM OF THE FUMONISINS

Toxicological experiments with culture material of *F. moniliforme* and the fumonisins in rats showed that apart from the kidney, the liver is the main target organ of the fumonisins

(Gelderblom *et al.*, 1991; Voss *et al.*, 1993). Chronic feeding studies in BD IX rats showed that hepatocarcinogenesis developed against a background of a chronic toxic hepatitis that culminates in cirrhosis (Gelderblom *et al.*, 1991). The role of hepatotoxic effects has been established recently as non-toxic dosages failed to initiate cancer in short-term experiments in Fischer 344 male rats (Gelderblom *et al.*, 1994). This was confirmed in a long-term study in rats where dietary levels up to 25 mg FB_1/kg, which did not induce excessive toxic effects, also failed to induce liver cancer (unpublished data). The mild hepatotoxic effects in the rats that received 25 mg FB_1/kg diet, included single cell necrosis, bile duct proliferation and fibrosis that slightly distorted the liver in some cases, while in one rat a large area of cholangiofibrosis was present. Lipid accumulation and mild to prominent anisonucleosis occurred frequently in the high dosage group. In the rats fed the diet which contained 10 mg FB_1/kg, these changes were far less prominent. The only "pre-neoplastic" effect noticed was the induction of hepatocyte nodules and basophilic foci in some of the rats. There was a marked increase in the size and number of $GSTP^+$ foci in the liver of the rats treated with the 10 and 25 mg FB_1/kg diets compared to the control. Large confluent areas of some liver sections of the high dose group also stained positively for GGT. Therefore, a mild hepatotoxic effect induced by the fumonisins only resulted in the induction of early lesions that can be related to cancer development in the liver. As suggested previously, it would appear that a chronic hepatotoxic effect is a prerequisite for cancer development in the rat (Gelderblom *et al.*, 1994), implying the existence of a threshold value. The importance of a certain threshold value concerning fumonisins in cancer induction and the implications regarding the assessment of the risk to humans will be discussed further on.

(i) *In Vitro* Cytotoxicity and Metabolism

The fumonisins are not very toxic to primary hepatocytes in culture with a CD_{50} dosage of 1000μM and 500μM for FB_1 and FB_2 respectively (Gelderblom *et al.*, 1993). The water solubility of the fumonisins appears to be the reason for the low cytotoxicity as the more polar FB_1 is less cytotoxic than the less polar FB_2 and the hydrolyzed aminopolyol products of FB_1 and FB_2. Binding studies using radiolabelled FB_1 indicated that the compound is associated with both the soluble and insoluble or membraneous compartments in the cells (Cawood *et al.*, 1994). Although very little of the labelled FB_1 interacts with the hepatocytes (<0.01%) the compound is associated tightly with the membraneous compartment. As the fumonisin molecule exhibits both hydrophobic and hydrophilic properties it may readily be associated with cellular membranes.

Subsequent studies regarding the possible metabolism of the fumonisins by various enzyme preparations, including the microsomal cytochrome P450 and esterases and the hepatic triglyceride lipases, showed that the fumonisins are not substrates for these enzymes (Cawood *et al.*, 1994). Fractionation of the incubation medium of primary hepatocytes treated with fumonisins also failed to indicate the presence of any metabolites. Investigations regarding the structure-activity relationships of the fumonisins (Gelderblom *et al.*, 1993) indicated that the intact molecule is responsible for the biological activity of the fumonisins.

(ii) *In Vivo* Toxicity and Metabolism

Short-term studies with fumonisins indicated that they are not very hepatotoxic (Gelderblom *et al.*, 1988; 1992; 1994). Dietary levels as high as 50mg FB_1/kg over a period of 21 days induced only scattered single cell necrosis. However, chronic feeding of the same dietary level over a period of 24 to 26 months caused cirrhosis in the liver. These data seem to imply that the onset of fumonisin toxicity is very slow and that events that precede cell death occur at a far slower rate during chronic feeding at low dietary levels. The irreversible nature of the interaction of fumonisins with cellular membranes (Cawood *et al.*, 1994) suggests that there could be, as a function of time, a slow accumulation of fumonisins in the cell which eventually precipitates hepatotoxicity. Toxicokinetic studies using radiolabelled FB_1 indicated that after a single gavage dosage, all the radiolabel was recovered in the feces as unmetabolized FB_1 (Shephard *et al.*, 1992). Only trace amounts of the radiolabelled parent molecule were found in the liver, kidneys and red blood cells. This again supports the *in vitro* studies that fumonisins are not readily metabolized by the liver.

(iii) Dose-Response Relationships Between FB_1-Induced Toxicity and Carcinogenesis

With respect to dose response effects in cancer initiation, promotion and the induction of the ultimate cancer, the following need to be considered:

- 1)In short-term cancer initiation/promotion experiments cancer initiation occurs at relatively high dietary (250 mg FB_1/kg) levels which also induce toxic effects in the liver (see above). Contrary to this, a dietary level of 50 mg FB_1/kg that lacks cancer initiation activity, effects cancer promotion.
- 2)In long-term feeding studies liver cancer is induced in rats at a dietary level of 50 mg FB_1/kg after 18 to 24 months. As discussed above a dietary level of 25 mg FB_1/kg failed to induce liver cancer when fed to rats over a period of 24 months. However, cancer promotion and initiation were effected in rats fed dietary levels of 10 and 25 mg FB_1/kg for 2 years, resulting in the formation of hepatocyte nodules in the absence of any liver cancer. Of these dosage levels only the 50 mg FB_1/kg diet exhibited any carcinogenic activity (promotion) during short-term studies (see above).

Therefore, the kinetics for the induction of the early events differ depending on the dosage used and the length of exposure. It became apparent that high-dose/short-term effects cannot be extrapolated to those effects obtained with low dose exposure for longer periods of time. It is known that the incidence and latency period of the development of cancer with genotoxic and non-genotoxic carcinogens differ and depend on the type of carcinogen and the dosage used (Rao *et al.*, 1991). Another perspective is that the stage of carcinogenesis beyond the transformation phase of a fraction of cells is controlled by the dose, as well as the length of exposure (Jones, 1978). This time-dose relationship is a general phenomenon of most carcinogens. With respect to the relationship between toxicity and carcinogenicity in rat liver, no consistent data are available (Melnick and Huff, 1993). It has been shown that the majority of chemicals tested in laboratory animals, even mutagens, showed no relationship between carcinogenesis and chronic toxicity (Hoel *et al.*, 1988). Cancer induction with the trihalomethanes in female mice revealed a striking dose response curve in the absence of hepatocellular necrosis (Melnick and Huff, 1993). In this case overt toxicity and hyperplasia were not the major driving force for cancer induction. It has been argued that if toxic effects such as irritation and inflammation, cellular degeneration and/or regeneration, cytoplasmic alterations, hyperplasia, metaplasia and dysplasia were associated with carcinogenesis, then all chemicals that induce these lesions have to cause cancer. Despite the above arguments a chronic hepatotoxic effect seems to be a prerequisite for FB_1-induced cancer initiation (Gelderblom *et al.*, 1992) and the development of the ultimate cancer (Gelderblom *et al.*, 1991). In the study of Hoel *et al.*, (1988) it was noticed that a few chemicals exhibited toxic effects that may be related to the increase in tumor incidence. They have classified these compounds as "secondary carcinogens" as they produce carcinogenic effects via a mechanism that includes cytotoxicity and compensatory cell proliferation. This type of classification of carcinogens as "primary" (mutagens) and "secondary" carcinogens is in agreement with that of Cohen and Ellwein (1990) who used similar criteria to classify carcinogens as genotoxins and non-genotoxins. At present it would appear that the fumonisins should be classified as secondary or non-genotoxic carcinogens.

ADVANCES IN THE ELUCIDATION OF MECHANISMS OF FUMONISIN-INDUCED HEPATOTOXIC AND -CARCINOGENIC EFFECTS

(i) Interruption of Sphingolipid Biosynthesis

The interruption of sphingolipid biosynthesis by fumonisins and the toxicological and carcinogenic effects thereof have been discussed in detail elsewhere (Riley et al., 1994; Merrill et al., 1993; Schroeder et al., 1994). Experiments have been described that strongly suggest that the disruption of sphingolipid metabolism leads to cell death in a pig kidney renal epithelial cell line (Yoo et al., 1992). A recent study has implicated the mitogenic effect of FB_1 in Swiss 3T3 cells in the hepatocarcinogenicity of the mycotoxins (Schroeder et al., 1994). The FB_1-induced accumulation of sphinganine was associated with the mitogenic effect as both sphingosine and sphinganine induced mitogenesis when added exogenously to Swiss 3T3 cells.

However, an in vitro study in primary hepatocyte cultures indicated that disruption of sphingolipid biosynthesis is apparently not involved in the hepatotoxicity and cancer promoting activity of the fumonisins (Gelderblom et al., 1995). In this regard the following need to be considered:

- 1) The sphinganine/sphingosine (Sa/So) ratio is maximally altered in primary hepatocytes at a concentration of 1 μM FB_1 while cytotoxicity is effected at 250 μM.
- 2) Neither FB_1, Sa or So exhibited a mitogenic effect in primary hepatocytes.
- 3) Sa and/or So does not stimulate the EGF mitogenic response in primary hepatocytes.
- 4) The disruption of sphingolipid biosynthesis is not involved in the mitoinhibitory response of FB_1 on the EGF mitogenic response.
- 5) The mitoinhibitory effect of FB_1 in primary hepatocyte cultures is reversible while the accumulation of Sa is not (unpublished data).

In vivo studies indicated that the disruption of the Sa/So ratio occurs below the level that effects cancer initiation, while a slight increase (not significant) was observed at the lowest dietary level of FB_1 that induces cancer promotion and inhibition of cell proliferation (unpublished data). As the in vitro data also did not implicate the sphingolipids in the mitoinhibitory effect of the fumonisins, it is unlikely that their accumulation plays a role during cancer promotion.

(ii) Effect on in Vitro and in Vivo Lipid Biosynthesis

Studies in primary hepatocyte cultures treated with toxic (500 μM) and non-toxic (150 μM) concentrations revealed that FB_1 alters the incorporation of ^{14}C palmitic acid into cellular lipids indicating that, apart from the effect on sphingolipid biosynthesis, the synthesis of cellular lipids is also affected (Gelderblom *et al.*, 1996). As expected the radiolabelling of sphingomyelin (SM) was decreased as a result of the reduced synthesis of the phospholipid. In contrast, the radiolabelling of both phosphatidylcholine (PC) and phosphatidylethanolamine (PEA) was enhanced as a result of the increase in their respective concentrations. Fatty acid (FA) analysis of the major phospholipids (PC and PEA) and the neutral lipid triacylglyceride (TAG) showed marked alterations with respect to the n-6 fatty acid profiles. At the highest and cytotoxic dosage (500 µM/dish) C18:2 was markedly increased in PC, PEA and TAG while C20:4 was increased in TAG and PC. At the lower and non-cytotoxic dosage (150 µM) both C18:2 and C20:4 were increased in TAG while C18:2 was increased in PC only. In addition to these changes the free cholesterol (membrane associated) was also markedly decreased in hepatocytes treated with the highest dosage of FB$_1$. These data seem to imply that, apart from the effect on sphingolipid biosynthesis, FB$_1$ has important effects on the structure of the major membrane components, the FA storage pool (TAG-FA's) and the accumulation of long chain FA's within the cell. In this regard fat accumulation, noticed histochemically in rat liver treated with toxic dietary levels (Gelderblom *et al.*, 1994) of the fumonisins, is of relevance and could have important implications regarding the toxicity of these compounds. Changes in the membrane structure and function could also eventually lead to the disintegration of membrane continuity and eventually result in cell death.

The lipid profiles of rat liver, subjected to different dietary levels of FB$_1$, were also monitored *in vivo* in short-term and long-term experiments (unpublished data). In the short-term experiments three different dietary levels (50, 100 and 250 mg/kg) of FB$_1$ were fed to rats over a period of 21 days. Only mild to moderate toxic effects were noticed at the low dietary levels while more prominent toxic lesions such as single cell necrosis, bile ductule proliferation and early signs of fibrosis were observed at the high dosage level. Of the fumonisin containing diets only the 250 mg FB$_1$/kg diet initiated cancer while all three dietary levels exhibited cancer promoting activity (Gelderblom *et al.*, unpublished data). In the long-term studies three different diets, containing 1, 10 and 25 mg FB$_1$/kg, were fed over a period of 24 months. The major toxicological effects obtained in the liver were similar to those described above. The major changes observed in the short-term studies were noticed in the PEA phospholipid fraction in which the level of C18:2 was significantly increased. As a result of the negative feedback of C18:2 on the delta-4-desaturase the level of C22:5 was decreased. The levels of C18:2 and C22:5 were also markedly effected in a similar manner in PC but the differences were not significant. Analyses of the FA profiles in the livers (total FA and individually of PC and PEA) of the rats fed low dietary levels of FB$_1$ over a period of 24 months (see above) showed that C18:2 was again markedly increased.

In contrast to the *in vitro* studies, no effect was noticed on the arachidonic acid (AA) content in either the PEA and PC phospholipid fractions, presumably due to the fact that no excessive toxic effects were noticed as compared to the *in vitro* studies where 30 to 40 % cell death was recorded. Another difference was that *in vitro* exposure of hepatocytes to FB$_1$ significantly decreased the total cholesterol by reducing the concentration of free cholesterol. In the *in vivo* study, the highest dietary level (250 mg FB$_1$/kg) significantly increased the serum and the total cholesterol in the liver.

(iii) Fatty Acid Accumulation and Cell Proliferation

It has been suggested that cell proliferation plays a determining role during cancer initiation and promotion (Gelderblom *et al.*, 1994; 1995). In this regard, the inhibitory effect of FB$_1$ on cell proliferation is likely to be the prominent determinant during cancer promotion (see above)). *In vitro* studies in primary hepatocytes indicated that FB$_1$ inhibited the EGF growth response, thereby mimicking many cancer promoters such as AAF and PB (Gelderblom et al., 1995). The exact mechanism of this inhibition is not known at present. As discussed above the increase in polyunsaturated fatty acids (linoleic acid [C18:2] and AA) in phospholipid and neutral lipid fractions in primary hepatocytes after treatment with FB$_1$ indicated that the FA biosynthesis is altered. Short-term feeding studies in rats also showed that the concentration of PEA was increased while the level of C18:2 was increased markedly in PEA and PC and no effect was observed on AA (see above). As mentioned above, in long-term feeding studies, low dietary levels of FB$_1$ also markedly alter the C18:2 FA profiles of PEA. The accumulation of C18:2 suggests that FB$_1$ alters the n-6 FA metabolic pathway, presumably by affecting the activity of the delta-6-desaturase.

Regarding the disruption of FA metabolism it is known that long chain fatty acids control cell proliferation in different cell culture types via their modulation of prostaglandin levels (Cornwell and Norisaki, 1984). It was shown that, depending on the cell type, prostaglandins of the E series can either inhibit and/or stimulate cell proliferation. *In vitro* studies using Balb/c 3T3 cells, have shown that AA metabolism is required for the mitogenic response of the EGF (Nolan et al., 1988; Handler et al., 1990). Indomethacin inhibits the response while this inhibitory effect is overcome by the addition of prostaglandins, specifically prostaglandin F2$_\alpha$. Therefore, the disruption of the prostaglandin levels within the cell regulates the EGF response. Our studies (unpublished data) have shown that ibuprofen, a non-steroidal anti-inflammatory drug, also inhibits the EGF mitogenic response; again implicating the role of AA metabolism. The addition of PGE$_2$ counteracts the inhibitory effect of FB$_1$ indicating that the fumonisins also interfere with AA metabolism. The increased levels of AA within primary hepatocytes (see above) suggest that FB$_1$ inhibits the release and/or the metabolism of AA in primary hepatocytes, an effect that is greatly enhanced at a cytotoxic dosage of FB$_1$.

(iv) Fumonisin-Induced Toxicity and Oxidative Stress

Chronic inflammation increased the risk of liver cancer in humans as indicated by the fact that both alcohol and hepatitis B infection are important risk factors as they chronically damaged the liver. In the absence of these agents the incidence of liver cancer in humans is low (Ames and Gold, 1990). It has been suggested that the phagocytes liberate oxidants (superoxide) while destroying dead cells which resulted from the chronic infection and/or exposure to a toxic agent (Halliwell, 1994). Another physiological free radical that could be potentially harmful during chronic hepatic injury is nitric oxide which could be involved in the formation of carcinogenic nitrosamines (Bartsch et al., 1992). These oxidants are similar to the process of ionizing radiation and thus could affect the cell by damaging cellular membranes and/or DNA (Ward et al., 1987). It is known that tissue injury leads to oxidative stress when the antioxidant defenses become depleted. Exposure of cells to hepatotoxicants leads to oxidative stress due to various causes (Halliwell, 1994) and increased the probability to further cell injury. Cancer initiation and the induction of the ultimate cancer by FB$_1$ occur during the development of a chronic active hepatitis that eventually culminates in cirrhosis (Gelderblom et al., 1991; 1994). It has been indicated that primary hepatocytes exposed to FB$_1$ showed an accumulation of polyunsaturated fatty acids (PUFA) (Gelderblom et al., 1996). The cytotoxic effects of PUFA when added to both normal and cancer cells have been studied extensively and has been associated with an increase in the extent of lipid peroxidation (Gavino et al., 1981; Begin, et al., 1988). These studies imply that FB$_1$-treated cells could be more susceptible to lipid peroxidation than normal cells. Investigations regarding the effect of the fumonisins on the oxidative mechanisms within the cell are in progress.

(v) The Fumonisins as Peroxisome Proliferators

As the fumonisins appear to be non-genotoxic the possibility that they belong to another class of non-genotoxic carcinogens, the peroxisome proliferators, was investigated (Huber *et al*, unpublished data). The activities of two marker enzymes for peroxisomes, i.e. peroxisomal β-oxidation and acetyl-carnitin-transferase were monitored (Huber *et al*., 1992) in the liver of rats exposed to 3 dietary levels (1, 10, 25 mg FB_1/kg) of FB_1 over a period of 24 months. As discussed above the high dosage levels (10 and 25 mg FB_1/kg) did not cause cancer but induced both GGT^+ and $GSTP^+$ foci and/or nodules in the liver. However, with respect to the parameters measured, no changes were observed above those of the groups receiving the control and low FB_1-containing (1 mg/kg) diets. With respect to the induction of early hepatic lesions associated with liver cancer development, the carcinogenicity of FB_1 is not associated with peroxisome proliferation.

(vi) The Role of p53-Associated Changes in FB_1-Induced Hepatocarcinogenesis

Epidemiological and molecular studies on patients have shown a close association between aflatoxin B (AFB), hepatitus B virus (HBV) infection and hepatocellular carcinoma as well as an association with a mutation at codon 249 in the p53 gene (Bressac, 1991). The synergistic role of FB_1 that has been shown to co-occur with aflatoxin B_1 under natural conditions (Chamberlain *et al.*, 1993; Chu and Li, 1994) needs to be elucidated. Studies, currently in progress, are investigating the role of p53-associated mutations in early hepatocyte lesions and the ultimate cancer induced by FB_1 in rats fed a FB_1-containing diet (50 mg FB_1/kg) chronically for two years. DNA from rat liver nodules was amplified by PCR with gene specific primers, corresponding to codon 249 of the human gene. Aflatoxin B_1 (AFB_1), used as a positive control, induced mutations at a very low frequency at codon 243, the equivalent gene to codon 249 in humans, whereas no mutations were detected after cancer initiation with FB_1. In the long-term study with FB_1, only one rat fed 50 mg/kg for 12 months, showed a mutation at codon 243 while none of the rats with hepatic carcinomas after 18 to 26 months showed any mutations. Single strand conformation polymorphism (SSCP) analysis (Sheffield *et al.*, 1993) also failed to indicate any additional mutations in the vicinity (110 base pairs) of this codon. The single mutation detected with FB_1 and the low frequency of mutations induced by AFB_1 may not represent primary oncogenic effects. Subsequent studies using immunohistochemical staining procedures with monoclonal antibodies against the wild type and mutant protein will be used to assess whether the p53 protein is altered during FB_1-induced hepatocarcinogenesis.

IMPLICATIONS FOR ESTIMATION OF RISK FOR HUMAN EXPOSURE

(i) Exposure Assessment

Very little information is currently available regarding human exposure to fumonisins at a population level, i.e. estimates obtained from the levels that occur in food combined with the consumption pattern based on dietary recall. In a recent report, a fumonisin (total) intake profile was calculated based on the estimated consumption of "healthy" (14 μg/kg body weight/day) and "moldy" (440 μg/kg body weight/day) home-grown corn in high and low areas for esophageal cancer in southern Africa (Thiel *et al.*, 1992). More accurate data that include the actual daily corn consumption patterns of the populations at risk need to be obtained, since moldy corn does not normally form part of the daily food rations, but rather as an important component of the brewing of traditional beer or used in animal feeds. Therefore, studies need to be performed to monitor fumonisin-intake profiles (daily meals and traditionally brewed beer) at an individual level, over an extended period of time.

Numerous studies have been performed to investigate the natural contamination of corn and corn-based foods and feeds with the fumonisin B (FB) mycotoxins worldwide. High levels of fumonisin contamination of home-grown "moldy" and "healthy" corn have been recorded in the high esophageal cancer risk regions in Transkei, southern Africa (Rheeder et al., 1992; Sydenham et al., 1991) and in Linxian, China (Chu and Li, 1994). In Transkei, mean values of 53.7 (ranging between 3.3-117.5) and 13.7 (ranging between 0.8-23.0) mg/kg for FB_1 and FB_2, respectively, were recorded in homegrown "moldy" corn during the 1989 crop year, giving a total fumonisin content (mean) of 67.4 mg/kg. The corresponding level (total content) in home-grown corn harvested during the 1985 crop year was 31.5 mg FB/kg. The mean total values of the FB contamination of "healthy" corn during the 1985 and 1989 crop years were 9.0 and 5.1 mg FB/kg, respectively (Rheeder et al., 1992). In Linxian and Cinxian, high esophageal cancer areas in China, FB_1 levels up to 155 mg/kg were recorded in "moldy" corn The mean FB_1 levels in the "moldy" and healthy corn were 74 and 35.3 mg/kg, respectively (Chu and Li, 1994).

The mean levels of total FB recorded in commercial corn-based foodstuffs (cornmeal) from the USA was in the order of 1.3 mg FB/kg (Sydenham et al., 1991). The corresponding mean total levels of the fumonisins (FB_1 + FB_2) in South African commercial products were in the order of 0.2 mg/kg. A survey done in South African corn harvested in different locations during the 1989 to 1991 crop years indicated that the bulk of the samples (75%) contained FB (total) levels of 0 to 0.5 mg/kg, while less than 15% have levels of 0.5 to 1 mg/kg and the remaining samples contained levels below 2 mg/kg (Viljoen et al., 1993).

When the total FB level, found in commercial South African corn (0.3 mg/kg), is used and extrapolated to a rural black population that uses corn as a staple diet (70 kg person consuming 460 g corn/day, Thiel et al., 1992) a probable daily intake (PDI) of 2 µg FB/kg body weight/day can be estimated (Table 1).

In exported South African and American corn, containing FB levels of 0.4 and 1.1 mg/kg, respectively, corresponding PDI values are 2.6 and 7.3 µg FB/kg body weight/day respectively. When considering the levels of the fumonisins (total) that occur in commercial corn meal products in SA and the USA (Sydenham et al., 1991), with a mean level of 0.2 and 1.5 mg FB/kg, respectively, the corresponding PDI values are 1.3 and 9.9 µg FB/kg body weight/day (Table 1). The corn consumption pattern of urban black people (Bourne et al., unpublished data) is almost 50 % (276g/70kg body weight/day) when compared to the rural area, (Table 1). In developed countries, such as the European Community, consumption (Smith et al., 1994) of corn products is much lower (7.2g/70kg body weight/day) with PDI values that fall well into the ng FB/kg body weight/day range (Table 1). The corresponding PDI values obtained when using home-grown corn are between 20- to 200-fold higher when compared to the values mentioned above.

(ii) Hazard Assessment

Long-term studies in rats have indicated that the hepatotoxicity of FB_1 is a pre-requisite for the hepatocarcinogenic effect of the compound (Gelderblom et al., 1991). Additional studies on the cancer initiating/promoting potential of the fumonisins indicated that the control of cell proliferation appears to be the key aspect with regard to cancer induction of the fumonisins (Gelderblom et al., 1994). The stimulation of cell proliferation (regenerative) presumably occurs via its hepatotoxicity while the compound inhibits cell

Table 1. Fumonisin (FB) contamination (mean values of FB_1 and FB_2) of corn and corn products in different localities, including high incidence areas for esophageal cancer

Origin	Product	No. of samples	Mean FB level (μg/kg)	PDI (μg/kg body weight/day)		
South Africa (SA)	Exported corn	68[#]	400	2.6	1.6	0.04
	Commmercial corn	209[***]	300	2.0	1.2	0.03
	Corn meal	52[##]	200	1.3	0.8	0.02
SA (Transkei)	Healthy corn	18[*]	7100	46.6	28.0	0.73
	Moldy corn	18[*]	54000	354.9	212.9	5.54
China (Linxian & Cinxian)	Healthy corn	15[**]	35300	231.9	139.2	3.60
	Moldy corn	19[**]	74000	486.2	291.8	7.60
United States of America	Exported corn	1682[***]	1100	7.2	4.3	0.11
	Corn meal	16[##]	1500	9.9	5.4	0.15
Corn consumption profiles/70 kg body weight				460g	276	7.18

TDI (μg/kg/day): NOEL/5000 = 0.16; NOEL/1000 = 0.8; NOEL/100 = 8.0.
[*]1985 and 1989 crop years (Rheeder et al., 1992). [**]Based only on FB_1 content (Chu and Li, 1994).
[***]1991 and 1992 crop years (Viljoen et al., 1993). [#]1989 crop year (Rheeder et al., 1994). [##]Sydenham et al., 1991). PDI values calculated on the basis that a person (70 kg) in a rural area consumes 460 g (Thiel et al., 1992) , 267 g in an urban area (Bourne et al., 1990, BRISK Study, MRC, Tygerberg) and 7.18g maize/70 kg body weight/day in the European community(Smith et al., 1994).

proliferation via the disruption of growth effects related to normal cell growth processes (Gelderblom *et al.*, 1995). The inhibitory effect on cell proliferation, an important determinant for cancer promotion, is effected at non-toxic dosage levels both *in vitro* and *in vivo* (see above). Short-term studies (21 days) in Fischer rats showed that cancer initiation is effected at a dosage level of 0.7<effective dosage level(EDL)<1.5 mg FB_1/100 g body weight/day while cancer promotion and inhibition of cell proliferation are effected at a dosage level of 0.17<EDL<0.3 mg FB_1/100 g body weight/day (Gelderblom *et al.*, unpublished data). Long-term studies using BD IX rats indicated that cancer develops at a daily dietary dosage of 0.09<EDL<0.16 mg FB_1/100 g body weight/day while cancer initiation and promotion occur at a dietary dosage of 0.03 mg FB_1/100 g body weight/day and above. Comparison of the data obtained from the short-term and the long-term feeding studies with respect to the dosages required for cancer initiation and promotion, indicated differences of 50- and 10-fold, respectively.

In the long-term feeding studies only the rats that received the 50 mg FB_1/kg diet for a period of 26 months developed hepatocellular carcinoma. Apart from the induction of hepatocyte nodules, no cancers were observed in the long-term feeding experiments in which FB_1-containing diets (1, 10 and 25 mg FB_1/kg) were fed for 24 months. The dietary level of 25 mg FB_1/kg diet was considered as the no observed effect level (NOEL) with respect to cancer induction. Since only the highest dietary level induces cancer, the level at which 50 % of the rats would be expected to develop cancer, the TD_{50} value (44 mg FB_1/kg) was determined by linear extrapolation which reflects a numerical estimate of carcinogenic potency (Kuiper-Goodman, 1990). Although 5 animals died during the final stages of the experiment, the 50 mg FB_1/kg diet is below the maximum tolerated dose (MTD) of FB_1 in BD IX male rats. These data represent an intake profile of 800 μg FB_1/kg body weight/day for NOEL and 1548 µg FB_1/kg body weight/day for the TD_{50}. To determine the potential for human disease two different approaches can be followed, i.e. NOEL/safety factor (SF) using a scale factor of 1000-5000, and the TD_{50}-SF using a scale factor of 50000 (Kuiper-Goodman, 1990).

Using this approach, the estimated tolerable daily intake (TDI) based on the NOEL in carcinogenicity studies (divided by 5000), is 160 ng/kg body weight/day while based on the TD_{50}/50 000 is 31 ng/kg body weight/day. The corresponding TDI values for aflatoxin B_1, as determined in male Fischer rats are 0.15 ng/kg body weight/day for NOEL/SF and 0.023 ng/kg body weight/day for TD_{50}/SF. For ochratoxin A, values of 4.2 (NOEL/SF) and 1.5 (TD_{50}/SF) ng/kg body weight/day were reported in male Fischer rats (Kuiper-Goodman, 1990). Based on these data the estimated TDI values for FB_1 are more than 1100 X (NOEL/SF) and 1350 X (TD_{50}/SF) higher than those of aflatoxin B_1 and 40 X (NOEL/SF) and 20 X (TD_{50}/SF) as compared to ochratoxin A. However, since cancer induction by the fumonisins is dependent on FB_1-induced toxicity, safety factors used for genotoxic carcinogens, such as aflatoxins, cannot be applied directly to the fumonisins. It is suggested that safety factors used for toxic compounds may be more realistic. When a value of 1000, which is the borderline value for differentiating between toxic and carcinogenic effects (Kuiper-Goodman, 1990), is used (safety factors for NOEL are 100-1000), the calculated TDI for NOEL/SF is 800 ng/kg/day. Similarly, with a safety factor of 100, the TDI is 8000 ng/kg/day.

In the case of TDI based on the NOEL/1000, the corresponding acceptable level of total fumonisins in corn is 0.122 mg/kg which is slightly lower than the actual level (0.2 mg/kg) in South African corn meal (Table 1). When NOEL/100 is used, the acceptable value is 1.22 mg/kg, which is well within or higher than the actual mean FB levels in commercial South African corn and corn products (0.2 - 0.4 mg/kg). In contrast, the NOEL/5000 results in a tolerable level of 0.024 mg FB/kg, which is well below FB levels that are known to occur in commercial corn and corn products.

(iii) Thresholds in FB_1 Induced Carcinogenesis and Implications for Assessing Risk to Humans

Most mathematical models treat all carcinogens as mutagens (genotoxins). They assume that even at low doses, DNA reactive molecules could escape the cell's detoxifying mechanisms and induce a mutation in a critical site on the DNA. As a result, many regulatory policies of various countries rely upon the outcome of these models. However, oversimplified speculations on mechanisms of carcinogenesis induced by non-genotoxic carcinogens, such as the fumonisins, should therefore not serve as the basis for risk assessment procedures. Compounds, specifically cancer promoters that act through specific receptors, tend to be active at low doses and it is unclear whether a no-effect threshold exists. On the other hand, compounds that act through a cytotoxic mechanism would be expected to have a no-effect threshold (Cohen and Ellwein, 1990). Below the threshold, cytotoxicity and increased cell proliferation would not occur and thus not increase the tumor risk. Recent studies concerning two compounds, uracil and melamine, that are carcinogenic in the urinary bladder, indicated that urothelial proliferation is a prerequisite for the formation of calculi and tumors (Cohen and Ellwein, 1991). Although these two compounds are carcinogenic in animals, dose-related considerations suggest that they are obviously not carcinogenic since humans are only exposed to doses that are unable to induce urothelial proliferation. Therefore, apart from the DNA-interactive role of genotoxins, other parameters including pharmacokinetics and mode of action as well as the induction of cell proliferation need to be incorporated into the risk assessment (Butterworth and Goldsworth, 1991). FB_1 is carcinogenic at a dosage level that causes adverse toxic lesions culminating in cirrhosis, therefore suggesting the requirement of a cytotoxic-proliferative response for cancer induction. This was confirmed in a study where relatively low dietary levels of FB_1, that failed to induce excessive toxic effects, also failed to induce any cancers in the liver of the rats. A recent study on the cancer initiating potential of FB_1, indicated that cancer initiation is also effected by dietary levels of FB_1 that induce hepatotoxic effects over a period of 21 days. A threshold for the induction of liver cancer by the fumonisins becomes apparent, and therefore the level of concern regarding the fumonisins as potential human carcinogens, may become irrelevant below doses that fail to induce a cytotoxic-proliferative response.

Thus it is reasonable to accept that with respect to the induction of liver cancer, it is unlikely that the fumonisins can be regarded as human carcinogens at dietary levels below the threshold level that could effect a cytotoxic-proliferative response. This can be deduced when comparing the PDI's (μg/kg body weight/day) of the fumonisins in a population using corn as a dietary staple and the TDI for humans calculated from the NOEL/SF value using a safety factor of 1000 (Table 1). Based on these calculations the consumption of South African commercial corn and corn products yielded PDI values of 1.3 to 2.0 (rural area) and 0.8 to 1.8 (urban area) μg/kg/day are in the same order or slighty higher than the TDI value of 0.8 μg/kg/day (Table 1). The PDI values are based on the assumption that a person (70 kg) in a rural and urban area consumes 460 and 276 g corn per day, respectively. The corresponding PDI values for corn and corn products from the USA are about 3 to 6 fold higher, respectively and more closely resembled the NOEL/100 value. However, the PDI values of populations in the high risk areas for esophageal cancer in Transkei and China are well above the TDI value obtained from NOEL/100 (Table 1). As indicated previously the PDI values in developed countries are likely to be overestimated as the corn consumption in these countries is well below 460 g/day/ person of 70 kg (Thiel at al., 1992) as was indicated for the European Community countries (Smith *et al.*, 1994). This again emphasises the need to perform more accurate studies that will closely monitor the corn intake profiles and consumption patterns of different population groups.

Alternative risk assessment models need to be considered for the fumonisins and other non-genotoxic carcinogens, using NOEL for induced cytotoxic-proliferation responses with a safety factor different from that used for the genotoxic carcinogens where there is no apparent threshold (Butterworth and Goldsworthy, 1991). The safety factors used in the case of the fumonisins, need to consider the fact that cancer induction is likely to occur solely via a cytotoxic mechanism and that a no-effect threshold is likely to exist. On the other hand a possible synergistic interaction with other important carcinogenic stimuli for humans, such as hepatitis B virus (HBV) and AFB, could also eventually affect the potential risk of low dietary levels of the fumonisins. Recent reports indicated that the fumonisins co-occur with aflatoxin in corn, suggesting possible synergism (Chamberlain *et al.*, 1993; Chu and Li, 1994). A recent prospective nested case control study by Ross et al (1990), involving analyses of 18 000 urine samples, indicated a relative risk of 2, 5 and 60, respectively, for AFB, HBV and when the combined effect was considered as the causative principle(s) in the development of liver cancer. Comparable data for the fumonisins in combination with other risk factors, including HBV and AFB, are urgently required.

ACKNOWLEGDEMENTS

The authors are indebted to Prof NPJ Kriek, Department of Veterinary Pathology, University of Pretoria for the interpretation of the histopathological data. Members of the Experimetal Biology Programme, Medical Research Council, P.O. Box 19070, Tygerberg for carrying out the rat experiments. Ms P Bakkes for the isolation of fumonisin B_1. The South African Maize Board for financial support.

REFERENCES

Ames, B.N.; Gold, L.S. Dietary carcinogens, environmental pollution, and cancer: some misconceptions. *Med. Oncol. & Tumor Pharmacother.* **1990**, *7*, 69-85.

Barstch, H.; Ohshima, H.; Pignatelli, B.; Calmels, S. Endogenously formed N-nitroso compounds and nitrosating agents in human etiology. *Pharmacogenetics* **1992**, *2*, 272-277.

Begin, M.E.; Ells, G.; Horribin, D.F. Polyunsaturated fatty acid-induced cytotoxicity against tumor cells and its relationship to lipid peroxidation. *J. Natl. Cancer Inst.* **1992**, *80*, 188-194.

Bishop, J.M. The molecular genetics of cancer. *Science* **1987**, *235*, 305-311.

Bressac, B.; Kew, M.; Wands, J.; Ozturk, M. Selective G to T mutations of the p53 gene in hepatocellular carcinoma from South Africa. *Nature* **1991**, *350*, 429-432.

Butterworth, B.E.; Goldsworthy, T.L. The role of cell proliferation in multistage carcinogenesis. *Proc. Soc. Exp. Biol. Med.* **1991**, 683-687.

Chamberlain, W.J.; Bacon, C.W.; Norred, W.P.; Voss, K.A. Levels of fumonisin B_1 in maize naturally contaminated with aflatoxins. *Fd. Chem. Toxic.* **1993**, *31*, 995-998.

Cawood, M.E.; Gelderblom, W.C.A.; Alberts, J.F.; Snyman, S.D. Interaction of ^{14}C-labelled fumonisin B mycotoxins with primary hepatocytes cultures. *Fd Chem. Toxicol.* **1994**, *32*, 627-632.

Cohen, S.M.; Ellwein, L.B. Cell proliferation in carcinogenesis. *Science* **1990**, *249*, 1007-1011.

Cohen, S.M.; Ellwein, L.B. Carcinogenesis mechanisms: The debate continues. *Science* **1991**, *252*, 902-904.

Columbano, A.; Ledda-Columbano, G.M.; Coni, P.; Pichiri-coni, G.; Curto, M.; Pani, P. Chemically induced cell proliferation and carcinogenesis: differential effect of compensatory cell proliferation and mitogen-induced direct hyperplasia on hepatocarcinogenesis in the rat. *Progress. Clin. Biol. Res.* **1991**, *369*, 217-225.

Chu, F.S.; Li, G.Y. Simultaneous occurence of fumonisin B_1 and other mycotoxins in moldy maize collected from People's Republic of China in regions with high incidences of esophageal cancer. *Appl. Environ. Microbiol.* **1994**, *60*, 847-852.

Farber, E. Clonal adaptation during carcinogenesis. *Biochem. Pharmacol.* **1983**, *39*, 1837-1846.

Farber, E.; Chen, Z-Y.; Harris, L.; Lee, G.; Rinaudo, J.S.; Perera, F.P. Molecular cancer epidemiology: A new tool on cancer prevention. *J. Natl. Cancer Inst.* **1987**, *78*, 887-897.

Farber, E.; Chen, Z-Y.; Harris, L.; Lee, G.; Rinaudo, J.S.; Roomi, W.M.; Rotstein, J.; Semple, E. The biochemical-molecular pathology of the stepwise development of liver cancer: new insights and problems. In *Liver Cell Carcinoma*; Bannasch, P., Keppler, D., Weber, G., Eds.; Kluwer Academic Publishers: Dordreccht,Boston,London, 1989; pp 273-291.

Farber, E. Hepatocyte proliferation in stepwise development of experimental liver cell cancer. *Digest. Disease and Science* **1991**, *36*, 973-978.

Farber, E.; Rubin H. Cellular adaptation in the origin and development of cancer. *Cancer Res.* **1991**, *51*, 2751-2761.

Farber, E. Is carcinogenesis fundamentally adversarial confrontational or physiologic-adaptive? *J. Invest.Dermatol.* **1993**, *100*, 251s-253s.

Gavino, V.C.; Miller, J.S.; Ikharebha, S.O.; Milo, G.E.; Maizewell, D.G. Effect of polyunsaturated fatty acids and antioxidants on lipid peroxidation in tissue cultures. *J. Lipid Res.* **1981**, *22*, 763-769.

Gelderblom, W.C.A.; Jaskiewics, K.; Marasas, W.F.O.; Thiel, P.G.; Horak, R.M.; Vleggaar, R.; Kriek, N.P.J. Fumonisins - novel cancer promoting activity produced by *Fusarium moniliforme. Appl. Environ. Microbiol.* **1988**, *54*, 1806-1811.

Gelderblom, W.C.A.; Snyman, S.D. Mutagenicity of potentially carcinogenic mycotoxins produced by *Fusarium moniliforme. Mycotoxin Res.* **1991**, *7*, 46-52.

Gelderblom, W.C.A.; Kriek, N.P.J.; Marasas, W.F.O.; Thiel, P.G. Toxicity and carcinogenicity of the *Fusarium moniliforme* metabolite, fumonisin B_1, in rats. *Carcinogenesis* **1991**, *12*, 1247-1251.

Gelderblom, W.C.A.; Semple, E.; Farber, E. The cancer initiating potential of the fumonisin mycotoxins produced by *Fusarium moniliforme. Carcinogenesis* **1992**, *13*, 433-437.

Gelderblom, W.C.A.; Cawood, M.E.; Snyman, S.D.; Vleggaar, R.; Marasas, W.F.O. Structure-activity relationships of fumonisins in short-term carcinogenesis and cytotoxicity assays. *Fd. Chem. Toxicol.* **1993**, *31*, 407-414.

Gelderblom, W.C.A.; Cawood, M.E.; Snyman, S.D.; Marasas, W.F.O. Fumonisin B_1 dosimetry in relation to cancer initiation in rat liver. *Carcinogenesis* **1994**, *15*, 209-214.

Gelderblom, W.C.A.; Snyman, S.D.; Van der Westhuizen, L.; Marasas, W.F.O. Mitoinhibitory effect of fumonisin B_1 on rat hepatocytes in primary culture. *Carcinogenesis* **1995**, *16*, 625-631.

Gelderblom, W.C.A.; Snyman, S.D.; Cawood, M.E.; Smuts C.M.,; Abel, S. and Van der Westhuizen, L. Effect of fumonisin B$_1$ on protein and lipid synthesis in primary hepatocytes. *Food Chem Toxicol.* **1996**, In press.

Ghoshal, A.K.; Rushmore, T.H.; Farber, E. Initiation of carcinogenesis by a dietary deficiency of choline in the absence of added carcinogens. *Cancer Lett.* **1987**, *36*, 289-296.

Grasl-Kraupp, B.; Waldhor, T.; Huber, W.; Schulte-Herman, R. Glutathione S-transferase isoenzymes patterns in different subtypes of enzyme-altered rat liver foci treated with the peroxisome proliferator nafenopin or with phenobarbital. *Carcinogenesis* **1993**, *14*, 2407-2412.

Halliwell, B. Free radicals, antioxidants, and human disease: curiosity, cause, or consequence. *The Lancet* **1994**, *344*, 721-724.

Handler, J.A.; Danilowics, M.; Eling, T.E. Mitogenic signalling by epidermal growth factor requires arachidonic acid metabolism in BALB/c 3T3 cells. *J. Biol. Chem.* **1990**, *265*, 3669-3673.

Hoel, D.G.; Haseman, J.K.; Hogan, M.D.; Huff, J.; McConnell, E. The impact of toxicity on carcinogenicity studies: implications for risk assessment. *Carcinogenesis* **1988**, *9*, 2045-2052.

Huber, W.W.; Kraupp-Grasl, B.; Gschwentner, C. Influence of cicadian rhythm, feeding rhythm and fat charge on lipid peroxidation under peroxisome proliferation by nafenopin. Naunyn-Schmiedeberg's Arch. *Pharmacol.* **1992**, *345* Suppl. 153

Kato, T.; Imaida, K.; Ogawa, K.; Hesegawa, R.; Shirai, T.; Tatematsu, M. Three-dimensional analysis of glutathione S-transferase placental form-positive lesion development in early stages of rat hepatocarcinogenesis. *Jpn. J .Cancer Res.* **1993**, *84*, 1252-1257.

Kraupp-Grasl, B.; Huber, W.; Taper, H.; Schulte-Hermann, R. Increased susceptibility of aged rats to hepatocarcinogenesis by the peroxisome proliferator nafenopin and the possible involvement of altered liver foci occurring spontaneously. *Cancer Res.* **1991**, *51*, 666-671.

Kuiper-Goodman, T. Uncertainties in the risk assessment of three mycotoxins: aflatoxin, ochratoxin, and zearalenone. *Can.J. Physiol. Pharmacol.* **1990**, *68*, 1017-1024.

Manjeshwar, S.; Rao, P.M.; Rajalakshmi, S.; Sarma, D.S.R. Inhibition of DNA synthesis by phenobarbital in primary cultures of hepatocytes from normal liver and from hepatic nodules. *Carcinogenesis* **1992**, *13*, 2287-2291.

Maizewell, D.G.; Morisaki, N. Fatty acid paradoxes in the control of cell proliferation: prostaglandins, lipid peroxides, and co-oxidation reactions. In *Free Radicals in Biology VI*; Pryor, W.A., Ed.; Academic Press INC: Orlando, Florida, USA, London, 1984; pp 95-148.

Melnick, R.L.; Huff, J. Liver carcinogenesis is not a predictive outcome of chemically induced hepatocyte proliferation. *Toxicol. Indust. Health* **1993**, *9*, 415-438.

Morisaki, N.; Lindsey, J.A.; Milo, G.E.; Maizewell, D.G. Fatty acid metabolism and cell proliferation. Effect of prostaglandin biosynthesis either from exogenous fatty acid metabolism or endogenous fatty acid release with hydralazine. *Lipids* **1983**, *18*, 349-352.

Nagai, M.K.; Armstrong, D.; Farber, E. Induction of resistant hepatocytes by clofibrate, a non-genotoxic carcinogen. *Proc. Am. Ass. Cancer Res.* **1993**, *34*, 164.

Nolan, R.D.; Danliowicz, R.M.; Eling, T.E. Role of arachidonic acid metabolism in the mitogenic response of Balb/c 3T3 fibroblasts to epidermal growth factor. *Molecular Pharmacol.* **1988**, *33*, 650-656.

Norred W.P.; Plattner, R.D.; Vesonder, R.F.; Bacon, C.W.; Voss, K.A. Effects of selected secondary metabolites of *Fusarium moniliforme* on unscheduled synthesis of DNA by primary rat hepatocytes. *Fd. Chem. Toxicol.* **1992**, *30*, 233-237.

Pichiri-coni, G.; Coni, P.; Laconi, E.; Schwarze, P.E.; Seglen, P.O.; Rao, P.M.; Rajalakshimi, S.; Sarma, D.S.R. Studies on the mitoinhibitory effect of orotic acid on hepatocytes in primary culture. *Carcinogenesis* **1990**, *11*, 981-984.

Rao, M.S.; Reddy, J.K. An overview of peroxisome proliferator-induced hepatocarcinogenesis. *Environ. Health Perspect.* **1991**, *93*, 205-209.

Reifferscheid, G.; Heil, J.; Oda, Y.; Zahn, R.K. A microplate version of SOS/umu-test for rapid detection of genotoxins and genotoxic potential of environmental samples. *Mutat Res.* **1991**, *253*, 215-222.

Rheeder, J.P.; Marasas, W.F.O.; Thiel, P.G.; Sydenham, E.W.; Shephard, G.S.; Van Schalkwyk, D.J. *Fusarium moniliforme* and fumonisins in maize in relation to human esophageal cancer in Transkei. *Phytopathology* **1992**, *82*, 353-357.

Rheeder, J.P.; Sydenham, E.W.; Marasas, W.F.O.; Thiel, P.G.; Shephard, G.S.; Schlecther, M.; Stockenström, S.; Cronje, D.E.; Viljoen, T.H. Ear-rot fungi and mycotoxins in South Africa maize of the 1989 crop exported to Taiwan. *Mycopathologia* **1994**, *127*, 35-41.

Roomi, W.M.; Rotstein, J.; Semple, E. The biochemical-molecular pathology of the stepwise development of liver cancer: new insights and problems. In *The molecular biology of carcinogenesis*; Bannasch, P., Keppler, D., Pitot, H.C. *Cancer* **1993**, Suppl. *72*, 962-970.

Ross, R.K.,; Yaun, J.M.,; Yu, M.C.,; Wogan, G.N., Qian, G.S. Urinary aflatoxin biomarkers and risk of hepatocellular carcinoma. Lancet **1992**, *339*, 9343-9346.

Satoh, K.; Hatayama, I.; Tateoka, N.; Tamai, K.; Shimizu, T.; Tatematsu, M.; Ito, N.; Sato, K. Transient induction of single GST-P positive hepatocytes by DEN. *Carcinogenesis* **1989**, *10*, 2107-2111.

Sheffield, V.C.; Beck, J.S.; Kwitek, A.E.; Sandstrom, D.W.; Stone, E.M. The sensitivity of single-strand conformation polymorphism analyses for the detection of single base substitutions. *Genomics* **1993**, *16*, 325-332.

Shephard, G.S.; Thiel, P.G.; Sydenham, E.W.; Alberts, J.F.; Gelderblom, W.C.A. Fate of single dose of the ^{14}C-labelled mycotoxin, fumonisin B$_1$, in rats. *Toxicon* **1992**, *30*, 768-770.

Smith, J.E.; Lewis, C.W.; Anderson, J.G.; Solomons, G.L. Eds.; Mycotoxins in human nutrition and health. Directorate-General X11, *Science, Research and Development*, EUR 16048 EN, **1994**, pp 43-48.

Sydenham, E.W.; Shephard, G.S.; Thiel, P.G.; Marasas, W.F.O.; Stokenström, S. Fumonisin contamination of commercial maize-based human foodstuffs. *J. Agric. Food Chem.* **1991**, *39*, 2014-2018.

Thiel, P.G.; Marasas, W.F.O.; Sydenham, E.W.; Shephard, G.S.; Gelderblom, W.C.A. The implications of naturally occurring levels of fumonisins in maize for human and animal health. *Mycopathologia* **1992**, *177*, 3-9.

Viljoen, J.H.; Marasas, W.F.O.; Thiel P.G. Fungal infection and mycotoxin contamination of commercial maize. In *Cereal Science and Technology: Impact on a Changing South Africa*; Taylor, J.R.N., Randall, P.G., Viljoen, J.H., Eds.; CSIR, Pretoria, **1993**.

Voss, K.A.; Chamberlain, W.J.; Bacon, C.W.; Norred, W.P. A preliminary investigation on renal and hepatic toxicity in rats fed purified fumonisin B$_1$. *Natural Toxins* **1993**, *1*, 222-228.

Ward, J.M.; Henneman, J.R. Naturally-occurring age-dependent glutathione S-transferase P immunoreactive hepatocytes in aging female F344 rat liver as potential promotable targets for non-genotoxic carcinogens. *Cancer Lett.* **1990**, *52*, 187-195.

Ward, J.F.; Limoli, C.L.; Calabro-Jones, P.; Evans, J.W. Radiation vs. chemical damage to DNA. In Anticarcinogenesis and Radiation Protection; Cerutti P.A., Nygaard, O.F., Simic, M.G., Eds.; Plenum Press: New York, **1987**.

Weinstein, I.B. Nonmutagenic mechanism in carcinogenesis: role of protein kinase C in signal transduction and growth control. *Environ. Health Perspec.* **1991**, *93*, 175-179.

Yoo, H.; Norred, W.P.; Wang, E.; Merrill A.H.(Jr); Riley, R.T. Fumonisin inhibition of de novo sphingolipid biosynthesis and cytotoxicity are correlated in LLC-PK$_1$ cells. *Toxicol. Appl. Pharmacol.* **1992**, *114*, 9-15.

25

FUMONISIN TOXICITY AND SPHINGOLIPID BIOSYNTHESIS

A. H. Merrill, Jr. [*][1], E. Wang,[1] T. R. Vales,[1] E. R. Smith,[1] J. J. Schroeder,[1]
D. S. Menaldino,[2] C. Alexander,[2] H. M. Crane,[2] J. Xia,[2] D. C. Liotta,[2]
F. I. Meredith,[3] and R. T. Riley[3]

Departments of [1]Biochemistry and [2]Chemistry
Emory University
Atlanta, GA 30322; and
[3] U. S. Department of Agriculture
   Agriculture Research Service, Toxicology and Mycotoxins Research Unit
   Athens, GA 30613
[*] Author to whom correspondence should be addressed (telephone
      404-727-5978; telefax 404-727-3954; e-mail amerril@unix.cc.emory.edu

ABSTRACT

Fumonisins are inhibitors of sphinganine (sphingosine) N-acyltransferase (ceramide synthase) *in vitro*, and exhibit competitive-type inhibition with respect to both substrates of this enzyme (sphinganine and fatty acyl-CoA). Removal of the tricarballylic acids from fumonisin B_1 reduces the potency by at least 10 fold; and fumonisin A_1 (which is acetylated on the amino group) is essentially inactive. Studies with diverse types of cells (hepatocytes, neurons, kidney cells, fibroblasts, macrophages, and plant cells) have established that fumonisin B_1 not only blocks the biosynthesis of complex sphingolipids; but also, causes sphinganine to accumulate. Some of the sphinganine is metabolized to the 1-phosphate and degraded to hexadecanal and ethanolamine phosphate, which is incorporated into phosphatidylethanolamine. Sphinganine is also released from cells and, because it appears in blood and urine, can be used as a biomarker for exposure. The accumulation of these bioactive compounds, as well as the depletion of complex sphingolipids, may account for the toxicity, and perhaps the carcinogenicity, of fumonisins.

INTRODUCTION

Fumonisins are a family of mycotoxins that are produced by some strains of *Fusarium moniliforme* (Bezuidenhout et al., 1988) and have been shown to be toxic and carcinogenic for animals and, perhaps, humans (for a recent review see Riley et al., 1993b). It is not yet

Fumonisins in Food, Edited by L. Jackson *et al.*
Plenum Press, New York, 1996

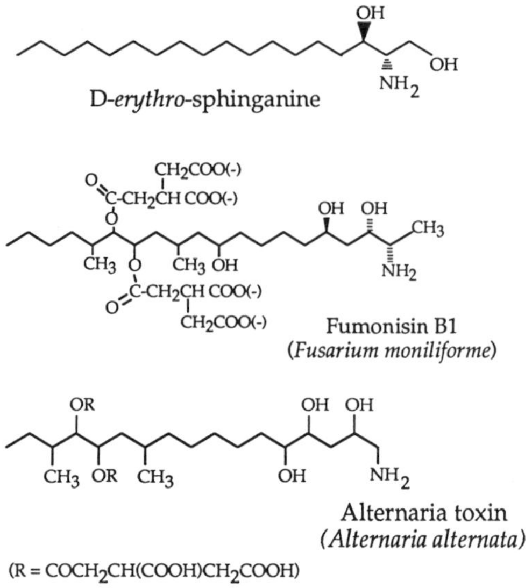

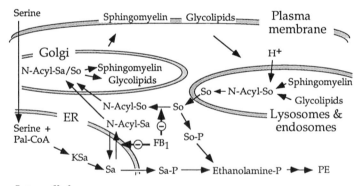

D-*erythro*-sphinganine

Fumonisin B1
(*Fusarium moniliforme*)

Alternaria toxin
(*Alternaria alternata*)

(R = COCH₂CH(COOH)CH₂COOH)

Figure 1. Structures of sphinganine, fumonisin B₁, and alternaria toxin.

Figure 2. Sphingolipid biosynthesis and turnover. The first steps occur in the endoplasmic reticulum (ER), where serine is condensed with palmitoyl-CoA (Pal-CoA) to yield 3-ketosphinganine (KSa), which is reduced to sphinganine (Sa). Sphinganine is acylated to dihydroceramides (also called N-acyl-sphinganines, N-Acyl-Sa) by ceramide synthase using various long-chain fatty acyl-CoA's. Headgroups (e.g., phosphorylcholine, glucose, galactose, and hundreds of more complex polysaccharides) and the 4,5-*trans*-double bond (of sphingosine, So) are subsequently added to the 1-hydroxyl group in the Golgi and plasma membranes. Sphingolipid turnover is thought to involve the internalization of the sphingolipids, followed by their hydrolysis in acidic compartments (lysosomes and endosomes) to ceramides (N-Acyl-So), then to sphingosine (So), as shown. Sphingosine is either reacylated or phosphorylated (to So-P) and cleaved to a fatty aldehyde and ethanolamine phosphate, which is incorporated into phosphatidylethanolamine (PE). In the presence of fumonisin B₁ (FB₁), sphinganine accumulates and there is also an increase in sphinganine 1-phosphate (Sa-P).

known how many different fungi produce fumonisins; however, Alternaria toxins are structurally similar (Bottini et al., 1981), which makes it likely that additional fungal metabolites with some of these structural features will be found.

Fumonisins are elaborations of a backbone that is characteristic of sphinganine (Figure 1), an intermediate in the biosynthesis of complex sphingolipids (Sweeley, 1991, Bell et al., 1993). This similarity led us to hypothesize that fumonisins might interact with enzyme(s) of sphingolipid metabolism or interfere with some of the functions of sphingolipids (or both).

Studies of sphingolipid biosynthesis by rat hepatocytes in culture (Wang et al., 1991) revealed that fumonisin B$_1$ is a potent inhibitor of ceramide synthase, the enzyme that is responsible for the acylation of sphinganine in the *de novo* biosynthetic pathway for sphingolipids as well as the re-acylation of sphingosine that is released upon sphingolipid turnover (Figure 2). This article will summarize some of the properties of this inhibition *in vitro*, and the results of various studies that have explored the implications of this inhibition for the toxicity and carcinogenicity of fumonisins.

INHIBITION OF CERAMIDE SYNTHASE *IN VITRO*

Fumonisins of the "B" series (which have structures similar to the one shown in Figure 1) have, thus far, all been shown to inhibit ceramide synthase at low or sub-micro-molar concentrations (Wang et al., 1991; Merrill et al., 1993b,c). In recent assays of fumonisin A$_1$, which has the same structure as fumonisin B$_1$ except that there is an acetyl group on the nitrogen, we have found <2 % inhibition of ceramide synthase at 10 µM; therefore, acetylation appears to reduce the potency by more than 50 fold. Removal of the tricarballylic acid sidechains reduces the potency *in vitro* by approximately 10 fold (Merrill et al., 1993c).

Fumonisin B$_1$ inhibits ceramide synthase with either sphinganine or sphingosine as the substrate, and with a variety of fatty acyl-coenzyme A's as the co-substrate (Wang et al., 1991; Merrill et al., 1993a and 1993b). The inhibition is competitive with both the long-chain (sphingoid) base and fatty-acyl-CoA (Merrill et al., 1993c), which indicates that fumonisins may inhibit ceramide synthase by interacting with both the binding site for sphinganine (sphingosine) and the site for the fatty acyl-CoA's, as envisioned by the model in Figure 3.

In recent studies of a series of 2-amino-,3,5-dihydroxyoctadecanes (Figure 4) (T. R. Vales et al., manuscript in preparation) that were synthesized as sphinganine and fumonisin analogs, all of the compounds with the *erythro*-configuration were utilized by ceramide synthase almost as well as the natural substrate D-*erythro*-sphinganine, despite removal of the 1-hydroxyl group and addition of an extra hydroxyl at position 5.

The *threo*-analogs were also acylated, but were much poorer substrates and inhibitors. Fumonisins have been found to have the *threo*-configuration (ApSimon et al., 1994; Harmange et al., 1994; Hoye et al., 1994; Poch et al., 1994); therefore, it is tempting to speculate that fumonisins may be acylated by ceramide synthase since the *threo*-analogs were, albeit poor, substrates for this enzyme. All of the analogs in Figure 4 are cytotoxic for various cell lines in culture (data not shown), as has been seen before (Krauss et al., 1992); however, this toxicity may be only indirectly related to the mechanism of action of fumonisins because long-chain amines are generally cytotoxic (Stevens et al., 1990).

It is possible that fumonisins also inhibit other enzymes that interact with either long-chain bases or fatty acyl-CoA's. To test this possibility, fumonisin B$_1$ has been tested with sphingosine kinase (R. M. Bell, personal communication) and serine palmitoyltransferase (Wang et al., 1991), and neither enzyme was inhibited by the levels that result in complete inhibition of ceramide synthase.

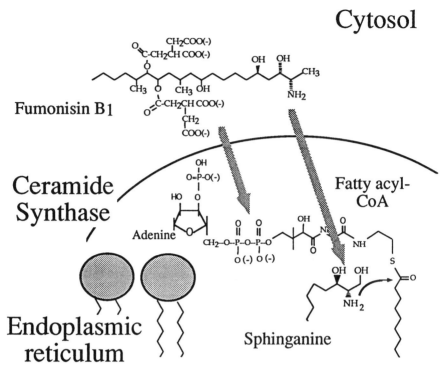

Figure 3. A hypothetical scheme illustrating the interaction of ceramide synthase with its substrates, sphinganine and fatty acyl-CoA's, and a possible mechanism for inhibition by fumonisins.

INHIBITION OF SPHINGOLIPID BIOSYNTHESIS IN CELL SYSTEMS

Fumonisins are potent inhibitors of sphingolipid biosynthesis by rat hepatocytes, with an IC_{50} of approximately 0.1 μM for fumonisin B_1 and B_2 (Wang et al., 1991; Merrill et al., 1993c). Inhibition has also been demonstrated with a variety of cell types (as described below); however, higher fumonisin concentrations have typically been required for inhibition. Fumonisin B_1 blocks the formation of more complex sphingolipids (Wang et al., 1991, Merrill et al., 1993b, Riley et al., 1994a); in cultured mouse cerebellar neurons (where the individual species of sphingolipids have been analyzed), the IC_{50} for inhibition of sphingomyelin biosynthesis was 10-fold lower than for glycolipid biosynthesis (Merrill et al., 1993b).

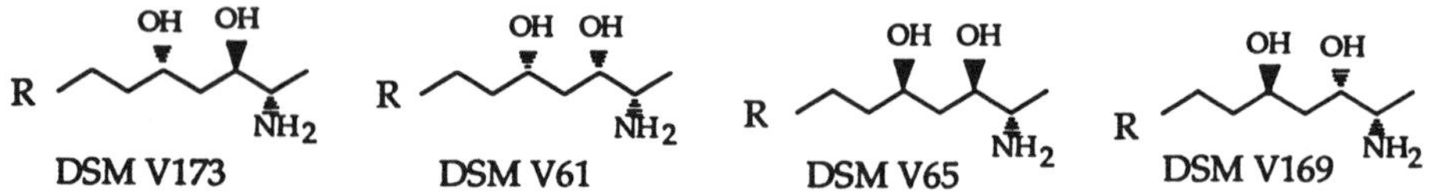

Figure 4. Structures of synthetic 1-deoxy-5-hydroxy-sphinganines as fumonisin analogs. The R group represents the same alkyl sidechain, $CH_3(CH_2)_{10}$-, that is found in C18-sphingosine and sphinganine.

The inhibition of ceramide synthase by fumonisins also causes sphinganine to accumulate, and at least a portion of the sphinganine undergoes phosphorylation and cleavage (Figure 2) (Merrill et al., 1993c). In recent studies of J774 cells (Smith and Merrill, submitted for publication), we have found that upon addition of fumonisin B_1, long-chain base turnover can account for about one-third of the ethanolamine that is used to synthesize phosphatidylethanolamine. This raises the possibility that fumonisins can affect the metabolism of phosphoglycerolipids via this indirect route.

In all cases studied to date, the concentration of sphinganine rises significantly when cells are treated with fumonisins. For example, sphinganine levels increased 110 fold (approximately 1 nmol/mg protein) when rat hepatocytes were incubated with 1 μM fumonisin B_1 for 4 days (Wang et al., 1991). In cell culture, there is usually little change (or decrease) in free sphingosine, consistent with this long-chain base arising from turnover rather than *de novo* biosynthesis. However, when cells begin to die, sphingosine levels would be expected to increase due to the breakdown of membrane lipids.

INHIBITION OF SPHINGOLIPID BIOSYNTHESIS *IN VIVO*

At least some of the sphinganine that accumulates in cells treated with fumonisins leaves the cells and appears in blood (Wang et al., 1992). As an example of typical changes, a pony that consumed a diet containing 44 ppm fumonisin B_1 exhibited a 2.7-fold increase in sphinganine by day 2, a 4.5-fold increase by day 7, and elevations in both sphinganine (13 fold) and sphingosine (2-fold) by day 10 (Wang et al., 1992). On day 10, there was a 7-fold increase in serum transaminase activities, which indicated that the pony had significant cellular injury. At this stage, at least, the elevation in serum sphinganine is reversible because when the pony refused to eat the contaminated feed the amount of sphinganine in serum began to return to normal. Upon resumption of the fumonisin-contaminated diet, there was another increase in sphinganine and the animal began to show clinical signs of equine leukoencephalomalacia and required euthanasia on day 45 (Wang et al., 1992). Thus, the level of sphinganine in serum can be used to detect consumption of fumonisins and, if found early enough, might allow a response before the mycotoxin has caused irreversible injury to the animal. Although sphingosine is also sometimes elevated—especially if there is cell death and membrane lipids are being turned over more rapidly, the sphinganine/sphingosine ratio also correlates well with exposure (Wang et al., 1992), and may be a more useful index of exposure (Riley et al., 1993c).

There is also a reduction in the amount of complex sphingolipids in serum of animals that consume fumonisins (Wang et al., 1992), as would be expected if the elevation in sphinganine is due to inhibition of *de novo* sphingolipid biosynthesis. One explanation for these changes might be that fumonisin B_1 blocks the secretion of sphingolipids with lipoproteins, since the liver is a major site of synthesis of sphingolipids (Merrill and Jones, 1990) and it secretes sphingolipids with very-low density lipoproteins (Merrill et al., 1995).

Similar analyses have been conducted with pigs (Riley et al., 1993a), chickens (Weibking et al., 1993), rats (Riley et al., 1994a) and other animals. As far as we are aware, there have been no reports where exposure of animals to toxic levels of fumonisins has not led to an elevation in these biomarkers; therefore, it appears that they can be used as early indicators of exposure.

POSSIBLE MECHANISMS FOR GROWTH INHIBITION AND TOXICITY OF FUMONISINS

The disruption of sphingolipid biosynthesis would be predicted to have profound effects on cells because these compounds have important roles in membrane and lipoprotein structure, cell-cell communication, interactions between cells and the extracellular matrix, regulation of growth factor receptors, and as second messengers for a wide range of factors, including tumor necrosis factor, interleukin 1, and nerve growth factor (for recent reviews, see Merrill et al., 1993a and other articles in the two volumes edited by Bell, Hannun, and Merrill, 1993a,c).

Based on current knowledge about some of the systems that have been discovered to be regulated by sphingolipids, a number of mechanisms can be proposed whereby disruption of sphingolipid metabolism by fumonisins could account for the toxicity (Figure 5) and carcinogenicity (Figure 6) of these mycotoxins.

Two of the most likely explanations for the cell death after inhibition of sphingolipid biosynthesis by fumonisins are:

1) The accumulation of free sphinganine (and possibly its metabolites, such as the 1-phosphate) is growth inhibitory and cytotoxic for the cells. Long-chain (sphingoid) bases are well known to be growth inhibitory and cytotoxic (Merrill, 1983; Stevens et al., 1990); therefore, the accumulation of sphinganine (and sometimes sphingosine) might lead to cell death. The cellular target that accounts for these effects of long-chain bases is unknown, but there are a number of candidates. Long-chain bases have been shown to inhibit protein kinase C, to activate phospholipase D, and to activate or inhibit other enzymes of lipid signalling pathways (such as phosphatidic acid phosphatase and phospholipase C), to inhibit the Na^+/K^+ ATPase, to induce dephosphorylation of retinoblastoma protein (a key regulator of the G to S transition of the cell cycle), to induce release of Ca^{2+} from intracellular stores (apparently via sphingosine 1-phosphate), and to affect a large number of other cell regulatory systems (for an overview see Merrill et al., 1993a). In addition, sphingosine has been observed to induce apoptosis in thymocytes (Bai et al., 1990) and neutrophils (Ohta et al., 1994).

2) The loss of complex sphingolipid biosynthesis would be expected to alter cell behavior, and could also lead to cell death based on findings with mutants in serine palmitoyltransferase, the initial enzyme of sphingolipid biosynthesis (Hanada et al., 1990, 1992).

Possible mechanisms for the cell behavior changes, growth inhibition, and cytotoxicity induced by fumonisins via disruption of sphingolipid metabolism

---

Accumulation of sphinganine & sphingosine and metabolites (such as the 1-phosphates)

Depletion of critical complex sphingolipids

Inhibitions of Na^+/K^+ ATPase

Inhibition of protein kinase C

Release of intracellular Ca^{2+}

Promotion of Rb dephosphorylation

Induction of apoptosis

Figure 5. Possible mechanisms for changes in cell behavior, growth inhibition, and cytotoxicity of fumonisins.

Possible mechanisms for promotion of
carcinogenesis by fumonisins via disruption
of sphingolipid metabolism

Activation of EGF receptor/MAP kinase

Activation of Phospholipase D and/or
    Inhibition of Phosphatidic acid phosphohydrolase

Release of intracellular calcium

Activation of AP-1

Cytotoxicity for normal cells

Loss of regulation of differentiation

Loss of regulation of apoptosis

Loss of lipid mediators of TNF

Figure 6. Possible mechanisms for the carcinogenicity of fumonisins.

There have only been a few studies that have attempted to relate the effects of fumonisins on sphingolipid metabolism to cell behavior. Yoo et al. (1992) have shown a close association between the inhibition of sphingolipid biosynthesis in LLC-PK$_1$ cells (a renal epithelial cell line) and the growth inhibition and toxicity of fumonisin B$_1$. In more recent studies (Yoo et al., manuscript in preparation), it has been found that there is a close association between the amount of sphinganine that accumulates in the cells, the depletion of complex sphingolipids, and cell death (as measured by reduction in cell protein and release of lactate dehydrogenase into the medium).

Studies with rat hippocampal neurons have shown that fumonisins inhibit neuron extension and synapse formation, and that this could be circumvented by addition of ceramide to the cells (Harel and Futerman, 1993). The ability to restore normal function by ceramide replacement shows that these effects of fumonisins are due to the loss of biosynthesis of a key sphingolipid (or sphingolipids).

In another recent study (Ramasamy et al., manuscript submitted), we have examined the effect of fumonisin B$_1$ on the barrier function of endothelial cells. For these experiments, endothelial cells from porcine pulmonary arteries were cultured on micropore filters and the effect of fumonisin B$_1$ (at 20 to 50 µM) on the movement of albumin across the monolayer was determined. The addition of fumonisin B$_1$ doubled the rate of albumin transfer without any apparent change in the number of viable cells, as measured by trypan blue exclusion and morphology. Therefore, these effects may be due to changes in cell-cell interaction due to a loss of complex sphingolipids, or to related changes due to the accumulation of sphinganine (or both). The finding that endothelial cells are affected by fumonisins may explain the vascular damage observed in pigs (Casteel et al., 1994) and may play a role in the leakage of fluid into the lung in porcine pulmonary edema syndrome. This may also contribute to the edema and perivascular hemorrhaging that have been noted in the brain of horses with fumonisin-induced equine leukoencephalomalacia (Marasas et al., 1988).

Some of these same mechanisms may account for the ability of fumonisins to serve as carcinogens (Figure 6). For a recent review of these hypotheses, see Riley et al., 1994b.

While fumonisin B$_1$ has been found to be capable of being a complete carcinogen at high doses, fumonisins were not found to be mutagenic in the Salmonella mutagenicity assay (Gelderblom et al., 1991, 1992), and do not appear to be genotoxic based on DNA repair assays using primary rat hepatocytes (Norred et al., 1992). Therefore, at low dosages, fumonisins appear to be mainly promoters (Gelderblom et al., 1988). Tumor promoters are often mitogens, which intrigued us because Spiegel and co-workers have shown that sphingosine and sphingosine 1-phosphate can induce DNA synthesis in growth-arrested Swiss 3T3 cells (Zhang et al. 1990 and 1991). Therefore, we hypothesized that fumonisins might induce DNA synthesis via the accumulation of sphinganine (Schroeder et al., 1994). Using Swiss 3T3 cells, we have demonstrated that addition of fumonisin B_1 to the cells elevates sphinganine and induces an increase in [^{3}H]thymidine incorporation into DNA. Furthermore, both were blocked by addition of an inhibitor of serine palmitoyltransferase (β-fluoro-L-alanine), which established that this effect of fumonisins is due to sphinganine accumulation, not the depletion of complex sphingolipids.

There are many possible ways in which fumonisins might affect carcinogenesis by altering sphingolipid metabolism (Figure 6) (for recent reviews see Merrill et al., 1993a,c and other chapters in Bell et al., 1993). These include the activation of an epidermal growth factor receptor kinase—probably via the mitogen activated protein kinases (MAP kinases), control of intracellular calcium (seemingly via the 1-phosphate), increases in AP-1—an early step in the growth of some cell types, *inter alia*. In addition, more complex sphingolipids are involved in vital cell functions that are part of the regulation of normal cell growth, differentiation, and apoptosis. Therefore, fumonisins might impact many of these processes, and will likely have a variety of effects depending on the cell type and other factors.

SUMMARY

Unless system(s) other than sphingolipid metabolism are discovered to be affected by fumonisins, the inhibition of ceramide synthase should be considered the initial step in the toxicity and carcinogenicity of these mycotoxins. Nonetheless, this represents only the starting point for understanding the pathological effects of these compounds. More must be learned about the cellular consequences of elevations in sphinganine (and various sphinganine metabolites) and the reductions in key complex sphingolipids caused by fumonisins. Such investigations should provide insight into not only the the mechanisms of diseases caused by mycotoxins; but also, into the biological roles of sphingolipids.

ACKNOWLEDGMENTS

The authors are grateful to several collaborators in various aspects of this work, namely Drs. C. W. Bacon, V. Beasley, W. Haschek, B. Hennig, B. Lagu, R. LaRocque, W. P. Norred, R. D. Plattner, S. Ramasamy, P. F. Ross, K. Sandhoff, P. Stancel, G. vanEchten-Deckert, K. A. Voss and T. Wilson. We also thank Ms. Winnie Scherer for help in preparing this text. This work was supported by funds from the USDA (including grant #91-37204-6684) and the NIH (grants GM33369 and GM46368).

REFERENCES

ApSimon, J. W.; Blackwell, B. A.; Edwards, O. E.; Fruchier, A. Relative configuration of the C-1 to C-5 fragment of fumonisin B_1. *Tetrahedron Lett.* **1994**, 35, 7703-7706.

Bai, C.; Aw, T. Y.; Wang, E.; Merrill, A. H., Jr.; Jones, D. P. Effect of sphingosine, gangliosides, cyclic AMP, and interferons on programmed cell death, *FASEB J.* **1990**, 4, 477.

Bell, R.M.; Hannun, Y.A.; Merrill, A.H. Jr., Eds. *Advances in Lipid Research: Sphingolipids and Their Metabolites*, Academic Press, Orlando, FL, 1993, volumes 25-26.

Bezuidenhout, C. S.; Gelderblom, W. C. A.; Gorstallman, C. P.; Horak, R. M.; Marasas, W. F. O.; Spiteller, G.; Vleggaar, R. Structure elucidation of the fumonisins, mycotoxins from *Fusarium moniliforme. J. Chem. Soc. Commun.* **1988**, 743-745.

Bottini, A. T.; Bowen, J. R.; Gilchrist, D. G. Phytotoxins II. Characterization of a phytotoxic fraction from *Alternaria alternata f. sp. lycopersici, Tetrahedron Lett.* **1981**, 22, 2723-2726.

Casteel, S. W.; Turk, J. R.; Rottinghaus, G. E. Chronic effects of dietary fumonisin on the heart and pulmonary vasculature of swine. *Fundam. Appl. Toxicol.* **1994**, 23, 518-524.

Gelderblom, W. C. A.; Jaskiewicz, K.; Marasas, W. F. O.; Thiel, P. G.; Horak, R. M.; Vleggaar, R.; Kriek, N. P. J. Cancer promoting potential of different strains of *Fusarium moniliforme* in a short-term cancer initiation/promotion assay. *Carcinogenesis* **1988**, 9, 1405-1409.

Gelderblom, W. C. A.; Kriek, N. P. J.; Marasas, W. F. O.; Thiel, P. G. Toxicity and carcinogenicity of the *Fusarium moniliforme* metabolite, fumonisin B₁, in rats. *Carcinogenesis* **1991**, 12, 1247-1251.

Gelderblom, W. C. A,; Semple E.; Marasas, W. F. O.; Farber, E. The cancer-initiating potential of the fumonisin-B mycotoxins. *Carcinogenesis* **1992**, 13, 433-437.

Hanada, K.; Nishijima, M.; Akamatsu, Y. A temperature-sensitive mammalian cell mutant with thermolabile serine palmitoyltransferase for the sphingolipid biosynthesis. *J. Biol. Chem.* **1990**, 265, 22137-22142.

Hanada, K.; Nishijima, M.; Kiso, H.; Hasegawa, A.; Fujita, S.; Ogawa, T.; Akamatsu, Y. Sphingolipids are essential for the growth of Chinese hamster ovary cells. Restoration of the growth of a mutant defective in sphingoid base biosynthesis with exogenous sphingolipids, *J. Biol. Chem.* **1992**, 267, 23527-23533.

Harel, R.; Futerman, A. H. Inhibition of sphingolipid synthesis affects axonal outgrowth in cultured hippocampal neurons, *J. Biol. Chem.* **1993**, 268, 14476-14482.

Harmange, J.-C.; Boyle, C. D.; Kishi, Y. Relative and absolute stereochemistry of the fumonisin B₂ backbone, *Tetrahedron Lett.* **1994**, 35, 6819-6822.

Hoye, T. R.; Jimenez, J. I.; Shier, W. T. Relative and absolute configuration of the fumonisin B₁ backbone, *J. Am. Chem. Soc.* **1994**, 116, 9409-9410.

Krauss, G. A.; Applegate, J. M.; Reynolds, D. Synthesis of analogs of fumonisin B₁. *J Agric. Food Chem.,* **1992**, 40, 2331-2332.

Marasas, W. F. O.; Kellerman, T. S.; Gelderblom, W. C. A.; Coetzer, J. A. W.; Thiel, P. G.; van der Lugt, J. J. Leukoencephalomalacia in a horse induced by fumonsin B₁ isolated from *Fusarium moniliforme.* *Onderstepoort J. Vet. Res.* **1988**, 55, 197-203.

Merrill, A. H., Jr. Characterization of serine palmitoyltransferase activity in Chinese hamster ovary cells. *Biochim. Biophys. Acta* **1983**, 754, 284-291.

Merrill, A. H., Jr.; Jones, D. D. An update of the enzymology and regulation of sphingomyelin metabolism. *Biochim. Biophys. Acta 1044;* **1990**, 1-12.

Merrill, A. H.; Jr; Hannun, Y. A.; Bell, R. M, Sphingolipids and their metabolites in cell regulation. In *Advances in Lipid Research: Sphingolipids and Their Metabolites,* Bell, R. M.; Hannun, Y. A.; Merrill, A. H. Jr, Eds., Academic Press, San Diego, CA., 1993a, 25, pp 1-24,

Merrill, A. H., Jr.; vanEchten, G.; Wang, E.; Sandhoff, K. Fumonisin B₁ inhibits sphingosine (sphinganine) N-actyltransferase and *de novo* sphingolipid biosynthesis in cultured neurons *in situ. J. Biol. Chem.,* **1993b**, 268, 27299-27306.

Merrill, A. H., Jr.; Wang, E.; Gilchrist, D. G.; Riley, R. T. Fumonisins and other inhibitors of de novo sphingolipid biosynthesis. In *Advances in Lipid Research: Sphingolipids and Their Metabolites,* Bell, R. M.; Hannun, Y. A.,; Merrill, A. H. Jr., Eds., Academic Press, San Diego, CA, 1993c, 26, pp 215-234.

Merrill, A. H., Jr.; Lingrell, S.; Wang, E.; Nikolova-Karakashian, M.; Vales, T.R.; and Vance, D. E. Sphingolipid biosynthesis *de novo* by rat hepatocytes in culture: Ceramide and sphingomyelin are associated with, but not required for, very low density lipoprotein secretion. *J Biol Chem,* **1995**, 270, 13834-13841.

Norred, W. P.; Plattner, R. D.; Vesonder, R. F.; Bacon, C. W.; Voss; K. A. Effects of selected secondary metabolites on unscheduled synthesis of DNA by rat primary hepatocytes. *Food. Chem. Toxicol.* , **1992a**, 26, 233-237.

Ohta, H.; Yatomi, Y.; Sweeney, E. A.; Hakomori, S. -I.; Igarashi, Y. A possible role of sphingosine in induction of apoptosis by tumor necrosis factor-α in human neutrophils. *FEBS Lett.* **1994**, 355, 267-270.

Poch, G. K.; Powell, R.G.; Plattner, R. D.; Weisleder, D. Relative stereochemistry of fumonisin B₁ at C-2 and C-3. *Tetrahedron Lett.* **1994**, 35, 7707-7710.

Riley, R. T.; An, N. H.; Showker, J. L.; Yoo, H-S.; Norred, W. P.; Chamberlain, W. J.; Wang, E.; Merrill, A. H., Jr.; Motelin, G.; Beasley, V. R.; Haschek, W. M. Alteration of tissue and serum sphinganine to sphingosine ratio: An early biomarker of exposure to fumonisin-containing feeds in pigs. *Toxicol. & Appl. Pharmacol* **1993a**, 118, 105-112.

Riley, R. T.; Norred, W. P.; Bacon, C. W. Fungal toxins in foods: Recent concerns. *Annu. Rev. Nutr.,* **1993b**, 13, 167-189.

Riley, R. T.; Wang, E.; Merrill, A. H., Jr. Liquid chromatography of sphinganine and sphingosine: Use of sphinganine to sphingosine ratio as a biomarker for consumption of fumonisins. *J. Assoc. Off. Anal. Chem.* **1993c**, 77, 533-540.

Riley, R. T.; Hinton, D. M.; Chamberlain, W. J.; Bacon, C. W.; Wang, E.; Merrill, A. H., Jr.; Voss, K. A. Dietary fumonisin B$_1$ induces disruption of sphingolipid metabolism in Sprague-Dawley rats: A new mechanism of nephrotoxicity. *J. Nutr.* **1994a**, 124, 594-603.

Riley, R. T.; Voss, K. A.; Yoo, H. -S.; Gelderblom, W. C. A.; Merrill, A. H., Jr. Mechanism of fumonisin toxicity and carcinogenicity. *J. Food Protect.*, **1994b**, 57, 638-645.

Schroeder, J. J.; Crane, H. M.; Xia, J.; Liotta, D. C.; Merrill, A. H., Jr. Disruption of sphingolipid metabolism and stimulation of DNA synthesis by fumonisin B$_1$: A molecular mechanism for carcinogenesis associated with *Fusarium moniliforme*. *J. Biol. Chem.*, **1994**, 269, 3475-3481.

Stevens, V. L.; Nimkar, S.; Jamison, W. C.; Liotta, D. C.; Merrill, A. H. Jr. Characteristics of the growth inhibition and cytotoxicity of long-chain (sphingoid) bases for Chinese hamster ovary cells: Evidence for an involvement of protein kinase C. *Biochim. Biophys. Acta* **1990**, 1051, 37-45.

Sweeley, C. C. Sphingolipids. In *Biochemistry of Lipids, Lipoproteins, and Membranes, Vance, D. E.; Vance, J. E., Eds., Elsevier Science Publ., Amsterdam, 1991, pp 327-361.*

Wang, E.; Norred, W. P.; Bacon, C. W.; Riley, R. T.; Merrill, A. H., Jr. Inhibition of sphingolipid biosynthesis by fumonisins: Implications for diseases associated with *Fusarium moniliforme*. *J. Biol. Chem.* **1991**, 266, 14486-14490.

Wang, E.; Ross, P. F.; Wilson, T. M.; Riley, R. T.; Merrill, A. H., Jr. Alteration of serum sphingolipids upon dietary exposure of ponies to fumonisins, mycotoxins produced by *Fusarium moniliforme*. *J. Nutr.* **1992**, 122, 1706-1716.

Weibking, T. S.; Ledoux, D. R.; Bermudez, A. J.; Turk, J. R.; Rottinghaus, G. E.; Wang, E. Merrill, A. H. Jr. Effects of feeding *Fusarium moniliforme* culture material, containing known levels of fumonisin B$_1$, on the young broiler chick. *Poultry Sci.* **1993**, 72, 456-466.

Yoo, H.; Norred, W. P.; Wang, E.; Merrill, A. H., Jr.; Riley, R. T. Fumonisin inhibition of *de novo* sphingolipid biosynthesis and cytotoxicity are correlated in LLC-PK$_1$ cells. *Toxicol. & Appl. Pharmacol.* **1992**, 114, 9-15.

Zhang, H.; Buckley, N. E.; Gibson, K.; Spiegel, S. Sphingosine stimulates cellular proliferation via a protein kinase C-independent pathway. *J. Biol. Chem.* **1990**, 265, 76-81.

Zhang, H.; Desai, N. N.; Olivera, A.; Seki, T.; Booker, G.; Spiegel, S. Sphingosine-1-phosphate, a novel lipid, involved in cellular proliferation. *J. Cell Biol.* **1991**, 114, 155-167.

26

CHEMISTRY AND BIOLOGICAL ACTIVITY OF AAL TOXINS

Carl K. Winter[1], David G. Gilchrist[2], Martin B. Dickman[3], Clinton Jones[4]

[1] Department of Food Science and Technology
[2] CEPRAP and Department of Plant Pathology
University of California, Davis
Davis, CA 95616
[3] Department of Plant Pathology
[4] Department of Veterinary and Biomedical Science and The Center for
    Biotechnology
University of Nebraska, Lincoln
Lincoln, NE 68583

ABSTRACT

AAL toxins and fumonisins comprise a family of highly reactive, chemically related mycotoxins that disrupt cellular homeostasis in both plant and animal tissues. Two critical issues to resolve are the detection of the entire family in food matricies and the mode of cellular disruption. Analysis of the entire set of chemical congeners in food matrices is difficult but has been achieved by a combination of different HPLC and mass spectrometry strategies. The mode of cellular disruption is unknown but likely involves changes associated with the inhibition of ceramide synthase in both plants and animals. Toxin treated cells exhibit morphological and biochemical changes characteristic of apoptosis. Further evaluation of the specific genetic and biochemical changes that occur during toxin-induced cell death may aid in understanding the mole of the action of these mycotoxins.

INTRODUCTION

Mycotoxins produced by the fungus *Alternaria alternata* f. sp. *lycopercici*, collectively known as the AAL toxins, were described initially as host-specific determinants of the Alternaria stem canker disease of tomato (Gilchrist and Grogan, 1976). Structural characterization of AAL toxins TA and TB indicated that each of the toxins existed as a pair of regioisomers of 1,2,3 propanetricarboxylic acid (tricarballylic acid) esterified to 1-amino-11,15-dimethylheptadeca-2,4,5,13,14-pentol (TA) and 1-

Fumonisins in Food, Edited by L. Jackson *et al.*
Plenum Press, New York, 1996

amino-11,15- dimethylheptadeca-2,4,13,14-tetrol (TB) (Bottini and Gilchrist, 1981; Bottini et al., 1981).

Interest in the AAL toxins increased following the South African reports confirming that fumonisins, mycotoxins produced by isolates of *Fusarium moniliforme*, were structurally related to the AAL toxins as bistricarballylic aminopolyol esters (Figure 1) (Bezuidenhout et al., 1988) and that animal (Marasas et al., 1988a; Gelderblom et al., 1988) and human diseases (Marasas et al., 1988b; Sydenham et al., 1990) associated with consumption of maize colonized by *F. moniliforme* were linked to levels of fumonisins present in the grain. AAL toxins and the fumonisins share similar toxicological mechanisms of action as both inhibit ceramide synthase in animal cells (Merrill et al., 1993), both inhibit cell proliferation in rat liver and dog kidney cells (Mirocha et al., 1992), and both induce cell death in tomato tissues and protoplasts (Gilchrist et al., 1992; Moussatos et al., 1993a). The toxicological properties and structural similarities of the AAL toxins and the fumonisins have raised concern about the potential effects of these mycotoxins on human and animal health. This review focuses on the chemistry and biological activity of the AAL toxins.

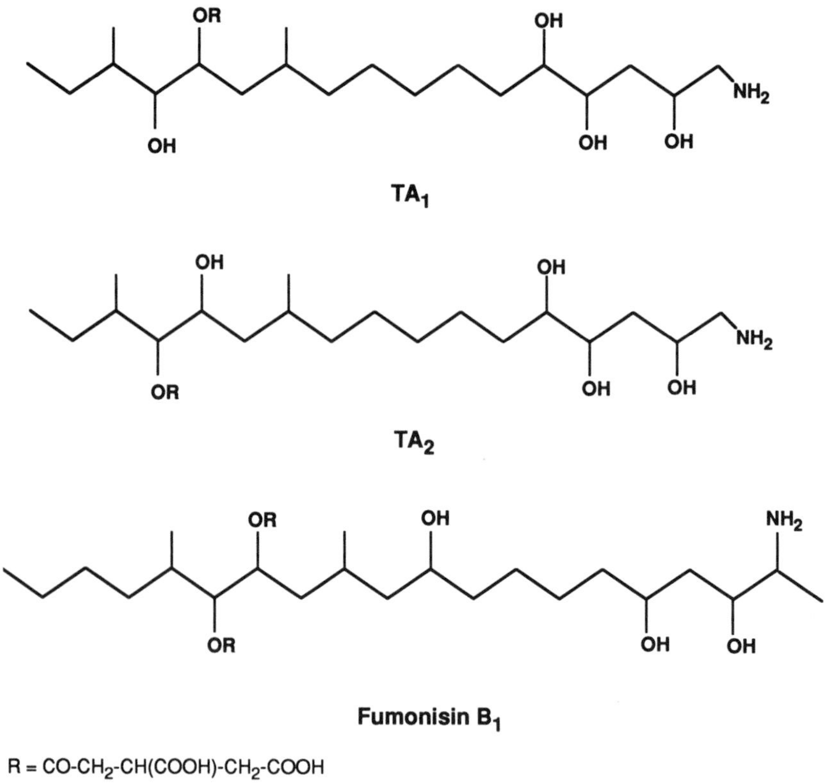

R = CO-CH$_2$-CH(COOH)-CH$_2$-COOH

Figure 1. Comparison of structures of AAL toxins TA$_1$ and TA$_2$ with that of fumonisin B$_1$.

CHEMISTRY OF AAL TOXINS

Isolation and Identification

AAL toxins TA and TB were first isolated from cell-free culture filtrates of *Alternaria alternata* f. sp. *lycopercisi* by Bottini et al. (1981) using isoelectric focusing at pH 4-5 and normal phase thin layer chromatography (TLC). Reaction of the terminal amino groups with ninhydrin allowed visualization of the toxin-ninhydrin complexes. The structure of TA was characterized following basic hydrolysis to yield 1,2,3-tricarboxylate and an aminodimethyl-heptadecapentol (Bottini and Gilchrist, 1981; Bottini et al., 1981). High resolution mass spectrometry studies confirmed the molecular weights and empirical formulas of the hydro-lyzed compounds. Proton and ^{13}C magnetic resonance studies, through multiplet analysis, band counting, and chemical shift data, provided the basis for hydroxylation at carbons 2, 4, and 5, methylation at carbons 11 and 15, and esterification with tricarballylic acid at carbon 13 (TA$_1$, with hydroxylation at carbon 14) or at carbon 14 (TA$_2$, with hydroxylation at carbon 13). Bottini et al. (1981) also indicated that AAL toxin TB consisted of two analogous regioisomers differing in the position of tricarballyic acid esterification (carbon 13 or 14) but lacking the carbon 5 hydroxyl, although no spectroscopic data were presented.

The discovery of the fumonisins in 1988 renewed interest into studies of AAL toxins, and Caldas et al. (1994) reported the discovery of three new pairs of regioisomeric toxins designated as AAL toxins TC, TD, and TE. The toxins were produced from liquid culture filtrates of *Alternaria alternata* f.sp. *lycopercici*, eluted from XAD-2 resin using methanol, and fractionated using silica gel column chromatography. Individual toxins were isolated using analytical normal phase TLC and a solvent system of ethyl acetate/acetic acid/water (6:3:1). Toxins were visualized using *p* -anisaldehyde. Ratio of the front values for TA, TB, TC, TD, and TE were 0.37, 0.48, 0.61, 0.73, and 0.80, respectively, while relative levels of toxins differed significantly (TA>TB>TE>TD>TC) . Toxins were further purified by column chromatography prior to characterization by mass spectrometry and magnetic resonance studies.

Under conditions of positive fast atom bombardment mass spectrometry (FABMS), each of the toxins produced abundant signals corresponding to the protonated molecular ions [M+H]$^+$ (Caldas et al., 1994). Daughter ion tandem mass spectrometry (MS/MS) spectra generated from the [M+H]$^+$ ions yielded abundant fragments of [M+H - 176]$^+$, corresponding to the loss of the tricarballylic acid group. In addition, the spectra also contained evidence of the number of hydroxyl groups indicated by the presence of a succession of fragments attributed to [M + H - 176 - n H$_2$O], where n is an integer that ranges from 1 to the number of hydroxyl groups. Peaks in the daugher ion spectra of [M+H]$^+$ for TD and TE also suggested the presence of *N* -acetyl groups due to losses of tricarballylic acid, n H$_2$O, and an additional loss of 42 daltons due to the loss of ketene. Subsequent daughter ion MS/MS experiments for ions at m/z 370 and m/z 356 obtained via *in situ* hydrolysis of TD and TE, yielded characteristic fragments at m/z 60, presumably corresponding to [CH$_3$CONH$_2$ + H]$^+$, which also supported *N* -acetylation of TD and TE.

Proton and ^{13}C magnetic resonance studies of the toxins (Tables 1 and 2), including COSY experiments, supported the findings of Bottini et al. (1981) for TA and TB, and provided, in combination with mass spectrometry findings, the basis for the structural assignments for TC, TD, and TE. Each toxin was represented by a pair of regioisomers esterified at either carbon 13 or 14. AAL toxin TC resembled TA but lacked hydroxyls at carbons 4 and 5, while TD represented *N* -acetylated TB and TE represented *N* -acetylated TC. The structures for AAL toxins TA-TE are provided in Figure 2.

Table 1. ^{1}H NMR Chemical Shifts for AAL toxins (relative to CD$_3$OD = 3.30ppm)

	TA$_1$/TA$_2$	TB$_1$/TB$_2$	TC$_1$/TC$_2$	TD$_1$/TD$_2$	TE$_1$/TE$_2$
H$_1$	3.05	3.03	3.01	3.25	3.26
H$_{1'}$	2.82	2.82	2.75	3.12	3.06
H$_2$	4.01	4.02	3.77	3.82	3.60
H$_3$	1.72	1.57	1.70	1.48	1.36
H$_{3'}$	1.51	1.44	1.46	1.45	1.33
H$_4$	3.67	3.79		3.76	
H$_5$	3.44	1.53		1.39	
H$_6$	1.36				

	TA$_1$	TA$_2$	TB$_1$	TB$_2$	TC$_1$	TC$_2$	TD$_1$	TD$_2$	TE$_1$	TE$_2$
H$_{12}$	1.68	1.31	1.63	1.31	1.67	1.31	1.60	1.31	1.60	1.31
H$_{13}$	5.12	3.75	5.11	3.76	5.02	3.74	5.07	3.78	5.11	3.77
H$_{14}$	3.39	4.75	3.39	4.77	3.39	4.75	3.44	4.77	3.36	4.78
H$_{15}$	1.34	1.77	1.37	1.74	1.37	1.75	1.36	1.74	1.36	1.74

Adapted from Caldas, et al (1994)

An additional AAL toxin corresponding to N-acetylated TA was identified by Caldas et al. (1995) following electrospray mass spectrometry (ESMS), ESMS/MS, and proton magnetic resonance studies. From a biosynthetic standpoint, it is curious that N-acetylated TA is present at much lower levels than other N-acetylated AAL toxins (TD and TE) while TA is the most abundant of the AAL toxins produced from *Alternaria alternata* f.sp. *lycopercici* .

The relative stereochemistry of AAL toxin TA for the carbon 1-5 region was initially proposed by Bottini and Gilchrist (1981) to be $2S$, $4S$, and $5S$ or its antipode based upon analysis of proton magnetic resonance coupling constants. A subsequent study (Oikawa et al., 1994 a) confirmed the stereochemistry for the same region as $2S$, $4S$, and $5R$. Further studies by Oikawa et al. (1994b) and by Boyle et al. (1994) identified the absolute configuration of the TA backbone, while the absolute configurations of the the tricarballylic acid side chains for TA and fumonisin B$_1$ were found to be equivalent (S) at carbon 3' (Shier et al., 1995).

Analysis of AAL Toxins

The first quantitative high performance liquid chromatography (HPLC) technique for AAL toxin analysis was developed by Siler and Gilchrist (1982) which involved reacting

Table 2. ^{13}C NMR Shifts of Carbons Attached to the OH, Tricarballylic Acid, and Nitrogen of TA, TB, TD, and TE (relative to CD$_3$OD = 49.00 ppm)

	TA$_1$/TA$_2$	TB$_1$/TB$_2$	TC$_1$/TC$_2$	TD$_1$/TD$_2$
C$_1$	45.28	46.50	46.13	46.71
C$_2$	64.97	66.27	68.36	69.49
C$_4$	70.46	68.58	69.49	
C$_5$	74.70			

	TA$_1$	TA$_2$	TB$_1$	TB$_2$	TD$_1$	TD$_2$	TE$_1$	TE$_2$
C$_{13}$	74.46	69.16	74.62	69.75	69.56	68.94	74.89	71.33
C$_{14}$	76.30	81.72	77.30	82.12	77.49	82.24	77.58	82.30

Adapted from Caldas, et al (1994)

TOXIN	R1	R2	R3	R4	R5
TA1	H	CO-CH2-CH(COOH)-CH2-COOH	OH	OH	H
TA2	CO-CH2-CH(COOH)-CH2-COOH	H	OH	OH	H
TA1-NAc	H	CO-CH2-CH(COOH)-CH2-COOH	OH	OH	C(=O)CH3
TA2-NAc	CO-CH2-CH(COOH)-CH2-COOH	H	OH	OH	C(=O)CH3
TB1	H	CO-CH2-CH(COOH)-CH2-COOH	H	OH	H
TB2	CO-CH2-CH(COOH)-CH2-COOH	H	H	OH	H
TC1	H	CO-CH2-CH(COOH)-CH2-COOH	H	H	H
TC2	CO-CH2-CH(COOH)-CH2-COOH	H	H	H	H
TD1	H	CO-CH2-CH(COOH)-CH2-COOH	H	OH	C(=O)CH3
TD2	CO-CH2-CH(COOH)-CH2-COOH	H	H	OH	C(=O)CH3
TE1	H	CO-CH2-CH(COOH)-CH2-COOH	H	H	C(=O)CH3
TE2	CO-CH2-CH(COOH)-CH2-COOH	H	H	H	C(=O)CH3

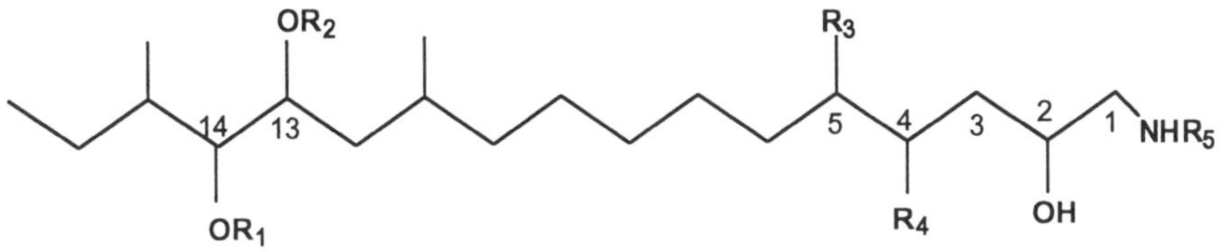

Figure 2. Chemical structures of all reported congeners of the AAL toxins produced in liquid culture by *Alternaria alternata* f. sp. *lycopersici.*

the primary amines of TA and TB with maleic anhydride to form an ultraviolet chromophore that absorbs at 250 nm. This simple and inexpensive method works well for high levels (parts per thousand) of mycotoxins but is not sensitive enough for analysis of naturally-occurring food samples.

A highly-sensitive but non-specific detached leaflet bioassay for leaf tissue of the *asc/asc* isoline of tomato was developed by Clouse and Gilchrist (1986) that was capable of detecting AAL toxins in intact tomato tissue at 20 nM concentrations. Using this approach, phytotoxic concentrations of TA, TB, TC, TD, and TE were determined to be 10, 10, 300, 4000, and 4000 ng/mL, respectively (Caldas et al., 1994).

Improvements in the sensitivity of HPLC-based analysis of AAL toxins have been made through derivatization of the primary amines with *o*-phthaldialdehyde (OPA) (Shephard et al., 1993) to yield fluorescent derivatives that enable detection of AAL toxins TA and TB from corn culture at levels of 2 ng of toxins per injection. While the OPA method is the most widely used technique in the quantitative analysis of fumonisins and was used in a major interlaboratory study of reproducibility characteristics (Thiel et al., 1993), the derivatives formed are relatively unstable and must be injected into the HPLC immediately after formation which limits the use of autosampling for maximizing throughput.

A technique superior to OPA derivatization involves formation of the more stable naphthalene 2,3-dicarboxaldehyde-potassium cyanide (NDA-KCN) derivatives. This approach has been applied for the analysis of fumonisins possessing free amino groups (Scott and Lawrence, 1992). Modifications of this procedure for the analysis of AAL toxins TA and TB provided a limit of detection of approximately 10 ppb (Gilchrist, 1995, unpublished data).

A major limitation of the HPLC flourescence techniques previously described is their inability to simultaneously detect all known toxin congeners since the flourescent reagents are unreactive toward the *N*-acetylated forms of the AAL toxins or fumonisins. Caldas et al. (1995) reported that positive and negative mode ESI was capable of detecting all known AAL toxin congeners in food samples but the detection limits ranged from 4 to 10 µg/mL injected. It was concluded that additional sample cleanup is required for detection of AAL toxins in foods at sub-parts per million levels using ESI. It is likely that innovative approaches such as affinity column purifications, mixed mode solid phase extractions, and carboxyl-specific derivatization may be necessary to achieve sensitive and specific detection of all known AAL toxins from foodstuffs.

Currently, no methods have been published for the immunochemical analysis of AAL toxins, although several methods for the immunochemical detection have been reported for the fumonisins (Azcona-Olivera et al., 1992a,b; Maragos and Richard, 1994; Abouzied and Pestka, 1994; Shelby et al., 1994; Usleber et al., 1994).

Biosynthesis of AAL Toxins

While a comprehensive understanding of the biosynthetic pathways for AAL toxin production is lacking, recent work (Caldas and Gilchrist, 1995, unpublished data) suggests that glycine is directly incorporated into the carbon 1 and nitrogen positions of TA. The use of ^{13}C-glycine also showed slight ^{13}C enrichment for all of the carbons and ^{13}C-labeled methionine was used to demonstrate the methionine origins of the carbon 7 and 11 methyl groups.

BIOLOGICAL ACTIVITY OF AAL TOXINS

The AAL toxins, as molecular determinants of the Alternaria stem canker disease of tomato and as potent mycotoxins, are suitable chemical tools to characterize the basis of disease-induced cell death in plants and animals. Briefly, the key features include: a) a single genetic locus (*Asc*) that regulates susceptibility to this pathogen in tomato (Clouse and Gilchrist, 1986), b) sensitivity to AAL-toxin (Gilchrist and Grogan, 1976), visualized in tomato as interveinal necrosis of leaflets, also regulated by the *Asc* locus and c) induction of cell death in rat liver (Mirocha et al., 1992), dog kidney (Mirocha et al., 1992), and African green monkey kidney (CV-1) cells (Jones, 1995, unpublished data) .

The macroscopic pattern of stem canker symptoms *in planta* includes initiation of necrosis within the interveinal areas, greatest sensitivity to AAL-toxins in the youngest sink leaflets, and induction of leaflet epinasty similar to a response to ethylene (Gilchrist and Grogan, 1976; Grogan et al., 1975). We interpret this pattern of cell death to suggest that the physiological consequence of AAL-toxin action may be mediated through the phloem by processes underlying the regulation of carbon allocation in the mesophyll and could require coordinate hormonal action. Published results (Meir et al., 1984; Moussatos et al., 1993a,b) suggest a temporal and metabolic linkage between the action of the *Asc* alleles, sucrose transport, ethylene biosynthesis, pyrimidine metabolism and cell death. Natural differences in sucrose influx rates occurring between the AAL-toxin-resistant (*Asc/Asc*) and sensitive (*asc/asc*) genotypes also suggest a relationship between the normal physiological function of the *Asc* gene product and sucrose influx (Moussatos et al., 1993b).

The current evidence also suggests that dihydroorotic acid treatment (a pyrimidine precursor) alters steady state levels of 1-aminocyclopropane-1-carboxylic acid, (ACC) (ethylene precursor) and interferes with toxin-induced cell death (Moussatos et al., 1994). That pyrimidine precursor treatment also interferes with the necrosis response to AAL-toxin suggests that dihydroorotic acid may negatively regulate ACC levels. Additional genetic and metabolic studies are needed to establish such a regulation point. However, it would seem likely that these relationships suggested by the toxin interaction are not limited to the stress induced by the toxin but may be part of a natural order of metabolic interlock among these key pathways.

In addition, fumonisin B_1 also induces cell death at nM concentrations against *asc/asc* tomato tissues and protoplasts (Gilchrist et al., 1992) as well as against both rat liver, dog kidney (Mirocha et al., 1992) and african monkey kidney (CV-1) cells (Jones, 1995, unpublished data). Both fumonisins and AAL-toxins inhibit ceramide synthase from rat hepatocytes (Merrill et al., 1993). We also have confirmed that ceramide synthase activity in microsomal preparations from green tomato fruit and leaf tissue is inhibited by both AAL-toxin TA and fumonisin B_1 (Gilchrist and Wang, 1995, unpublished data). In tomato there was a significant and equivalent inhibition of the enzyme at 20 nmolar concentrations with an I_{50} in the range of 35-40 nmolar for both AAL TA and FB_1. However, the physiological connection between inhibition of the synthesis of sphingoid bases and the symptoms (cell death or neoplasms) associated with these molecules is unknown. However, disruption of sphingolipid synthesis by FB_1 has been associated with a stimulation in DNA synthesis in animal cells (Schroeder et al., 1994).

There are reports in the literature indicating that ceramides and sphingosine derivatives are potent second messengers which trigger apoptosis or programmed cell death in animals. In addition, there are a number of genes now known to regulate the balance between the induction and the suppression of apoptosis leading to either death or proliferation (Tomei and Cope, 1991). A relevant question, in terms of the fumonisins and AAL-toxins action in both plants and animals, relates to the possible effect of these sphingoid base-related mycotoxins on apoptosis, either as second messenger mimics or by altering the normal level of key regulatory intermediates in the cell.

Inhibition of ceramide synthase in animal cells by both fumonisins and AAL-toxins (Wang et al., 1990; Merrill et al., 1993) suggested a molecular target for the molecules in animals leading to perturbation of biosynthesis of sphingoid bases and potential changes in production of ceramides. Physiological studies have confirmed that animals fed fumonisins at toxic concentrations show changes in levels of sphinganine and sphingosine expected if ceramide synthase was inhibited *in vivo* (Wang et al., 1992). Similar observations have been reported for plant cells (Abbas et al., 1994) which we have confirmed in preliminary studies with additional sphinganine analog mycotoxin (SAM) congeners. Altered sphingolipid metabolism also is known to be involved in apoptosis in animals but it is not known at this time if there is a direct relationship between the inhibition of ceramide synthase by SAMs and the induction of apoptosis. It is apparent that the molecules possess high specific biological activity and disrupt cell homeostasis though a fundamental mechanism. The fact that sphingosine-related metabolic intermediates are now recognized as potent signaling molecules in the regulation of cell cycle, protein phosphorylation, and gene expression underscores the need for a cell to maintain integrity of ceramide biosynthesis

Specifically, we currently are studying the question of whether common signal transduction pathways or specifically ordered metabolic changes occur as a consequence of developmental, environmental, and pathogen triggered events leading to necrosis. A related, but equally crucial, question is whether plants possess mechanisms of programmed cell death analogous to apoptosis in animal systems (Gerschenson and Totello, 1992; Lamb, 1994; Martin, 1993; Tomei and Cope, 1991) which could be triggered (or repressed) during infection by signals secreted by the pathogen as appears to be the case in several animal diseases. Preliminary evidence from our laboratories indicates that key hallmarks of apoptosis in animals (Figure 3) are expressed by both plant and animal cells exposed to AAL toxins and fumonisin B_1 (Jones, Dickman, and Gilchrist, 1995, unpublished data).

The hallmark of apoptosis is the orderly degradation of nuclear DNA by a calcium-dependent endonuclease that leads to cleavage at the nucleosomal linker (Barr and Tomei, 1994). The internucleosomal regions occur in 200 bp stretches and, when resolved by gel electrophoresis, appear as a ladder of bands increasing in size by about 200 bp (200, 400,

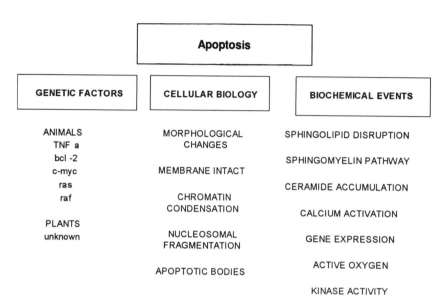

Figure 3. Genetic, cellular and biochemical hallmarks of apoptosis.

600 bp, etc). DNA fragmentation also can be detected histologically by immunochemical reagents that react with the exposed 3' hydroxyls of cleaved DNA.

In conclusion, it is apparent that the toxin molecules possess high specific biological activity and disrupt cell homeostasis though a fundamental mechanism. We believe that the SAMs are extremely interesting molecules, especially as chemical tools to investigate fundamental mechanisms that regulate cell stability and developmental processes. Lastly, studies of the SAM biosynthesis and development of multianalyte techniques to detect all SAM congeners in food products are needed urgently.

ACKNOWLEDGMENTS

Research (DGG) was supported in part by NSF Cooperative Agreement BIR-8920216 to CEPRAP, a NSF Science and Technology Center, by CEPRAP corporate associates Calgene, Inc. Ciba Biotechnology Coporation, Sandoz Seeds, and Zeneca Seeds, and (CKW and DGG) by the NIEHS Center for Environmental Health Sciences (1P30ES05707)

REFERENCES

Abbas, H.K.; Tanaka, T.; Duke, S.O.; Porter, J.K.; Wray, E.M.; Hodges, L.; Sessions, A.E.; Wang, E.; Merrill, A.H.; Riley, R.T. Fumonisin and AAL-toxin induced disruption of sphingolipid metabolism with accumulation of free sphingoid bases. *Plant Physiology* **1994,** *106,* 1085-93.

Abouzied, M.M.; Pestka, J.J. Simultaneous screening of fumonisin FB$_1$, aflatoxin B$_1$, and zearalenone by line immunoblot: A computer-assisted multianalyte assay system. *JAOAC International* **1994,** *77,* 495-500.

Azcona-Olivera, J.I.; Abouzied, M.M.; Plattner, R.D.; Norred, W.P.; Pestka, J.J. Generation of antibodies reactive with fumonisins B$_1$, B$_2$, and B$_3$ by using cholera toxin as the carrier-adjuvant. *App. Environ. Microbiol.* **1992a,** *58,* 169-173.

Azcona-Olivera, J.I.; Abouzied, M.M.; Plattner, R.D.; Norred, W.P.; Pestka, J.J. Production of monoclonal antibodies to the mycotoxins fumonisins B_1, B_2, and B_3. *J. Agric. Food Chem.* **1992b**, *40*, 531-534.

Barr, P.J.; Tomei, L.D. Apoptosis and its role in human disease. *Biotechnology* **1994**, *12*, 487-493.

Bezuidenhout, C.S.; Gelderblom, W.C.A.; Gorstallman, C.P.; Horak, R.M.; Marasas, W.F.O.; Spiteller, G.; Vleggaar, R. Structure elucidation of the fumonisins, mycotoxins from Fusarium moniliforme. *J. Chem. Soc. Commun.* **1988**, 743-745.

Bottini, A.T.; Bowen, J.R.; Gilchrist, D.G. Phytotoxins. II. Characterization of a phytotoxic fraction from *Alternaria alternata* f.sp. *lycopercici* . *Tetrahedron Letters* **1981**, *22*, 2723-2726.

Bottini, A.T.; Gilchrist, D.G. Phytotoxins. I. A 1-aminodimethylheptadecapentol from *Alternaria alternata* f.sp. *lycopercici* . *Tetrahedron Letters* **1981**, *22*, 2719-2722.

Boyle, C.D.; Harmange, J.-C.; Kishi, Y. Novel structure elucidation of AAL toxin TA backbone. *J. Am. Chem. Soc.* **1994**, *116*, 4995-4996.

Caldas, E.D.;Jones, A.D.; Ward, B.; Winter, C.K.; Gilchrist, D.G. Structural characterization of three new AAL toxins produced by *Alternaria alternata* f.sp. *lycopercici* . *J. Agric. Food Chem.* **1994**, *42*, 327-333.

Caldas, E.D.; Jones, A.D.; Winter, C.K.; Ward, B.; Gilchrist, D.G. Electrospray ionization mass spectrometry of sphinganine analog mycotoxins. *Anal. Chem.* **1995**, *67*, 196-207.

Clouse, S.D.; Gilchrist, D.G. Interaction of the *asc* locus in F8 paired lines of tomato with *Alternaria alternata* f.sp. *lycopercici* and AAL toxin. *Phytopathology* **1986**, *77*, 80-82.

Gelderblom, W.C.A.; Marasas, W.F.O.; Jaskiewicz, K.; Combrinck, S.; van Schalkwyk, D.J. Cancer promoting potential of different strains of *Fusarium moniliforme* in a short-term cancer initiation/promotion assay. *Carcinogenesis* **1988**, *9*, 1405-1409.

Gerschenson, L. E.; Totello, R. J. Apoptosis: a different type of cell death. *The FASEB Journal* **1992**, *6*, 2450-2455.

Gilchrist, D.G; Grogan R.G. Production and nature of a host-specific toxin from *Alternaria alternata* f. sp. *lycopersici*. *Phytopathology* **1976**, *66*, 165-171.

Gilchrist, D.G.; Ward, B.; Moussatos, V; Mirocha C.J. Genetic and physiological response to fumonisins and AAL-toxin by intact tissue of a higher plant. *Mycopathologia* **1992**, *117*, 57-64.

Grogan, R.G.; Kimble, K.A.; Misaghi, I. A stem canker disease of tomato caused by *Alternaria alternata* f. sp. *lycopersici*. *Phytopathology* **1975**, *65*, 880-886.

Lamb, C.J. Plant disease resistance genes in signal perception and transduction *Cell* **1994**, *76*, 419-22.

Maragos, C.M.; Richard, J.L. Quantitation and stability of fumonisins B_1 and B_2 in milk. *JAOAC International* **1994**, *77*, 1162-1167.

Marasas, W.F.O.; Kellerman, T.S.; Gelderblom, W.C.A.; Coetzer, J.A.W.; Thiel, P.G.; van der Lugt, J.J. Leukoencephalomacia in a horse induced by fumonisin B_1, isolated from *Fusarium moniliforme* . *Onderstepoort J. Vet. Res.* **1988a**, *55*, 197-203.

Marasas, W.F.O.; Jaskiewicz, K.; Venter, F.S.; van Schalkwyk, D.J. *Fusarium moniliforme* contamination of maize in esophageal cancer areas in Transkei. *S. Afr. Med. J.* **1988b**, *74*, 110-114.

Martin, S.J. Apoptosis: suicide, execution or murder? *Trends in Cell Biology* **1993**, *3*, 141-144.

Meir, S.; Philosoph-Hadas, S.; Epstein, E.; Aharoni, N. Role of sucrose in the metabolism of IAA-conjugates as related to ethylene production by tobacco leaf discs. In:*Ethylene: Biochemical, Physiological and Applied Aspects*. Fuchs, Y; Chalutz, E, eds. Martinus Nijhoff/Dr. W. Junk, publishers, Boston. **1984**, pp 97-98.

Merrill, A. H.; Wang, E.; Gilchrist, D. G.; Riley, R. T. Fumonisins and other inhibitors of *de nova* sphingolipid biosynthesis. *Advances in Lipid Research* **1993**, *26*, 215-234.

Mirocha, C.J.; Gilchrist, D.G.; Shier, W.T.; Abbas, H.K.; Wen, Y.J.; Vesonder, R.F. AAL toxins, fumonisin (biology and chemistry) and host-specificity concepts. *Mycopathologia* **1992**, *117*, 47-56.

Moussatos, V.V.; Witsenboer H.; Hille J.; and Gilchrist D. Behavior of the disease resistance gene *Asc* in protoplasts of *Lycopersicon esculentum* Mill. *Physiological and Molecular Plant Pathology* **1993a**, *43*, 255-263.

Moussatos, V.V.; Lucas, W.J.; Gilchrist, D.G. AAL-toxin-induced physiological changes in *Lycopersicon esculentum* Mill: Differential sucrose transport in tomato lines isogenic for the *Asc* locus. *Physiological and Molecular Plant Pathology* **1993b**, *42*, 359-371.

Moussatos, V.V.; Yang, S.F.; Ward, B.W.; Gilchrist, D.G. AAL-toxin induced physiological changes in *Lycopersicum esculentum* Mill: Roles of ethylene and pyrimidine biosynthesis intermediates in the necrosis response. *Physiological and Molecular Plant Pathology* **1994**, *44*, 455-468.

Oikawa, H.; Matsuda, I.; Ichihara, A.; Kohmoto, K. Absolute configuration of C(1)-C(5) fragment of AAL-toxin: Conformationally rigid acyclic aminotriol moiety. *Tetrahedron Letters* **1994a**, *35*, 1223-1226.

Oikawa, H.; Matsuda, I.; Kagawa, T.; Ichihara, A.; Kohmoto, K. Absolute configuration of main chain of AAL-toxins. *Tetrahedron* **1994b**, *50*, 13347-13368.

Schroeder, J.J.; Crane, H.M.; Xia, J.; Liotta, D.C.; Merrill, A.H. Disruption of sphingolipid metabolism and stimulation of DNA synthesis by fumonisin B_1. A molecular mechanism for carcinogenesis associated with *Fusarium moniliforme. J. Biol. Chem.* **1994**, *269 (5)*, 3475-81.

Scott, P.M.; Lawrence, G.A. Liquid chromatographic determination of fumonisins with 4-fluoro-7-nitrobenzofuran. *JAOAC International* **1992**, *75*, 829-834.

Shelby, R.A.; Rottinghaus, G.E.; Minor, H.C. Comparison of thin-layer chromatography and competitive immunoassay for detecting fumonisin on maize. *J. Agric. Food Chem.* **1994**, *42*, 2064-2067.

Shephard, G.S.; Thiel, P.G.; Marasas, W.F.O.; Sydenham, E.W.; Vleggaar, R. Isolation and determination of the AAL phytotoxins from corn cultures of the fungus *Alternaria alternata* f.sp. *lycopercici* . *J. Chromatogr.* **1993**, *641(1)*, 95-100.

Shier, W.T.; Abbas, H.K.; Badria, F.A. Complete structures of the sphingosine analog mycotoxins fumonisin B_1 and AAL toxin TA: Absolute configuration of the side chains. *Tetrahedron Letters* **1995**, *36*, 1571-1574.

Siler, D.J.; Gilchrist, D.G. Determination of host-selective phytotoxins from *Alternaria alternata* f.sp. *lycopercici* as their maleyl derivatives by high-performance liquid chromatography. *J. Chromatogr.* **1982**, *238*, 167-173.

Sydenham, E.W.; Thiel, P.G.; Marasas, W.F.O.; Shephard, G.S.; van Schalkwyk, D.J.; Koch, K.R. Natural occurrence of some Fusarium mycotoxins in corn from low and high esophageal cancer prevalence areas of the Transkei southern Africa. *J. Agric. Food Chem.* **1990**, *38*, 1900-1903.

Thiel, P.G.; Sydenham, E.W.; Shephard, G.S.; van Schalkwyk, D.J. Study of the reproducibility characteristics of a liquid chromatographic method for the determination of fumonisins B_1 and B_2 in corn: IUPAC collaborative study. *JAOAC International* **1993**, *76*, 361-366.

Tomei, L.D.; Cope, F.O., eds. *Apoptosis: The Molecular Basis of Cell Death. Current Communications in Cell and Molecular Biology*. Cold Spring Harbor Laboratory Press. Cold Spring Harbor, NY. **1991**, 246 pp.

Usleber, E., Straka, M., Terplan, G. Enzyme immunoassay for fumonisin B_1 applied to corn-based food. *J. Agric. Food Chem.* **1994**, *42*, 1392-96.

Wang, E.; Norred, W.P.; Bacon, C.W.; Riley, R.T.; Merrill, A.H Inhibition of sphingolipid biosynthesis by fumonisins. *J. Biol. Chem.* **1990**, 266, 14486-90.

Wang, E.; Ross, P.E.; Wilson, T.M.; Riley, R.T.; Merrill, A.H Alteration of serum sphingolipids upon exposure of ponies to feed containing fumonisins, mycotoxins produced by *Fusarium moniliforme J. Nutr.* **1992**, *122*, 1706-16.

DISTRIBUTION OF FUMONISINS IN FOOD AND FEED PRODUCTS PREPARED FROM CONTAMINATED CORN

Glenn A. Bennett,[1] John L. Richard,[1] and Steve R. Eckhoff[2]

[1] Mycotoxin Research
National Center for Agricultural Utilization Research, USDA/ARS
1815 N. University Street
Peoria, IL 61604
[2] Department of Agricultural Engineering
University of Illinois
Champaign-Urbana, IL 61801

ABSTRACT

The fate and distribution of the fumonisins B_1 (FB_1) and B_2 (FB_2) were determined in products obtained from naturally contaminated corn used for ethanol fermentation and wet milling operations. Fumonisins are stable to the conditions used in ethanol fermentations and tend to concentrate in the distillers dried grain, a fraction generally used for animal feed. No toxin was found in the ethanol. Starch from wet milling of corn, naturally contaminated at 13.9 µg fumonisin B_1/g, was free of detectable toxin. The other fractions contained fumonisins at the following levels: gluten (5.1-5.8 µg FB_1/g, 4.7-4.9 µg FB_2/g); fiber (2.7-5.7 µg FB_1/g, 2.1-3.1 µg FB_2/g); and germ (1.3-3.1 µg FB_1/g, 0.7-1.6 µg FB_2/g). The steep water and process water contained 22% of the recoverable fumonisins. A combination of analytical methodologies was required to determine fumonisins in the different products from the wet milling process.

INTRODUCTION

Since their discovery and characterization in 1988 (Gelderblom et al., 1988; Bezuidenhout et al., 1988), the fumonisins have been the focus of numerous investigations around the world. This intense interest was generated because this family of compounds have been shown to have cancer-promoting activity in rats (Gelderblom et al., 1991), to exhibit toxic effects in animals, especially horses and pigs (Marasas et al., 1988; Harrison et al., 1990), and to be associated with the elevated incidence of esophageal cancer in the Transkei region of South Africa (Sydenham et al., 1990). Fumonisins are produced by *Fusarium* species

Fumonisins in Food, Edited by L. Jackson *et al.*
Plenum Press, New York, 1996

which occur world wide (Nelson et al., 1993), and can persist in corn-based human foods (Sydenham et al., 1991). Although fumonisins are routinely found in most corn crops at low levels (<1 μg/g), drought conditions, such as those that existed in 1989 and 1990, can result in levels which are associated with equine leukoencephalomalacia (ELEM) and porcine pulmonary edema (PPE) (Ross et al., 1990). These animal disorders are usually related to the use of contaminated corn screenings which often contains ten times the level of fumonisins found in whole kernel corn. Unfortunately, physical appearance of corn is not an indicator of fumonisin levels and sound, visibly undamaged corn can contain >10 μg/g of total fumonisins. A survey of midwestern corn for the crop years 1989-1991 showed that the majority of the samples had FB_1 levels of 1-5 μg/g and the average concentrations of the total fumonisins were similar for these crop years (Murphy et al., 1993). Since the fumonisins are present in corn in most crop years and the potential exists for high levels to accumulate during conditions which favor disease development by *F. moniliforme* and *F. proliferatum*, we have conducted experiments to determine the fate and distribution of these toxins during ethanol fermentations and the wet milling process.

MATERIALS AND METHODS

Ethanol Fermentations

Two lots of naturally contaminated corn (1989 yellow corn and 1990 corn screenings) and one lot of Grade No. 2 yellow dent corn (control) were used for replicate ethanol fermentations (Bothast et al., 1992). The fumonisins FB_1 and FB_2 (Figure 1) levels in the original corn and the resulting fermentation products were determined by HPLC of the naphthalene-2,3-dicarboxaldehyde derivatives of extracts cleaned up on C_{18} solid phase extraction (SPE) columns (Bennett and Richard, 1994).

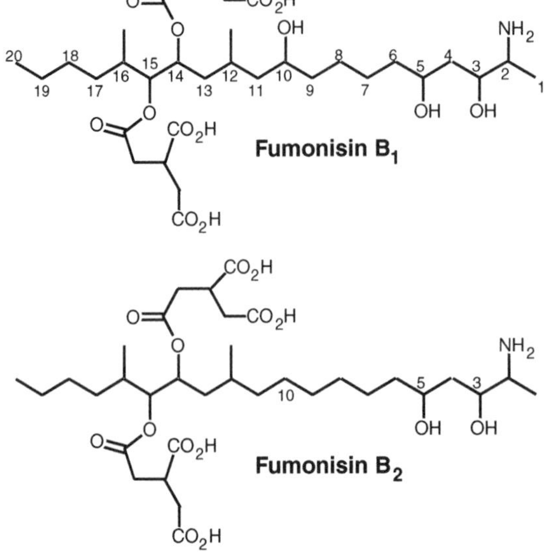

Figure 1. Structures of fumonisin FB_1 and FB_2.

Wet Milling Process for Fusarium Damaged Corn

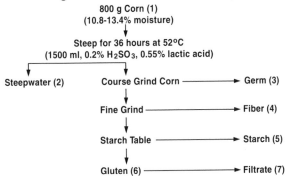

Figure 2. Flow diagram of laboratory scale wet milling procedure used in this study. Numbers in ( ) identify fractions assayed for fumonisins.

Wet Milling. Laboratory-scale milling of control and fumonisin contaminated corn was conducted as outlined in Figure 2. Replicate runs of 800 g of each corn lot were conducted and the product yields (% dry basis) determined. The whole kernel corn was steeped for 36 hours in 1500 ml of a solution containing 0.2% sulfurous acid and 0.55% lactic acid. After steeping, the steep water (995-1070 ml) was drained and retained for analysis. The corn was coarsely ground with distilled water to release the germ which was subsequently skimmed from the slurry. The remaining slurry was finely ground and passed through a fiber shaker to collect the fiber fraction. The specific gravity of the remaining slurry was adjusted to 1.04 and passed over a starch table at 300 ml/min. Starch was collected and dried. The gluten in the starch table overflow was collected by filtration. Finally, the process water (6900-7600 ml) was collected and retained for analysis. Analysis of the corn and the wet milling fractions was done as previously described (Bennett and Richard, 1994) with some modifications dictated by the nature of the different milling products. Germ and gluten fractions were assayed without difficulty. However, starch and fiber fractions were extracted with 2X the normal volume of acetonitrile-water (1+1) due to adsorption of solvent by the matrix. The pH of the steep water and process water samples was adjusted to 6.0-6.5 to accomplish clean up on SAX columns. These fractions have high ionic strength which effects retention of fumonisins on the SAX column. Therefore, samples (2 ml) of these fractions were diluted 2X with water, then mixed with 8 ml methanol-water (3:1) before addition to the column. Confirmation of FB_1 levels in selected fractions (gluten, starch, steep, and process waters) was accomplished by GC/MS (Plattner and Branham, 1994). In addition, starch fractions were spiked with deuterium-labeled (D_6) fumonisin FB_1 to determine the efficiency of the extraction and clean up procedures used.

RESULTS AND DISCUSSION

Ethanol Fermentations

The distribution of fumonisin B_1 in products from ethanol fermentations using sound corn (control) and naturally contaminated corn (15 µg/g) is shown in Table 1. These corn samples were similar in moisture, protein, ash and starch content and the final ethanol concentration in the mash was 8.8% (wt/vol) for both samples. Since fumonisin B_1 is water soluble, it was not surprising that 54% of the fumonisin B_1 in the starting corn was extracted into the whole stillage. Despite solubility in the stillage, higher levels of toxin were detected in the distillers dried grains than were in the original corn (19-25 µg/g vs. 15 µg/g). A similar

Table 1. Distribution of fumonisin B$_1$ in products obtained from ethanol fermentation of contaminated corn. (from Bothast et al., 1992)

Sample	Fumonisin B$_1$ (μg/g)[a]	
	Control Corn	Contaminated Corn
Starting corn	ND[b]	15
Fermented mash (72 hr)	ND - 0.4	1.8 - 2.7
Distillers dried grains	4.0 - 5.0	19.2 - 25.3
Thin stillage	ND - 0.6	1.5 - 1.7
Centrifuge solids	ND	ND
Distillers solubles	ND - 0.2	1.3 - 1.7
Distilled ethanol	ND	ND

[a] μg/g or μg/ml and each value is average of duplicate assays.
[b] ND = none detected, less than 0.1 μg/g.

distribution pattern (data not shown) was obtained when corn screenings containing 36 μg/g FB$_1$ were used as a substrate for ethanol fermentation.

A surprising result of the study was the detection of fumonisin B$_1$ in distillers dried grains from the control corn which originally contained <0.1 μg/g. Insufficient sensitivity of the assay method or sampling error in the starting corn may be causes for this discrepancy. The amount of recoverable fumonisin increased as the mash underwent the 72 hour fermentation. Total recoverable fumonisin B$_1$ was 85% of the level in the original corn (15 μg/g), clearly showing that the process did not significantly degrade the toxin. Additional studies are ongoing to determine if modifications in the washing procedures for distillers dried grains, a valuable feed co-product, can be used to reduce the toxin levels below the levels of concern. This study indicates that detoxification or removal strategies must be developed before ethanol fermentations can be used as a process for utilization of highly contaminated corn.

Wet Milling Studies. Product yields (% dry basis) for control (1 μg FB$_1$/g) and fumonisin contaminated corn (13.9 μg FB$_1$/g) are shown in Table 2. Although the level of fumonisins in the contaminated corn was high, the physical appearance of this corn was very similar to that of the No. 2 Grade control corn. Also, product yields from the contaminated corn were comparable to those from the control corn. However contaminated corn yielded

Table 2. Yields (%, d.b.) of products obtained from wet milling of fumonisin contaminated corn.

Product	Control Corn (1 μg FB$_1$/g)			Fumonisin Contaminated Corn (13.9 μg FB$_1$/g)		
	Rep A	Rep B	Average	Rep A	Rep B	Average
Germ	6.75	6.91	6.83	7.17	6.78	6.98
Fiber	7.93	9.87	8.90	5.17	11.56	8.37
Starch	67.02	66.09	66.56	65.00	62.74	63.87
Gluten	10.74	10.35	10.55	12.76	10.99	11.88
Solid in steepwater	4.93	4.31	4.62	4.75	4.62	4.69
Solid in process water	2.43	1.85	2.14	2.02	2.10	2.06
Total	99.80	99.33	99.57	96.87	98.79	97.83

Table 3. Results of HPLC analyses for fumonisins in wet-milled fractions of contaminated corn

Fraction	Replicate	FB$_1$ (µg/g or ml)	FB$_2$ (µg/g or ml)
Fiber	1	5.7	2.1
	2	2.7	3.1
Germ	1	1.3	0.7
	2	3.1	1.6
Gluten	1	5.8 (5.0)[b]	4.7
	2	5.1	4.9
Starch	1	< 0.1	ND[c]
	2	< 0.1	ND
Steep water	1	2.1 (1.71)	2.5
	2	0.3 (0.46)	—
Process water	1	ND (0.09)	ND
	2	ND (0.01)	ND

[a] Contaminated at 13.9 µg FB$_1$/g
[b] Value in ( ) determined by GC/MS
[c] ND = none detected

slightly less starch and more gluten than the control corn. The only milling variation for the two corn lots was that the water-coarse ground corn slurry (for germ release) was darker for the contaminated corn than for the control corn.

The levels of fumonisin FB$_1$ and FB$_2$ in fractions from duplicate runs of each corn lot are given in Table 3. Fumonisins could not be detected in any fraction, except steep water, from the 1 µg/g control corn. Considerable variations occurred in the levels of FB$_1$ and FB$_2$ in replicate runs of the contaminated corn. Also, the ratio of FB$_1$ to FB$_2$ was lower (more FB$_2$) in product fractions than in the original corn. These results indicate that the FB$_2$ may be more readily extracted from the products than from the original corn or that the FB$_1$ becomes less extractable due to heating of the fractions during the drying process. Verification of the measured FB$_1$ was confirmed by GC/MS analysis of gluten, steep water and process water. Starch fractions, the major wet milling product, did not contain measurable fumonisin B$_1$ by the HPLC method used which had a detection limit of 0.1 µg/g FB$_1$. Trace levels (<0.03 µg/g FB$_1$) were detected by FAB/MS/MS in samples spiked with D$_6$-labeled fumonisin to determine efficiency of the SPE clean up procedure used. Natural FB$_1$ contamination and spiked FB$_1$ (labeled) are easily distinguished by the MH$^+$ ions of 728 (D$_6$-FB$_1$) and 722 (FB$_1$). Twenty-two percent of the recoverable FB$_1$ was found in the steep and process waters. These data demonstrate that the fumonisins persist in products from wet milling of contaminated corn. Although water soluble in purified form, fumonisin B$_1$ remains in the fiber, gluten and germ fractions at 10-40% the level found in the starting corn. Also, this study showed the urgent need to develop specific assay procedures for the determination of fumonisins in matrices other than whole corn. Immunoaffinity columns appear to be adaptable to this problem and milling studies are being repeated to determine, more accurately, the levels of fumonisins in mill fractions. Low temperature drying and exhaustive extractions of the various fractions are two techniques which will be incorporated into this continuing study.

ACKNOWLEDGMENT

We express our appreciation to C. Tso, J. Wu, and K. Yeptanco for technical assistance in wet milling and B. Davis for assistance in fumonisin assays.

REFERENCES

Bennett, G. A.; Richard, J. L. Liquid chromatographic method for analysis of the naphthalene dicarboxaldehyde derivative of fumonisins. *J. AOAC Int.* **1994**, *77*, 501-506.

Bezuidenhout, S. C.; Gelderblom, W. C. A.; Gorst-Allman, C. P.; Horak, R. M.; Marasas, W. F. O.; Spiteller, G.; Vleggaar, R. Structure elucidation of the fumonisins, mycotoxins from *Fusarium moniliforme. J. Chem. Soc. Chem. Commun.* **1988**, *11*, 743.

Bothast, R. J.; Bennett, G. A.; VanCauwenberge, J. E.; Richard, J. L. Fate of fumonisin B_1 in naturally contaminated corn during ethanol fermentation. *Appl. Environ. Microbiol.* **1992**, *58*, 233-236.

Gelderblom, W.C.A.; Jaskiewicz, K.; Marasas, W.F.O.; Thiel, P.G.; Horak, R.M.; Vleggaar, R.; Kriek, N.P.J. Fumonisins - novel mycotoxins with cancer-promoting activity produced by Fusarium moniliforme. *Appl. Environ. Microbiol.* **1988**, 54, 1806-1811.

Gelderblom, W. C. A.; Kriek, N. P. J.; Marasas, W. F. O.; Thiel, P. G. Toxicity and carcinogenicity of the *Fusarium moniliforme* metabolite, fumonisin B_1 in rats. *Carcinogenesis* **1991**, *12*, 1247-1251.

Harrison, L. H.; Colvin, B. M.; Greene, J. T.; Newman, L. E.; Cole, Jr., J. R. Pulmonary edema and hydrothorax in swine produced by fumonisin B_1, a toxic metabolite of *Fusarium moniliforme. J. Vet. Diagn. Invest.* **1990**, *2*, 217-221.

Marasas, W. F. O.; Kellerman, T. S.; Gelderblom, W. C. A.; Coetzer, J. A. W.; Thiel, P. G.; van der Lugt, J. J. Leukoencephalomalacia in a horse induced by fumonisin B_1 isolated from *Fusarium moniliforme* Sheldon. *Onderstepoort J. Vet. Res.* **1988**, *55*, 197-203.

Murphy, P. A.; Rice, L. G.; Ross, P. F. Fumonisin B_1, B_2, and B_3 content of Iowa, Wisconsin, and Illinois corn and corn screenings. *J. Agric. Food Chem.* **1993**, *41*, 263-268.

Nelson, P. E.; Desjardins, A. E.; Plattner, R. D. Fumonisins, mycotoxins produced by *Fusarium* species: Biology, chemistry, and significance. *Ann. Rev. Phytopathol.* **1993**, *31*, 233-252.

Plattner, R. D.; Branham, B. E. Labeled fumonisins: Production and use of fumonisin B_1 containing stable isotopes. *J. AOAC Int.* **1994**, *77*, 525-532.

Ross, P. F.; Nelson, P. E.; Richard, J. L.; Osweiler, G. D.; Rice, L. G.; Plattner, R. D.; Wilson, T. M. Production of fumonisins by *Fusarium moniliforme* and *Fusarium proliferatum* isolated associated with equine leukoencephalomalacia and a pulmonary edema syndrome in swine. *Appl. Environ. Microbiol.* **1990**, *56*, 3225.

Sydenham, E. W.; Thiel, P. G.; Marasas, W. F. O.; Shephard, G. S.; Van Schalkwyk, D. J. Natural occurrence of some *Fusarium* mycotoxins in corn from low and high esophageal cancer prevalent areas of the Transkei, southern Africa. *J. Agric. Food Chem.* **1990**, *38*, 1900-1903.

Sydenham, E. W.; Shephard, G. S.; Thiel, P. G.; Marasas, W. F. O.; Stockenstrom, S. Fumonisin contamination of commercial corn-based human foodstuffs. *J. Agric. Food Chem.* **1991**, *39*, 2014-2018.

28

EFFECT OF PROCESSING ON FUMONISIN CONTENT OF CORN

Patricia A. Murphy, Suzanne Hendrich, Ellen C. Hopmans,
Cathy C. Hauck, Zhibin Lu, Gwendolyn Buseman, and Gary Munkvold[1]

Food Science and Human Nutrition Department
[1] Plant Pathology Department
Iowa State University
Ames, IA 50011

ABSTRACT

Fumonisins (FBs) are a family of mycotoxins produced by *Fusarium moniliforme* and *F. proliferatum*, predominant corn pathogens, and are found in most corn-containing foods. The FBs are heat stable, resistant to ammoniation, and unlike most mycotoxins, are water-soluble. The levels in corn and corn-containing foods will be presented ranging from <20 ppb to >2 ppm. Washing of contaminated FB-corn with water did not reduce the measured FB levels of significantly. The traditional processing step to make tortilla flour, nixtamalization [$Ca(OH)_2$ cooking] to produce masa, reduced FB levels but produced hydrolyzed FB which was almost as toxic as FB. Retorting sweet corn in brine apparently produced hydrolyzed FB. Fermentation of corn to ethanol did not alter FB levels but distillation yielded FB-free ethanol. Attempts to enzymatically modify FB with several enzymes were unsuccessful. Reactions between FB and reducing sugars (glucose or fructose) to produce Schiff's bases yielded products that were not toxic. The effects of these processing treatments must be evaluated both chemically and biologically.

INTRODUCTION

The fumonisins (FBs) are found in human and companion animal foods that contain corn. Although the levels are fortunately much lower than usually found in livestock feeds and intact corn, FB levels in foods are of concern given the epidemiological association between FB-contaminated corn consumption and certain types of cancer in humans (Marasas et al., 1988; Yoshizawa et al., 1994; Hopmans & Murphy, 1993). Regulatory agencies have not given action levels for these mycotoxins but advisory levels have been published and FB_1 has been identified as a class 2B carcinogen, a probable human carcinogen (J.D. Miller, personal communication). FBs are unusual mycotoxins because they are very soluble in water. Furthermore, FB levels can be relatively high in corn that appears normal. Visual inspection cannot reliably identify FB-contaminated corn. This paper will review the FB contents of human and pet foods and describe the effect of processing on this family of mycotoxins.

FB ACTIVE GROUPS

The main active groups in the FB structure that are of concern to the analytical chemist and the toxicologist are: the primary amine group, the hydroxyl groups, the tricarboxylic acid groups and the aliphatic backbone. Currently, most analytical methods involve either determination of a derivative of the primary amine or the mass of the aliphatic backbone. Other groups may be involved in metabolism of these mycotoxins and may also present new analytical challenges (Hopmans et al., 1995).

METHODS OF ANALYSIS OF FBs

Any evaluation of the effects of processing on FB levels in corn-containing foods must take into account the analytical method(s) used to measure the FB or its reaction products. The methods used to evaluate FB concentrations can be divided into: derivatization of a free amine group with uv absorbing or fluorescing reagents; gas- or high performance liquid- chromatography coupled with mass spectrometry; and antibody recognition (Table 1).

Unfortunately, the structures of the FBs do not contain any useful chromophores naturally. The premise for most thin-layer and HPLC methods for FB analysis rely on a free amine group that can be derivatized with one of the reagents listed in Table 1. If there are any interactions between the free amine and the other constituents in the food matrix, the amount of estimated FB will be lower than the actual concentration. Hydrolysis of the sample to produce the hydrolyzed, and more hydrophobic versions, of FBs can circumvent some of the these problems. Hydrolysis followed by derivatization has been the approach used for gas chromatographic (gc) methods. These potential analytical problems have not been of

Table 1. Methods of analysis for Fumonisins

Primary Amine	o-Phthaldialdehyde Derivatives
	Maleyl Derivatives
	4-Fluor-7-Nitrobenzofurazan
	Derivatives
	Naphthalene Dicarboxaldehyde
	derivatives
	Fluorescamine Derivatives
Mass Detection	Gas Chromatography-Mass
	Spectroscopy
	HPLC-Thermospray Mass
	Spectroscopy
	Thermospray Mass Spectroscopy
Immunoassay	ELISA
	Line Immunoblot

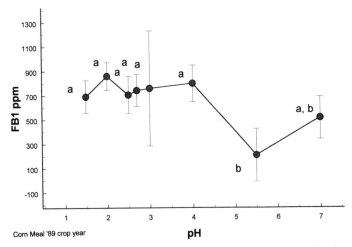

Figure 1. Effect of pH on fumonisin extraction for corn-containing food.

concern for analysis of corn or corn culture material. However, when more highly processed foods are examined, the availability of the amine should be of concern.

The effect of pH on FB extraction for mixed corn foods is more pronounced than in corn and minimally processed corn foods. Acidification of the food and extraction solvents prior to extraction improves FB recovery (Figure 1). The buffering capacity of corn-containing foods with multiple ingredients appears to have a major impact on FB extraction. In a comparison of corn-containing foods (purchased in 1992 and 1993) extracted with 50% acetonitrile (ACN) and with pH 2.7 adjusted 50% ACN, countering the pH buffering capacity of the food leads to greater FB recovery (Table 2). Pet foods, which seem to contain additional highly pigmented components, may have higher FB recovery but further clean-up procedures need to be developed for those products.

When FB containing foods do not show significant amine reactivity, an analysis of the FB backbone can circumvent this problem using the same derivatization protocol. Hydrolysis of components blocking the amine group can be accomplished by refluxing the

Table 2. Effect of pH on fumonisin extraction from foods

	Fumonisin (ng/g food)	
Food	50% ACN	pH 2.7, 50% ACN
Masa-A 92	0	120
Masa-B 92	t	60
Masa-C 92	17	180
Corn Maze 92	10	360
Corn Pasta 92	90	246
Corn meal 93	42	290
Tortilla chips 93	0	90
Corn Sugar 93	0	88
Cornmeal mix 93	t	601
Blue Corn Chips 93	84	409
Corn Bran 93	2742	1974

t = trace

food sample in 1 N KOH-50% ACN for 1 h. The refluxed sample can then be acidified and extracted as the intact FB. The free amine group is ready for derivatization and detection.

Thin-layer chromatography is useful for screening corn samples (≥ 500 ng/g) but it usually does not have the sensitivity for FB levels found in foods. HPLC detection limits are dictated by the sensitivity of the detectors and the concentration of other amine reactive constituents in the food product.

Direct mass spectroscopy (MS) with or without coupling with GC (Plattner et al. 1994) or with HPLC (Thakur & Smith, 1994) have been reported as analytical methods for FBs. The GC or GC-MS methods involve considerable work-up of the sample to volatilize the mycotoxins as trifluoro-derivatives. A negative-ion thermospray MS procedure requires less sample preparation and yields distinctive fragmentation fingerprints of the FBs. This procedure can be coupled with HPLC.

Immunoassays have been developed for FB_1 and FB_2 and commercial clean-up immunoaffinity columns are available (Pestka et al, 1994; Abouzied & Pestka, 1994). Pestka et al. (1994) reported correlations between an ELISA method for FB_1 and GC-MS or HPLC which were 0.478 and 0.512, respectively. Correlation between the two chemical assays was 0.946. While all reported correlations were significant at $p<0.05$, the lack of good correlation between antibody recognition and chemically generated indicators was explained by Pestka et al. (1994) in several ways. Pestka et al. (1994) proposed: that the recovery FB in sample workup for the GC and HPLC methods were less than for immunoassay; or that the food matrix may enhance ELISA inhibition response and increase the FB estimates; or that the FB precursors or metabolites in food cross-react with the ELISA and increase FB estimates. Interestingly, in minimally processed corn foods, the ELISA yielded higher FB concentrations than the other two methods. In more processed or reformulated corn-containing foods, ELISA gave concentrations lower than detected by GC-MS or HPLC. Further studies should help to resolve these differences.

LEVELS IN FOODS

The concentration of FBs expected in foods can be anticipated by the FB levels observed in the current corn crop year. Average concentrations of FB observed in Iowa corn from 1988 to 1994 show considerable variability (Figure 2). Since Iowa is usually first or second in the U.S. in terms of corn production, we should be able to use these averages to predict problems in foods made from these crop years. Some of these data have already been reported in Murphy et al. (1993) and Murphy et al. (1995) and are reproduced for comparison. No correlation can be made between the climatic conditions and FB levels found in corn. For example, 1993 was the flood year in Iowa and had essentially the same FB levels as 1994 which was a bumper crop year. The climatic conditions in 1992 were approximately similar to 1991 but FB average contents were drastically different. Further research is underway to determine the pattern of field infection and toxin production.

There appears to be no correlation between the level of fungal contamination and FB concentration found in corn (Figure 3). Corn was heavily contaminated with *Fusarium moniliforme* and *F. proliferatum* in 1993, yet the levels of FB found in these corn samples were extremely low.

Screenings, the broken corn kernel material, are mentioned here because they are used as one of the main forms of corn used in companion animal foods as a function of least cost formulation. Since corn screenings typically contain FB concentrations that are ten times higher than in intact corn, monitoring of pet foods containing corn should be advised. Size segregation of corn screenings showed no correlation with FB contents (Murphy et al, 1993).

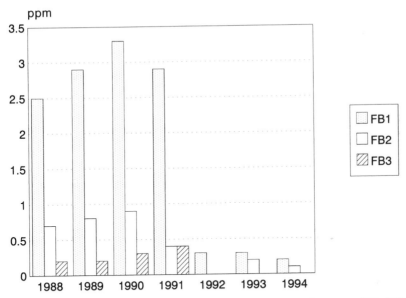

Figure 2. Fumonisin content of Iowa corn 1988-1994. Each bar represents the average FB1, FB2 and FB3 levels (in ppm). The number of samples from each crop year are as follows: 1988 n=22; 1989 n=44; 1990 n=59; 1991 n=50; 1992 n=81; 1993 n=50; 1994 n=43.

Foods produced from some of these crop years were been analyzed using the methods of Hopmans & Murphy (1993) with the pH modification indicated above. Generally, FB levels in human corn-containing foods were somewhat higher in the 1989 and 1990 corn crop foods than those produced from the 1992 and 1993 crop (Table 3). However, the levels detected in foods from the latter years were not as low as would have been predicted from the average levels found in Iowa corn of those years. For some foods such as masa, FB levels were higher in the 1992 and 1993 masas than the 1989 masa. FB content of tortilla chips were highly variable. A few foods from 1992 and 1993 contain FB levels exceeding 1000 ng/g. High lysine cornmeals contained more FB than regular cornmeals sampled in 1992 and 1993. Corn brans had relatively high FB levels in 1992 and 1993. Blue corn products sampled had relatively high FB concentrations for these years.

DRY AND WET MILLING

The effects of milling corn are reported elsewhere in this symposium by Bennett et al. (1995). We have found high levels (>10 µg/ml) of FB in corn steep liquor sold as a feed supplement or microbial media supplement.

FERMENTATION

FBs survive the fermentation process used to produce ethanol from corn (Bothast et al., 1992). Ethanol distilled from the whole stillage did not contain any FB, however, all the other products produced from this fermentation contained FB (Figure 4). Sydenham et al. (1991) suggested an epidemiological relationship between FB consumption and human

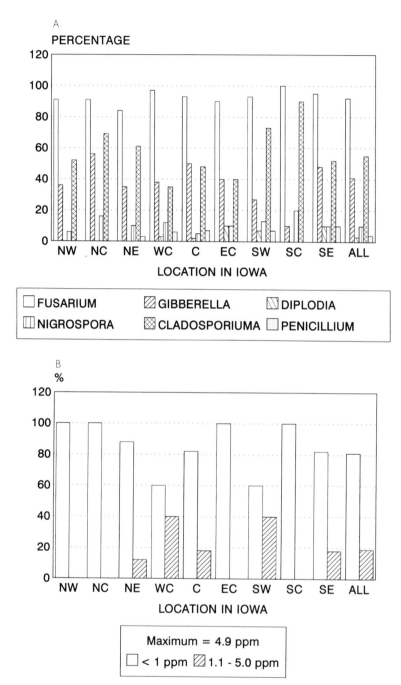

Figure 3. Fungal species and fumonisin levels in 1993 Iowa corn from regional locations. A. Percentage infection of corn by fungal group. B. Fumonisin level in same corn samples.

Table 3. Fumonisin content found in foods in 1992 & 1993

		ng/g food	
Food	n	FB$_1$	HFB$_1$
Masa-A 92	3	36±16	
Masa-B 92	3	84±53	
Corn Maze 92	4	200±56	1
Corn Pasta 92	4	246±34	1
Corn meal, high lysine 92	1	448	
Corn Masa 93	3	25±21	
Corn Meal 93	5	161±33	
Corn meal, high lysine 93	4	370±90	
Tortilla Chips A 93	3	29±33	
Tortilla Chips B 93	3	111±161	
Tortilla Chips C 93	3	11±11	
Tortilla Chips D 93	3	25±31	
Tortilla Chips E 93	2	15±3	
Corn Chips 93	3	17±15	
Corn Sugar 93	3	101±31	1
Cornmeal Mix 93	3	101±31	1
Blue Corn Chips 93	4	409±115	1
Corn Bran 93	4	1974±632	1
Corn Grits-93	3	143±56	
Blue Corn Flakes, Organic 93	3	17±15	
Blue Corn Pancake Mix 93 4	4	939±205	
Corn Puffs, Organic 93	3	86±59	
Popcorn, yellow or white 93	1	ND	

ND = not detected

FERMENTATION

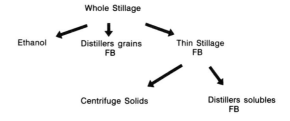

Figure 4. Fumonisin-containing fractions from ethanol fermentation of contaminated corn adapted from Bothast et al.(1992).

esophageal cancers. A principle source of the FB in these diets was probably beer produced from heavily FB-contaminated-corn. Levels found in beer made from FB contaminated corn could be significant.

ENZYME ACTIVITY WITH AMINE GROUP

We have examined the potential for detoxifying FB during processing through the use of enzymes targeting the primary amine group. Commercial monoamine oxidase, diamine oxidase, L-amino acid oxidase and D-amino acid oxidase (Sigma Chemical Co., St. Louis, MO) were incubated with FB_1 under published conditions. No loss of o-phthaldial-dehyde (OPA) reactivity was detected following incubation with any of these enzymes, although the reference substrates (benzylamine, putrescine, L-leucine, D-alanine, respectively) were totally active with each enzyme. FB reactivity to OPA did not change following incubation with a cytochrome P-450 monooxygenase (Cawood et al., 1994).

NIXTAMALIZATION

The traditional process to produce masa, or tortilla flour, called nixtamalization, was evaluated as a method to detoxify FB (Hendrich et al., 1993). A flow diagram for this process is presented in figure 5. This process has been used for centuries in Central America for corn to improve the nutritional value by increasing niacin bioavailability in the masa. We expected the alkaline cooking and steeping step would hydrolyze FB. When corn fermented with *F. proliferatum* to produce 50 µg/g FB_1 was nixtamalized, FB_1 concentrations were reduced to 0.4 µg/g in the finished product and 10 hydrolyzed FB_1 (HFB) µg/g was produced and retained in the masa. Treatment of FB-corn with the same process minus $Ca(OH)_2$ yielded no change in FB or loss of FB_1 due to washing/leaching.

To evaluate if this traditional process decreased the toxicity of *F. proliferatum* fermented corn, male F344/N rats were initiated with 15 mg/kg diethylnitrosamine at 10 d. At weaning, animals were randomly assigned to 8 groups of 6 rats to evaluate the effect of

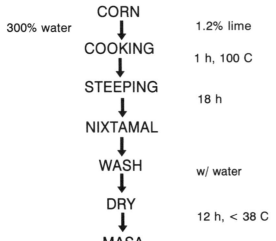

Figure 5. Nixtamalization flow diagram for corn processing.

F. proliferatum fermented corn, with and without nixtamalization and nutrient supplementation. The animals were fed for 4 weeks. Nixtamalization did not reduce the toxicity observed compared to the effects of *F. proliferatum* fermented corn. Animals fed nixtamalized *F. proliferatum* fermented corn containing 10 µg HFB/g had decreased body weight, increased relative liver weight, plasma cholesterol and plasma glutamate-pyruvate transaminase, however, these changes were less than those observed in animals fed *F. proliferatum* fermented corn containing 50 µg FB_1/g. Animals fed the nixtamalized diet supplemented with nutrients developed similar levels of neoplasia as animals that received the *F. proliferatum* diet without nutrient supplementation. These data suggest that the effects of FBs may be nutrient related and that nixtamalization is not a useful detoxification strategy for *F. moniliforme* or *F. proliferatum* contaminated corn consumed by populations with good nutritional status.

HFB has been detected in commercial masa, tortilla chips and canned sweet corn. HFB in the two former products probably formed due to the nixtamalization process. The canned corn was packed in a $CaCl_2$ brine to maintain crispness. The alkalinity of this brine coupled with the very high heat treatment for retorting were probably sufficient for hydrolysis of FB in the sweet corn.

HEATING

Dry heating of corn or corn meal probably does not significantly destroy FB. Dupuy et al. (1993) reported apparent first order losses of FB_1 at high temperatures. Only the loss of primary amine reactivity was evaluated in this study. No analysis of the FB backbone was reported. It is possible that the FB amine was chemically blocked in the heating process. Without an evaluation of the biological activity of FB, it is difficult to conclude whether the higher temperature processes reduced the toxicity of FB or only resulted in the loss of the primary amine group.

The heating of semi-moist samples of suspected FB-contaminated corn by various veterinary diagnostic laboratories has resulted in lower levels of FB than expected as assayed by OPA derivatives (Bordson et al., 1995). Higher heating rates and/or temperatures, resulted in lower FB recoveries in these field samples. Hydrolysis of the heated samples resulted in recovery of the expected levels of the FB backbone. These observations suggested that some kind of amine blocking reaction occurred as these feed samples were dried.

Model pasteurization (62°C/30 min or boiling) did not result in significant loss of FBs from spiked milk samples (Maragos & Richard, 1994; Scott et al., 1994). Storage of milk for greater than 1 week at 4°C did not result in losses of FB spikes. Levels of FB in milk from cows fed 5 mg FB_1/kg body weight or given 200 µg/kg body weight intravenously were not detectable (Scott et al., 1994). This dose would represent consumption of feed

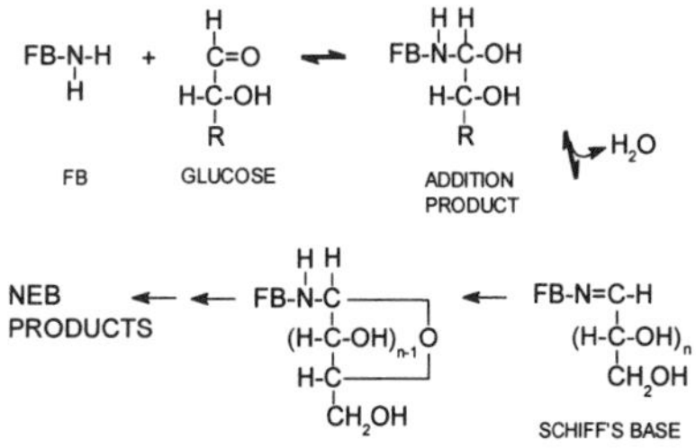

Figure 6. Proposed Schiff's base formation between fumonisin and a reducing sugar.

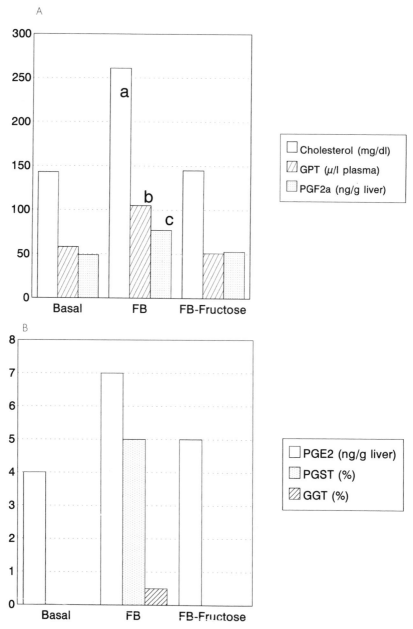

Figure 7. Effect on control, fumonisin and fructose-fumonisin fed rats on: A) plasma enzymes, prostaglandin and cholesterol levels; B) prostaglandin levels and altered hepatic foci percentages (p<0.05).

contaminated with 125 ppm FB_1. The effects of producing and storing nonfat dry milk contaminated with FBs would be useful given the results of the nonenzymatic browning study discussed below.

NONENZYMATIC BROWNING (NEB) REACTION

The NEB reaction is initiated by formation of a Schiff's base (figure 6). The requirements for this reaction are: a primary amine; a reducing sugar such as glucose, fructose or lactose; and water at pH > 7. FBs in the presence of reducing sugars and water, such as the field feed samples described above, fit the requirements for this reaction. Nonenzymatic browning would explain the increasing loss of FB detected in these samples with increasing heating times and temperatures. Once a Schiff's base undergoes the dehydration step, the adducts are committed to NEB products and the amine cannot be regenerated in its original form. This led us to consider if the formation of this adduct and its reaction products would render FB nontoxic. Murphy et al. (1995) reported that 100 mM fructose or glucose in a model system with 5 µg/ml (69.3 µM) FB_1 in 50 mM K phosphate, pH 7.0 for 48 h at 80°C resulted in the apparent first order rate loss of FB as measured by OPA reactivity. FB_1 treated in a similar manner without sugar resulted in no loss of FB amine reactivity. Preliminary cell tissue culture tests (Kraus et al., 1992) suggested that the adducts were less toxic than FB. Kraus et al. (1992) also reported that FB analogues without the primary amine group were not toxic in their test systems.

Male F344/N rats were initiated with 15 mg/kg diethylnitrosamine were fed diets containing 69.3 µmole FB_1/kg diet or 69.3 µmole FB_1-fructose reaction product/kg diet for 4 weeks (Lu et al., 1995). Compared to rats fed control diets or the fructose-FB_1 reaction diet, rats fed FB_1 diets had increased plasma glutamate/pyruvate transaminase and endogenous plasma prostaglandin $F_{2\alpha}$ (figure 7). Placental glutathione S-transferase-positive and γ-glutamyl-transferase-positive altered hepatic foci occurred in rats fed FB_1 diets while no foci developed in rats fed the fructose-FB reaction product diet or control diet. It appears that modifying FB_1 with a reducing sugar such as fructose eliminated FB_1 hepatocarcinogenicity in rats.

ACKNOWLEDGEMENTS

This work was supported in part by Pioneer Hibred International and the Center for Advanced Technology Development, Iowa State University. This is Journal Paper J-16286 of the Iowa Agricultural and Home Economics Experiment Station, Ames, IA, Projects 2955 and 2406, the latter a contributing project to North Central Regional Project NC-129.

REFERENCES

Abouzied, M.M., Pestka, J.J. Simultaneous screening of fumonisin B_1, aflatoxin B_1, and zearalenone by line immunoblot: a computer-assisted multianalyte assay system. *J. AOAC Int.* **1994**, *77*, 495-500.

Bennett, G.A.; Richard, J.L.; Eckhoff, S.R. Distribution of fumonisins in food and feed products prepared from contaminated corn. American Chemical Society National Meeting, Anaheim, CA; April 2-7, 1995.

Bordson, G.O., Meerdink, G.L., Tumbleson, M.E. Effect of drying temperature on fumonisin recovery from feeds. *J. AOAC Int.* **1995**, 1183-1187.

Bothast, R.J., Bennett, G.A., Vancauwenberge, J.E., Richard, J.L. Fate of fumonisin B_1 in naturally contaminated corn during ethanol fermentation. *Appl. Environ. Microbiol.* **1992**, *58*, 233-236.

Cawood, M.E., Gelderblom, W.C.A., Alberts, J.F., Snyman, S.D. Interaction of ^{14}C-labeled fumonisin B mycotoxins with primary rat hepatocyte cultures. *Food Chem. Toxicol.* **1994**, *32*, 627-632.

Dupuy, J., Le Bars, P., Boudra, H., Le Bars, J. Thermostability of fumonisin B_1, a mycotoxin from *Fusarium moniliforme*, in corn. *Appl. Environ. Microbiol.* **1993**, *59*, 2864-2867.

Hendrich, S. Miller, K.A., Wilson, T.A., Murphy, P.A. Toxicity of *Fusarium proliferatum*-fermented nixtamalizated corn-based diets fed to rats: effects of nutritional status. *J. Agric. Food Chem.* **1993**, *41*, 1649-1654.

Hopmans, E.C. & Murphy, P.A. Detection of fumonisins B_1, B_2, and B_3 and hydrolyzed fumonisin B_1 in corn-containing foods. *J. Agric. Food Chem.* **1993**, *41*, 1655-1658.

Hopmans, E.C., Murphy, P.A., Hendrich, S. Bioavailability of fumonisin B_1, hydrolyzed fumonisin B_1 and fructose-fumonisin B_1 adduct. unpublished data, 1995.

Kraus, G.A., Applegate, J.M., Reynolds, D. Synthesis of analogs of fumonisin B_1. *J. Agric. Food Chem.* **1992**, *40*, 2331-2332.

Lu, Z., Hopmans, E.C., Prisk, V., Murphy, P.A., Hendrich, S. Reaction with fructose detoxifies fumonisins B_1. *J. Agric. Food Chem.* **1995**, submitted.

Marasas, W.F.O., Jaskiewicz, K., Venter, F.S., Van Schalkwyk, D.J. *Fusarium moniliforme* contamination of maize in oesophageal cancer areas in Transkei. *S. Afr. Med. J.* **1988**, *74*, 110-114.

Margos, C.M., Richards, J.L. Quantitation and stability of fumonisins B_1 and B_2 in milk. *J. AOAC Int.* **1994**, *77*, 1162-1167.

Murphy, P.A., Rice, L.D., Ross, P.F. Fumonisin B_1, B_2 and B_3 content of Iowa, Wisconsin and Illinois corn and corn screening. *J. Agric. Food Chem.* **1993**, *4*, 263-266.

Murphy, P.A., Hopmans, E.C., Miller, K., Hendrich, S. Can fumonisins in foods be detoxified? in *Natural Protectants and Natural Toxicants in Food, Vol. 1*, W.R. Bidlack & S.T. Omaye, Ed; Technomic Publishing Co., Lancaster, PA, 1995, pp. 105-117.

Pestka, J.J., Azcona-Olivera, J.I., Plattner, R.D., Minervini, F., Doko, M.B., Visconti, A. Comparison assessment of fumonisin in grain-based foods by ELISA, GC-MS and HPLC. *J. Food Prot.* **1994**, *57*, 169-172.

Plattner, R.D. & Branham, B.E. Labeled fumonisins: production and use of fumonisin B_1 containing stable isotopes. *J. Assoc. Off. Anal. Chem. Int.* **1994**, *77*, 525-532.

Scott, P.M., Delgada, T., Prelusky, D.B., Trenholm, H.L., Miller, J.D. Determination of fumonisins in milk. *J. Environ. Sci. Health.* **1994**, *B29*, 989-998.

Sydenham, E.W., Shephard, G.S., Thiel, P.G., Marasas, W.F.O., Stockenstrom, S. Fumonisin contamination of commercial corn-based human foodstuffs, *J. Agric. Food Chem.* **1991**, *39*, 2014-2018.

Thakur, R.A., Smith, J.S. Analysis of fumonisin B_1 by negative-ion thermospray mass spectroscopy. *Rapid Commun. Mass Spectro.* **1994**, *8*, 82-88.

Yoshizawa, T., Yamashita, A., Luo, Y. Fumonisin occurrence in corn from high- and low-risk areas for human esophageal cancer in China. *Appl. Environ. Microbiol.* **1994**, *60*, 1626-1629.

29

REDUCTION OF RISKS ASSOCIATED WITH FUMONISIN CONTAMINATION IN CORN

Douglas L. Park,[1,2] Rebeca López-García,[1,2] Socrates Trujillo-Preciado,[1,2] Ralph L. Price[1]

[1] Department of Nutritional Sciences
University of Arizona
Tucson, Arizona 85721
[2] Current Address: Department of Food Science
Louisiana State University
Baton Rouge, Louisiana, 70803

ABSTRACT

Fumonisins, produced by *Fusarium moniliforme*, have been recognized as an important group of chemicals which cause health risks in domestic animals and humans. Decontamination procedures for fumonisin B$_1$ (FB$_1$) were evaluated to determine chemical modification and reduction in toxic/carcinogenic potentials. Ammoniation, a procedure used for decontamination of aflatoxins, yielded a 79% reduction in FB$_1$ levels in naturally contaminated corn. Authentic FB$_1$ and FB$_1$-contaminated corn were exposed to alternative treatments containing various combinations of Ca(OH)$_2$, NaHCO$_3$, and H$_2$O$_2$ simulating a modified nixtamalization procedure. Treatments also included NH$_4$Cl alone or in combination with H$_2$O$_2$ or horseradish peroxidase. The brine shrimp assay (*Artemia* spp.) was used to monitor toxicity of reaction products and the *Salmonella*/microsomal mutagenicity assay, using tester strains TA-100 and TA-102, was used to evaluate mutagenicity. Treatments of FB$_1$-contaminated corn simulating modified nixtamalization (Ca(OH)$_2$ alone or with NaHCO$_3$ + H$_2$O$_2$) gave 100% reduction of FB$_1$ and reduced brine shrimp toxicity by ca. 40%. The positive mutagenic potential (without S-9) for extracts of corn naturally contaminated with FB$_1$ was eliminated following exposure to modified nixtamalization. Reaction products formed when pure FB$_1$ was treated with Ca(OH)$_2$ and H$_2$O$_2$/NaHCO$_3$ were inhibitory to *Bacillus cereus*, *B. subtilis*, and *B. megaterium*. No inhibitory potential was evident for contaminated corn extracts following the chemical treatments.

Fumonisins in Food, Edited by L. Jackson *et al.*
Plenum Press, New York, 1996

INTRODUCTION

The interest of the scientific and regulatory communities in both aflatoxins and fumonisins has grown in response to the reports of the incidence of diseases in farm animals of non-microbial origin and the increased awareness of potential human risk. Aflatoxins and fumonisins are hepatotoxic and carcinogenic metabolites produced by *Aspergillus flavus* and *Fusarium moniliforme*, respectively. *A. flavus* and *F. moniliforme* are reported to infect corn by different routes. *A. flavus*, a non pathogenic fungus, colonizes corn kernels through the silk scars and cracks within the outer seed covering, the pericarp. The occurrence of this fungus on corn is random, and primarily surface born. *F. moniliforme*, a pathogenic fungus of corn, invades the kernel through the pedicle to the internal space distal to the tip cap, and is primarily an internally seed-borne fungus. Because an additive or synergistic toxic response is possible in animals or humans on simultaneous exposure to genotoxic and non-genotoxic carcinogens, Chamberlain and co-workers (1993) analyzed aflatoxin-contaminated corn for the presence of fumonisin. The study suggested that the two carcinogenic mycotoxins naturally coexist on corn and that toxicological investigations of simultaneous exposure to these compounds were warranted. When conditions within the kernels favor mycotoxin production, both aflatoxins and fumonisins can accumulate.

Treatment with ammonia has been successfully used to decontaminate aflatoxin-contaminated feed and foodstuffs, including corn (Brekke *et al.*, 1977), peanut, and cottonseed meals (Gardner, 1971). Several studies have shown that ammonia treatment prevents both acute and chronic aflatoxicoses in animals (Frayssinet and Lafarage-Frayssinet, 1990; Fremy and Quillardet, 1985; Hughes, *et al.*, 1979; Manson and Neal, 1987; Norred, 1979 and 1982; Norred and Morrisey, 1983; Shroeder *et al.*, 1985), and the process is generally believed to be the most effective method for the decontamination of aflatoxin-containing foodstuffs (Park *et al.*, 1988). Norred and coworkers (1991) recorded partial reduction of fumonisin levels by ammoniation, but the toxicity of the treated material was retained. Park and co-workers (1992), however, observed 80% reductions in FB_1 following exposure to ammoniation.

Price and Jorgensen (1985) reported that the nixtamalization process, used in tortilla manufacture, involving boiling and soaking of corn in lime water, reduced levels of aflatoxin in tortillas; however, reformation occurred in acidic conditions. Therefore, the process may not necessarily yield a product safe for human consumption as originally presumed (Stoloff, 1979). According to Sydenham (1994), the nixtamalization process hydrolyzes fumonisins into its two tricarballylic branches and the aminopentol backbone; however, the aminopentol backbone has still proven to be toxic. Further destruction of the fumonisin molecule could be achieved with an oxidizing agent such as hydrogen peroxide. Since hydrogen peroxide treatments under basic conditions have been widely explored for the decontamination of aflatoxin in human food and animal feed, the effects of treatments involving modified nixtamalization, *i.e.*, $Ca(OH)_2$ in combination with H_2O_2 and $NaHCO_3$ on naturally contaminated corn were studied for the decontamination of both toxins.

The fermentation process is used for the production of ethanol from corn. Bothast and co-workers (1992) reported that fumonisin was concentrated in the distillers' grains following the normal fermentation process, and there was no significant reduction in alcohol production. A modified fermentation (fermentation in presence of ammonia-based compounds) may have an effect on the structure of fumonisin. Ester bonds in the presence of ethanol, ammonia, and heat can be disrupted. The addition of several FDA approved additives during the process of fermentation could result in the reduction of aflatoxin-fumonisin contamination levels in corn. Price and co-workers (University of Arizona, personal communication) have developed a modified fermentation procedure to obtain ethanol from aflatoxin-contaminated corn. This procedure was used on pure fumonisin B_1 standard to determine its effect on the basic structure of fumonisin.

To fully evaluate the effectiveness and safety of decontamination treatments, toxicity testing of reaction products is necessary. The development of meaningful *in vitro* assays has made it possible to determine toxic effects that minor components have on biological systems (Dombrink *et al.*, 1994). The toxic potential of several mycotoxins has been evaluated utilizing disc diffusion assays using different strains of microorganisms. Fumonisin has not been evaluated in these assays. Disc diffusion assays using *Bacillus cereus, B. megaterium, and B. subtilis* presented potential alternatives for screening for the presence and toxicity of fumonisins and decontamination reaction products. Additionally, brine shrimp larvae (*Artemia* sp.) have been used extensively for determining the toxicity potential (Panigrahi, 1993). The brine shrimp assay is a simple bioassay that yields acute toxicity information within 48 hours; therefore, the evaluation of the sensitivity of this assay to fumonisins might prove useful for the evaluation of fumonisin contamination and the effectiveness of decontamination treatments.

Park and co-workers (1992) reported that fumonisins are not mutagenic in the *Salmonella* mutagenicity test using tester strain, TA-100. In the current study, fumonisin was exposed to an oxidative agent. A new *Salmonella* tester strain, TA-102, that has higher sensitivity and detects a variety of oxidants and other agents as mutagens was used to evaluate the mutagenic potentials of fumonisin B_1 and reaction products. TA-102 differs from the other standard tester strains used in the *Salmonella*/microsomal mutagenicity assay in that it has A:T base pairs at the site of reversion, whereas all the other tester strains have G:C base pairs. It is likely that this difference is responsible for the unique sensitivity of TA-102 to reversion, especially by chemical oxidants (Levin *et al.*, 1982).

MATERIALS AND METHODS

Samples

Naturally contaminated corn involved in a fumonisin poisoning outbreak of 60 horses (16 died from equine leucoencephalomalacia) in 1989 was obtained from Winslow, Arizona. The corn was stored in tightly sealed plastic bags under refrigeration (4°C). Pure fumonisin B_1 standard was obtained from the Center for Food Safety and Applied Nutrition, Food and Drug Administration, and stored dry in amber vials in a freezer (ca -10°C).

Fumonisin Analysis

Corn sample preparation and extractions were performed by the method of Bennett and Richard (1994). A composite of three corn extracts was used for the determinative test and toxicity testing. High performance liquid chromatography (HPLC) analyses were performed according to the method reported by Sydenham (1994). Accuracy and precision of the analytical method was confirmed with standardized solutions. A sample of aflatoxin-fumonisin-free corn served as a control.

Aflatoxin Analysis

Aflatoxin levels in corn were determined using AOAC method 991.31 (AOAC International, 1990).

Decontamination Procedures

Modified Nixtamalization. Solutions of 0.01 M Ca(OH)$_2$ (final pH =11.63), 2% NaHCO$_3$ (final pH=7.92), and 3% H$_2$O$_2$ were prepared with analytical grade reagents and distilled deionized water. Reactions of these solutions with pure FB$_1$ were done in amber vials at 23°C (Table 1). Each treatment was run in triplicate. A total of 100 ml of treatment solutions was added to each bag of corn to make a slurry. After 12 hours, the treatment using only Ca(OH)$_2$ was neutralized (pH=7) with 1 N HCl to stop the reaction and frozen with dry ice. For the other treatments, 100 ml of H$_2$O$_2$ and/or NaHCO$_3$ was added. Following the reaction, all samples were frozen with dry ice, freeze-dried, weighed and stored at ca 4°C.

Modified Fermentation. A 1% (w/w) NH$_4$Cl solution was used alone or in combination with a 1% (w/v) H$_2$O$_2$ solution or 220 units of horseradish peroxidase for the modified fermentation treatments. Pure FB$_1$ standard (100 g) was diluted with 5 mL of the solutions mentioned above. The pH was adjusted to 6.0 with glacial acetic acid or a solution of sodium acetate (99 %). Samples were held at 60°C for 2 hours. The pH was then adjusted to 4.2, then incubated at 34°C for 1 hour. Samples were frozen overnight at -70°C, then freeze-dried. Fermentative microorganisms were not used in this study. However, heat treatments in combination with chemicals normally used during fermentation procedures were evaluated to observe any possible fumonisin structure modification during the fermentation procedure. Treatments were done in duplicate.

Toxic/Mutagenic Potential Determination

Brine Shrimp Assay. The brine shrimp (*Artemia* sp.) assay was performed according to the method reported by Park *et al.* (1986). Brine shrimp sensitivity to FB$_1$ (0.001, 0.01, 0.1, 1.0, 10, and 100 μg) untreated and following the various treatments was evaluated. Negative (methanol) and positive (okadaic acid, 250 ng) controls and an extract of aflatoxin-fumonisin-free corn were included. Tubes containing 1, 10, 100 μL of the extract or 1, 10,100 μg of fumonisin and/or its reaction products were prepared and evaporated to dryness under N$_2$ in a water bath at 60°C.

Table 1. Modified nixtamalization chemical decontamination treatments evaluated for their ability to reduce fumonisin levels naturally occurring in corn

Treatment #	Reagent I	Time I	Reagent II	Time II
0 Control (-) Water	Distilled water	12 hours	Distilled water	1.0 hour
Ca(OH) H$_2$O$_2$/NaHCO$_{32}$	Ca(OH)$_2$ (0.01 M)	12 hours	H$_2$O$_2$ (3%) / NaHCO$_3$ (2%)	1.0 hour
Ca(OH)$_2$	Ca(OH)$_2$ (0.01 M)	12 hours	—	—
H$_2$O$_2$/NaHCO$_3$	—	—	H$_2$O$_2$ (3%) / NaHCO$_3$ (2%)	1.0 hour
H$_2$O$_2$	—	—	H$_2$O$_2$ (3%)	1.0 hour
NaHCO$_3$	—	—	NaHCO$_3$ (2%)	1.0 hour
Ca(OH)$_2$ H$_2$O$_2$	Ca(OH)$_2$ (0.01 M)	12 hours	H$_2$O$_2$ (3%)	1.0 hour
Ca(OH)$_2$ NaHCO$_3$	Ca(OH)$_2$ (0.01 M)	12 hours	NaHCO$_3$ (2%)	1.0 hour

Table 2. Chemical treatments used in conduction with modified fermentation that were evaluated for their ability to decompose pure fumonisin B_1

Treatment	(A)	H_2O_2 (1)	Peroxidase (2)
A	+	-	-
A1	+	+	-
A2	+	-	+
A12	+	+	+

+: Reagent was added
-: Reagent was not added

Bacterial Assays. Bacillus cereus, B. subtilis, and B. megaterium were used to screen for toxicity using the disc diffusion assay following the method described by Madhyastha and co-workers (1994).

Salmonella/microsomal mutagenicity assay. The *Salmonella*/microsomal mutagenicity assay was conducted essentially as described by Maron and Ames (1983). *Salmonella* tester strains TA-100 and TA-102 were used. Pure aflatoxin B_1 and fumonisin B_1 controls were run (0.01, 0.1, 1.0, 10, and 100 μg/plate). Mutagenicity was evaluated with and without metabolic activation (S-9). Composites from the corn extracts were prepared to obtain a representative sample for each treatment. Dilutions were made from each sample extract to obtain ca. 0.0025, 0.025, 0.25, 2.5 μg of fumonisin/reaction products per plate. The fumonisin/aflatoxin/reaction products were extracted from corn with acetonitrile-water (50+50, v/v). The extracts were evaporated to dryness under N_2 and the residues taken up in 1 mL of dimethylsulfoxide (DMSO). DMSO-water (50+50, v/v) and an extract of aflatoxin-fumonisin-free corn served as controls.

RESULTS AND DISCUSSION

The results for the modified nixtamalization procedure showed that the levels of chemically modified FB_1 increased with time. Samples treated with $Ca(OH)_2$ alone showed ca. 50% reduction in FB_1 levels. However, pure FB_1 and FB_1 contaminated corn exposed to the modified nixtamalization procedure, *i.e.*, treatment with $Ca(OH)_2$ and/or H_2O_2 and $NaHCO_3$, etc. had 81 and 100% losses of FB_1, respectively (Figure 1). HPLC analyses showed several new peaks on the chromatograms which corresponded to the R_f values reported for partially hydrolyzed fumonisin molecules (PH_1 and AP_1) (Sydenham, 1994). No PH_1 and AP_1 standards were available to confirm the identity of these peaks.

The most effective decontamination treatments for contaminated corn were the modified nixtamalization procedure ($Ca(OH)_2$, H_2O_2 and $NaHCO_3$) and the treatment with $H_2O_2/NaHCO_3$. Treatments on the pure toxin and contaminated corn showed different responses. This may be due to a matrix effect on the corn that aided in fumonisin destruction or to the lower fumonisin recovery in the purification procedures (Figure 1).

With respect to brine shrimp toxic potential, positive toxicity responses were obtained at 0.1 μg. At the higher concentrations (10.0 and 100 g), the response was more evident. When comparing chemical modification of fumonisin based on the HPLC analysis and toxic potential for both pure toxin and corn extracts, a relationship between chemical modification and mortality reduction was noticed. The greater the % loss of FB_1, the lower the brine shrimp

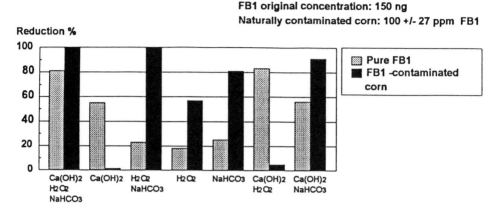

Figure 1. Chemical modification of pure fumonisin B₁ and naturally contaminated corn exposed to the modified nixtamalization procedure.

mortality (Figure 2). Although two treatments resulted in 100 % FB₁ reduction, the treatment utilizing H₂O₂/NaHCO₃ showed the best reduction in brine shrimp mortality.

In the modified fermentation procedure, all treatments resulted in 70-100% reduction in fumonisin B₁ levels as determined by HPLC. With the exception of NH₄Cl/H₂O₂, all treatments resulted in a significant reduction in brine shrimp (*Artemia* sp.) mortality (Figure 3).

Bacillus cereus, B. megaterium and *B. subtilis* sensitivity to fumonisin B₁ was tested using a disc diffusion assay at the 0.001-100 g concentration range. None of the three bacteria used showed sensitivity to FB₁. On the other hand, 20 ng of aflatoxin caused a 2 mm inhibition zone. The disc diffusion assay was used to determine if toxic reaction products formed during treatment of pure FB₁ and corn extracts with chemical solutions. Growth

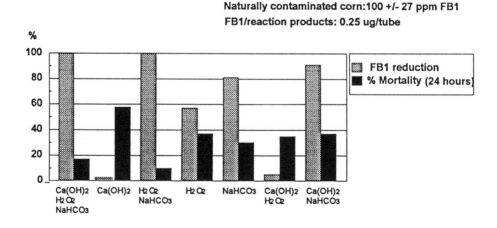

Figure 2. Comparison between chemically modified fumonisin B₁ and brine shrimp (*Artemia* sp.) mortality for FB₁ contaminated corn exposed to the modified nixtamalization procedure and selected chemicals.

# of Revertants

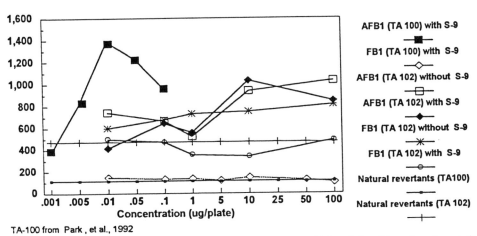

TA-100 from Park , et al., 1992

Figure 3. Comparison between chemically modified fumonisin B$_1$ and brine shrimp (*Artemia* sp.) mortality for pure FB$_1$ exposed to the modified fermentation procedure and selected chemicals.

inhibition zones (1.5-3.5 mm) were observed for pure FB$_1$ exposed to treatments containing a combination of Ca(OH)$_2$ and H$_2$O$_2$/NaHCO$_3$. None of the contaminated corn extracts presented bacterial growth inhibition. This effect may be due to the presence of a high concentration of residual salts in the FB$_1$ treatments. The corn matrix may have inhibited these salts causing a decrease in toxicity to the bacteria.

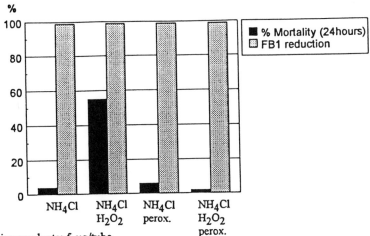

FB$_1$/reaction products: 5 µg/tube
perox.: horseradish peroxidase
Controls of 5, and 10 µg authentic FB$_1$ resulted in 13 and 47% mortality (24 hours) to brine shrimp respectively.

Figure 4. Comparison between tester strains TA-100 and TA-102 with and without metabolic activation for aflatoxin B$_1$ and fumonisin B$_1$ using the *Salmonella*/microsomal mutagenicity assay.

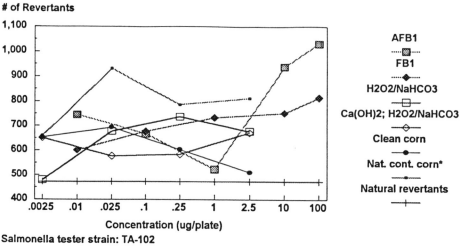

# of Revertants

Salmonella tester strain: TA-102

* Total Aflatoxin: 65 ppb

Fumonisin B: 100 $\pm$ 27 ppm

Figure 5. Mutagenic potential of selected corn extracts from fumonisin/aflatoxin contaminated corn exposed to the modified nixtamalization procedure and selected chemicals using tester strain TA-102 without metabolic activation.

For the *Salmonella*/microsomal mutagenicity assay, sample revertant counts greater than or equal to two times the spontaneous reversion rate are considered a positive mutagenic response using this assay. The mutagenic response of *Salmonella* TA-102 to FB$_1$ showed a positive mutagenic tendency without metabolic activation; however, a positive mutagenic activity was not confirmed. In the same concentration range (0.01-100 µg/plate), the mutagenic response for aflatoxin was more evident (Figure 4). Due to the high spontaneous revertant rate (ca. 400/plate), this tester strain is difficult to use.

The naturally contaminated corn extracts without metabolic activation presented positive mutagenic responses (Figure 5). Further evaluation of this sample, found that it contained 65 ng/g total aflatoxin which would explain the positive mutagenic response. None of the chemically treated corn extracts presented a mutagenic event (with and without metabolic activation) (Figures 5 and 6). These results suggest that the chemical treatments evaluated in this study may also be effective for aflatoxin decontamination. Further research will be needed to confirm the effect of these treatments on aflatoxin.

The most effective treatments for FB$_1$ chemical modification, *i.e.*, the modified nixtamalization procedures (Ca(OH)$_2$ with H$_2$O$_2$ and NaHCO$_3$, and H$_2$O$_2$/NaHCO$_3$) did not show positive mutagenic potentials. This may be due to the absence of toxic/mutagenic oxidized forms of fumonisin following chemical treatment (Figures 5 and 6).

CONCLUSIONS

Fumonisins are common contaminants in corn and corn products worldwide. Decontamination procedures appear to be crucial for controlling the fumonisin content of corn. Treatment of FB$_1$ with Ca(OH)$_2$ hydrolyzed FB$_1$ to compounds that were still toxic. Further decontamination and detoxification of FB$_1$ contaminated corn was achieved by using the modified nixtamalization procedure. However, the best reaction involved treating corn with

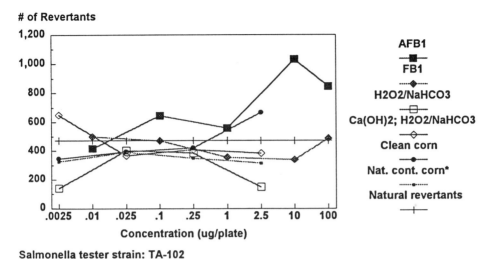

# of Revertants

AFB1
FB1
H2O2/NaHCO3
Ca(OH)2; H2O2/NaHCO3
Clean corn
Nat. cont. corn*
Natural revertants

Concentration (ug/plate)

Salmonella tester strain: TA-102

* Total Aflatoxin: 65 ppb

Fumonisin B: 100 $\pm$ 27 ppm

Figure 6. Mutagenic potential of selected corn extracts from fumonisin/aflatoxin contaminated corn exposed to the modified nixtamalization procedure and selected chemicals using tester strain TA-102 with metabolic activation.

a combination of $H_2O_2/NaHCO_3$ for one hour. This is a simple treatment that could be incorporated into industrial processing procedures for corn. Further studies are needed to determine the identity of the reaction products as well as the effects of the treatment on the nutritional and functional properties of corn.

Brine shrimp (*Artemia* sp.) were sensitive to FB_1 at 0.01 μg. *Bacillus cereus, B. subtilis and B. megaterium* were not sensitive to FB_1 when exposed up to 100 μg FB_1. In the *Salmonella*/microsomal mutagenicity assay, tester strain TA-102 presented a tendency to a positive mutagenic event in the 0.01-100 μg FB_1/plate concentration range; however, the mutagenic potential was not confirmed. No mutagenic reaction products were detected with and without metabolic activation even when it was shown that the samples contained originally a high concentration of aflatoxins. More research is needed to determine the effect of these treatments on aflatoxin.

Since fumonisins and aflatoxins have been reported to be co-contaminants in corn, the development of decontamination/detoxification treatments will be of great importance for control programs for both toxins.

REFERENCES

AOAC. Official Method of Analysis, method 991.31. AOAC International, Arlington, Virginia. 1990.

Bennett, G. A.; Richard, J. L. Liquid chromatographic method for analysis of the naphthalene dicarboxaldehyde derivative of fumonisins. *J. of AOAC International*. **1994**, 77, 501-506.

Bothast, R. J.; Bennett, G. A.; Vancauwenberge, J.E.; Richard, J.L. Fate of fumonisin B_1 in naturally contaminated corn during ethanol fermentation. Appl. Environ. Microbiol. 1992, 58, 233-236.

Brekke, O. L.; Sinnhuber, R. O.; Peplinski, A.J.; Wales, H.H.; Putman, G. B.; Lee, D. J.; Ciegler, A. Aflatoxin in corn: ammonia inactivation and bioassay with rainbow trout. Appl. Environ. Microbiol. 1977, 34, 34-37.

Chamberlain, W. J.; Bacon, C. W.; Norred, W. P.; Voss, K. A. Levels of fumonisin B₁ in corn naturally contaminated with aflatoxins. Food Chem. Toxicol. **1993**, 31, 995-998.

Dombrink-Kurtzman, M. A.; Bennet, G. A.; Richard, J. L. An optimized MTT bioassay for determination of cytotoxicity of fumonisins in turkey lymphocytes. J. AOAC International. **1994**, 77, 512-516.

Frayssinet, C.; Lafarge-Frayssinet, C. Effects of ammoniation on the carcinogenicity of aflatoxin-contaminated groundnut oil cakes: long-term feeding study in the rat. Food Add. Contam. **1990**, 7, 63-68.

Fremy, J. M.; Quillardet P. The carry over of aflatoxin into milk of cows fed ammoniated rations: Use of an HPLC method and genotoxicity test for determining milk safety. Food Add. Contam. **1985**, 2, 201-207.

Gardner H. K. Inactivation of aflatoxins in peanut and cottonseed meals by ammoniation. J. Am. Oil Chem. Soc. **1971**, 48, 70-73.

Hughes B. L.; Barnett B. D.; Jones, J. E.; Dick, J. W.; Norred, W. P. Safety of feeding aflatoxin-inactivated corn to white leghorn layer breeders. Poultry Sci. **1979**, 58, 1202-1209.

Levin, D. E.; Hollstein, M.; Christman, M. F.; Schwiers, E. A.; Ames, B. N. A new *Salmonella* tester strain (TA-102) with A-T base pairs at the site of mutation detects oxidative mutagens. Proc. Natl. Acad. Sci. U.S.A. **1982**, 79, 7445-7449.

Madhyastha, M. S.; Marquardt, R. R. Comparison of toxicity of different mycotoxins to several species of bacteria and yeasts: Use of *Bacillus brevis* in a disc diffusion assay. J. Food Prot. **1994**, 57, 48-53.

Manson, M. M.; Neal, G. E. The effect of feeding aflatoxin-contaminated groundnut meal, with or without ammoniation, on the development of gamma-glutamyl transferase lesions in the livers of Fischer 344 rats. Food Add. Contam. **1987**, 4, 141-147.

Maron, D. M; Ames, B. N. Revised methods for the *Salmonella* mutagenicity test. Mutation Res. **1983**, 113, 173-215.

Norred, W. P. Effect of ammoniation on the toxicity of corn artificially contaminated with aflatoxin B₁. Toxicol. Appl. Pharm. **1979**, 51, 411-416.

Norred, W. P. Ammonia treatment to destroy aflatoxins in corn. J. Food Prot. **1982**, 45, 972-976.

Norred, W. P.; Morrisey, R. E. Effects of long-term feeding of ammoniated, aflatoxin contaminated corn to Fischer 344 rats. Toxicol. Appl. Pharm. **1983**, 70, 96-104.

Norred, W. P.; Voss, K. A.; Bacon, C. W.; Riley, R. T. Effectiveness of ammonia treatment in detoxification of fumonisin-contaminated corn. Food Chem. Toxicol. **1991**, 29, 815-819.

Panigrahi S. Bioassay of mycotoxins using terrestrial and aquatic animal species. Food Chem. Toxicol. **1993**, 31, 167-190.

Park, D. L.; Aguirre-Flores, I.; Scott, W. F.; Alterman, E. Evaluation of chicken embryo, brine shrimp, and bacterial bioassays for saxitoxin. J. of Tox. and Env. Health. **1986**, 18, 589-594.

Park, D. L.; Lee, L. S.; Price, R. L.; Pohland, A. E. Review of decontamination of aflatoxins by ammoniation: Current status and regulation. J. AOAC. **1988**, 71, 685-703.

Park, D. L.; Rua, S. M.; Mirocha C. J.; Abd-Alla El Sayed, A. M.; Weng, C. Y. Mutagenic potentials of fumonisins contaminated corn following ammonia decontamination procedure. Mycopathol. **1992**, 117, 105-108.

Price, R. L.; Jorgensen, K. V. Effects of processing on aflatoxin levels and on mutagenic potential of tortillas made from naturally contaminated corn. J. Food Sci. **1985**, 50, 347-349 & 357.

Shroeder, T.; Zweifel, U.; Sagelsdorff, P.; Friendich, U.; Luthy, J.; Schlatter, C. Ammoniation of aflatoxin-containing corn: Distribution, *in vivo* covalent deoxyribonucleic acid binding, and mutagenicity of reaction products. J. Agr. Food Chem. **1985**, 38, 1900-1903.

Stoloff, L. The three eras of fungal toxin research. J. Am. Oil Chem. Soc. **1979**, 56, 784.

Sydenham, E. W. Fumonisins: Chromatographic methodology and their role in human and animal health. Ph. D. dissertation. University of Cape Town, Capetown, South Africa, 1994.

30

EFFECT OF THERMAL PROCESSING ON THE STABILITY OF FUMONISINS

Lauren S. Jackson[1], Jason J. Hlywka[2], Kannaki R. Senthil[3], and Lloyd B. Bullerman[2]

[1] U.S. Food and Drug Administration
National Center for Food Safety and Technology
Summit-Argo, IL 60501
[2] Department of Food Science and Technology
University of Nebraska
Lincoln, NE 68583
[3] Illinois Institute of Technology
National Center for Food Safety and Technology
Summit-Argo, IL 60501

ABSTRACT

Fumonisins, a group of mycotoxins produced by *Fusarium moniliforme* in corn, have been implicated in several animal and human diseases. *F. moniliforme* and the fumonisins are an area of increasing concern for corn producers and consumers. Consequently, there is interest in reducing human and animal exposure to these fungal toxins. Studies of the effects of biological, chemical, and physical treatments on the reduction of fumonisin levels in food have shown variable results. Work was conducted at the U.S. Food and Drug Administration, National Center for Food Safety and Technology, to determine the effects of thermal processing on fumonisins B_1 (FB_1) and B_2 (FB_2) in an aqueous buffer. Parameters that were studied included processing time (0-60 min), processing temperature (100-235°C), and buffer pH (4, 7, and 10). The rate and extent of fumonisin decomposition increased with processing temperature. Less than 27% of FB_1 and less than 20% of FB_2 were lost when processing temperatures were less than or equal to 125°C for 60 min. After 60 min at 150°C, losses of FB_1 and FB_2 were 80-90% at pH 4, 18-30% at pH 7, and 40-52% at pH 10. At temperatures greater than or equal to 175°C, more than 80% of FB_1 and FB_2 was lost after 60 min. These results indicate that foods reaching temperatures greater than 150°C during processing may have lower fumonisin levels. More work is needed to quantitate the effects of different processing operations (baking, extrusion, frying) on the fumonisin content of corn-based foods.

Fumonisins in Food, Edited by L. Jackson *et al.*
Plenum Press, New York, 1996

INTRODUCTION

The fumonisins are a family of mycotoxins produced by *Fusarium moniliforme* and *F. proliferatum*, two of the most prevalent molds associated with corn and other cereal grains. Since the early 1900s it was strongly suspected that metabolites produced by *F. moniliforme* and related species were responsible for the onset of a variety of animal and human diseases. Peters (1904) reported that ingestion of moldy corn contaminated with *F. moniliforme* led to sloughing of hooves in cattle and horses, loss of feathers in chickens, and convulsions and death in a variety of animals. More recently, feeding culture material of *F. moniliforme* or purified fumonisin B_1 (FB_1) to experimental animals resulted in hepatic cirrhosis and hyperplasia in rats (Kriek et al., 1981a,b; Gelderblom et al., 1991, 1992), leukoencephalomalacia (LEM) and toxic hepatosis in horses (Kriek, et al, 1981a,b; Marasas et al., 1988; Sydenham et al., 1992; Wilson et al., 1992), pulmonary edema (PE) in swine (Harrison et al., 1990), and acute congestive heart failure in baboons and monkeys (Kriek et al., 1981a,b; Fincham et al., 1992). In humans, ingestion of home-grown corn infected with *F. moniliforme* has been implicated in the high incidence of esophageal cancer in the Transkei region of South Africa (Sydenham et al., 1991) and in the Linxian region of China (Cheng et al., 1985).

Fumonisins have been found in corn grown in all regions of the world. This is not unexpected since *F. moniliforme* and *F. proliferatum* are found worldwide, thriving under various environmental conditions. The fungi are important ear and stalk rot pathogens of corn and can form systemic infections which are asymptomatic (Bacon and Nelson, 1994). Therefore, ears of corn may have high fumonisin content yet appear normal (i.e., without moldy or discolored kernels).

Corn implicated in outbreaks of diseases in humans and animals contained high levels of fumonisins. In the Transkei region of South Africa, where the incidence of esophageal cancer is high, the fumonisin content of corn reached 140 ppm (Rheeder et al., 1992). Feeds associated with field cases of equine LEM and porcine PE in the United States contained from 8 to 117 ppm FB_1 and from 20 to 360 ppm FB_1, respectively (Ross et al., 1992). Most "good quality" corn, or corn not implicated in disease outbreaks, contains <1 ppm fumonisin (Sydenham et al., 1991; Pittet et al., 1992).

Data on the toxicity and carcinogenicity of the fumonisins suggest that these compounds should be evaluated as potential risks to human and animal health. FB_1 and fumonisin B_2 (FB_2), two of the most abundant forms of fumonisin in food, have been classified as possible human carcinogens by the International Agency for Research on Cancer (IARC, 1993). Work is under way at the U.S. Food and Drug Administration's (FDA) National Center for Toxicological Research (NCTR) and other government agencies and academic institutions to determine the health implications of chronic consumption of fumonisins. In addition, the FDA is collecting data to determine the occurrence and levels of fumonisins in human foods. Toxicity and survey data may be used by the FDA to establish residue levels for fumonisin in corn-based foods and feeds.

METHODS FOR REDUCING THE FUMONISIN CONTENT OF FOOD

Contamination of foods and feeds with *F. moniliforme* and the fumonisins is an increasing concern that may threaten human and animal health and reduce the marketability of corn. Currently, there has been increased attention directed at reducing the human and animal exposure to these fungal toxins. Because of the widespread presence of *F. moniliforme* on corn, its control may require the breeding of corn for resistance to *F. moniliforme* (Riley et al., 1993) or the use of soil and endophytic bacteria, such as *Enterobacter cloacae*, to inhibit the growth of *F. moniliforme* and other corn pathogens (Bacon and Nelson, 1994). In a study of postharvest control of fumonisin production, Le Bars et al. (1994) found that fumonisin production was reduced during storage of low-moisture corn (<22%) under modified atmospheres (i.e., N_2 or CO_2). Biological, chemical, and physical processes have been explored to salvage fumonisin-contaminated corn. These processes and, in particular, thermal processing will be discussed in detail in the following sections.

Biological Treatments

Bothast et al. (1992) studied fermentation as a means to salvage fumonisin-contaminated corn. They found that only a limited amount of FB_1 was degraded during the fermentation of contaminated corn into ethanol. Toxin in the ethanol was removed by distillation. However, the distillers' grains, thin stillage, and distillers' solubles, which contained significant quantities of fumonisin, required detoxification before they could be used as animal feed.

Chemical Treatments

Norred et al. (1991) found that atmospheric pressure/ambient temperature ammoniation reduced the fumonisin content of *F. moniliforme* culture material but did not reduce the toxicity of the material when fed to rats. Park et al. (1992) reported 79% reduction of fumonisin levels in corn after high-pressure/ambient-temperature ammoniation followed by a low-pressure/high-temperature treatment. However, they did not measure the toxicity of the treated corn. A survey by Sydenham et al. (1991) showed that corn products (masa, tortillas) from South America and the United States that were treated with lime water and heat had very low levels of fumonisins. Hendrich et al. (1993) reported that treating corn with lime water and heat hydrolyzed FB_1 to the aminopentol backbone and tricarballylic acid. When the corn was fed to rats, it was found that the toxicity was not reduced by the treatment.

Physical Methods

Various physical methods for decontaminating fumonisin- containing corn have been investigated. Sydenham et al. (1994) reported that physically removing the fines or screenings from the bulk shipments of corn reduced the fumonisin content by 26-69%. Studies conducted by Bennett and Richard (1995) at the U.S. Department of Agriculture National Center for Agricultural Utilization Research showed that starch prepared from contaminated corn by wet milling was fumonisin-free. However, the gluten and fiber fractions contained considerable amounts of fumonisin and required further decontamination before they could be used for animal feed.

Thermal Processing

Surveys (Stack and Eppley, 1992; Pittet et al., 1992; Doko and Visconti, 1994) have shown that thermally processed corn products (e.g., canned corn, tortillas, ready-to-eat cereals) generally have a lower concentration of FB_1 than do unprocessed products (e.g., corn meal and grits) (Figure 1). Since most corn-based foods receive some sort of thermal treatment before they are eaten, there has been interest in determining the effects of thermal processing on the fumonisin content of food.

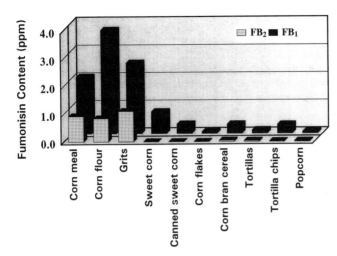

Figure 1. Fumonisin B_1 and fumonisin B_2 concentrations in commercial corn-based human foods from retail outlets in several countries. Data were obtained from surveys by Doko and Visconti (1994), Pittet et al. (1992), Stack and Eppley (1992), and Sydenham et al. (1991). Bars represent the range of fumonisin levels in each corn-based food.

Alberts et al. (1990) reported that boiling culture material of *F. moniliforme* in water for 30 min followed by oven drying at 60°C for 24 h resulted in no loss of fumonisin or the cancer-promoting (gamma-glutamyl-transpeptidase-positive) activity of the culture material. Subjecting raw milk spiked with FB_1 and FB_2 to pasteurization conditions (62°C, 30 min) resulted in no loss of either toxin (Maragos and Richard, 1994). Scott and Lawrence (1994) observed losses of FB_1 and FB_2 exceeding 70% in dry corn meal heated to 190 and 220°C, for 60 and 25 min, respectively. Similarly, fumonisin content was partially reduced in corn meal muffins baked at 220°C for 25 min (Scott and Lawrence, 1994).

Dupuy et al. (1993) reported that the decomposition of FB_1 in dry corn heated at 100-150°C followed a first order decay. They also found that substantial losses of FB_1 occurred only after dry corn was heated at 150°C for 40 min. Lower temperatures (50, 75, 100, and 125°C) resulted in small losses of FB_1.

Reports by Bordson et al. (1993) and Scott and Lawrence (1994) suggest that data from surveys and thermal processing studies should be interpreted with caution. Bordson et al. (1993) found that the extraction efficiency of fumonisin decreased when corn and feed were heated. Similarly, Scott and Lawrence (1994) reported that in some heated corn-based foods, fumonisin appeared to be bound to the food matrix rather than chemically decomposed. More work is needed to develop analytical methodology that distinguishes between binding and matrix effects and true losses of fumonisin (Bullerman and Tsai, 1994).

Work was initiated at the FDA, National Center for Food Safety and Technology (NCFST), to eliminate matrix-related recovery problems by studying the effects of an aqueous buffer (citrate-phosphate-borate) at pH 4, 7, and 10 (Jackson et al., 1995a) on the stability of FB_1 and FB_2 (5 ppm) during thermal processing (processing time, 0-60 min; temperature, 100-235°C). Fumonisin solutions were processed in a Parr reactor that was heated with an electric heating mantle. A proportional controller was used to maintain the solutions at the desired temperature while they were agitated at a constant speed. All processing runs were done in triplicate.

Figure 2 shows the time-temperature profiles recorded for the fumonisin solutions during processing. The come-up times, i.e., the length of time necessary for the fumonisin solutions to reach the desired processing temperatures, were 19, 28, 33, 39, 44, and 53 min for temperatures of 100, 125, 150, 175, 200, and 235°C, respectively.

Aliquots of the fumonisin solutions were removed during the come-up time and at 10-min intervals for 60 min once the processing temperature was attained. The solutions

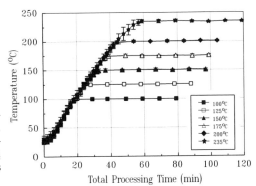

Figure 2. Temperatures recorded in aqueous solutions of FB$_1$ (5 ppm) during thermal processing. The solutions were processed for 1 h at temperatures of 100-235°C in a 1-L Parr pressure reactor. Points indicate the average of nine runs. Error bars indicate one standard deviation of the mean.

were analyzed for FB$_1$ content by using the method of Shephard et al. (1990). Because of the absence of matrix effects and interferences in the processed solutions, steps normally used to extract and purify fumonisins were omitted. Consequently, the fumonisin solutions required minimal preparation for analysis by high-performance liquid chromatography (HPLC). Liquid chromatography-mass spectrometry (LC-MS) was used to identify decomposition products resulting from some of the thermal processing runs involving FB$_1$.

LC-MS analysis of the thermally processed FB$_1$ solutions indicated that at least three decomposition products existed: two partially hydrolyzed products and a fully hydrolyzed form. Structures of FB$_1$ and its hydrolysis products are shown in Figure 3. The pH had no apparent effect on the types of decomposition products found in the processed FB$_1$ solutions. However, it was noted that at pH 10, the major decomposition species was fully hydrolyzed FB$_1$, whereas at pH 4 and 7, partially hydrolyzed FB$_1$ was also present. It is interesting to note that FB$_1$ and its hydrolysis products were not found in the pH 4 solution after 60 min of processing at temperatures ≥175°C. These results suggest that the decomposition products formed at these temperatures could not be derivatized by the *o*-phthaldialdehyde (OPA) reagent and, thus, could not be measured by HPLC. In addition, these decomposition products may not have been in the mass range scanned by LC-MS.

Figures 4-6 illustrate that the decomposition of FB$_1$ during thermal processing depended on the pH of the solution. Overall, FB$_1$ appeared least stable at pH 4 and most stable at pH 7. The rate and extent of FB$_1$ decomposition increased with processing temperature; less than 27% of FB$_1$ was lost after processing at temperatures ≤125°C for 60 min (Jackson et al., 1995a). After 60 min at 150°C, loss of FB$_1$ ranged from 11 to 90%, depending on buffer pH. At temperatures ≥175°C, more than 80% of FB$_1$ was lost after 60 min of processing.

In general, the stability of FB$_2$ during processing was similar to that of FB$_1$ (Jackson et al., 1995b). The rate and extent of decomposition of FB$_2$ increased with processing temperature and length of processing time. No losses of FB$_2$ were seen at 100°C and minimal losses (<20%) occurred at 125°C. After 60 min at 150°C, loss of FB$_2$ ranged from 30 to 80%, depending on buffer pH. At temperatures ≥175°C, more than 90% of FB$_2$ was lost after the 60-min processing time. At 200°C, all FB$_2$ was lost after 30 min of processing. In general, FB$_2$ was least stable at pH 4 and most stable at pH 7.

These results (Jackson et al., 1995a,b) as well as those reported by Alberts et al. (1990), Dupuy et al. (1993), and Scott and Lawrence (1994) indicate that the fumonisins are fairly heat-stable compounds. In general, the loss of FB$_1$ and FB$_2$ was more rapid and extensive in alkaline or acidic environments than at neutral pH. Thermal processing operations such as boiling or retorting, which occur at temperatures <125°C, would be expected

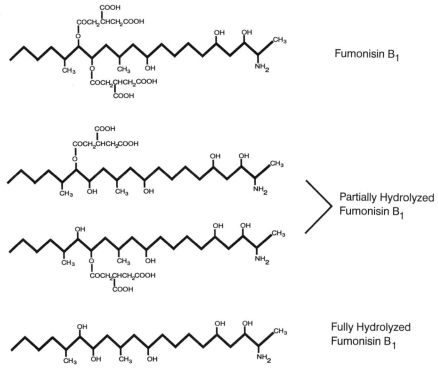

Figure 3. Chemical structures of FB$_1$ and its hydrolysis products.

to have little effect on the fumonisin content of food. However, foods that reach temperatures >150°C during processing may have significant losses of fumonisins.

Although the decomposition of FB$_1$ and FB$_2$ was evident during this study, detoxification of these compounds cannot be assumed. Hendrich et al. (1993) found that corn treated with lime and heat was more toxic than untreated corn. The toxicity of the treated corn was attributed to the presence of hydrolysis and other breakdown products of FB$_1$. Studies by Wang et al. (1991) and Norred and Voss (1994) suggest that the aminopentol backbone of FB$_1$ is responsible for the alterations in sphingolipid synthesis in animals which exhibit fumonisin toxicity. Work is under way at the University of Nebraska to determine the toxicity of the thermally processed fumonisin solutions.

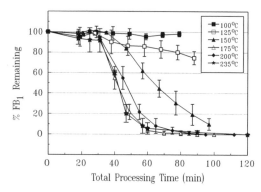

Figure 4. Effects of processing temperature and time on the decomposition of FB$_1$ in an aqueous buffer at pH 4. Each point represents the average of three replicates, and error bars indicate one standard deviation of the mean.

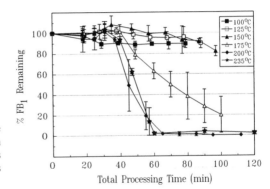

Figure 5. Effects of processing temperature and time on the decomposition of FB₁ in an aqueous buffer at pH 7. Each point represents the average of three replicates, and error bars indicate one standard deviation of the mean.

CONCLUSIONS

Surveys have shown that only a small fraction of corn grown in the world (<17% of the corn grown in the United States) is consumed by humans (Riley et al., 1993). However, corn is consumed worldwide and is the staple grain for some populations. Consequently, more work is needed to estimate the human exposure to fumonisins from the diet and to devise methods for reducing the fumonisin content of corn.

Physical parameters such as processing time and temperature are factors that affect the fumonisin content of food. Published studies suggest that when foods are heated at temperatures encountered in boiling or retorting foods (100-125°C), little change in fumonisin content would be expected. Foods that reach temperatures of >150°C during processing (baking or frying) may have some losses of fumonisins. More work is needed to further define the effects of thermal processing operations (baking, frying, extrusion) on the stability of fumonisins in contaminated corn.

ACKNOWLEDGMENTS

The authors would like to thank Dr. Robert Eppley (FDA) for donating the FB₁ for the processing studies and Dr. Steven Musser (FDA) for analyzing the processed fumonisin solutions by LC-MS. This publication was supported (or partially supported) by cooperative agreement no. FD-000431 from the U.S. Food and Drug Administration (and the National Center for Food Safety and Technology). Its contents are solely the opinions of the authors and do not necessarily represent official views of the U.S. Food and Drug Administration.

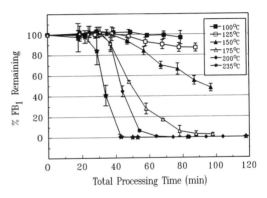

Figure 6. Effects of processing temperature and time on the decomposition of FB₁ in an aqueous buffer at pH 10. Each point represents the average of three replicates, and error bars indicate one standard deviation of the mean.

REFERENCES

Alberts, J.F.; Gelderblom, W.C.A.; Thiel, P.G.; Marasas, W.F.O.; van Schalwyk, D.J.; Behrend, Y. Effects of temperature and incubation period on production of fumonisin B_1 by *Fusarium moniliforme*. Appl. Environ. Microbiol. 1990, 56, 1729-1733.

Bacon, C.W.; Nelson, P.E. Fumonisin production in corn by toxigenic strains of *Fusarium moniliforme* and *Fusarium proliferatum*. J. Food Prot. 1994, 57, 514-521.

Bennett, G.A.; Richard, J.L. Distribution of fumonisins in food and feed products prepared from contaminated corn. 1995, Accompanying chapter in this volume.

Bordson, G.; Meerdink, G.; Bauer, K.; Tumbleson, M. Fumonisin recovery from samples dried at various temperatures. Midwest Section meeting of AOAC INTERNATIONAL, Des Moines, IA, 1993.

Bothast, R.J.; Benett, G.A.; Vancauwenberge, J.E.; Richard, J.L. Fate of fumonisin B_1 in naturally contaminated corn during ethanol fermentation. Appl. Environ. Microbiol. 1992, 58, 233-236.

Bullerman, L.B.; Tsai, W.-Y. Incidence and levels of *Fusarium moniliforme, Fusarium proliferatum* and fumonisins in corn and corn-based foods and feeds. J. Food Prot. 1994, 57, 541-546.

Cheng, S.J.; Jiang, Y.Z.; Li, M.H.; Lo, H.Z. A mutagenic metabolite produced by Fusarium moniliforme isolated from Linxian county, China. Carcinogenesis 1985, 6, 903-905.

Doko, M.B.; Visconti, A. Occurrence of fumonsins B_1 and B_2 in corn and corn-based human foodstuffs in Italy. Food Addit. Contam. 1994, 11, 433-439.

Dupuy, J.; Le Bars, P.; Boudra, H.; Le Bars, J. Thermostability of fumonisin B_1, a mycotoxin from *Fusarium moniliforme*, in corn. Appl. Environ. Microbiol. 1993, 59, 2864-2867.

Fincham, J.E; Marasas, W.F.O.; Taljaard, J.J.F.; Kriek, N.P.J.; Badenhorst, C.J.; Gelderlom, W.C.A.; Seier, J.V.; Smuts, C.M.; Faber, M.; Weight, M.J.; Slazus, W.; Woodroof, C.W.; van Wyk, M.J.; Kruger, M.; Thiel, P.G. Atherogenic effects in a non-human primate of *Fusarium moniliforme* cultures added to a carbohydrate diet. Atherosclerosis 1992, 94, 13-25.

Gelderblom, W.C.W.; Kriek, N.P.J.; Marasas, W.F.O.; Thiel, P.G. Toxicity and carcinogenicity of the *F. moniliforme* metabolite, FB_1, in rats. Appl. Environ. Microbiol. 1991, 12, 1247-1251.

Gelderblom, W.C.A.; Marasas, W.F.O.; Vleggar, R.; Cawood, M.E.; Thiel, P.G. Fumonisins: Isolation, chemical characterization and biological effects. Mycopathologia 1992, 117, 11-16.

Harrison, L.R.; Colvin, B.M.; Greene, T.J.; Newman, L.E.; Cole, R.J. Pulmonary edema and hydrothorax in swine produced by fumonisin B_1, a toxic metabolite of *Fusarium moniliforme*. J. Vet. Diagn. Invest. 1990, 2, 217-221.

Hendrich, S.; Miller, K.A.; Wilson, T.M.; Murphy, P.A. Toxicity of Fusarium proliferatum-fermented nixtamalized corn-based diets fed to rats: Effect of nutritional status. J. Agric. Food Chem. 1993, 41, 1649-1654.

International Agency for Research on Cancer (IARC). Toxins derived from *Fusarium monilforme*: Fumonisin B_1 and B_2 and fusarin C. IARC Monogr. Eval. Carcinog. Risk Chem. Hum. 1993, 56, 445-466.

Jackson, L.S.; Hlywka, J.J.; Senthil, K.R.; Bullerman, L.B.; Musser, S.M. The effects of time, temperature and pH on the stability of fumonisin B_1 in an aqueous model system. J. Agric. Food Chem. 1995a, Submitted.

Jackson, L.S.; Hlywka, J.J.; Senthil, K.R.; Bullerman, L.B. The effects of thermal processing on the stability of fumonisin B_2 in an aqueous model system. 1995b, Unpublished data.

Kriek, N.P.J.; Kellerman, T.S.; Marasas, W.F.O. A comparative study of the toxicity of *Fusarium verticillioides* (=*F. moniliforme*) to horses, primates, pigs, sheep, and rats. Onderstepoort J. Vet. Res. 1981a, 48, 129-131.

Kriek, N.P.J.; Marasas, W.F.O.; Thiel, P.G. Hepato- and cardiotoxicity of *Fusarium verticillioides* (*F. moniliforme*) isolates from southern African maize. Food Cosmet. Toxicol. 1981b, 19, 447-456.

Le Bars, J.; Le Bars, P.; Dupuy, J.; Boudra, H. Biotic and abiotic factors in fumonisin B_1 production and stability. J. AOAC Int. 1994, 77, 517-521.

Maragos, C.M.; Richard, J.L. Quantitation and stability of fumonisins B_1 and B_2 in milk. J. AOAC Int. 1994, 77, 1162-1167.

Marasas, W.F.O.; Kellerman, T.S.; Gelderblom, W.C.A.; Coetzer, J.A.W.; Thiel, P.G.; van der Lugt, J.J. Leukoencephalomalacia in a horse induced by fumonisin B_1 isolated from *Fusarium moniliforme*. Onderstepoort J. Vet. Res. 1988, 55, 197-203.

Norred, W.P.; Voss, K.A.; Bacon, C.W.; Riley, R.T. Effectiveness of ammonia treatment in detoxification of fumonisin-contaminated corn. Food Chem. Toxicol. 1991, 29, 815-819.

Norred, W.P. ; Voss, K.A. Toxicity and role of fumonisins in animal diseases and human esophageal cancer. J. Food Prot. 1994, 57, 522-527.

Park, D.L.; Rua, S.M., Jr.; Mirocha, C.J.; Abd-Alla, E.A.M.; Weng, C.Y. Mutagenic potentials of fumonisin contaminated corn following ammonia decontamination procedure. Mycopathologia 1992, 117, 73-78.

Peters, A.T. A fungus disease in corn. Annu. Rep. Agric. Exp. Stn. Nebr. 1904, 17, 13-22.

Pittet, A.; Parisod, V.; Schellenberg, M. Occurrence of fumonisins B_1 and B_2 in corn-based products fromt the Swiss market. J. Agric. Food Chem. 1992, 40, 1352-1354.

Rheeder, J.P.; Marasas, W.F.O.; Thiel, P.G.; Sydenham, E.W.; Shephard, G.S.; van Schalkwyk, D.J. *Fusarium moniliforme* and fumonisins in corn in relation to human esophageal cancer in Transkei. Phytopathology 1992, 82, 353-357.

Riley, R.T.; Norred, W.P.; Bacon, C.W. Fungal toxins in foods: Recent concerns. Annu. Rev. Nutr. 1993, 13, 167-189.

Ross, P.F.; Rice, L.G.; Osweiler, G.D.; Nelson, P.E.; Richard, J.L.; Wilson, T.M. A review and update of animal toxicoses associated with fumonisin-contaminated feeds and production of fumonisins by Fusarium isolates. Mycopathologia 1992, 117, 109-114.

Scott, P.M ; Lawrence, G.A. Stability and problems in recovery of fumonisins added to corn-based foods. J. AOAC Int. 1994, 77, 541-545.

Shephard, G.S.; Sydenham, E.W.; Thiel, P.G.; Gelderblom, W.C.A. Quantitative determination of fumonisin B_1 and B_2 by high performance liquid chromatography with fluorescence detection. J. Liq. Chromatogr. 1990, 13, 2077-2087.

Stack, M.E.; Eppley, R.M. Liquid chromatographic determination of fumonisins B_1 and B_2 in corn and corn products. J. AOAC Int. 1992, 75, 834-837.

Sydenham, E.W.; Shephard, G.S.; Thiel, P.G.; Marasas, W.F.O.; Stockenstrom, S. Fumonisin contamination of commercial corn-based human foodstuffs. J. Agric. Food Chem. 1991, 25, 767-771.

Sydenham, E.W.; Marasas, W.F.O.; Shephard, G.S.; Thiel, P.G.; Hirooka, E.Y. Fumonisin concentrations in Brazilian feeds associated with field outbreaks of confirmed and suspected animal mycotoxicoses. J. Agric. Food Chem. 1992, 40, 994-997.

Sydenham, E.W.; Van der Westhuizen, L.; Stockenstrom, S.; Shephard, G.S.; Thiel, P.G. Fumonisin-contaminated maise: Physical treatment for the partial decontamination of bulk shipments. Food Addit. Contam. 1994, 11, 25-32.

Wang, E.; Norred, W.P.; Bacon, C.W.; Riley, R.T.; Merrill, A.H., Jr. Inhibition of sphingolipid biosynthesis by fumonisins: Implications for diseases associated with *Fusarium moniliforme*. J. Biol. Chem. 1991, 266, 1706-1716.

Wilson, T.M.; Ross, P.F.; Owens, D.L.; Rice, L.G.; Green, S.A.; Jenkins, S.J.; Nelson, H.A. Experimental reproduction of ELEM- a study to determine the minimum toxic dose in ponies. Mycopathologia 1992, 117, 115-120.

31

REGULATORY ASPECTS OF FUMONISINS IN THE UNITED STATES

Terry C. Troxell

Division of Programs and Enforcement Policy
Office of Plant and Dairy Foods and Beverages
Center for Food Safety and Applied Nutrition
Food and Drug Administration, Washington, DC 20204

ABSTRACT

The hazards and risks from fumonisins, a relatively recently discovered class of mycotoxins, are in the process of being characterized. Any risk management approach must consider the uncertainties in the risk characterization and practicalities of control options. This paper addresses risk management alternatives, especially in the context of the Food and Drug Administration's legal authorities, and the potential impacts of the alternatives.

DISCUSSION

During this symposium, the results of a great deal of research on the toxicity and occurrence of fumonisins, methods for their determination, and the potential for reducing fumonisin levels during processing have been presented. Because all of this information and the substantial work in progress will be needed to effectively manage the potential risks to consumers, it forms an essential element in the risk assessment/risk management process.

The issue before us is more than the regulatory aspects of fumonisins. It is the appropriate use of risk assessment and risk management approaches, in the context of the societal level of protection deemed appropriate in the U.S., to arrive at an appropriate response to this potential human health threat. The relevant societal level of protection is, to a large extent, determined by Congress and, in this case, is set forth in the provisions of the Federal Food, Drug, and Cosmetic Act (FDC Act). However, the societal level of protection is determined by other factors, including consumer perception of risk, which is influenced by statements from consumer organizations, scientific organizations, trade associations, industry, and government bodies as conveyed by the media.

There are no simple answers to questions relating to these matters, but I would like to consider the issues dealing with contaminants in food from a broader perspective than regulatory aspects alone.

Fumonisins in Food, Edited by L. Jackson *et al.*
Plenum Press, New York, 1996

In a simple world, the risk manager's goal would be to determine a specific level of intake of fumonisin that is safe, i.e., a level that would result in negligible risk or a reasonable certainty of no harm. The latter is the meaning of safe in the food additives legislation. For food additives which are intentionally added, that goal can be achieved. Food additives must be shown to be safe for their intended use, and a regulation must be issued to permit that use before they may be legally added to foods. Thus, the intake can be limited as much as necessary until there is high confidence that the risk is negligible. To achieve this high confidence, the sponsor of the additive must conduct a set of toxicology studies. If the initial set of studies raises concerns, additional work may be needed to resolve the outstanding issues. Agency scientists then use a time-honored safety assessment approach to determine a safe level, the Acceptable Daily Intake (ADI). This approach generally involves applying a series of tenfold safety factors to the lowest credible no-observed-adverse-effect level to account for animal-to-man extrapolation, heterogeneity of the human response, and conversion from short-term study results to long-term potential exposures of humans. The result of the process is a well-controlled situation with high confidence that no harm will result, which is as it should be for a purposely added ingredient.

The safety assessment approach is a good screening method for food additive clearance and some contaminant situations. On the other hand, while the safety factors have survived the test of time, they also embody some science policy decisions that constitute risk management. They create "bright lines" that demarcate "safe" from "not safe" or "not shown to be safe," thus hiding the implicit uncertainty of the results. The single number ADI or its contaminant relative, the Tolerable Daily Intake (TDI), cannot possibly describe the risk to a diverse population with its varying sensitivities. Worse yet, how can a single number allow for an adequate assessment of the gains that may be expected from the use of alternative risk management options?

Although food additive decisions are by no means simple, the complex problems presented by contaminants such as fumonisins, aflatoxins, lead, and methyl mercury often require the use of risk assessment, which provides a full description of the probability of harm for the key adverse effects as well as an evaluation of the uncertainties. Unlike food additives, contaminants do not wait for permission to be present in food; therefore, to protect health they must be dealt with on the basis of the available data. Rarely is a "complete" set of data available, so for many contaminants the degree of uncertainty is larger than for food additives. Generally, contaminant issues involve a significant degree of "decision making in uncertainty." The research discussed at this symposium and under way will help to reduce the uncertainty in the risk assessment of fumonisins. Having the best risk assessment possible is crucial for addressing major contaminant problems because control options invariably involve nontrivial costs.

Risk assessment and cost-benefit analysis are now important issues in Congress. For example, H.R. 1022, passed by the House of Representatives, would have agencies carry out scientifically objective, unbiased best estimates of risk. It would require an assessment of substitution risks and of incremental costs and benefits of alternative risk management options. This legislation would require that agencies certify "that the incremental risk reduction or other benefits of any strategy chosen will be likely to justify, and be reasonably related to, the incremental costs incurred..." The Food and Drug Administration (FDA) supports the appropriate use of risk assessment and believes that it exercises its regulatory discretion to minimize the impact on affected industries while acting to protect the public health. Also, President Clinton's Executive Order 12866 (Fed. Regist. 1993) requires that agencies assess the costs and benefits of available regulatory alternatives and choose the one that maximizes net benefits unless a statute requires another regulatory approach. The risk assessment legislation is mentioned for two reasons. First, there is a growing concern that regulation be cost-effective, and FDA supports that concept for environmental contaminants found in foods. Second, implicit in this legislation and arguments put forth by some is the assumption that the risk and the costs and benefits of the risk management options can be quantitated with minimal uncertainty in the values derived. On the contrary, for many contaminants the uncertainties can be large and the resulting estimates, "soft". Because of the uncertainties in both the risk and the cost-benefit analyses for the risk management options, it is likely that such calculations will often yield a wide range of benefit-to-cost ratios. The "softness" of these cost-benefit calculations will also spawn some intense arguments and court cases about specific rulemaking. Nevertheless, calculations can and should be done. FDA should be making public policy decisions with the best available description of the risk and control costs, including a description of the uncertainties.

Irrespective of the outcome of the risk assessment debate in Congress, contaminants such as the fumonisins present some of the most difficult and complex problems for risk managers. From a risk manager's perspective, the goal with respect to contaminants should be to protect the public health while minimizing the impact on the affected industry. An alternative way of saying this is to reduce exposure to the contaminant to the extent feasible or practicable and consistent with adequate public health protection. Simply described, it is the "risk versus avoidability" concept, that is, the effort to remove a contaminant should be commensurate with the risk from the exposure or, better yet, the risk reduction achieved. This approach is crippled by "bright lines" like TDIs. TDIs suggest that unlimited effort is required to bring the intake beneath this level of exposure. Implicit in the concept of risk versus avoidability is the recognition that risk is not a step function of intake; it is a continuum. We do not magically have zero risk below a TDI (or ADI). So the issue becomes how much risk should be tolerated, and that depends on the avoidability, i.e., the feasibility to eliminate or reduce it. Thus, several risk levels can be consistent with protecting the public health, with various associated "margins of safety."

No matter how rationally, deliberately, and carefully such risk management decisions are reached, they may be of little utility if the public perceives itself at significant risk from exposure to the contaminant. It can be argued that in the debate over risk assessment and cost-benefit analysis we must never lose sight of the importance of consumer confidence in the safety of the food supply. We all have a responsibility to communicate risk to consumers in an understandable way and especially to characterize situations accurately and without exaggeration. To do otherwise may irresponsibly undermine consumer confidence.

Whose responsibility is it to protect consumers from fumonisins, contaminants, or other health risks from foods? There is no doubt that the primary food mission of FDA is to protect the public health. However, it is clearly not FDA's responsibility alone. The FDC Act states that distributing adulterated food is a prohibited act. In fact, the Act recognizes that it is the responsibility of the producer to ensure that food products are safe, wholesome, and truthfully labeled. FDA's responsibilities under the FDC Act are to monitor foods and take enforcement actions such as seizures of adulterated foods in domestic status and detentions of imports in accordance with the provisions of the FDC Act. Others with major responsibilities are researchers, both academic and government, who are providing the data and means for reducing the risk, such as development of corn cultivars that resist *Fusarium moniliforme* and agricultural practices that reduce contamination by the fumonisins. Of course, consumers also have a major responsibility, especially in minimizing risk from microbiological sources by properly handling and preparing vulnerable foods. Because FDA has limited resources to test lots of food and inspect plants, and because end-testing by industry is not the most effective way to maximize food safety, FDA is considering the use of Hazard Analysis of Critical Control Points (HACCP) as a problem prevention tool to establish critical control points for likely hazards. Under HACCP, dry millers might establish a critical control point for incoming corn and require it to meet an appropriate fumonisin B_1 level prior to processing.

FDA carries out its food mission by various means, including enforcement actions, providing guidance to industry, encouraging industry action to address problems, working with the states and foreign countries, and public education. However, FDA's actual regulatory powers flow from and are limited by the FDC Act and several other related statutes. The following discussion outlines the pertinent provisions of the FDC Act related to contaminants. These provisions do not entirely mesh with the simple risk management principle of risk versus avoidability.

The main provisions that apply to contaminants in foods are Section 402(a)(1) and Section 406. Section 402(a)(1) provides that "A food shall be deemed to be adulterated if it bears or contains any poisonous or deleterious substance which **may** render it injurious to health; but in case the substance is not an added substance such food shall not be considered adulterated under this clause if the quantity of such substance in such food does not **ordinarily** render it injurious to health" (emphasis added). Clearly, fumonisin B_1 is a poisonous substance, so it passes the first criterion for consideration under Section 402(a)(1). But then it gets interesting. What is the difference between "may render" and "ordinarily render" injurious? In the 1938 Act, which replaced the original 1906 Food and Drugs Act, Congress included the "ordinarily render" terminology in recognition that foods contain natural toxicants that may be hazardous at high enough intakes. Thus, Congress inserted the "ordinarily render" language to require the FDA to prove that these naturally occurring toxicants were harmful at ordinary intake levels. Congress apparently wanted to capture poisonous mushrooms and mussels in this net, but not coffee or tea (Merrill & Hutt, 1980).

In 1977, FDA finalized regulations to further interpret the meaning of "added". These regulations are in Part 109 of Title 21 of the Code of Federal Regulations (1994). In these regulations, FDA defined a "naturally occurring poisonous or deleterious substance" as "a poisonous or deleterious substance that is an inherent natural constituent of a food and is not the result of environmental, agricultural, industrial, or other contamination". This definition appears consistent with the legislative history of the "ordinarily render" provision. Furthermore, the definition for "added" in the regulations views a naturally occurring substance as added to the extent that it is increased to abnormal levels through mishandling or other intervening acts.

However, FDA's interpretation was not followed in a Fifth Circuit decision in the Anderson Seafoods case in 1980 (U.S. v. Anderson Seafoods, Inc., U.S. Court of Appeals, Fifth Circuit, 1980, 622 F.2d 157). This case resulted from seizure actions due to excessive methyl mercury levels in swordfish. Relevant to our discussion here is the belief of the Court that FDA's interpretation, i.e., nonadded substances were only those that were inherent natural constituents of the food, was too restrictive. The judge argued that FDA's definition was inconsistent with the legislative history and a Supreme Court opinion. The judge stated, "The term 'added' ... means artificially introduced, or attributable to the acts or intervention of man" (U.S. v. Anderson Seafoods, Inc., 622 F.2d 157, 1980). Thus, this decision draws the distinction between acts of man and acts of nature. On the other hand, the court said FDA needed only to prove that some portion of the mercury in swordfish could be traced to man's activities in order to use the added substance "may render" standard. The judge also stated that "may" connotes "a reasonable possibility". FDA has not acquiesced in this decision and still believes that the better view of "added" is embodied in its definition.

Clearly, fumonisin B_1 is not an inherent component of corn, and thus, is "added" under FDA's definition. If FDA could not sustain the position that fumonisin B_1 is added, the alternative is an "ordinarily injurious" charge. Considering the potential seriousness of the adverse effects, we would not want to wait until the data were strong enough to prove that fumonisin B_1 ordinarily rendered the food injurious to health. This would probably require a high certainty of some injury to otherwise healthy humans from nonextreme intakes. As a matter of public health, we believe action needs to be taken to control fumonisin

B_1 well below intakes that border on "ordinarily injurious". Not doing so would likely also undermine public confidence in the affected products and the federal food safety system. Another strategy, one that has been used with aflatoxin, might be to demonstrate that fumonisin B_1 increased to some extent during storage. That would provide the hand-of-man linkage described by the judge in the Anderson Seafood case. However, in the case of fumonisins, that does not normally appear to be the case.

It is important also to emphasize that in all of these descriptions FDA bears the burden of proof in enforcement actions. Thus, for emerging contaminant problems, it is not enough for FDA to have general safety concerns; those concerns must also be supported by adequate, substantial science to sustain a charge if legal action is needed.

Section 402(a)(1) is also difficult in another way. The provision refers to "a food" being deemed adulterated. FDA seizes foods that meet the above criteria. In court, the issue will be whether the seized food is hazardous to health. That issue is simple for acute hazards such as pathogens which cause immediate illness from the food in question. However, this provision does not clearly delineate how to view hazards from contaminants that have multiple sources, whether in the context of several specific foods (such as aflatoxin in corn and peanuts), specific foods and specific environmental sources (such as lead in paint), or specific foods relative to other background exposures. Even given all these factors, it would be wrong to think that Section 402(a)(1) necessarily sets such a high standard that the government cannot readily address public health problems. The standard it sets is that the government must show a reasonable possibility of injury from the food for a vulnerable population.

As part of the 1938 Act, Congress also enacted another standard in Section 406 that addresses added poisonous or deleterious substances that are either required in the production of food or are unavoidable under good manufacturing practices. This section is the basis for tolerances. Congress recognized that sometimes it is necessary to make a trade-off between protection and avoidability. In Section 406, Congress provided for regulations that limit the substance enough to protect the public health, but also account for the unavoidability of the substance and "other ways in which the consumer may be affected by the same or other ...substance." This standard would permit the consideration of all exposures to a contaminant, whether from food or not, in setting tolerances. However, because this provision applies only to added substances, it also is affected by the Anderson case.

Finally, tolerances established under Section 406 are promulgated through formal rulemaking, which requires an administrative hearing if there are objections to the final rule. The administrative hearing process can lead to protracted rulemaking, which is resource-intensive for the government. Because of objections, when FDA issued tolerances for polychlorinated biphenyls (PCBs) in various foods, it took 11 and 7 years to establish the current tolerances for PCBs in paper food packaging and fish, respectively.

The focus thus far has been on the statutory provisions that address contaminants. As indicated, they can be applied directly in enforcement actions, such as seizures, injunctions, and import detentions to control contamination problems. The establishment of a tolerance by formal rulemaking to control levels of added contaminants has also been mentioned. To take an enforcement action once a tolerance is established, FDA would need only to prove that the prescribed limit was exceeded, and would not need to justify the hazard each time. A tolerance would not be established if the matter was in a state of flux (21 CFR 109.4 and 109.6). A tolerance also might not be established because formal rulemaking is resource-intensive. A judgment would need to be made on the basis of available government resources whether to control a problem by case-by-case enforcement or establishment of a tolerance.

For years FDA has used action levels in lieu of issuing tolerances, especially in those situations where information on control and toxicology was still evolving (21 CFR 109.4

and 109.6). For example, FDA has an action level of 20 ppb for aflatoxins in most human foods. Action levels are not issued by rulemaking; therefore, they do not have the force and effect of law. They are not binding. They are used as informal guidelines by FDA's field staff on when to consider enforcement action. They can be issued relatively quickly. However, unlike tolerances, FDA must prove all aspects of the statutory violation in each enforcement action.

Finally, FDA more recently has provided for the establishment of regulatory limits based on the Section 402(a)(1) provision (Fed. Regist. 1990; 21 CFR 109.4 and 109.6). These limits would be established by notice-comment rulemaking and would not be delayed by administrative hearings. Regulatory limits would have the force and effect of law, and FDA would need to prove only that the limit was exceeded to take an enforcement action.

FDA also occasionally uses advisories to provide guidance to industry on contaminant levels. For example, after the extremely wet conditions in the spring of 1993, the domestic wheat crop was substantially contaminated with deoxynivalenol (vomitoxin, DON). In response to inquiries from industry about the applicability of an advisory issued by FDA in 1982, FDA rapidly updated its scientific review of this mycotoxin. The driving adverse effect was acute gastrointestinal illness. Within roughly one month FDA issued a revised advisory. For human use the advisory retained the 1 ppm DON level for milled products, such as flour and bran, but removed the 2 ppm level for raw wheat. FDA stated that a DON level of 1 ppm or lower would not appear to present a public health hazard. In considering this problem, FDA learned that milling methods varied and could result in a roughly two- to eightfold reduction of DON in the milled product. Since public health goals could be satisfied with the 1 ppm level in the milled product, FDA believed it was unnecessary to retain the level for raw wheat at 2 ppm when methods were available to the milling industry that could satisfactorily reduce levels of 4, 6, and even 8 ppm DON in the wheat.

As discussed earlier, industry and the regulators collectively have a range of risk management options to control intake of fumonisins. Considering the serious adverse effects associated with fumonisins, it would be prudent to minimize exposure to them. However, we need a better understanding of the science, including year-to-year variability in crops, before any long-term risk management decisions are made.

FDA has used and will continue to use establishment of regulatory standards as a compliance option, and it may soon be appropriate to establish an interim level. Limits are in some ways a crude risk management tool. One can vary the products (raw, intermediate, or final) to which they are applied and the numeric values of the limits. Usually, there are a few major sources of contamination. Certainly, judiciously applied limits can minimize contaminant exposures. A limit applied to the corn coming into a dry milling operation would divert unacceptable product to wet milling production, animal feed, or ethanol production. For efficiency, a rapid screening analytical method would be needed. This approach seems practical because only about 2-3% of the corn crop is used in dry milled products. However, it may be deceptive because of the transportation expenses involved in finding and hauling acceptable product, especially in bad years. These expenses have not been characterized yet.

Also, setting a limit at the high end of the contamination distribution curve really does not result in elimination of a substantial percentage of the fumonisins from the food supply, which might be desirable if we ultimately conclude that they are a significant human health risk that needs to be addressed. If the adverse effects indicate that a limit needs to be set at a lower level on the contamination distribution curve, the limit will clearly prevent a large amount of the contaminant from reaching the human food supply and may force industry to develop control procedures. However, such a limit may have a significant adverse impact on the food supply and/or cause significant disruption in the marketplace. These effects would be exacerbated if such limits were imposed suddenly.

A preferred risk management approach is to develop control methods, which would depend on the efforts of researchers and industry. Developing cultivars that are resistant to *Fusarium moniliforme*, safe treatments that inactivate the fumonisins, and related technologies could reduce the levels of fumonisins in the food supply while minimally affecting the marketplace. Consider, for example, the major gains in reduction of lead intakes from foods by the elimination of most leaded gasoline and lead-soldered cans. Dietary intake of lead as measured through FDA's Total Diet Study dropped by a factor of 10 from 1978 to 1990. Source control, or other means of prevention, is often more effective than enforcement limits. Therefore, we strongly encourage efforts to develop control methodologies and recommend that these efforts be pursued vigorously.

In summary, risk assessment which describes the probability of harm to the vulnerable populations and the uncertainties in the estimates should be used for substantial contaminant problems. Risk managers should consider the risk versus avoidability, and consequently the costs, in targeting controls. The best risk management approaches are often source controls and processing controls rather than regulatory limits, which may be more costly because of rejection of product. It is important to vigorously research control options for fumonisins now in order to have practical means available, to the extent needed, when more is known about potential human effects. However, even the present knowledge about the hazards of fumonisins indicates that it would be prudent to minimize exposure immediately.

ACKNOWLEDGMENT

The author acknowledges the assistance of Norma J. Yess in the preparation of this manuscript.

REFERENCES

Code of Federal Regulations, Title 21; U.S. Government Printing Office: Washington, DC, 1994.
Fed. Regist. May 21, **1990**, 55, 20785.
Fed. Regist. October 4, **1993**, 58, 51735.
Merrill, R. A., and Hutt, P. B. *Food and Drug Law: Cases and Materials*; The Foundation Press: Mineola, New York, **1980**, pp. 54-62.

32

REGULATORY ASPECTS OF FUMONISINS WITH RESPECT TO ANIMAL FEED

Animal Derived Residues in Foods

Margaret A. Miller, John P. Honstead, and Randall A. Lovell

Center for Veterinary Medicine, USFDA
7500 Standish Place
Rockville, MD 20855

ABSTRACT

The fumonisins are a recently discovered class of mycotoxins produced primarily by *Fusarium (F.) moniliforme* and *F. proliferatum*. Fumonisins present in mycotoxin-contaminated feed have been identified as the causative agent of equine leukoencephalomalacia and porcine pulmonary edema. To prevent these diseases, FDA has utilized informal guidance levels for fumonisins in feed and initiated a surveillance program for fumonisins in feed corn and corn by-products during FY 93 and 94. Natural contaminants present in animal feed can enter the human food supply as residues present in animal tissues and other animal derived products. Although fumonisin guidance levels were originally set based on animal safety, FDA also ensures the human food safety of animal products from animals fed mycotoxin-contaminated feed. Recent pharmacokinetic studies in food-producing animals as well as statutory requirements for regulating natural toxins will be discussed in light of FDA's human food safety mandate.

INTRODUCTION

The United States Food and Drug Administration (FDA) is the primary federal agency responsible for the regulation of food products intended for human and animal consumption. FDA's authority, as set out in the Federal Food, Drug, and Cosmetic (FD&C) Act, and related aspects of the Public Health Service Act, includes the safety of direct food additives and of indirect food additives and contaminants. While the Center for Food Safety and Applied Nutrition regulates the vast majority of human food, the Center for Veterinary Medicine (CVM) ensures the human food safety of animal drugs, devices, feed additives and feed contaminants which enter the human food supply in milk, meat and eggs. In carrying out its

Fumonisins in Food, Edited by L. Jackson *et al.*
Plenum Press, New York, 1996

public health mission, CVM is concerned about animal safety, animal productivity, and the safety of animal-derived products consumed by humans.

Mycotoxins are generally considered unavoidable contaminants in food and feed under the FD&C Act, and the regulation of mycotoxins differs markedly from the regulation of feed additives and animal drugs. For animal drugs and feed additives, the Agency has premarket authority. The sponsor must demonstrate that the product is safe, efficacious and can be manufactured to ensure product purity, strength and identity, before the product enters the marketplace. In establishing the safety of animal drugs and feed additives, CVM requires the product's sponsor to conduct extensive target animal, human food and environmental safety studies. For example, the amount of drug residues which can be safely consumed in edible products from treated animals is established based on a battery of toxicology and metabolism studies performed with the purified product.

The human food safety concerns for drug residues in food focus on the toxicological potential of chronic low level exposure. Generally, the concentration of drug residues in animal tissues is not high enough to cause acute toxicity in humans. The results of the toxicology studies are used as a basis for establishing an acceptable daily intake (ADI) for residues. The ADI is the level of residue which can be safely consumed daily for a lifetime. The advantage of this approach is that the exposure limit is established based on safety studies and not on the sensitivity of the analytical methods used for monitoring.

SAFETY ASSESSMENT OF FUMONISINS IN ANIMAL FEED

Although the safety concerns for feed contaminants, including mycotoxins, are similar to those for drugs, the regulatory authority, data quantity and quality, product quantity and purity, and economic considerations are vastly different for feed contaminants. The Agency has no premarket authority over natural occurring contaminants such as mycotoxins. To assess the safety of natural contaminants, the Agency relies on literature studies, epidemiological data, and anecdotal information. In most cases, these data are incomplete. Often the purity, strength and identity of the contaminants are not well characterized. In examining the regulatory aspects of fumonisins in animal feed, CVM is concerned about animal safety, animal productivity and human food safety.

Equine leukoencephalomalacia (ELEM) or "moldy corn poisoning" is a syndrome of acute illness and death in horses that is associated with feeding horses moldy corn. Necropsy of these horses often reveals a softening or liquefaction of the white matter in the brain known as leukoencephalomalacia. Studies demonstrate that fumonisin B_1, a mycotoxin produced principally by the fungi *Fusarium moniliforme* and *Fusarium proliferatum*, is the causative agent in ELEM. Fumonisin B_1 is also responsible for natural outbreaks of porcine pulmonary edema and for toxicoses in cattle, sheep, chickens, turkeys and ducks in controlled studies. While the acute toxicity of fumonisin has been described in several species, the chronic effects of lower doses on animal health and loss of productivity are less well defined.

Fusarium moniliforme and *Fusarium proliferatum* contamination occurs primarily on corn-based products. Corn is an important agricultural commodity in the United States because of its extensive use in animal feed. Elimination of corn from the diet of animals is not practical because corn represents approximately 60 percent of the grain fed to animals. Feed corn is used extensively for dairy cattle, feedlot cattle, pigs, poultry and other livestock. For some feedlot cattle, feed corn can represent up to 90% of the daily ration.

Corn can be added to animal feed in various forms. Corn by-products are also often integrated into the animal feeding program. The composition of feed varies widely because of differences in climate, soil conditions, maturity, variety, management and processing. Similarly, mycotoxin contamination varies widely with these conditions. In some cases, several mycotoxins and their metabolites may exist together in feed which may alter their toxicity.

Unlike humans who consume a varied diet, high corn rations can be fed to animals daily throughout their lifetime. This type of continuous exposure increases the potential for contaminants present in feed entering the human food supply as residues in meat, milk or eggs. The human food safety concern for fumonisin residues in meat, milk and eggs is not from an acute toxic effect. The concern for fumonisin residues in food-producing animals focuses on the potential toxic effect of repeated exposure to low levels. Since fumonisins have been associated with the induction of tumors in laboratory animals, cancer is the main public health concern for chronic exposure to fumonisin residues in animal tissues.

INFORMAL GUIDANCE FOR FUMONISIN CONTENT IN ANIMAL FEED

Because most mycotoxins are unavoidable, naturally occurring compounds, limiting exposure provides an important tool in controlling the quality of the animal feed. Theoretically, all the safety concerns identified for fumonisin could be eliminated by controlling exposure to fumonisin in animal feed. FDA has utilized informal guidance levels for fumonisins in animal feed as a mechanism to help prevent the animal diseases caused by fumonisin and to help control fumonisin residues in food derived from animals.

In 1993, the Mycotoxin Committee of the American Association of Veterinary Laboratory Diagnosticians (AAVLD), Inc. recommended the following guidance to assist the animal industry in responding to the presence of fumonisin in animal feeds. The committee recommended that animal feed contain fumonisin B_1 below the following:

- 1) For equine species, it is recommended that the non-roughage portion of the feed be less than 5 ppm fumonisin B_1.
- 2) For porcine species, it is recommended that the total diet contain less than 10 ppm fumonisin B_1.
- 3) For beef cattle, the non-roughage portion of the feed is recommended to be less than 50 ppm fumonisin B_1.
- 4) For dairy cattle, no recommendation was agreed upon (or simply restated, it was recommended that no level be recommended).
- 5) For poultry, it is recommended that the complete feed be less than 50 ppm (fumonisin B_1). Turkeys are of more concern than chickens and there is concern about transfer into eggs.

FDA's DOMESTIC SURVEILLANCE FOR FUMONISIN

The regulation of animal feed is a joint state-federal endeavor. FDA's jurisdiction over feed is established if the feed or any of its ingredients has entered interstate commerce. FDA's regulatory authority includes establishing the kinds and amounts of drugs or feed additives that may be used in the manufacture of medicated feed and regulating potential or actual contamination, particularly with respect to pesticides, industrial chemicals, mycotoxins and similar adulterants. To gain more information on these contaminants, the FDA collects and analyzes many samples each year as part of its Feed Contaminants Program.

The FDA collects four types of feed samples: domestic surveillance, domestic compliance, import surveillance and import compliance. Surveillance samples are collected on a routine basis where there is no inspectional or other evidence of a problem with the product. Compliance samples are collected on a selective basis as the result of an inspection, complaint or other evidence that there may be a problem with the product. Domestic samples are samples of products of domestic origin (including all 50 States, the District of Columbia, and all U.S. territories), or of foreign origin if the product is in the domestic channels of trade. Import samples are samples of products from foreign countries at ports of entry.

Fumonisins were added to the Feed Contamination Program in fiscal year 1993 (FY 93). Table 1 summarizes the data from FDA's domestic surveillance samples collected and analyzed in FY 93 and FY 94. For the most part, fumonisins did not appear to be a major problem in U.S. corn and corn by-products during FY 93 and 94.

RECENT EQUINE LEUKOENCEPHALOMALACIA OUTBREAKS

There have been two separate outbreaks of ELEM related to fumonisin contaminated corn in FY 95. In January 1995, a veterinarian at Murray State University informed CVM of horse deaths in western Kentucky following consumption of white corn screenings from

Table 1. FDA's domestic surveillance for fumonisins in FY 93 and FY 94

Commodity and Year	Total No. Samples	Fumonisin B$_1$ + Fumonisin B$_2$ Levels*				
		None Detected No. (%)	Trace No. (%)	1.0 - 4.9 ppm No. (%)	5.0 - 9.9 ppm No. (%)	≥ 10.0 ppm No. (%)
Corn—FY 93	69	63 (91.3)	3 (4.3)	3 (4.3)	0 (0.0)	0 (0.0)
Corn—FY 94	33	19 (57.6)	3 (9.1)	9 (27.3)	1 (3.0)	1° (3.0)
Corn By-products•—FY 93	11	9 (81.8)	0 (0.0)	2 (18.2)	0 (0.0)	0 (0.0)
Corn By-products••—FY 94	5	5 (100.0)	0 (0.0)	0 (0.0)	0 (0.0)	0 (0.0)
Mixed Feed Rations—FY 93	9	8 (88.9)	1 (11.1)	0 (0.0)	0 (0.0)	0 (0.0)
Mixed Feed Rations—FY 94	4	2 (50.0)	1 (25.0)	0 (0.0)	0 (0.0)	1°° (25.0)
Totals	131**	106 (80.9)	8 (6.1)	14 (10.7)	1 (0.8)	2 (1.5)

*all samples were analyzed for fumonisins by extraction using methanol and subsequent liquid-liquid partitioning into acidified chloroform, derivatization with naphthalene 2,3 dicarboxaldehyde reagent in the presence of cyanide ion at pH 9.5, and determination by HPLC with fluorescence detection (Laboratory Information Bulletin, December 1991, Vol. 7, No. 12, 3621).

°34.1 ppm in shelled corn from SC

• n=4 for corn meal; n=3 for corn screenings; n=2 for corn on-the-cob; n=1 for corn gluten meal; n=1 for hominy.

•• n=2 for corn meal; n=2 for corn screenings; n=1 for hominy.

°°15.3 ppm in a cattle ration from MS

**FY 93 distribution = 16 from WI, 11 from LA, 10 from ND, 8 from TX, 6 from GA, 5 from IA, 4 each from VA, IL, MN, and SD, 3 each from MS, FL and CA, 2 each from NC and MO, and 1 each from SC, VT, ME and AL. FY 94 distribution = 17 from SC, 6 from CA, 5 from IL, 3 each from GA and FL, 2 each from NC and LA, and 1 each from VT, ND, MS and NJ.

one feed manufacturer. The horses had classic signs of ELEM, and samples of the white corn screenings showed the presence of fumonisin at 2 to 40 ppm.

The FDA's Cincinnati District Office investigated the incident and found that the corn in question was white corn harvested in 1994 from farms in Illinois, Kentucky and Indiana. The district collected four samples of corn and corn flour intended for human consumption and found levels of fumonisin ranging between 26 and185 ppb.

The Cincinnati District collected one sample of corn screenings intended for animal feed and it contained 14 ppm of fumonisin B_1 and 5.2 ppm of fumonisin B_2. They reported that 19 horses have died on 3 farms with at least 200 other horses at risk from exposure to the contaminated corn.

Murray State University released a statement to veterinarians cautioning against feeding corn screenings to horses. The feed manufacturer agreed not to sell corn screenings to anyone feeding horses and will attempt to get signed statements from the purchaser that they will not feed corn screenings to horses.

The second incident occurred in Virginia. In early February 1995, a diagnostician from Virginia informed CVM of several horse deaths from ELEM in southeastern Virginia. The regional diagnostic laboratories in Virginia had 18 suspect ELEM cases between October 1994 and February 1995. These cases came from 7 owners and the majority of the cases were reported in January. Seven of the 18 cases have been confirmed by pathology findings and the other 11 are considered likely based on clinical signs, gross necropsy finding by the attending veterinarian, and/or fumonisin analysis of feed. The corn in question appears to represent corn grown in 1994 from the tidewater area of Virginia and North Carolina. Three of the 7 horse owners likely purchased the suspect corn from one feed manufacturer in North Carolina. The FDA collected 5 surveillance samples of feed corn from elevators in the tidewater area in Virginia and sent them to the New Orleans Laboratory for fumonisin analysis. The levels of fumonisin B_1 and B_2 in these 5 samples were 2.1 ppm, 5.5 ppm, 14.8 ppm, 15.9 ppm and 21.2 ppm. The analysis of the 6 samples collected in the tidewater area of North Carolina has not been completed.

The Virginia Department of Agriculture and Consumer Services issued a press release on February 6, 1995 advising horse owners and swine producers to have their corn tested for fumonisin B_1 prior to feeding.

HUMAN FOOD SAFETY CONCERNS FOR FUMONISIN RESIDUES

Ideally, the guidance levels in animal feed should be protective of both animal health and human consumers of animal products. Unfortunately, as indicated in the AAVLD recommendations, the data currently available do not permit a definitive statement that the levels of fumonisin suggested for animal feed are sufficient to protect consumers from fumonisin residues. While a definitive safe level for fumonisin residues cannot be made at this time, the toxicological and biochemical properties of fumonisin and the metabolism studies conducted to date suggest that limiting exposure to fumonisin in animal feed will protect consumers from unsafe residues.

Fumonisins are water soluble. Generally, water soluble compounds do not bioaccumulate in animal tissues and are less likely to leave residues than lipid-soluble compounds. Fumonisin B_1 is structurally related to the endogenous compounds, sphinganine and sphingosine. Compounds which are structurally related to endogenous substrates are generally metabolized by the animal to less toxic metabolites. Finally, residues are usually present only in organ tissues, kidney and liver. The human consumption of these tissues is much less than the consumption of muscle.

The metabolism studies reported in the literature indicate that fumonisin residues in animal tissues are low. For example, when an oral dose of 5 mg of FB_1/kg b.w. was administered to four dairy cows, no detectable FB_1 residues were present in milk (Scott et al., 1994). The analytical detection limits were approximately 3-7 ng/ml for FB_1 and 20-25 ng/ml for AP_1. Maragos and Richard (1994) monitored 155 samples of unpasteurized, unhomogenized milk throughout Wisconsin because of a severe mold problem reported in the 1992 Wisconsin corn crop. Only one sample contained detectable fumonisin and that was only 1.29 ppb of FB_1. In the same study, ten samples of whole homogenized milk contained less that 5 ppb of FB_1. These studies suggest that exposure of humans to fumonisin in milk is very low.

The studies by Vudathala et al. (1994) on the metabolism of ^{14}C-fumonisin B_1 in laying hens showed that a single dose of 2 mg/kg b.w. produced no detectable residues in blood, esophagus, proventriculus, gizzard, lung, spleen, heart, brain, abdominal fat, muscle, ovary, oviduct and largest ova. Only trace residues were detectable in the crop, liver, small intestine, cecum and kidney 24 hr post-dosing. No residues were found in eggs laid during the 24 hr post-dosing time period. In a similar study, Prelusky et al. (1994) dosed 5 barrows with 0.5 mg ^{14}C-fumonisin B_1/kg b.w. Only a very small fraction of the dose remained in tissue at 72 hours. The highest levels were in the liver (90 ppb), large intestine (39 ppb) and kidney (31 ppb).

Based on these studies and our knowledge of the biological properties of fumonisin, we believe that fumonisin residues in edible products are low and do not represent a public health concern. In other words, we do not consider food products derived from animals exposed to fumonisins at or below the guidance limits set by AAVLD as adulterated under the Act because, at this time, these products do not appear to bear or contain any poisonous or deleterious substance which may render the food injurious to health.

The data collected to date do not provide enough information for us to determine a "safe level" for fumonisin residues in milk, meat and eggs. The "safe level" cannot be determined based solely on monitoring exposure. It must be based on risk assessment. Assuming cancer is the endpoint of toxicological concern for humans exposed to low doses of fumonisin in the diet, an estimation of the "safe level" should be possible after the National Center for Toxicological Research completes the fumonisin bioassay and mechanism studies. Based on the results of these studies, CVM will be able to perform quantitative risk assessment that predicts the risk to humans and the level at which the risk is likely to occur.

REFERENCES

Maragos, C.M.; Richard, J.L. Quantitation and Stability of Fumonisins B_1 and B_2 in Milk. *J. AOAC Int.*, **1994**, 77, 1162-1167.

Prelusky, D.B.; Trenholm, H.L.; Savard, M.E. Pharmacokinetic Fate of ^{14}C-Labelled Fumonisin B_1 in Swine. *Natural Toxins*, **1994**, 2, 73-80.

Scott, P.M.; Delgado, T.; Prelusky, D.B.; Trenholm, H.L.; Miller, J.D. Determination of Fumonisin in Milk. *J. Environ. Sci. Health*, **1994**, B29, 989-998.

Vudathala, D.K.; Prelusky, D.B.; Ayroud, M.; Trenholm, H.L.; Miller, J.D. Pharmacokinetic Fate and Pathological Effects of ^{14}C-Fumonisin B_1 in Laying Hens. *Natural Toxins*, **1994**, 2, 81-88.

33

APPROACHES TO THE RISK ASSESSMENT OF FUMONISINS IN CORN-BASED FOODS IN CANADA

T. Kuiper-Goodman, P. M. Scott, N. P. McEwen, G. A. Lombaert,[1] and W. Ng[2]

Health Canada, Food Directorate
Ottawa, Ontario, K1A 0L2
[1] Health Protection Branch, Winnipeg
Manitoba, R2J 3Y1
[2] Health Protection Branch
Scarborough, Ontario, M1P 4R7

ABSTRACT

The presence of fumonisins and associated mycotoxins from *Fusarium moniliforme* in corn-based foods has recently become a concern in North America and elsewhere. Monitoring of various corn based foods and food commodities for fumonisins is ongoing in both the USA and Canada, and the results can be used for preliminary exposure assessments. The role of *Fusarium moniliforme* and the fumonisins in some diseases of livestock has been established. Considerable information is available on the mechanism of action of the fumonisins. With the availability of increased quantities of pure fumonisins, several sub-chronic toxicity studies, designed to establish dose response characteristics in rodents have now been completed. However, since concerns about the chronic toxicity of the fumonisins have not yet been adequately addressed, a tolerable daily intake cannot be established at this time. With the information at hand it is, nevertheless, possible to arrive at an interim risk assessment, which can be used to make interim risk management decisions. A total of 361 samples, covering 4 years of a Canadian survey, have been analyzed to date. Of these, 64 contained ≥ 0.1 μg/g fumonisin B_1, and 10 contained ≥ 1 μg/g. The 'all persons' estimate for the intake of fumonisins from these foods was < 0.089 μg/kg bw for 5-11 year-old children, and lower for other age groups. Based on an assessment of the available information on the toxicity of fumonisins, it can be concluded that these estimated intakes are unlikely to pose a health risk.

Fumonisins in Food, Edited by L. Jackson *et al.*
Plenum Press, New York, 1996

INTRODUCTION

The fumonisins (fumonisins B_1, B_2, B_3, B_4, A_1, A_2) are a group of mycotoxins, produced by *Fusarium moniliforme*, which have recently been isolated and chemically characterized. Research with *F. moniliforme* toxins was initiated in South Africa in order to isolate the chemical substances responsible for the diseases in livestock and possibly humans associated with the presence of *F. moniliforme* in corn-based feeds and foods.

Because of the ubiquitous nature of *F. moniliforme*, a 5-year survey for the presence of fumonisin B_1 (FB_1) and fumonisin B_2 (FB_2) in Canadian corn and corn-based food items is being conducted by the Health Protection Branch. In this paper results from the first three years and part of the fourth year of the survey will be presented and compared with those from surveys in other countries, especially the USA. To assess whether the levels of fumonisins found pose a health risk, an interim exposure assessment for Canadians, based on the survey data, and an interim health hazard and risk assessment will be made.

EXPOSURE ASSESSMENT

Canadian Survey

A total of 361 samples (0.5 to 1 kg) of corn or corn-based foods, sold or produced in Canada, were surveyed over a 4-year period (1991-1995) by the Health Protection Branch.

Determination of FB_1 and FB_2

During 1991-92 and (in part) 1992-93, samples were analysed for FB_1 and FB_2 by a modification of the method of Scott and Lawrence (1992). Extraction was with methanol-water (3 + 1, v/v), cleanup of 0.1 g sample equivalent was on a 1 ml strong anion exchange (SAX) solid phase extraction column, and quantitation was by reverse phase liquid chromatography (LC) of the fluorescent derivatives formed with naphthalene-2,3-dicarboxaldehyde (NDA) and potassium cyanide (KCN). The mobile phase was acetonitrile-methanol-water-acetic acid (40 + 40 + 20 + 1, or 25+55+20+1 v/v). Subsequently, methanol-water (8 + 2, v/v) containing 5% sodium chloride was used to extract the samples and the equivalent of 1 g sample was added to a Fumonitest® immunoaffinity column for cleanup. Fumonisins were eluted from the column with methanol following column washes with a solution of 2.5% sodium chloride, 0.5% sodium bicarbonate and 0.01% Tween 20, then water. o-Phthaldialdehyde (OPA) and mercaptoethanol (MCE) (Shephard et al., 1990) or NDA/KCN were used as primary derivatization reagents for reverse phase LC; the mobile phase was methanol-water-acetic acid (81 + 19 + 1, v/v) or acetonitrile-methanol-water-acetic acid (40 + 40 + 20 + 1, v/v). The alternative derivative was used for confirmation purposes (1993-94, 1994-95, and in part 1992-93). The detection limit (DL) was 100 ng/g sample for each fumonisin.

For a study specifically on calcium hydroxide processed corn foods (Scott and Lawrence, 1994b), samples were extracted with methanol-water (8 + 2, v/v) and 0.2 g sample equivalent was cleaned up on a 1 ml SAX column. Reverse phase LC of the OPA/MCE derivative of FB_1 was carried out with a gradient of the mobile phases methanol-0.05 M NaH_2PO_4 (55 + 45, v/v) adjusted to pH 3.3 with H_3PO_4, and acetonitrile-water (80 + 20, v/v). The DL was 30 ng/g. 4-Fluoro-7-nitrobenzofurazan (NBD-F) was used as derivatizing reagent for confirmatory LC (Scott and Lawrence, 1992).

Table 1. Recoveries of fumonisins from samples

Method	Lab.[a]	Year(s)	Food/food-stuff	Spiking levels (µg/g) FB$_1$	FB$_2$	n	Mean recoveries (%) ± SD and range FB$_1$	FB$_2$
SAX, NDA/KCN	O,C	1992-93 1991-92	Corn	1.0-5.0	1.0-4.0	20	78±22 (52-126)	63±21 (20-115)
SAX, NDA/KCN	C	1991-92	Corn flour, corn meal	0.5-4.2	0.5-3.0	13	69±25 (26-112)	74±25 (43-124)
SAX, NDA/KCN	C	1991-92	Corn snacks/cereals	0.5-4.2	0.5-3.0	12	59±26 (24-110)	44±30 (17-102)
Fumonitest, OPA/MCE	C	1992-93 1993-94	Corn	0.5-1.0	0.5-1.0	11	88±12 (69-110)	82±11 (68-105)
Fumonitest, OPA/MCE	C	1992-93 1993-94	Corn flour, corn meal	0.5-1.0	0.5-1.0	7	88±16 (67-111)	91±14 (69-108)
Fumonitest, OPA/MCE	C	1992-93	Tortillas/cereals	0.9-1.0	0.9-1.0	4	82±11 (75-91)	83±8 (73-92)
Fumonitest, NDA/KCN	C	1993-94	Corn	0.5-0.7	0.5-0.7	2	77 (77,77)	82 (76,87)
Fumonitest, NDA/KCN	O,C	1993-94	Corn flour, corn meal	0.5-1.5	0.5-1.0	11	79±12 (68-106)	80±11 (61-98)
Fumonitest, NDA/KCN	O,C	1993-94	Corn snacks/cereals	0.5-1.5	0.5-1.0	11	70±17 (40-100)	80±21 (39-107)
SAX, OPA/MCE		1994	Corn snacks	0.1	0.1	18,15	85±28 (40-115)	77±24 (23-109)

[a] O, Ontario; C, Central Region

Recoveries of Fumonisins B$_1$ and B$_2$

Analyses were carried out using two cleanup procedures - strong anion exchange solid phase extraction (SAX) or an immunoaffinity column (Fumonitest™). Derivatization was with naphthalene-2,3-dicarboxaldehyde (NDA) - potassium cyanide (KCN) or o-phthaldialdehyde (OPA) - mercaptoethanol (MCE). There were therefore four methods for which recovery data have been obtained; recoveries have been further segregated by matrix to some extent (Table 1). Recoveries at different spiking levels have been averaged, although this may not be considered to be entirely appropriate. Generally, recoveries were more variable and in some cases, lower, with the SAX cleanup procedure than with the immunoaffinity column cleanup. There also appeared to be lower recoveries with corn snacks/cereals than with other commodities; with the Fumonitest™ methods, this was evident only in the corn snacks component of this category (data not shown). Where direct comparisons were made with the same matrix and spiking level (7 samples), NDA/KCN and OPA/MCE derivatizations both afforded good recovery values (immunoaffinity column cleanup). Similar results for the two derivatives were also obtained on 29 actual positive samples (data not shown).

Food Intake Estimates

Corn and corn product consumption rates were obtained in most instances from the 1970-72 Nutrition Canada Survey (Health and Welfare Canada, 1976). For some food

Table 2. Food intake estimates of Canadians

Commodity	Eaters Only g/ person/ day			All Persons g/ person/ day		
	5-11 yr[a]	12-19 yr M	Adult	5-11 yr	12-19 yr M	Adult
Fresh corn	128	105	118	4.2	3.1	3.8
Corn, dried corn, corn meal, flour, semolina	18	27	36	1.3	2	2.6
Popcorn	10	23	26	0.9	1.2	1.6
Tortillas/taco shells	25	37.5	50	1.8	2.7	3.6
Chips (tortilla, taco)	28.6	40.4	29.5	5.2	9.6	3.3
Corn flakes, other breakfast cereals	35.7	37	29	3.8	3.6	2.2

[a] Body weights are 26.4, 53.8, and 60 kg for 5-11 year-old children, 12-19 year old males, and adults, respectively.

commodities this information was not known, and either other sources (USDA, 1982) were utilized or some assumptions were made regarding the intake of a given commodity, usually by comparing it to the intake of a similar type of food, for which consumption data was available. In other instances, the product analyzed would not be consumed as such (e.g. corn meal, corn flour), and estimated intakes of the corn portion of the finished product were made. For the fumonisin intake estimates, mean consumption rates for 'eaters only' and 'all persons' were used (Table 2). In addition to intake estimates for the 60 kg adult, estimates were also made for the 5-11 year old child and the 12-19 year old adolescent male, as some corn products, especially snacks, are consumed regularly by children.

Survey Results

The results of the survey are shown in Table 3 which indicates the incidence of samples $\geq 0.1 \mu g/g$ for each year of the survey, the incidence of samples $\geq 1 \mu g/g$ over the 4-year period, the range of the values obtained for each of the commodity groupings, and the overall mean content of fumonisins. Over the 4-year period, 64/361 and 10/361 samples, mostly in the dried corn and corn meal commodities, contained $\geq 0.1 \mu g$ FB_1/g or 1 μg FB_1/g, respectively. Nine of the 10 samples that contained more than 1 μg FB_1/g were found during the first year of the survey, and these were found to contain relatively low levels of FB_2. The sample with the highest value (3.5 $\mu g /g$) came from Italy.

The mean FB_1 content was calculated by assuming that the level of FB_1 in samples below the detection limit was either equal to zero (ND=0), or equal to the detection limit (ND=DL). The true mean is somewhere between these two estimates, which, among other things, depends on the number of samples below the detection limit and the distribution of the values (Kuiper-Goodman et al., 1993). A value of 20% was added for the overall level of FB_2, based on the overall ratio of the levels of FB_1 to FB_2, typically seen here and elsewhere. The dried corn/corn meal/corn flour commodities had the highest levels of fumonisins, which ranged from 0.20 to 0.28 $\mu g/g$ for ND=0 and ND=DL, respectively.

Fumonisin Intake Estimates

Since fumonisins are relatively heat-stable (Alberts et al., 1990; Dupuy et al., 1993; Le Bars et al., 1994; Scott and Lawrence, 1994a), it was assumed that there would be no loss of fumonisins either during processing or cooking of fresh or dried corn. However losses can occur during calcium hydroxide processing of corn (nixtamalization) with the formation of the aminopentol (AP_1 or HFB_1) (Hendrich et al., 1993; Sydenham et al., 1992) and in the

Table 3. Presence of fumonisins in corn and corn foods sold in Canada and analyzed by the Health Protection Branch

Product	1991/92	1992/93	1993/94	1994/95 in part	Total analyzed	Total ≥ 0.1 μg/g FB_1	Total ≥ 1 μg/g FB_1	Range μg/g FB_1	Mean FB_1+FB_2[a] μg/g	
									ND^b=0	ND=DL
Fresh corn, frozen corn	0/0	0/2	1/1	0/4	7	1	0	<0.1-0.12	0.021	0.12
Corn, dried corn, corn meal, flour, semolina	19/51	12/41	15/48	0/20	160	46	9	<0.1-3.5	0.20	0.28
Popcorn, cheesies	0/2	0/5	0/8	0/6	21	0	0	<0.1	0	0.12
Tortillas, tacos	1/7	2/7	0/6[c]	0/1	21	3	1	<0.03-1.2	0.08	0.17
Chips (corn, tortilla, taco) and other corn based snacks	0/3	0/18	6/28[d]	0/13	62	2	0	<0.03-0.22	0.010	0.11
Corn flakes and other corn based breakfast cereals	6/12	3/23	2/17	0/11	63	11	0	<0.1-0.32	0.028	0.13
Other corn based foods	1/3	0/2	0/2	0/20	27	1	0	<0.1-0.8	0.04	0.15
Total	27/78	17/98	24/110[c,d]	0/75	361	64	10	<0.03-3.5		

[a] FB_2 level was estimated as 20% of FB_1 level, except for dried corn, where actual values were used for the high samples; [b] In determining the mean, samples below the level of detection (ND) were taken as either 0 or the detection limit (DL); [c] includes 3 samples analyzed with a lower detection limit; [d] includes 14 samples analyzed with a lower detection limit.

manufacture of corn starch by wet milling, since FB_1 is water soluble (Bullerman and Tsai, 1994).

For some commodities the number of samples was very small, and the use of such data may lead to an erroneous estimate of intake, especially for the estimates for eaters, which indicate the highest consumption food items. Therefore estimates of fumonisin intake were only made for commodities for which >3 samples are available. For fresh or frozen corn, one positive sample (0.12 µg/g) was found among 7 samples. For this commodity the maximum intake estimate, based on ND=DL for 'eaters only' was omitted, since, because of the relatively high detection limit, it would lead to an overestimate; the intake from this commodity was however included in the all person estimate.

Table 4 is a summary of the range of intake estimates (for ND=0 to ND=DL) for the highest intakes from the 'eaters only' estimates. As is usual, the highest intakes, on a body weight basis, were seen in 5-11 year-old children (0.14 to <0.19 µg/kg bw/day) for the intake from dried corn. Table 4 also shows the 'all person' intake, which is the intake estimate from several corn-based foods at an average rate for all persons. This second intake estimate was made to address possible concerns about a chronic intake of fumonisins. For 5-11 year-old children the estimates ranged from 0.025 to <0.089 µg/kg bw/day, with the other age categories consuming less than this. As part of a risk management exercise, we have also shown in Table 4 the impact of removing all samples greater than 1 µg/g on the above fumonisin intake estimates.

Another source of fumonisins is from the intake of beer. Based on a recent survey, mean levels (ND=DL= 0.4 ng/ml) of FB_1 in 41 samples of Canadian and imported beer were 0.90 ng/ml (range <0.4 to 15 ng/ml) and 4.6 ng/ml (range <0.4 to 38 ng/ml), respectively (Scott and Lawrence, 1995). The FB_1 intake estimates for adult beer drinkers were 0.010 and 0.049 µg/kg bw, respectively, based on an average intake of 643 ml/day. Heavier beer

Table 4. Estimated intake of fumonisins by Canadians in µg/kg bw/day

	ND=0	ND=DL
a: for food commodities with more than 3 samples		
Eaters only [a]		
5-11 year-old children	0.14 (corn meal)	0.19 (corn meal)
12-19 year-old boys	0.10 (corn meal)	0.14 (corn meal)
adults	0.12 (corn meal)	0.17 (corn meal)
All Persons		
5-11 year-old children	0.025	0.089
12-19 year-old boys	0.017	0.057
adults	0.017	0.044
b: for food commodities with more than 3 samples, and minus 10 samples >1 ppm		
Eaters		
1-5 year-old children	0.058 (corn meal)	0.17 (breakfast cereal)
12-19 year-old boys	0.043 (corn meal)	0.089 (breakfast cereal)
adults	0.051 (corn meal)	0.10 (corn meal)
All Persons		
1-5 year-old children	0.015	0.079
12-19 year-old boys	0.009	0.049
adults	0.008	0.036

[a] The intake from fresh or frozen corn was excluded from the eaters only estimates, since only one positive sample of 0.12 µg/g was found; the high intake of fresh corn for eaters would distort the intake estimate, if the maximum value for the mean (0.103 µg/g) was used.

drinkers, consuming 950 ml/day (90th percentile) at the maximum value found, would have FB_1 intakes of 0.24 and 0.60 µg/kg bw for domestic and imported beers, respectively.

HAZARD ASSESSMENT OF FUMONISINS

Experimental studies with the fumonisins have been and are being conducted at a rapid pace, and much progress has been made since their isolation only a few years ago. Also of interest to those conducting a hazard assessment are the studies of natural outbreaks of disease associated with *F. moniliforme* toxins and fumonisins.

In the studies reported below, corn contaminated with *F. moniliforme*, *F. moniliforme* culture material, and pure fumonisins have been used. In the contaminated corn and culture material studies, additional metabolites, both identified and unidentified, may have played a role.

Epizootics and Experimental Studies in Livestock

The presence of *F. moniliforme* in feed has been related to at least two diseases in livestock that have a high fatality rate: equine leukoencephalomalacia (ELEM), which affects the brain, and which is known to have occurred in South Africa, the USA, Australia and Brazil (Ross et al., 1991; Wilson et al., 1990; Thiel et al., 1992; Christley et al., 1993) and porcine pulmonary edema, known to have occurred in the USA (Colvin and Harrison, 1992) (Table 5). In addition, a possible association has been observed between fumonisins and mystery swine disease (MSD), characterized by adverse effects on reproduction as well as respiratory disease, also occurring in the USA (Bane et al., 1992).

Of the several mycotoxins isolated from *F. moniliforme*, it has been possible to experimentally induce ELEM in young horses by oral administration of pure FB_1 at 1-4 mg/kg bw/day (20-21 doses over 29-33 days) (Kellerman et al., 1990) (Table 6). ELEM was also induced in one of four ponies that were fed feed naturally contaminated with 22 ppm FB_1, equal to 0.18 mg/kg bw/day for 159 days, interspersed with 2 periods of control feed for 30 and 7 days, and an initial feed containing 15 ppm FB_1 for 130 days (Wilson et al., 1992). The same authors found mild lesions in the brain, liver and kidney and behavioural changes after feeding a group of 5 ponies naturally contaminated feed at 8 ppm, equal to 0.06 mg/kg bw/day for 201 days followed by the same feed at 0.12 mg/kg bw/day for 58

Table 5. Diseases associated with *Fusarium moniliforme*

Disease	Species	Target organs	Area	Reference
equine leukoencephalomalacia (ELEM)	horse	brain	South Africa, USA, Australia and Brazil	Ross et al., 1991, Ross et al., 1992; Wilson et al., 1990; Thiel et al., 1992; Christley et al., 1993
porcine pulmonary edema (PPE)	pig	lung	USA	Colvin and Harrison, 1992
mystery swine disease (MSD)	pig	reproduction respiratory	USA	Bane et al., 1992
esophageal cancer	human	esophagus liver	Transkei, South Africa	Jaskiewicz et al., 1987a Marasas et al. 1988 Sydenham et al., 1990
			China	Zhen et al., 1984
			Northern Italy	Franceschi et al., 1990

Table 6. Oral administration studies in horses

	Feed	Duration	Level in feed (ppm)	Dose mg/kg bw/day	Effects	Reference
Young horses	FB₁	29-33d		1-4	ELEM	Kellerman et al., 1990
Ponies	naturalᵃ contam.	159 d	FB₁ 22	0.18	1/4 ELEM	Wilson et al., 1992
Ponies	naturalᵃ contam.	201 d + 58 d	FB₁ 8 + FB₁	0.06 FB₁ 0.12	brain, kidney, liver, behavioural	Wilson et al., 1992
Horse	epizootics 1984-1990 naturalᵃ contam.		FB₁>10 FB₁<6 5	0.1	89% ELEM 94% no ELEM taken as NOAEL	Ross et al., (1992)

ᵃ These feeds may contain other toxic factors besides fumonisin B₁

days. Since the effective oral doses of pure versus naturally occurring fumonisins differ, additional toxic factors may be involved.

Ross et al. (1992) reviewed 45 confirmed cases of ELEM that had occurred in the USA between 1984 and 1990, and found that 40 of these were associated with feed that contained >10ppm FB₁. Horse feeds that were not associated with health problems contained less than 6 ppm 94% of the time. Therefore, 5 ppm of FB₁ has been considered a 'no observed adverse effect level' (NOAEL) for horses. Based on an intake of 10 kg of feed for a 500 kg horse, this NOAEL is equivalent to 0.1 mg/kg bw/ day. In the USA (Arizona) a maximum residue level of 5 ppm has been set for the presence of FB₁ in the non-roughage portion of horse feed, and with the recommendation that corn screenings should be avoided (Frank Ross, personal communication). The American Federation of Veterinary Laboratory Diagnosticians supports these levels and has also made recommendations for maximum levels in swine feed, poultry feed and beef cattle feed (see chapter by Miller et al. in this volume).

Haschek et al. (1992) were able to induce porcine pulmonary edema after five days of oral administration of corn screenings naturally contaminated with fumonisins B₁ and B₂ to give a dose of 4.5-6.6 mg/kg bw/day (total fumonisins) (Table 7). Swine fed corn screenings containing FB₁ developed pulmonary edema at a dose level of 175 ppm (equal to 6.3 mg/kg bw/day total fumonisins and 4.6 mg/kg bw/day FB₁) after 4 to 6 days, accompanied by severe liver damage (Motelin et al., 1994) (Table 7). At lower dose levels (23, 39 and 101 ppm total fumonisins) fed for 14 days, the same authors found only liver damage. The NOAEL for liver damage was 5 ppm total fumonisins, equal to 0.17 mg/kg bw/day FB₁. Using regression analysis, the authors derived statistically a NOAEL (point estimate) equal to 0.41 mg/kg bw/day FB₁. The lower confidence limit on this value would, however, be less than 0.17 mg/kg bw.

In gavage studies with *F. moniliforme* culture material, Colvin et al. (1993) found liver failure at lower dosage levels (4-16 mg/kg bw/day) and pulmonary edema at higher dosage levels (16-64 mg/kg bw/day) in 12-16 kg pigs (Table 7). The same authors also conducted a feeding study (200 ppm FB₁), in which the pigs exhibited feed refusal, reduced body weight gain and developed liver disease as indicated by biochemical and pathological observation. These changes were reversed when two remaining pigs were placed on a control diet for 10 days.

Table 7. Oral administration studies in swine

Feed	Duration	Level FB$_1$ in feed (ppm)	Dose mg/kg bw/day	Effects or target organ	Reference
corn screenings[a]	5 d		4.5-6.6 total fumonisins	PPE	Haschek et al. (1992)
corn screenings[a]	4-6 d	175	FB$_1$ 4.6; tot.	PPE	Motelin et al., 1994
	14 d	101, 39, 23	6.3	liver	
		5	~0.17	NOAEL	
culture material[a]	45d	gavage	4	liver failure	Colvin et al., 1993
	8 d	"	8	liver failure	
	3-5d	"	16-64	PPE	
	21d	200	< 14	liver	
culture material + naturally contamin. corn[a]	7d+ 10-76 d	125 (total) + 237 (total)	~ 3 (total) + ~ 5 (total)	liver nodules[b] esophagus stomach	Casteel et al., 1993
	10 d + 205 d	131 (total) + 210 (total)	~ 4 (total)	vasculature[b], lung, heart liver nodules	Casteel et al., 1994

[a] These diets may contain other toxic factors besides fumonisin B$_1$
[b] No conccurrent control data were provided in these studies.

In two successive trials, six and five weanling pigs were fed diets containing fumonisin culture material and naturally contaminated corn. In the first trial, the diet contained 125 mg FB$_1$+FB$_2$/kg for the first 7 days and 237 mg FB$_1$+FB$_2$/kg, equivalent to 3 mg/kg bw/day, for the remaining 17 to 83 days; in the second trial the diet contained 131 mg FB$_1$ +FB$_2$ for the first 12 days followed by an average of 210 mg FB$_1$ +FB$_2$/kg for the remaining 205 days (Table 7). Control pigs were weight matched. All pigs developed varying degrees of hepatic necrosis, changes in biochemical serum parameters and after 93 days hepatic nodular hyperplasia. This latter lesion is seldom seen in pigs at slaughter. In addition, esophageal hyperplasia was reported in all of the pigs in the first trial, but not the second trial, and hypertrophy of the medial layer of the small and medium arterioles of the lung was observed in the second trial (Casteel et al., 1993; 1994). It should be noted that concurrent controls subjected to feed restriction were not available. It has not been possible to reproduce the esophageal lesions or nodular hyperplasia in the liver in pigs fed purified FB$_1$, so that these lesions must be attributed to other contaminants in the diet (Casteel, personal communication).

Epidemiology

In Transkei, South Africa, a high prevalence of esophageal cancer has been noted, accounting for almost half of the reported cancer cases (Jaskiewicz et al., 1987a). In this area liver cancer was the second most prevalent type of cancer.

Marasas et al. (1988) reported a significantly higher level of *F. moniliforme* contamination of home grown corn from households in which adults showed cytological evidence of premalignant esophageal lesions, in a district of Transkei with a high esophageal cancer rate, compared to "unaffected" households from a district with a lower esophageal cancer rate. An ecological study in Transkei (Sydenham et al., 1990) also showed a higher incidence of *F. moniliforme* contamination in "healthy" and "moldy" corn samples taken from a district that had higher rates of esophageal cancer compared to a district with a lower rate. Significant differences in FB$_1$ content were also noted for the "healthy" corn (1.6 versus 0.06 µg/g) and "moldy" corn (23.9 versus 6.5 μg/g) samples used for beer brewing from the 2 regions, suggesting an association between the incidence of esophageal cancer and the incidence of *F. moniliforme* and the presence of fumonisins in corn.

Table 8. Oral administration studies with rodents

Species	Feed/gav-age	Number	Duration	Level in diet ppm	Dose mg/kg bw/day	Effects	Reference
Rat, BDIX	gavage		3 d		FB$_1$ 237	3/4 death lesions in liver, heart, lungs, kidney, git	Gelderblom et al., 1988
Rat, BDIX	gavage or diet		21 d 33 d		FB$_1$ 60 FB$_1$ ~ 60	red. weight gain, lesions in liver, kidney	Gelderblom et al., 1988
Rat, SD, m,f	feed with pure FB$_1$	3/g	28 d	150 50 15	13.6 4.7 1.4	m,f liver, kidney m,f kidney m, kidney	Voss et al., 1993
Rat, SD, m	gavage	6/g	11 d		75 35 15 5 1	kidney, liver kidney, liver kidney, liver kidney kidney	Bondy et al., 1995
Rat, Fischer m,f	feed with pure FB$_1$	15/g	91 d	81 27 9 3 1	5.66; 6.35 1.92; 2.15 0.62; 0.73 0.21; 0.24 0.07; 0.08	m,f kidney m, kidney m, kidney m, NOAEL	Voss et al., 1995a
Mice, B6C3F1, m,f	feed with pure FB$_1$	15/g	91 d	81 27 9 3 1	23.1; 28.9 7.38; 9.71 2.44; 3.30 0.84; 1.00 0.30; 0.31	f, liver f, NOAEL	Voss et al., 1995a
Rat BDIX, m	corn based feed with pure FB$_1$	25/g	182 d 365 d 600 d 780 d	50	3.75	hepatitis cirrhosis 66% HCC	Gelderblom et al., 1991 Thiel et al., 1992
				0.5 (control)	0.04	no effects	

Acute, Subacute and Subchronic Toxicity in Rodents

The fumonisins are only moderately toxic in acute and subacute toxicity studies. Three out of 4 young male BD IX rats died after receiving 3 daily doses of 237 mg/kg bw of FB_1 by gavage. These rats developed toxic hepatitis, with single cell necrosis, as well as lesions in the heart (severe disseminated acute myocardial necrosis in 2 rats), lungs (severe pulmonary edema in 2 rats), kidney (proximal tubular necrosis) and gastrointestinal tract (necrosis in Peyer's patches and scattered epithelial necrosis) (Table 8) (Gelderblom et al., 1988). Two groups of rats that were administered about 1/4 of the above dose had a significant reduction in weight gain and developed hepatitis, which was more advanced in those rats receiving fumonisin in the diet for 33 days than in those dosed by gavage for 21 days. Kidney lesions were similar but less extensive than those seen in rats that died after 3 days. No other organs were affected (Gelderblom et al., 1988). The reduced weight gain and the pathological changes in the livers of rats caused by FB_1 were similar to those caused by *F. moniliforme* MRC 826 cultures.

Groups of 3 male and female Sprague Dawley rats, 5 weeks of age, were fed diets containing 0, 15, 50, and 150 ppm FB_1 (≥99% pure) for 28 days. These dietary levels provided FB_1 at 0, 1.4, 4.7, and 13.6 mg/kg bw/day to males and 0, 1.4, 4.1, and 13.0 mg/kg bw/day to females. No differences in survival, behaviour, and appearance, and no significant differences in body weight or food consumption were found between control and treated groups. Serum alanine aminotransferase, aspartate aminotransferase (males only), alkaline phosphatase, total cholesterol and triglycerides were significantly increased or outside the range found in the control group in high-dose males and females. Liver lesions, characterized by focal single cell hepatocellular necrosis, variability of nuclear size and staining, and hepatocellular cytoplasmic vacuolization were observed in all rats fed 150 ppm FB_1. Absolute and relative kidney weight was decreased in all male treatment groups and in the high dose female group. Kidney changes, consisting of focal cortical (proximal) tubular epithelial basophilia, hyperplasia, and focal cell necrosis, were found in all males fed 15 ppm or more of FB_1 and in all females fed 50 ppm or more of FB_1 (Table 8) (Voss et al., 1993

Groups of 15 male and female Fischer rats, 5 weeks of age, were fed diets containing 0, 1, 3, 9, 27, and 81 ppm FB_1 (≥98% pure) for 13 weeks. These dietary levels provided FB_1 at 0.07, 0.21, 0.62, 1.92, and 5.66 mg/kg bw/day to males and 0.08, 0.24, 0.73, 2.15, and 6.35 mg/kg bw/day to females. No differences in survival, behaviour, appearance, body weight, or food consumption were found between control and treated groups. There was a small dose related increase in serum creatinine levels in males (0.45 mg/dL at 0 ppm to 0.58 mg/dL at 81 ppm) and females (0.49 mg/dL at 0 ppm to 0.62 mg /dL at 81 ppm) at 13 weeks. This was accompanied by a dose related reduction in absolute and relative body weight of about 20% in males and 12% in females. Histopathologically, nephrosis was seen in 93% to 100% of males at dose levels of 9 ppm and above, with a severity that increased with time and that was dose related. The kidney changes involved individual proximal tubular cells, which became degenerate or necrotic. Some necrotic cells with pyknotic nuclei were sloughed into the tubular lumina. No glomerular or vascular lesions were observed. The same kidney changes were observed in females at a dose level of 81 ppm. At these dose levels, FB_1 was not hepatotoxic. Thus, under the conditions of this experiment, the NOAEL for kidney changes in males was 3 ppm, equal to 0.21 mg/kg bw/day of FB_1 (Table 8) (Voss et al., 1995a).

Groups of 15 male and female B6C3F1 mice, 5 weeks of age, were fed diets containing 0, 1, 3, 9, 27, and 81 ppm FB_1 (≥98% pure) for 13 weeks. These dietary levels provided FB_1 at 0.30, 0.84, 2.44, 7.38, and 23.1 mg/kg bw/day to males and 0.31, 1.00, 3.30, 9.71, and 28.9 mg/kg bw/day to females. No differences in survival, behaviour, appearance, body weight, or food consumption were found between control and treated groups. Serum levels of alanine aminotransferase, aspartate aminotransferase, alkaline phosphatase, lactate dehydrogenase, total protein and total bilirubin were significantly increased in high-dose females, and total serum cholesterol was significantly increased in females at 81 ppm and 27 ppm. Liver lesions, designated as mild hepatopathy were observed in all high dose females. The lesions were primarily centrilobular, with focal hepatocyte necrosis and cytomegaly, an increase in mitotic figures, mixed neutrophil and macrophage infiltrates, and macrophage pigmentation. Thus, under the conditions of this experiment, the NOAEL for liver damage in females was 27 ppm, equal to 9.71 mg/kg bw/day of FB_1 (Table 8) (Voss et al., 1995a).

Metabolic Disposition

In preliminary studies, the fate of a single dose of ^{14}C-labelled or unlabelled FB_1 was determined in 4 BD IX rats given 7.5 mg/kg bw by gavage. Within 24 hr, more than 99% of the FB_1 was eliminated unmetabolized in the feces, with only trace amounts in urine, liver, kidney and red blood cells (Shephard et al., 1992a,b). The same group found that ^{14}C-labelled FB_1, at a dose of 7.5 mg/kg bw administered by gavage to four male Wistar rats, was poorly absorbed from the gut with no appreciable enterohepatic circulation and no evidence of metabolic products (Shephard et al., 1994a). These results suggest that FB_1 is poorly absorbed by rats, and that the absorbed FB_1 is rapidly eliminated through biliary excretion.

Male Sprague Dawley rats were dosed intravenously or intragastrically with a single dose of ^{14}C labelled FB_1 (label at C_{22} and C_{23}), and the presence of radiolabel in various tissues was determined at various time intervals up to 96 hours in 3 rats per time interval. Up to 3% of radiolabel was excreted in the urine, indicating that some absorption had occurred. About 80% of radiolabel was excreted in the feces within 48 hours. Radiolabel in the liver, kidney and blood accounted for about 0.5% of the dose, and persisted for at least 96 hours. Tissue distribution was similar after i.v. dosing, with about 35% of the label excreted in the feces, indicating biliary excretion (Norred et. al., 1993). The same authors found an increase in specific radioactivity in liver and kidney but not blood after 3 doses of the same labelled material, 24 hours apart.

Administration of a single dose of ^{14}C-labelled FB_1 (8 mg/kg bw) to 2 Vervet female monkeys by gavage resulted in 76% recovery of radioactivity over a subsequent 3-day period with most of it in feces (61%) and intestinal contents (12%) and with 1.2% accounted for by urinary excretion. Over 90% of the recovered radioactivity was attributed to FB_1 or partially hydrolyzed FB_1 with the latter only found in the feces. Low levels (avg. 0.6% of the dose) of the fully hydrolyzed FB_1, the aminopentol AP_1 were detected in the feces. Residual radioactivity was recovered at low levels from skeletal muscle (1%), liver (0.4%), brain (0.2%), kidney, heart, plasma, red blood cells and bile (0.1% each). The authors concluded that, because of the rapid elimination of FB_1 and the low recoveries from plasma, levels of FB_1 and its metabolites in human blood are unsuitable as a biomarker of animal and human exposure (Shephard et al., 1994b). The same authors also administered a single dose of ^{14}C-labelled FB_1 (1.6 mg/kg bw) intravenously to 2 female Vervet monkeys. Elimination of FB_1 from the plasma was rapid with a mean half-life of 40 min. Over a 5-day period, about 47% of the radioactivity was recovered, 6-20% in the urine and 26-38% in the feces, mostly attributed to FB_1 and partially hydrolyzed FB_1. Trace levels (<0.1% of the dose) of AP_1 were found in the feces (Shephard et al., 1994b).

Male Yorkshire pigs were dosed intravenously or intragastrically with a single dose of ^{14}C labelled FB_1, and the presence of radiolabel in various tissues was determined at various time intervals up to 72 hours in 3 rats per time interval. Up to 1.6 % of radiolabel was excreted in the urine, and up to 95% was excreted in the feces, in the intragastrically dosed animals. Highest activities were measured in the liver and kidneys. Extensive enterohepatic recycling was evident and this could result in accumulation of residues after prolonged dosing (Prelusky et al., 1994). The same authors have conducted a multiple dosing study in which radiolabelled FB_1 was incorporated in the feed for 24 days, and in which increasing tissue residues were seen with time (Chapter by Prelusky et al., in this volume).

Carcinogenicity

FB_1 in the diet of BD IX rats at a level of 50 mg/kg bw/day was found to produce positive results in a short term (4-week) liver cancer initiation/promotion bioassay. In the bioassay, FB_1 induced gamma-glutamyl-transpeptidase-positive (GGT+) foci both in diethylnitrosamine (DEN) initiated rats and control (DMSO treated) rats. This bioassay was actually used as a monitoring system to isolate cancer promoting compounds from *F. moniliforme* culture material, and formed the basis for the isolation of the fumonisins (Gelderblom et al., 1988). The authors concluded that FB_1 was a promoter as well as possibly a "weak initiator". To further investigate the initiating potential of FB_1, a "resistant hepatocyte" model was used (Solt and Farber, 1976). This model measures the ability of cells to proliferate in an environment that inhibits the growth of normal hepatocytes. In this assay, FB_1 was found to be a "weak initiator" of carcinogenesis when incorporated at 0.1% in the diet (equivalent to 0.5 mg/kg bw) for 26 days or longer (Gelderblom et al., 1992b). Higher doses of FB_1 (total dose 50 to 250 mg/kg bw), given in one or two doses by gavage, were not effective at causing initiation in the model (Gelderblom et al., 1992b). These authors found that the effective dose to cause initiation (EDI) was a function of length of treatment and method of administration, with no initiation observed for exposures lasting less than 7 days. With an exposure of 14 days the EDI values were $17 \leq EDI \leq 24$ mg FB_1/kg bw/day for administration in the diet, and $3.9 \leq EDI \leq 8.3$ mg FB_1/kg bw/day for administration by gavage (Gelderblom et al., 1994).

Further studies have shown that fumonisins B_2 and B_3 closely mimic the toxic and tumor promoting effects seen with FB_1 in the short term liver bioassay; similar to FB_1, they are able to induce "resistant hepatocytes" (Gelderblom et al., 1992a). In contrast, under identical experimental conditions, no adverse effects could be detected for fumonisin A_1, the tricarballylic acid moiety, or the two hydrolysis products of FB_1 and FB_2 (Gelderblom et al., 1992a; Gelderblom et al., 1993).

A carcinogenicity study, considered as preliminary, has been completed with 2 groups of 25 BD IX rats given semi-purified corn-based diets to which purified FB_1 ($\geq 90\%$ pure) was added at 0 or 50 mg/kg, equal to 3.75 mg/kg bw/day (Table 8). The actual level of FB_1 in the control diet was 0.5 mg/kg. Five rats per group were sacrificed at 6, 12, 20 and 26 months (Gelderblom et al., 1991; Thiel et al., 1992). The major target organ of FB_1 was the liver. All FB_1-treated rats that died (5 rats) or were sacrificed (10 rats) from 18 months onward suffered from "chronic toxic hepatitis" which had progressed to cirrhosis. Ten out of 15 of these rats had developed primary hepatocellular carcinoma, with metastases to the heart, lungs or kidneys in 4 rats with hepatocellular carcinoma. Large areas of cholangiofibrosis were present at the hilus of the liver. All fumonisin treated livers contained regenerative nodules as well as GGT+ foci. Relative liver weight, probably due to the presence of tumours and regenerative nodules, was significantly increased in treated rats by almost 4-fold at 26 weeks. No neoplastic changes were observed in control rats, and their survival rate was 96%. Serum levels of aspartate aminotransferase, gamma-glutamyltranspeptidase and bilirubin (total, conjugated and unconjugated) in treated rats were significantly higher at 20 and 26 months. The kidneys of the fumonisin-treated rats were markedly affected at 26 months. The lesions included fibrosis, retention cysts and proximal tubular necrosis. Contrary to observations in rats fed culture material of *F. moniliforme* (MRC 826) (Jaskiewicz et al., 1987b; Marasas et al., 1984), there were no lesions in the esophagus, heart or forestomach.

Chronic Toxicity other than Cancer

A borderline NOAEL or 'lowest observed adverse effect level' (LOAEL) of approximately 0.15 mg FB_1/kg bw has been found in a 1631 day study, that is still ongoing, with 9 Vervet monkeys (4 males, 5 females) which were fed a carbohydrate diet low in fat to which *Fusarium moniliforme* culture material, containing FB_1, was added. At the next highest dose of approximately 0.35 mg FB_1/kg bw/day, changes in liver enzymes, elevated total cholesterol, mild fibrosis in liver biopsies and an atherogenic plasma lipid profile were observed (Fincham et al., 1992).

Studies on Reproductive Effects and Teratology

Diets containing *F. moniliforme* culture material to provide 0, 1, 10 or 55 ppm FB_1 were fed to male and female rats beginning 9 and 2 weeks, respectively, before mating (Table 9). Although kidney lesions (nephrosis) were seen in males fed 10 and females fed 55 ppm FB_1, there were no significant reproductive effects in males and dams and in fetuses examined on gestation day 15, or dams and litters examined on day 21 *post partum*. Litter weight gain of groups fed 10 or 55 ppm FB_1 was slightly decreased, but gross litter weight was unaffected. Although this study did not include a teratology component, there were no obvious abnormalities in the offspring (Voss et al., 1995b).

Timed-bred Syrian hamsters were dosed with FB_1 at levels of 0, 1.0, 2.0, 3.0, 6.0, and 12.0 mg/kg bw from day 8-10 or day 8-12 of gestation (Table 9). No clinical signs of maternal toxicity were observed, although pregnant animals had reduced weight gains compared to non-pregnant animals. There was no evidence of liver damage in maternal animals based on serum levels of aspartate aminotransferase or bilirubin, but histologic changes were observed in the kidneys, livers (dose related increase in karyomegaly) and placentas (especially at the highest dose level) A dose-related increase in the numbers of prenatal deaths and resorptions was seen in hamsters that were dosed at 3 mg FB_1/kg bw or higher. The incidence ranged from 3/6 litters affected with 9% dead and resorbed fetuses for

Table 9. Reproductive effects of fumonisins in rodents

Species	Feed/gavage	Number	Duration	Level in diet ppm	Dose mg/kg bw/day	Effects	Reference
Hamster	gavage	6-12/g	d 8-12 of gestation		12	100%, dead/resorbed	Floss et al., 1994
					6	21%, fetuses	
					3	9%, LOAEL	
					2	3%, NOAEL	
					1	4%	
					0.25-0.5	0%	
Rat	diet, culture material	12/g	2 weeks prior mating + gestation	55 FB_1 10 1	~2.7	No effect on mating, fertility, offspring	Voss et al., 1995b

animals dosed at 3 mg/kg bw to all litters affected and 100% dead and resorbed fetuses for animals dosed at 12 mg/kg bw. These changes indicate that FB_1 is a developmental toxicant in the hamster. No consistent structural defects were observed, suggesting that FB_1 is not a teratogen in the hamster (Floss et al., 1994).

Genotoxicity

FB_1 and FB_2 were found to be negative in the Ames test (Gelderblom and Snyman, 1991; Park et al., 1992). FB_1 (at concentrations of 0.04-250 µM) and FB_2 (at 0.04-40 µM) also lacked genotoxic effects in the *in vitro* DNA repair assays in primary rat hepatocytes (Gelderblom et al., 1992b; Norred et al., 1992a). Similarly, no DNA repair was induced in rat hepatocytes that were obtained 13-14 hr after rats had been orally administered FB_1 (Gelderblom et al., 1992b).

Cytotoxicity

In vitro studies, measuring release of lactate dehydrogenase by primary hepatocytes, showed that FB_1 and FB_2 are cytotoxic only at high dose levels (175 µM and 87 µM, respectively) when compared with aflatoxin B_1 (<1 µM) (Gelderblom et al., 1992b). In this bioassay, the aminopentol hydrolysis products of FB_1 and FB_2, AP_1 and AP_2 exhibited greater cytotoxicity than the respective parent compounds (Gelderblom et al., 1993). In another more sensitive bioassay, measuring cell viability in a hepatoma cell line (H4TG) and in the MDCK dog kidney epithelial cell line, FB_1 and FB_2, on a molar basis, after 4 days of culture, were about 1000 to 3000 times less toxic than T-2 toxin, with inhibitory concentrations (IC_{50}'s) in MDCK cells of 3.5 µM FB_1, 2.8 µM FB_2 and 1 nM T2 toxin (Shier et al., 1991).

Studies into the Mechanism of Action

The fumonisins are structurally similar to the long chain base sphingosine, a component of the long chain backbone of sphingolipids. It is therefore not surprising that FB_1 and FB_2 are naturally occurring specific inhibitors of *de novo* sphingolipid biosynthesis and sphingolipid turnover. The site of inhibition is at the formation of ceramides catalyzed by sphingosine- and sphinganine N-acetyl-transferase (ceramide synthase). In primary rat hepatocytes this inhibition occurs at concentrations of fumonisin that are not toxic to the cells (IC_{50} = 0.1 µM) (Wang et al., 1991; Norred et al., 1992b). Similarly, FB_1 and FB_2 inhibited *de novo* sphingosine biosynthesis (IC_{50} = 10-15 µM) in a proliferating cell line of pig kidney cells (LLC-PK1) and caused a remarkable (128 fold) increase in cellular levels of sphinganine (Yoo et al., 1992; Norred et al., 1992b). In these cells inhibition of sphingolipid biosynthesis was an early event in the toxicity of fumonisins, and preceded inhibition of cell proliferation and cytotoxicity. Inhibition of sphingolipid biosynthesis was also observed in yeast cells (Kaneshiro et al., 1992).

In pigs dosed orally with fumonisin-containing culture material (equivalent to 0.9 to 1.2 mg fumonisins/kg bw/day), or intravenously with 4.5-6.6 mg FB_1/kg bw/day, the primary target organ was the liver (Haschek et al., 1992). Based on electron-microscopic observations, the authors hypothesized that altered sphingolipid metabolism resulted in hepatocellular damage, leading to the release of membranous material into circulation. It was further hypothesized that this material is engulfed by pulmonary macrophages, where it could trigger the release of mediators, leading to porcine pulmonary edema.

Inhibition of sphingolipid biosynthesis may be the primary effect of fumonisin toxicity in ELEM and porcine pulmonary edema. It may also be responsible for the tumor-promoting ability of the fumonisins. In the latter, fumonisin-induced inhibition of sphingosine biosynthesis may lead to a deregulation of protein kinase C, which could lead to the proliferation of initiated cells (Norred et al., 1992b). In addition, stimulation of DNA synthesis by FB_1, at concentrations of 10 µM to 100 µM, in Swiss 3T3 fibroblasts was attributed to the mitogenic effect of the sphingoid bases that accumulated as a result of a disruption of sphingolipid metabolism (Schroeder et al., 1994).

The ratio of free sphinganine to free sphingosine in serum and tissues has been found to increase when rats, ponies and pigs were exposed to fumonisins in their feed. In rats, ponies and pigs fed diets that contained < 1ppm of fumonisins the ratio was <0.35. Levels as low as 5 ppm in the feed of pigs caused statistically significant increases, so that this ratio has been suggested as a specific biomarker for fumonisin exposure (Riley et al., 1994). Elevations in this ratio were seen before there was evidence of tissue damage as indicated by serum biochemical parameters or histopathology. The implications of persistent elevations in these ratios in the development of fumonisin associated diseases are of interest.

Other *F. moniliforme* Toxins

A number of other mycotoxins have been isolated from *F. moniliforme* cultures, and have been assessed for their possible role in *F. moniliforme* associated diseases. Fusarin C was found to be mutagenic in the Ames test (Gelderblom and Snyman, 1991), but was not positive in a short term initiation/promotion bioassay in rat liver (Gelderblom et al., 1986). Rapid conjugation with glutathione and excretion are possibly related to its lack of carcinogenicity (Norred et al., 1992a). Inconclusive results with fusarin C were obtained with a DNA repair assay in primary hepatocytes (Norred et al., 1992a). Moniliformin and bikaverin were not genotoxic in a DNA repair assay with primary hepatocytes (Norred et al., 1992a). Additional mycotoxins continue to be isolated from *F. moniliforme* cultures. The possibility for interaction between the various *F. moniliforme* toxins in the genesis of both acute and chronic effects needs to be considered in the overall risk assessment and future regulation of this group of toxins.

Summary of LOAELS and NOAELS

Table 10 summarizes the results of the various studies that can be used to determine LOAELS and NOAELS of toxicological endpoints in a variety of species. The horse appears to be the most sensitive species, and based on epizootics of ELEM the NOAEL for naturally contaminated feed is 5 ppm, equivalent to 0.1 mg/kg bw of FB_1. It should be noted that experimentally induced ELEM requires higher doses of pure FB_1.

Similarly, based on epizootics, the NOAEL for porcine pulmonary edema (PPE) for naturally contaminated feed for pigs is 20 ppm, equivalent to 0.8 mg/kg bw. In a short term feeding study with *Fusarium moniliforme* culture material, the NOAEL for histopathological damage (liver) was 5 ppm (total $B_1 + B_2$) in feed, equal to 0.17 mg FB_1/kg bw.

In 28-day and 90-day studies conducted with rodents, several strains of rats (BDIX, Sprague-Dawley, Fischer) were more sensitive than mice (B6C3F₁). In both Sprauge-Dawley and Fischer rats the kidney was affected sooner and at lower doses than the liver, and males were more sensitive than females. Female mice were more sensitive with respect to effects on the liver, but this could be related to the slightly higher intake of fumonisins by females. In a 28-day study in Sprague Dawley rats, no liver changes were observed below 150 ppm. No evidence of hepatotoxicity was observed at dose levels up to 81 ppm in a 90 day study with FB_1 (≥98% pure) in Fischer rats. In that study a NOAEL of 3 ppm, equal to 0.21 mg FB_1/kg bw/day, for kidney damage was established.

Table 10. Summary of studies and parameters effected

Species	Duration	Effect	LOAEL mg/kg bw/day	NOAEL mg/kg bw/day	Margin of Safety[a]	Reference
Mouse	91d	liver	28.8	9.71		Voss et al., 1995
Rat	91d	kidney	0.62	0.21 (m)	2400	Voss et al., 1995
Rat	2 yr	liver cancer (inc. 66%)	3.75	>0.04 (control)	42,000	Gelderblom et al., 1988
Pig	4-6 d	PPE	4.6			Haschek et al., 1992
Pig	epizootics	PPE		0.8 FB$_1$ (natural)		Ross et al., 1992
Pig	14 d	liver	0.78	0.17 (FB$_1$+FB$_2$)	1900	Motelin et al. 1994
Horse	epizootics	ELEM		0.1 FB$_1$	1100	Ross et al., 1992
Vervet (Monkey)	1631 d	liver	0.35	0.15	1700	Fincham et al., 1992
Hamster	8-10 d	fetal death/ resorptions	3	2		Floss et al., 1994

[a] Expressed relative to an intake of 0.089 µg/kg bw/day and the various NOAELS, except for liver cancer

The preliminary carcinogenicity study conducted with BD IX rats in South Africa showed that FB$_1$ (≥90% pure) induced liver tumors in 10/15 rats at a dietary level of 50 ppm (equal to 3.75 mg/kg bw/day) for 26 months. No LOAEL or NOAEL was established. Although there are many deficiencies in this study (such as only a few animals, only one dose level, purity of FB$_1$ only ≥ 90%, type of diet, little information on the strain of rat), it indicated that FB$_1$ is a hepatocarcinogen in the rat under the conditions of the experiment. This study will be validated by a FDA National Center for Toxicological Research (NCTR) study, now ongoing.

A NOAEL of approximately 0.15 mg FB$_1$/kg bw has been found in the chronic study with Vervet monkeys, which were fed a diet to which *Fusarium moniliforme* culture material, containing FB$_1$, was added.

Extrapolation from Animals to Humans

From the available information on the toxicity of fumonisins, summarized above, it is difficult to estimate an unconditional tolerable daily intake (TDI) for this group of mycotoxins, mainly because of uncertainties with respect to their tumorigenic potential. Attempts to set a preliminary/provisionary TDI would have to use additional safety/uncertainty factors to take into account these and other uncertainties in the existing data and database. In addition to the LOAELS, and NOAELS indicated above, the following points need to be considered in choosing appropriate safety factors, and in setting a TDI for fumonisins:

- Naturally *Fusarium moniliforme* contaminated food contains other contaminants besides FB$_1$, such as other fumonisins and other toxic *Fusarium moniliforme* metabolites.

- FB$_1$ was found to be carcinogenic in the rat, although there are shortcomings in the study.
- Studies with radiolabelled FB$_1$ in pigs suggest an accumulation of residues in liver and kidney.
- d. Inadequate information on comparative metabolic disposition or other factors that could be important in causing species differences in target toxicity and sensitivity.

RISK ASSESSMENT

As an alternative to setting a preliminary TDI, the observed Canadian fumonisin exposure estimates can be compared to the effects seen in animals (Table 10). For chronic effects the 'all person' intake by 5-11 year-old children of 0.025 to <0.089 µg FB$_1$/kg bw/day is still considered a conservative estimate, since this would likely decrease if detection limits were to be less than 0.1 µg/g. The maximum value of 0.089 µg FB$_1$/kg bw/day is:

- 1100 times less than the NOAEL for the induction of field cases of ELEM, a generally fatal disease;
- 2400 times less than the NOAEL of 0.21 mg/kg bw/day in the 90 day rat study; this study does not address possible carcinogenic effects because of its relative short duration of exposure;
- 1900 times less than the NOAEL of 0.17 mg/kg bw/day of FB$_1$ in the 14 day pig study; similar to the above rat study, this study does not address possible carcinogenic effects because of its short duration of exposure, and very few animals studied;
- 1700 times less than the NOAEL of 0.15 mg/kg bw/day in the chronic Vervet monkey study; with so few animals in the study, although of long duration, the carcinogenic potential in this species cannot be addressed;.
- 42,000 times less than the dose of 3.76 mg/kg bw/day in the 26 month BDIX rat study, which induced liver cancer in 66% of the animals tested. This study suffered from a number of insufficiencies, such as few animals studied, and only one dose level;

From these comparisons, it may be concluded that, based on the present Canadian exposure data, the intake of FB$_1$ and FB$_2$ in Canada, even for the most sensitive age group, is unlikely to pose a health hazard.

DISCUSSION

Several surveys of U.S. corn and corn-based foods have been published (Hopmans and Murphy, 1993; Murphy et al., 1993; Pestka et al., 1994; Ross, 1994; Stack and Eppley, 1992; Sydenham et al., 1991; Trucksess et al., 1995). These are in addition to surveys of feed corn and corn screenings associated with outbreaks of animal disease (Ross, 1994) which will not be considered here. The concentrations of FB$_1$ and FB$_2$ found so far in U.S. corn and corn foods are shown in Table 11. The commodities are listed to correspond to those analyzed in Canada. Only one example of U.S. dried corn data is given as the proportion of dried corn intended for human consumption is not known; Iowa corn in 1988-91 had much higher levels of fumonisins than corn/dried corn analyzed in Canada.

Levels of FB$_1$ in fresh corn (based on very limited sampling, Table 3) appeared to be similar to U.S. data for frozen sweet corn and canned corn (Table 11); further Canadian information is clearly needed. Incidence and levels of fumonisins in dried corn/corn meal/flour were lower than in the U.S.A.; the sample with 3500 ng FB_1/g which was analyzed in Canada actually came from Italy. Corn grits (3 samples only) appeared to contain less fumonisin than those in the U.S.A.

Few U.S. data are available for alkali-processed corn foods (corn tortillas, corn chips, tortilla chips, etc.). Comparison is difficult due to the generally higher Canadian detection limit, but maximum levels found were similar. Concentrations of fumonisins in popcorn and corn flakes were below 100 ng/g in both Canada and the U.S. and maximum levels found in other corn-based cereals were similar.

Although data are limited, maximum levels of FB_1 in Canadian fresh corn were surpassed by those in Italian sweet corn (canned, cobs), which ranged from 60 to 790 ng/g (DL= 10 ng/g) (Doko and Visconti, 1994). Levels of fumonisins in Italian corn in general and also in corn consumed in parts of China and southern Africa were much higher than in Canada (Doko and Visconti, 1994; Ueno et al., 1993; Chu and Li, 1994; Sydenham et al., 1990).

Corn meal, corn flour and dried corn averaged < 200 ng fumonisins/g (for ND=0), as did corn meal in Switzerland (Pittet et al., 1992) and South Africa (Sydenham et al., 1991) and corn flour in Spain and China (Sanchis et al., 1994; Ueno et al., 199). The maximum level of fumonisins found in corn flakes in Switzerland, Spain, Italy and South Africa was 100 ng/g (Pittet et al., 1992; Sanchis et al., 1994; Doko and Visconti, 1994; Sydenham et al., 1991), which was not exceeded in Canada (detection limit), although in Canada low levels of fumonisins were detected in some other corn-based breakfast cereals (100-320 ng/g of FB_1 in 5 bran cereals).

There is a need to continue with the surveys of fumonisins in corn-based food products being marketed in Canada. Moreover, the need for methods with detection limits below 0.1 µg/g is evident. Maximum food intake estimates, based on a mean that includes values equal to the DL for values below the detection limit, are especially exaggerated when only a few values slightly above the detection limit are encountered.

It is now well established that *Fusarium moniliforme* contaminated feed and food have been the cause of epizootics affecting horses and pigs, and have been associated with human disease, such as esopahgeal cancer. *F. moniliforme* culture material has been found to be harmful to horses, pigs, monkeys and rodents. Many of the effects have been duplicated with purified FB_1. It has become evident that some of the target organs can be quite species specific. Thus, for horses it is the brain and liver, and for the pig it is the lung, liver, and kidney. The liver has been found to be a target in most species, and the extent of damage is dose, duration and species specific. In many species (pig, rodent) the kidney is also involved. Since FB_1 has affected several target tissues in animal studies, it is difficult to predict what the major target tissues will be in humans.

Pharmacokinetic studies with ^{14}C labelled FB_1 (in rats and pigs) with single doses (i.v. or gavage) have shown low bioavailability, but an accumulation of radioactivity mainly in liver and kidney. With multiple dosing (3 days in the rat) and 24 days in pigs, no plateau of radioactive label was found in these two tissues. Thus, it appears probable that hepatotoxic or nephrotoxic effects will not become evident (by histopathology or changes in enzymes) until sufficient tissue residues have accumulated. For this reason, one needs to be concerned with chronic low dose exposure. So far, no studies have correlated tissue residues with associated damage.

In a review of the data, the International Agency for Research on Cancer (IARC) concluded that there was *inadequate evidence* in humans for the carcinogenicity of toxins derived from *F. moniliforme*, *sufficient evidence* for the carcinogenicity of *F. moniliforme* culture material in experimental animals, and *limited evidence* for the carcinogenicity of FB_1 in experimental animals. Overall, toxins derived from *F. moniliforme* are *possibly carcino-*

Table 11. Fumonisins in random samples of U.S. corn and corn foods

Product	Year	N	Fumonisin B₁				N	Fumonisin B₂				Total mean μg/g	References
			DL μg/g	Mean[a] μg/g	Range μg/g	No. >DL		DL μg/g	Mean[a] μg/g	Range μg/g	No. >DL		
Frozen sweet corn	1993	27	0.004	0.018	ND-0.35	9						0.018[b]	Trucksess et al., 1994
Canned corn	1990-93	71	0.004-0.010	0.012	ND-0.24	29	1	0.020	ND	—	0	0.12[b]	Hopmans and Murphy, 1993; Trucksess et al., 1994
Corn, dried (mainly Iowa)	1988-91	180	0.100	2.98	ND-37.9	148	180	(0.50)[c]	0.82	ND-12.3	?	4.09[d]	Murphy et al., 1993
Corn meal	1989-91	52	0.010-0.200[e]	1.05	ND-6.8	45	38	0.01-0.05	0.22	ND-0.920	35	1.21[b]	Stack and Eppley, 1992; Hopmans and Murphy, 1993; Sydenham et al., 1991; Pestka et al., 1994
Corn grits/hominy corn	1990-91	16	0.010-0.050	0.44	0.060-2.54	16	16	0.01-0.05	0.14	ND-1.07	10	0.58	Stack and Eppley, 1992; Sydenham et al., 1991
Corn tortillas/tortilla mix/masa	1989-91	7	0.010-0.200[e]	0.06	ND-0.120	4	6	0.01-0.05	0.005	ND-0.03	1	0.064[b]	Stack and Eppley, 1992; Sydenham et al., 1991; Pestka et al., 1994; Hopmans and Murphy, 1993
Corn/tortilla chips	1989-91	6	0.010	0.11	ND-0.32	3	6	0.01	ND	—	0	0.11	Stack and Eppley, 1992; Hopmans and Murphy, 1993
Popcorn	1990-91	2	0.010	0.035	0.010-0.06	2	2	0.01	ND	—	0	0.035	Stack and Eppley, 1992
Corn muffin mix	1990-91	5	0.010-0.200[e]	0.29	ND-1.21	3	2	0.01	0.005	ND-0.010	1	0.29[b]	Stack and Eppley, 1992; Pestka et al., 1994

Corn flakes	1990-91	8	0.010-0.050	0.002	ND-0.01	2	8	0.01-0.05	ND	—	0	0.002	Stack and Eppley, 1992; Sydenham et al., 1994
Other corn-based cereals	1990-91	14	0.010-0.200[e]	0.08	ND-0.330	7	11	0.01	0.015	ND-0.070	5	0.09[b]	Stack and Eppley, 1992; Pestka et al., 1994
Misc.	1990-91	4	0.050	0.41	0.085-0.70	4	4	0.05	0.111	ND-0.240	3	0.52	Sydenham et al., 1991

[a] Overall mean, including ND (not detected) taken as 0, [b] Fumonisin B_2 not measured in some or all samples, [c] Quantitation limit, [d] Includes fumonisin B_3, [e] ELISA detection limit

genic to humans (Group 2B) (IARC, 1993). Experimental carcinogenicity studies with the fumonisins have been hampered by a lack of pure toxins. Major efforts are now ongoing in Canada, the US and in South Africa to produce fumonisins on a large scale. Further carcinogenicity studies are being conducted in South Africa (using lower dosage levels) and by the US Department of Health and Human Services National Toxicology Program.

Further epidemiological studies at the level of the individual, such as case control or cohort studies, are required to more precisely define the role of *F. moniliforme* and its metabolites in the development of esophageal cancer in Transkei and other areas where the incidence of esophageal cancer is high, such as parts of China (Ueno et al., 1993; Chu and Li, 1994; Yoshizawa et al., 1994) and northern Italy (Franceschi et al., 1990). In several of these areas there is a high incidence of *F. moniliforme* in corn, high levels of fumonisins in corn have been detected, and a higher per capita corn consumption than in Canada. In a case-control study in northern Italy, heavy drinking (>42 drinks/week) together with frequent consumption of maize was found to be a risk factor for esophageal cancer (odds ratio=2.80) (Franceschi et al., 1990). As the use of corn in the processing of beer is becoming more common, this source of fumonisins could be of concern for heavy beer drinkers.

In conclusion, the low levels of fumonisins found to date in corn-based foods in Canada do not appear to pose a health risk, based on a comparison with effects seen at much higher levels of fumonisins in various animal species. There appears, therefore, to be no need to establish guidelines or action levels for fumonisins in corn based foods in Canada at the present time.

REFERENCES

Alberts, J. F.; Gelderblom, W. C. A.; Thiel, P. G.; Marasas, W. F. O.; van Schalkwyk, D. J.; Behrend, Y. Effects of temperature and incubation period on production of fumonisin-B_1 by *Fusarium moniliforme*. Appl. Environ. Microbiol. 1990, 56, 1729-1733.

Bane, D. P.; Neumann, E. J.; Hall, W. F.; Harlin, K. S.; Slife, R. L. N. Relationship between fumonisin contamination of feed and mystery swine disease - A case-control study. Mycopathologia 1992, 117, 121-124.

Bondy, G.; Barker, M.; Mueller, R.; Fernie, S.; Miller, J. D.; Armstrong, C.; Hierlihy, S. L.; Rowsell, P.; Suzuki, C. Fumonisin B_1 toxicity in male Sprague Dawley rats. American Chemical Society, Annual Meeting, April 2-6, Anaheim, Ca. 1995, in press.

Bullerman, L. B. and Tsai, W.Y.J. Incidence and levels of *Fusarium moniliforme*, *Fusarium proliferatum* and fumonisins in corn and corn-based foods and feeds. J. Food Prot. 1994, 57, 541-546.

Casteel, S. W.; Turk, J. R.; Cowart, R. P.; Rottinghaus, G. E. Chronic toxicity of fumonisin in weanling pigs. J. Vet. Diagn. Invest. 1993, 5, 413-417.

Casteel, S. W.; Turk, J. R.; Rottinghaus, G. E. Chronic effects of dietary fumonisin on the heart and pulmonary vasculature of swine. Fundam. Appl. Toxicol. 1994, 23, 518-524.

Christley, R. M.; Begg, A. P.; Hutchins, D. R.; Hodgson, D. R.; Bryden, W. L. Leukoencephalomalacia in horses. Aust. Vet. J. 1993, 70, 225-226.

Chu, F. S.; Li, G. Y. Simultaneous occurrence of fumonisin B_1 and other mycotoxins in moldy corn collected from the Peoples Republic of China in regions with high incidences of esophageal cancer. Appl. Environ. Microbiol. 1994, 60, 847-852.

Colvin, B. M.; Harrison, L. R. Fumonisin-induced pulmonary edema and hydrothorax in swine. Mycopathologia 1992, 117, 79-82.

Colvin, B. M.; Cooley, A. J.; Beaver, R. W. Fumonisin toxicosis in swine - clinical and pathologic findings. J. Vet. Diagn. Invest. 1993, 5, 232-241.

Doko, M. B.; Visconti, A. V. Occurrence of fumonisins B_1 and B_2 in corn and corn-based human food stuffs in Italy. Food Addit. Contam. 1994, 11, 433-439.

Dupuy, J.; Lebars, P.; Boudra, H.; Lebars, J. Thermostability of fumonisin B_1, a mycotoxin from *Fusarium moniliforme*, in corn. Applied and Environmental Microbiology 1993, 59, 2864-2867.

Fincham, J. E.; Marasas, W. F. O.; Taljaard, J. J. F.; Kriek, N. P. J.; Badenhorst, C. J.; Gelderblom, W. C. A.; Seier, J. V.; Smuts, C. M.; Faber, M.; Weight, M. J.; Slazus, W.; Woodroof, C. W.; Van Wijk, M. J.;

Kruger, M.; Thiel, P. G. Atherogenic effects in a non-human primate of *Fusarium-moniliforme* cultures added to a carbohydrate diet. Atherosclerosis 1992, 94, 13-25.

Floss, J. L.; Casteel, S. W.; Johnson, G. C.; Rottinghaus, G. E.; Krause, G. F. Developmental toxicity in hamsters of an aqueous extract of *Fusarium moniliforme* culture material containing known quantities of fumonisin B_1. Vet. Hum. Toxicol. 1994, 36, 5-10.

Franceschi, S.; Bidoli, E.; Baron, A. E.; La Vecchia, C. Maize and risks of cancers of the oral cavity, pharynx, and esophagus in northeastern Italy. J. Natl. Cancer Inst. 1990, 82, 1407-1411.

Gelderblom, W. C. A.; Thiel, P. G.; Jaskiewicz, K.; Marasas, W. F. O. Investigations on the carcinogenicity of fusarin C - a mutagenic metabolite of *Fusarium moniliforme*. Carcinogenesis 1986, 7, 1899-1901.

Gelderblom, W. C. A.; Jaskiewicz, K.; Marasas, W. F. O.; Thiel, P. G.; Horak, R. M.; Vleggaar, R.; Kriek, N. P. J. Fumonisins novel mycotoxins with cancer-promoting activity produced by *Fusarium moniliforme*. Appl. Environ. Microbiol. 1988, 54, 1806- 1811.

Gelderblom, W. C. A.; Kriek, N. P. J.; Marasas, W. F. O.; Thiel, P. G. Toxicity and carcinogenicity of the *Fusarium moniliforme* metabolite, fumonisin B_1, in rats. Carcinogenesis 1991, 12, 1247- 1251.

Gelderblom, W. C. A.; Snyman, S. D. Mutagenicity of potentially carcinogenic mycotoxins produced by *Fusarium moniliforme*. Mycotoxin Res. 1991, 7, 46-52.

Gelderblom, W. C. A.; Marasas, W. F. O.; Vleggaar, R.; Thiel, P. G.; Cawood, M. E. Fumonisins - isolation, chemical characterization and biological effects. Mycopathologia 1992a, 117, 11-16.

Gelderblom, W. C. A.; Semple, E.; Marasas, W. F. O.; Farber, E. The cancer-initiating potential of the fumonisin B mycotoxins. Carcinogenesis 1992b, 13, 433-437.

Gelderblom, W. C. A.; Cawood, M. E.; Snyman, S. D.; Vleggaar, R.; Marasas, W. F. O. Structure-activity relationships of fumonisins in short-term carcinogenesis and cytotoxicity assays. Food Chem. Toxicol. 1993, 31, 407-414.

Gelderblom, W. C. A.; Cawood, M. E.; Snyman, S. D.; Marasas, W. F. O. Fumonisin B_1 dosimetry in relation to cancer initiation in rat liver. Carcinogenesis 1994, 15, 209-214.

Haschek, W. M.; Motelin, G.; Ness, D. K.; Harlin, K. S.; Hall, W. F.; Vesonder, R. F.; Peterson, R. E.; Beasley, V. R. Characterization of fumonisin toxicity in orally and intravenously dosed swine. Mycopathologia 1992, 117, 83-96.

Health and Welfare Canada. Nutrition Canada. Food Consumption Patterns Report. Health and Welfare Canada: Ottawa, 1976.

Hendrich, S.; Miller, K. A.; Wilson, T. M.; Murphy, P. A. Toxicity of *Fusarium proliferatum*-fermented nixtamalized corn- based diets fed to rats: effect of nutritional status. J. Agric. Food Chem. 1993, 41, 1649-1654.

Hopmans, E. C.; Murphy, P. A. Detection of fumonisin-b(1), fumonisin-b(2), and fumonisin-b(3) and hydrolyzed fumonisin-b(1) in corn-containing foods. J. Agric. Food Chem. 1993, 41, 1655- 1658.

IARC. International Agency for Research on Cancer. Toxins derived from *Fusarium moniliforme*: fumonisins B_1 and B_2 and fusarin C. In Some Naturally occurring Substances: Food Items and Constituents, Heterocyclic Aromatic Amines and Mycotoxins. World Health Organization, IARC: Lyon, France, 1993, 56, 445-466; 537; 565- 567.

Jaskiewicz, K.; Marasas, W. F. O.; van der Walt, F. E. Oesophageal and other main cancer patterns in four districts of Transkei, 1981 - 1984. S. Afr. Med. J. 1987a, 72, 27-30.

Jaskiewicz, K.; van Rensburg, S. J.; Marasas, W. F. O.; Gelderblom, W. C. A. Carcinogenicity of *Fusarium moniliforme* culture material in rats. J. Natl. Cancer Inst. 1987b, 78, 321- 325.

Kaneshiro, T.; Vesonder, R. F.; Peterson, R. E. Fumonisin-stimulated N-acetyldihydrosphingosine, n-acetyl-phytosphingosine, and phytosphingosine products of *Pichia-(Hansenula)-ciferri*, NRRL Y-1031. Current Microbiol. 1992, 24, 319-324.

Kellerman, T. S.; Marasas, W. F. O.; Thiel, P. G.; Gelderblom, W. C. A.; Cawood, M.; Coetzer, J. A. W. Leukoencephalomalacia in two horses induced by oral dosing of fumonisin B_1. Onderstepoort J. Vet. Res. 1990, 57, 269-275.

Kuiper-Goodman, T.; Ominski, K.; Marquardt, R. R.; Malcolm, S.; McMullen, E.; Lombaert, G. A.; Morton, T. Estimating human exposure to ochratoxin A in Canada. In Human Ochratoxicosis and its Pathologies. Colloque INSERM vol. 231. Creppy, E. E.; Castegnaro, M.; Dirheimer, G., Eds.; John Libbey, Eurotext Ltd.: Montrouge, France, 1993, pp 167-174.

Lebars, J.; Lebars, P.; Dupuy, J.; Boudra, H.; Cassini, R. Biotic and abiotic factors in fumonisin B_1 production and stability. J. Assoc. Off. Anal. Chem. 1994, 77, 517-521.

Marasas, W. F. O.; Kriek, N. P. J.; Fincham, J. E.; van Rensburg, S. J. Primary liver cancer and oesophagaeal basal cell hyperplasia in rats caused by *Fusarium moniliforme*. Int. J. Cancer 1984, 34, 383-387.

Marasas, W. F. O.; Jaskiewicz, K.; Venter, F. S.; Van Schalkwyk, D. J. *Fusarium moniliforme* contamination of maize in oesophageal cancer areas in Transkei. S. Afr. Med. J. 1988, 74, 110-114.

Motelin, G. K.; Haschek, W. M.; Ness, D. K.; Hall, W. F.; Harlin, K. S.; Schaeffer, D. J.; Beasley, V. R. Temporal and dose- response features in swine fed corn screenings contaminated with fumonisin mycotoxins. Mycopathologia 1994, 126, 27-40.

Murphy, P. A.; Rice, L. G.; Ross, P. F. Fumonisin B$_1$, fumonisin B$_2$, and fumonisin B$_3$ content of Iowa, Wisconsin, and Illinois corn and corn screenings. J. Agric. Food Chem. 1993, 41, 263-266.

Norred, W. P.; Plattner, R. D.; Vesonder, R. F.; Bacon, C. W.; Voss, K. A. Effects of selected secondary metabolites of *Fusarium moniliforme* on unscheduled synthesis of DNA by rat primary hepatocytes. Food Chem. Toxicol. 1992a, 30, 233-237.

Norred, W. P.; Wang, E.; Yoo, H.; Riley, R. T.; Merrill, A. H. *In vitro* toxicology of fumonisins and the mechanistic implications. Mycopathologia 1992b, 117, 73-78.

Norred, W. P.; Plattner, R. D.; Chamberlain, W. J. Distribution and excretion of [14C] fumonisin B$_1$ in male Sprague-Dawley rats. Natural Toxins 1993, 1, 341-346.

Park, D. L.; Rua, S. M.; Mirocha, C. J.; Abdalla, E. S. A. M.; Cong, Y. W. Mutagenic potentials of fumonisin contaminated corn following ammonia decontamination procedure. Mycopathologia 1992, 117, 105-108.

Pestka, J. J.; Azconaolivera, J. I.; Plattner, R. D.; Minervini, F.; Doko, M. B.; Visconti, A. Comparative assessment of fumonisin in grain-based foods by ELISA, GC-MS, and HPLC. J. Food Prot. 1994, 57, 169-172.

Pittet, A.; Parisod, V.; Schellenberg, M. Occurrence of fumonisins B$_1$ and B$_2$ in corn-based products from the Swiss market. J. Agric. Food Chem. 1992, 40, 1352-1354.

Prelusky, D. B.; Trenholm, H. L.; Rotter, B. A.; Miller, J. D.; Savard, M. E.; Yeung, J. M.; Scott, P. M. Biological fate of fumonisin B$_1$ in food-producing animals. ACS meeting, April 2-6, Anaheim, Ca. 1995, in press.

Prelusky, D. B.; Trenholm, H. L.; Savard, M. E. Pharmacokinetic fate of ^{14}C-labelled fumonisin B$_1$ in swine. Natural Toxins 1994, 2, 73-80.

Riley, R. T.; Wang, E.; Merrill, A. H. Liquid chromatographic determination of sphinganine and sphingosine - use of the free sphinganine-to-sphingosine ratio as a biomarker for consumption of fumonisins. J.Assoc. Off. Anal. Chem. 1994, 77, 533-540.

Ross, P.F.; Rice, L. G.; Reagor, J. C.; Osweiler, G. D.; Wilson, T. M.; Nelson, H. A.; Owens, D. L.; Plattner, R. D.; Harlin, K. A.; Richard, J. L.; Colvin, B. M.; Banton, M. I. Fumonisin B$_1$ concentrations in feeds from 45 confirmed equine leukoencephalomalacia cases. J. Vet. Diagn. Invest. 1991, 3, 238- 241.

Ross, P. F.; Rice, L. G.; Osweiler, G. D.; Nelson, P. E.; Richard, J. L.; Wilson, T. M. A review and update of animal toxicoses associated with fumonisin-contaminated feeds and production of fumonisins by *Fusarium* isolates. Mycopathologia 1992, 117, 109-114.

Ross, P. F. What are we going to do with this dead horse. J. Assoc. Off. Anal. Chem. 1994, 77, 491-494.

Sanchis, V.; Abadias, M.; Oncins, L.; Sala, N.; Vinas, I.; Canela, R. Occurrence of fumonisins B$_1$ and B$_2$ in corn-based products from the Spanish market. Appl. Environ. Microbiol. 1994, 60, 2147-2148.

Schroeder, J. J.; Crane, H. M.; Xia, J.; Liotta, D. C.; Merrill, A. H. Disruption of sphingolipid metabolism and stimulation of DNA synthesis by fumonisin B$_1$ - a molecular mechanism for carcinogenesis associated with *Fusarium moniliforme*. J. Biol. Chem. 1994, 269, 3475-3481.

Scott, P. M.; Lawrence, G. A. Liquid chromatographic determination of fumonisins with 4-fluoro-7-nitroben-zofurazan. J. Assoc. Off. Anal. Chem. 1992, 75, 829-834.

Scott, P. M.; Lawrence, G. A. Determination of hydrolyzed fumonisin B$_1$ in alkali-processed corn foods. 108th AOAC International Annual Meeting, Sept 12-15, Portland, Oregon 1994a, p39.

Scott, P. M.; Lawrence, G. A. Stability and problems in recovery of fumonisins added to corn-based foods. J.Assoc. Off. Anal. Chem. 1994b, 77, 541-545.

Scott, P. M.; Lawrence, G. A. Analysis of beer for fumonisins. J. Food Prot. 1995, in press.

Shephard, G. S.; Sydenham, E. W.; Thiel, P. G.; Gelderblom, W. C. A. Quantitative determination of fumonisin B$_1$ and fumonisins B$_2$ by high-performance liquid chromatography with fluorescence detection. J. Liqu. Chromatogr. 1990, 13, 2077-2087.

Shephard, G. S.; Thiel, P. G.; Sydenham, E. W.; Alberts, J. F.; Gelderblom, W. C. Fate of a single dose of the ^{14}C-labeled mycotoxin, fumonisin B$_1$ in rats. Toxicon. 1992a, 30, 768-770.

Shephard, G. S.; Thiel, P. G.; Sydenham, E. W. Initial studies on the toxicokinetics of fumonisin B$_1$ in rats. Food Chem. Toxicol. 1992b, 30, 277-279.

Shephard, G. S.; Thiel, P. G.; Sydenham, E. W.; Alberts, J. F. Biliary excretion of the mycotoxin fumonisin B$_1$ in rats. Food Chem. Toxicol. 1994a, 32, 489-491.

Shephard, G. S.; Thiel, P. G.; Sydenham, E. W.; Alberts, J. F.; Cawood, M. E. Distribution and excretion of a single dose of the mycotoxin fumonisin B$_1$ in a non-human primate. Toxicon 1994b, 32, 735-741.

Shier, W. T.; Abbas, H. K.; Mirocha, C. J. Toxicity of mycotoxins fumonisins B$_1$ and B$_2$ and *Alternaria alternata* f. sp. *lycopersica* toxin (AAL) in cultured mammalian cells. Mycopathologia 1991, 116, 97-104.

Solt, D. B.; Farber, E. New principle for the analysis of chemical carcinogenesis. Nature 1976, 263, 702-703.

Stack, M. E.; Eppley, R. M. Liquid chromatographic determination of fumonisin B_1 and fumonisin B_2 in corn and corn products. J. Assoc. Off. Anal. Chem. 1992, 75, 834-837.

Sydenham, E. W.; Thiel, P. G.; Marasas, W. F. O.; Shephard, G. S.; van Schalwyk, D. J.; Koch, K. R. Natural occurrence of some *Fusarium* mycotoxins in corn from low and high esophageal cancer prevalence areas of the Transkei, Southern Africa. J. Agric. Food Chem. 1990, 38, 1900-1903.

Sydenham, E. W.; Shephard, G. S.; Thiel, P. G.; Marasas, W. F.; Stockenström, S. Fumonisin contamination of commercial corn-based human foodstuffs. J. Agric. Food Chem. 1991, 39, 2014-2018.

Sydenham, E. W.; Thiel, P. G.; Stockenström, S.; Shephard, G. S. $Ca(OH)_2$ treatment: its influence on fumonisin B_1 in corn [abstract]. VIII International IUPAC Symposium on Mycotoxins and Phycotoxins. November 6-13. 1992, p43.

Thiel, P. G.; Marasas, W. F. O.; Sydenham, E. W.; Shephard, G. S.; Gelderblom, W. C. A. The implications of naturally occurring levels of fumonisins in corn for human and animal health. Mycopathologia 1992, 117, 3-9.

Trucksess, M. W.; Stack, M. E.; Allen, S.; Barrion, N. Immunoaffinity column coupled with liquid chromatography for determination of fumonisin B_1 in canned and frozen sweet corn. J. Assoc. Off. Anal. Chem. 1995, 78, 705-710.

Ueno, Y.; Aoyama, S.; Sugiura, Y.; Wang, D. S.; Lee, U-S; Hirooka, E. Y.; Hara, S.; Karki, T.; Chen, G.; Yu S-Z. A limited survey of fumonisins in corn and corn-based products in Asian countries. Mycotoxin Res. 1993, 9, 27-34.

USDA. Foods commonly eaten by individuals: amounts per day and per eating occasion. Pao, E. M.; Fleury, K. H.; Gunther, P. M.; Mickle, S. G., Eds.; US Department of Agriculture: Hyattsville, Md, 1982,

Voss, K. A.; Chamberlain, W. J.; Bacon, C. W.; Norred, W. P. A preliminary investigation on renal and hepatic toxicity in rats fed purified fumonisin B_1. Natural Toxins 1993, 1, 222-228.

Voss, K. A.; Chamberlain, W. J.; Bacon, C. W.; Herbert, R. A.; Walters, D. B.; Norred, W. P. Subchronic feeding study of the mycotoxin fumonisin B_1 in B6C3F1 mice and Fischer 344 rats. Fund. Appl. Toxicol. 1995a, 24, 102-110.

Voss, K. A.; Bacon, C. W.; Norred, W. P.; Chapin, R. E.; Chamberlain, W. J. Reproductive toxicity study of *Fusarium moniliforme* culture material in rats [abstract]. Toxicologist 1995b, 15, 215.

Wang, E. et al. Inhibition of sphingolipid biosynthesis by fumonisins: Implications for diseases associated with *Fusarium moniliforme*. J. Biol. Chem. 1991, 26, 14486-14490.

Wilson, T. M.; Ross, P. F.; Rice, L. G.; Osweiler, G. D.; Nelson, H. A.; Owens, D. L.; Plattner, R. D.; Reggiardo, C.; Noon, T. H.; Pickrell, J. W. Fumonisin B_1 levels associated with an epizootic of equine leukoencephalomalacia. J. Vet. Diagnos. Invest. 1990, 2, 213-216.

Wilson, T. M.; Ross, P. F.; Owens, D. L.; Rice, L. G.; Green, S. A.; Jenkins, S. J.; Nelson, H. A. Experimental reproduction of ELEM - A study to determine the minimum toxic dose in ponies. Mycopathologia 1992, 117, 115-120.

Yoo, H. S.; Norred, W. P.; Wang, E.; Merrill, A. H.; Riley, R. T. Fumonisin inhibition of de novo sphingolipid biosynthesis and cytotoxicity are correlated in LLC-PK1 cells. Toxicol. Appl. Pharmacol. 1992, 114, 9-15.

Yoshizawa, T.; Yamashita, A.; Luo, Y. Fumonisin occurrence in corn from high- and low-risk areas for human esophageal cancer in China. Appl. Environ. Microbiol. 1994, 60, 1626-1629.

Zhen, B.; Yang, S.; Ding, L.; Han, F.; Yang, W.; Liu, Q. The culture and isolation of fungi from the cereals in five high and three low incidence counties of oesophageal cancer in Henan province (China). Chin. J. Oncol. 1984, 6, 27-29.

INDEX